Engineering Design Optimization

Based on course-tested material, this rigorous yet accessible graduate textbook covers both fundamental and advanced optimization theory and algorithms. It covers a wide range of numerical methods and topics, including both gradient-based and gradient-free algorithms, multidisciplinary design optimization, and uncertainty, with instruction on how to determine which algorithm should be used for a given application. It also provides an overview of models and how to prepare them for use with numerical optimization, including derivative computation. Over 400 high-quality visualizations and numerous examples facilitate understanding of the theory, and practical tips address common issues encountered in practical engineering design optimization and how to address them. Numerous end-of-chapter homework problems, progressing in difficulty, help put knowledge into practice.

Accompanied online by a solutions manual for instructors and source code for problems, this is ideal for a one- or two-semester graduate course on optimization in aerospace, civil, mechanical, electrical, and chemical engineering departments.

Joaquim R. R. A. Martins is a Professor of Aerospace Engineering at the University of Michigan. He is a fellow of the American Institute for Aeronautics and Astronautics, and the Royal Aeronautical Society.

Andrew Ning is an Associate Professor of Mechanical Engineering at Brigham Young University, and has previously worked at the National Renewable Energy Laboratory (NREL) as a Senior Engineer.

Engineering Design Optimization

Joaquim R. R. A. Martins
University of Michigan, Ann Arbor

Andrew Ning
Brigham Young University, Utah

Shaftesbury Road, Cambridge CB2 8EA, United Kingdom

One Liberty Plaza, 20th Floor, New York, NY 10006, USA

477 Williamstown Road, Port Melbourne, VIC 3207, Australia

314–321, 3rd Floor, Plot 3, Splendor Forum, Jasola District Centre, New Delhi – 110025, India

103 Penang Road, #05–06/07, Visioncrest Commercial, Singapore 238467

Cambridge University Press is part of Cambridge University Press & Assessment, a department of the University of Cambridge.

We share the University's mission to contribute to society through the pursuit of education, learning and research at the highest international levels of excellence.

www.cambridge.org
Information on this title: www.cambridge.org/9781108833417

DOI: 10.1017/9781108980647

© Joaquim R. R. A. Martins and Andrew Ning 2022

This publication is in copyright. Subject to statutory exception and to the provisions of relevant collective licensing agreements, no reproduction of any part may take place without the written permission of Cambridge University Press & Assessment.

First published 2022 (version 2, May 2024)

Printed in Great Britain by CPI Group (UK) Ltd, Croydon CR0 4YY, May 2024

A catalogue record for this publication is available from the British Library

Library of Congress Cataloging-in-Publication data
Names: Martins, Joaquim R. R. A., author. | Ning, S. Andrew (Simeon Andrew), author.
Title: Engineering design optimization / Joaquim R. R. A. Martins, Andrew Ning.
Description: Cambridge ; New York, NY : Cambridge University Press, 2021. |
  Includes bibliographical references and index.
Identifiers: LCCN 2021024825 (print) | LCCN 2021024826 (ebook) | ISBN
  9781108833417 (hardback) | ISBN 9781108833417 (ebook)
Subjects: LCSH: Engineering design – Mathematical models. | Mathematical
  optimization. | Multidisciplinary design optimization. | BISAC:
  MATHEMATICS / Optimization
Classification: LCC TA174 .M354 2021 (print) | LCC TA174 (ebook) | DDC 620/.0042–dc23
LC record available at https://lccn.loc.gov/2021024825
LC ebook record available at https://lccn.loc.gov/2021024826

ISBN 978-1-108-83341-7 Hardback

Additional resources for this publication at www.cambridge.org/martins-ning

Cambridge University Press & Assessment has no responsibility for the persistence or accuracy of URLs for external or third-party internet websites referred to in this publication and does not guarantee that any content on such websites is, or will remain, accurate or appropriate.

Contents

Contents   v

Preface   xi

Acknowledgements   xiii

1 Introduction   1
   1.1   Design Optimization Process   2
   1.2   Optimization Problem Formulation   6
   1.3   Optimization Problem Classification   17
   1.4   Optimization Algorithms   21
   1.5   Selecting an Optimization Approach   26
   1.6   Notation   28
   1.7   Summary   29
        Problems   30

2 A Short History of Optimization   33
   2.1   The First Problems: Optimizing Length and Area   33
   2.2   Optimization Revolution: Derivatives and Calculus   34
   2.3   The Birth of Optimization Algorithms   36
   2.4   The Last Decades   39
   2.5   Toward a Diverse Future   43
   2.6   Summary   45

3 Numerical Models and Solvers   47
   3.1   Model Development for Analysis versus Optimization   47
   3.2   Modeling Process and Types of Errors   48
   3.3   Numerical Models as Residual Equations   50
   3.4   Discretization of Differential Equations   52
   3.5   Numerical Errors   53
   3.6   Overview of Solvers   61
   3.7   Rate of Convergence   63
   3.8   Newton-Based Solvers   66
   3.9   Models and the Optimization Problem   70

    3.10 Summary 73
          Problems 75

4   Unconstrained Gradient-Based Optimization 79
    4.1   Fundamentals 80
    4.2   Two Overall Approaches to Finding an Optimum 94
    4.3   Line Search 96
    4.4   Search Direction 110
    4.5   Trust-Region Methods 139
    4.6   Summary 147
          Problems 149

5   Constrained Gradient-Based Optimization 153
    5.1   Constrained Problem Formulation 154
    5.2   Understanding n-Dimensional Space 156
    5.3   Optimality Conditions 158
    5.4   Penalty Methods 175
    5.5   Sequential Quadratic Programming 187
    5.6   Interior-Point Methods 204
    5.7   Constraint Aggregation 211
    5.8   Summary 214
          Problems 215

6   Computing Derivatives 223
    6.1   Derivatives, Gradients, and Jacobians 223
    6.2   Overview of Methods for Computing Derivatives 225
    6.3   Symbolic Differentiation 226
    6.4   Finite Differences 227
    6.5   Complex Step 232
    6.6   Algorithmic Differentiation 237
    6.7   Implicit Analytic Methods—Direct and Adjoint 252
    6.8   Sparse Jacobians and Graph Coloring 262
    6.9   Unified Derivatives Equation 265
    6.10 Summary 275
          Problems 277

7   Gradient-Free Optimization 281
    7.1   When to Use Gradient-Free Algorithms 281
    7.2   Classification of Gradient-Free Algorithms 284
    7.3   Nelder–Mead Algorithm 287
    7.4   Generalized Pattern Search 292
    7.5   DIRECT Algorithm 298
    7.6   Genetic Algorithms 306

7.7 Particle Swarm Optimization   316
7.8 Summary   321
     Problems   323

8 Discrete Optimization   327
8.1 Binary, Integer, and Discrete Variables   327
8.2 Avoiding Discrete Variables   328
8.3 Branch and Bound   330
8.4 Greedy Algorithms   337
8.5 Dynamic Programming   339
8.6 Simulated Annealing   347
8.7 Binary Genetic Algorithms   351
8.8 Summary   351
     Problems   352

9 Multiobjective Optimization   355
9.1 Multiple Objectives   355
9.2 Pareto Optimality   357
9.3 Solution Methods   358
9.4 Summary   369
     Problems   370

10 Surrogate-Based Optimization   373
10.1 When to Use a Surrogate Model   374
10.2 Sampling   375
10.3 Constructing a Surrogate   384
10.4 Kriging   400
10.5 Deep Neural Networks   408
10.6 Optimization and Infill   414
10.7 Summary   418
      Problems   420

11 Convex Optimization   423
11.1 Introduction   423
11.2 Linear Programming   425
11.3 Quadratic Programming   427
11.4 Second-Order Cone Programming   429
11.5 Disciplined Convex Optimization   430
11.6 Geometric Programming   434
11.7 Summary   437
      Problems   438

12 Optimization Under Uncertainty   441
- 12.1 Robust Design   442
- 12.2 Reliable Design   447
- 12.3 Forward Propagation   448
- 12.4 Summary   469
- Problems   471

13 Multidisciplinary Design Optimization   475
- 13.1 The Need for MDO   475
- 13.2 Coupled Models   478
- 13.3 Coupled Derivatives Computation   501
- 13.4 Monolithic MDO Architectures   510
- 13.5 Distributed MDO Architectures   519
- 13.6 Summary   533
- Problems   535

A Mathematics Background   539
- A.1 Taylor Series Expansion   539
- A.2 Chain Rule, Total Derivatives, and Differentials   541
- A.3 Matrix Multiplication   544
- A.4 Four Fundamental Subspaces in Linear Algebra   547
- A.5 Vector and Matrix Norms   548
- A.6 Matrix Types   550
- A.7 Matrix Derivatives   552
- A.8 Eigenvalues and Eigenvectors   553
- A.9 Random Variables   554

B Linear Solvers   559
- B.1 Systems of Linear Equations   559
- B.2 Conditioning   560
- B.3 Direct Methods   560
- B.4 Iterative Methods   562

C Quasi-Newton Methods   571
- C.1 Broyden's Method   571
- C.2 Additional Quasi-Newton Approximations   572
- C.3 Sherman–Morrison–Woodbury Formula   576

D Test Problems   579
- D.1 Unconstrained Problems   579
- D.2 Constrained Problems   586

Bibliography 591

Index 615

Preface

Despite its usefulness, design optimization remains underused in industry. One of the reasons for this is the shortage of design optimization courses in undergraduate and graduate curricula. This is changing; today, most top aerospace and mechanical engineering departments include at least one graduate-level course on numerical optimization. We have also seen design optimization increasingly used in an expanding number of industries.

The word *engineering* in the title reflects the types of problems and algorithms we focus on, even though the methods are applicable beyond engineering. In contrast to explicit analytic mathematical functions, most engineering problems are implemented in complex multidisciplinary codes that involve implicit functions. Such problems might require hierarchical solvers and coupled derivative computation. Furthermore, engineering problems often involve many design variables and constraints, requiring scalable methods.

The target audience for this book is advanced undergraduate and beginning graduate students in science and engineering. No previous exposure to optimization is assumed. Knowledge of linear algebra, multivariable calculus, and numerical methods is helpful. However, these subjects' core concepts are reviewed in an appendix and as needed in the text. The content of the book spans approximately two semester-length university courses. Our approach is to start from the most general case problem and then explain special cases. The first half of the book covers the fundamentals (along with an optional history chapter). In contrast, the second half, from Chapter 8 onward, covers more specialized or advanced topics.

Our philosophy in the exposition is to provide a detailed enough explanation and analysis of optimization algorithms so that readers can implement a basic working version. Although we do not encourage readers to use their implementations instead of existing software for solving optimization problems, implementing a method is crucial in understanding the method and its behavior.* A deeper knowledge of these methods is useful for developers, researchers, and those who want to use numerical optimization more effectively. The problems at

---

*In the words of Donald Knuth: *"The ultimate test of whether I understand something is if I can explain it to a computer. I can say something to you and you'll nod your head, but I'm not sure that I explained it well. But the computer doesn't nod its head. It repeats back exactly what I tell it. In most of life, you can bluff, but not with computers."*

the end of each chapter are designed to provide a gradual progression in difficulty and eventually require implementing the methods. Some of the problems are open-ended to encourage students to explore a given topic on their own. When discussing the various optimization techniques, we also explain how to avoid the potential pitfalls of using a particular method and how to employ it more effectively. Practical tips are included throughout the book to alert the reader to common issues encountered in engineering design optimization and how to address them.

We have created a repository with code, data, templates, and examples as a supplementary resource for this book: https://github.com/mdobook/resources. Some of the end-of-chapter exercises refer to code or data from this repository.

Go forth and optimize!

Acknowledgments

Our workflow was tremendously enhanced by the support of Edmund Lee and Aaron Lu, who took our sketches and plots and translated them to high-quality, consistently formatted figures. The layout of this book was greatly improved based in part on a template provided by Max Opgenoord. We are indebted to many students and colleagues who provided feedback and insightful questions on our concepts, examples, lectures, and manuscript drafts. At the risk of leaving out some contributors, we wish to express particular gratitude to the following individuals who helped create examples, problems, solutions, or content that was incorporated in the book: Tal Dohn, Xiaosong Du, Sicheng He, Jason Hicken, Donald Jones, Shugo Kaneko, Taylor McDonnell, Judd Mehr, Santiago Padrón, Sabet Seraj, P. J. Stanley, and Anil Yildirim. Additionally, the following individuals provided helpful suggestions and corrections to the manuscript: Eytan Adler, Josh Anibal, Eliot Aretskin-Hariton, Alexander Coppeans, Alec Gallimore, Philip Gill, Justin Gray, Christina Harvey, John Hwang, Kevin Jacobsen, Kai James, Eirikur Jonsson, Matthew Kramer, Alexander Kleb, Michael Kokkolaras, Yingqian Liao, Sandy Mader, Marco Mangano, Giuliana Mannarino, Yara Martins, Johannes Norheim, Bernardo Pacini, Malhar Prajapati, Michael Saunders, Nikhil Shetty, Tamás Terlaky, and Elizabeth Wong. We are grateful to peer reviewers who provided enthusiastic encouragement and helpful suggestions and wish to thank our editors at Cambridge University Press, who quickly and competently offered corrections. Finally, we express our deepest gratitude to our families for their loving support.

Joaquim Martins and Andrew Ning

Introduction 1

Optimization is a human instinct. People constantly seek to improve their lives and the systems that surround them. Optimization is intrinsic in biology, as exemplified by the evolution of species. Birds optimize their wings' shape in real time, and dogs have been shown to find optimal trajectories. Even more broadly, many laws of physics relate to optimization, such as the principle of minimum energy. As Leonhard Euler once wrote, "nothing at all takes place in the universe in which some rule of maximum or minimum does not appear."

The term *optimization* is often used to mean "improvement", but mathematically, it is a much more precise concept: finding the *best* possible solution by changing variables that can be controlled, often subject to constraints. Optimization has a broad appeal because it is applicable in all domains and because of the human desire to make things better. Any problem where a decision needs to be made can be cast as an optimization problem.

Although some simple optimization problems can be solved analytically, most practical problems of interest are too complex to be solved this way. The advent of numerical computing, together with the development of optimization algorithms, has enabled us to solve problems of increasing complexity.

> By the end of this chapter you should be able to:
> 1. Understand the design optimization process.
> 2. Formulate an optimization problem.
> 3. Identify key characteristics to classify optimization problems and optimization algorithms.
> 4. Select an appropriate algorithm for a given optimization problem.

Optimization problems occur in various areas, such as economics, political science, management, manufacturing, biology, physics, and engineering. This book focuses on the application of numerical optimization to the design of engineering systems. Numerical optimization first emerged in *operations research*, which deals with problems such as deciding on the price of a product, setting up a distribution network, scheduling, or suggesting routes. Other optimization areas include optimal control and machine learning. Although we do not cover these other areas specifically in this book, many of the methods we cover are useful in those areas.

Design optimization problems abound in the various engineering disciplines, such as wing design in aerospace engineering, process control in chemical engineering, structural design in civil engineering, circuit design in electrical engineering, and mechanism design in mechanical engineering. Most engineering systems rarely work in isolation and are linked to other systems. This gave rise to the field of *multidisciplinary design optimization* (MDO), which applies numerical optimization techniques to the design of engineering systems that involve multiple disciplines.

In the remainder of this chapter, we start by explaining the design optimization process and contrasting it with the conventional design process (Section 1.1). Then we explain how to formulate optimization problems and the different types of problems that can arise (Section 1.2). Because design optimization problems involve functions of different types, these are also briefly discussed (Section 1.3). (A more detailed discussion of the numerical models used to compute these functions is deferred to Chapter 3.) We then provide an overview of the different optimization algorithms, highlighting the algorithms covered in this book and linking to the relevant sections (Section 1.4). We connect algorithm types and problem types by providing guidelines for selecting the right algorithm for a given problem (Section 1.5). Finally, we introduce the notation used throughout the book (Section 1.6).

1.1 Design Optimization Process

Engineering design is an iterative process that engineers follow to develop a product that accomplishes a given task. For any product beyond a certain complexity, this process involves teams of engineers and multiple stages with many iterative loops that may be nested. The engineering teams are formed to tackle different aspects of the product at different stages.

The design process can be divided into the sequence of phases shown in Fig. 1.1. Before the design process begins, we must determine the requirements and specifications. This might involve market research, an analysis of current similar designs, and interviews with potential

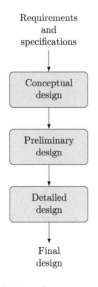

Fig. 1.1 Design phases.

customers. In the conceptual design phase, various concepts for the system are generated and considered. Because this phase should be short, it usually relies on simplified models and human intuition. For more complicated systems, the various subsystems are identified. In the preliminary design phase, a chosen concept and subsystems are refined by using better models to guide changes in the design, and the performance expectations are set. The detailed design phase seeks to complete the design down to every detail so that it can finally be manufactured. All of these phases require iteration within themselves. When severe issues are identified, it may be necessary to "go back to the drawing board" and regress to an earlier phase. This is just a high-level view; in practical design, each phase may require multiple iterative processes.

Design optimization is a tool that can replace an iterative design process to accelerate the design cycle and obtain better results. To understand the role of design optimization, consider a simplified version of the conventional engineering design process with only one iterative loop, as shown in Fig. 1.2 (top). In this process, engineers make decisions at every stage based on intuition and background knowledge.

Each of the steps in the conventional design process includes human decisions that are either challenging or impossible to program into computer code. Determining the product specifications requires engineers to define the problem and do background research. The design cycle must start with an initial design, which can be based on past designs or a new idea. In the conventional design process, this initial design is analyzed in some way to evaluate its performance. This could involve numerical modeling or actual building and testing. Engineers then evaluate the design and decide whether it is good enough or not based on the results.* If the answer is no—which is likely to be the case for at least the first few iterations—the engineer changes the design based on intuition, experience, or trade studies. When the design is finalized when it is deemed satisfactory.

*The evaluation of a given design in engineering is often called the *analysis*. Engineers and computer scientists also refer to it as *simulation*.

The design optimization process can be represented using a flow diagram similar to that for the conventional design process, as shown in Fig. 1.2 (bottom). The determination of the specifications and the initial design are no different from the conventional design process. However, design optimization requires a formal formulation of the optimization problem that includes the design variables that are to be changed, the objective to be minimized, and the constraints that need to be satisfied. The evaluation of the design is strictly based on numerical values for the objective and constraints. When a rigorous optimization algorithm is used, the decision to finalize the design is made only when the current

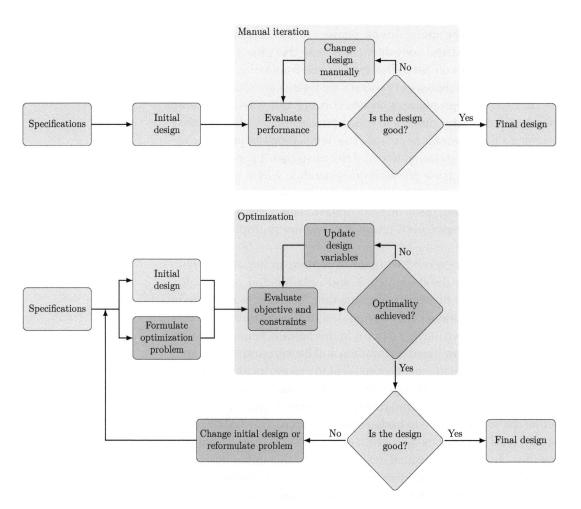

Fig. 1.2 Conventional (top) versus design optimization process (bottom).

design satisfies the optimality conditions that ensure that no other design "close by" is better. The design changes are made automatically by the optimization algorithm and do not require intervention from the designer.

This automated process does not usually provide a "push-button" solution; it requires human intervention and expertise (often more expertise than in the traditional process). Human decisions are still needed in the design optimization process. Before running an optimization, in addition to determining the specifications and initial design, engineers need to formulate the design problem. This requires expertise in both the subject area and numerical optimization. The designer must decide what the objective is, which parameters can be changed, and which constraints must be enforced. These decisions have profound effects on the outcome, so it is crucial that the designer formulates the optimization problem well.

After running the optimization, engineers must assess the design because it is unlikely that the first formulation yields a valid and practical design. After evaluating the optimal design, engineers might decide to reformulate the optimization problem by changing the objective function, adding or removing constraints, or changing the set of design variables. Engineers might also decide to increase the models' fidelity if they fail to consider critical physical phenomena, or they might decide to decrease the fidelity if the models are too expensive to evaluate in an optimization iteration.

Post-optimality studies are often performed to interpret the optimal design and the design trends. This might be done by performing parameter studies, where design variables or other parameters are varied to quantify their effect on the objective and constraints. Validation of the result can be done by evaluating the design with higher-fidelity simulation tools, by performing experiments, or both. It is also possible to compute post-optimality sensitivities to evaluate which design variables are the most influential or which constraints drive the design. These sensitivities can inform where engineers might best allocate resources to alleviate the driving constraints in future designs.

Design optimization can be used in any of the design phases shown in Fig. 1.1, where each phase could involve running one or more design optimizations. We illustrate several advantages of design optimization in Fig. 1.3, which shows the notional variations of system performance, cost, and uncertainty as a function of time in design. When using optimization, the system performance increases more rapidly compared with the conventional process, achieving a better end result in a shorter total time. As a result, the cost of the design process is lower. Finally, the uncertainty in the performance reduces more rapidly as well.

Considering multiple disciplines or components using MDO amplifies the advantages illustrated in Fig. 1.3. The central idea of MDO is to consider the interactions between components using coupled models while simultaneously optimizing the design variables from the various components. In contrast, sequential optimization optimizes one component at a time. Even when interactions are considered, sequential optimization might converge to a suboptimal result (see Section 13.1 for more details and examples).

In this book, we tend to frame problems and discussions in the context of engineering design. However, the optimization methods are general and are used in other applications that may not be design problems, such as optimal control, machine learning, and regression. In other words, we mean "design" in a general sense, where variables are changed to optimize an objective.

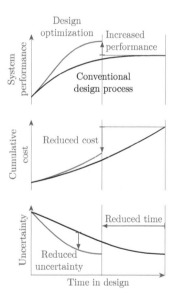

Fig. 1.3 Compared with the conventional design process, MDO increases the system performance, decreases the design time, reduces the total cost, and reduces the uncertainty at a given point in time.

1.2 Optimization Problem Formulation

The design optimization process requires the designer to translate their intent to a mathematical statement that can then be solved by an optimization algorithm. Developing this statement has the added benefit that it helps the designer better understand the problem. Being methodical in the formulation of the optimization problem is vital because *the optimizer tends to exploit any weaknesses you might have in your formulation or model.* An inadequate problem formulation can either cause the optimization to fail or cause it to converge to a mathematical optimum that is undesirable or unrealistic from an engineering point of view—the proverbial "right answer to the wrong question".

To formulate design optimization problems, we follow the procedure outlined in Fig. 1.4. The first step requires writing a description of the design problem, including a description of the system, and a statement of all the goals and requirements. At this point, the description does not necessarily involve optimization concepts and is often vague.

The next step is to gather as much data and information as possible about the problem. Some information is already specified in the problem statement, but more research is usually required to find all the relevant data on the performance requirements and expectations. Raw data might need to be processed and organized to gather the information required for the design problem. The more familiar practitioners are with the problem, the better prepared they will be to develop a sound formulation to identify eventual issues in the solutions.

At this stage, it is also essential to identify the analysis procedure and gather information on that as well. The analysis might consist of a simple model or a set of elaborate tools. All the possible inputs and outputs of the analysis should be identified, and its limitations should be understood. The computational time for the analysis needs to be considered because optimization requires repeated analysis.

It is usually impossible to learn everything about the problem before proceeding to the next steps, where we define the design variables, objective, and constraints. Therefore, information gathering and refinement are ongoing processes in problem formulation.

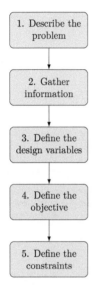

Fig. 1.4 Steps in optimization problem formulation.

1.2.1 Design Variables

The next step is to identify the variables that describe the system, the *design variables,** which we represent by the column vector:

$$x = [x_1, x_2, \ldots, x_{n_x}] \ . \tag{1.1}$$

*Some texts call these *decision variables* or simply *variables*.

1.2 Optimization Problem Formulation

This vector defines a given design, so different vectors x correspond to different designs. The number of variables, n_x, determines the problem's dimensionality.

The design variables must not depend on each other or any other parameter, and the optimizer must be free to choose the elements of x independently. This means that in the analysis of a given design, the variables must be input parameters that remain fixed throughout the analysis process. Otherwise, the optimizer does not have absolute control of the design variables. Another possible pitfall is to define a design variable that happens to be a linear combination of other variables, which results in an ill-defined optimization problem with an infinite number of combinations of design variable values that correspond to the same design.

The choice of variables is usually not unique. For example, a square shape can be parametrized by the length of its side or by its area, and different unit systems can be used. The choice of units affects the problem's scaling but not the functional form of the problem.

The choice of design variables can affect the functional form of the objective and constraints. For example, some nonlinear relationships can be converted to linear ones through a change of variables. It is also possible to introduce or eliminate discontinuities through the choice of design variables.

A given set of design variable values defines the system's design, but whether this system satisfies all the requirements is a separate question that will be addressed with the constraints in a later step. However, it is possible and advisable to define the space of allowable values for the design variables based on the design problem's specifications and physical limitations.

The first consideration in the definition of the allowable design variable values is whether the design variables are *continuous* or *discrete*. Continuous design variables are real numbers that are allowed to vary continuously within a specified range with no gaps, which we write as

$$\underline{x}_i \leq x_i \leq \overline{x}_i, \quad i = 1, \ldots, n_x, \tag{1.2}$$

where $\underline{x}$ and $\overline{x}$ are lower and upper bounds on the design variables, respectively. These are also known as *bound constraints* or *side constraints*. Some design variables may be unbounded or bounded on only one side.

When all the design variables are continuous, the optimization problem is said to be continuous.[†] Most of this book focuses on algorithms that assume continuous design variables.

[†] This is not to be confused with the continuity of the objective and constraint functions, which we discuss in Section 1.3.

When one or more variables are allowed to have discrete values, whether real or integer, we have a discrete optimization problem. An example of a discrete design variable is structural sizing, where only components of specific thicknesses or cross-sectional areas are available. Integer design variables are a special case of discrete variables where the values are integers, such as the number of wheels on a vehicle. Optimization algorithms that handle discrete variables are discussed in Chapter 8.

We distinguish the design variable bounds from constraints because the optimizer has direct control over their values, and they benefit from a different numerical treatment when solving an optimization problem. When defining these bounds, we must take care not to unnecessarily constrain the design space, which would prevent the optimizer from achieving a better design that is realizable. A smaller allowable range in the design variable values should make the optimization easier. However, design variable bounds should be based on actual physical constraints instead of being artificially limited. An example of a physical constraint is a lower bound on structural thickness in a weight minimization problem, where otherwise, the optimizer will discover that negative sizes yield negative weight. Whenever a design variable converges to the bound at the optimum, the designer should reconsider the reasoning for that bound and make sure it is valid. This is because designers sometimes set bounds that limit the optimization from obtaining a better objective.

At the formulation stage, we should strive to list as many independent design variables as possible. However, it is advisable to start with a small set of variables when solving a problem for the first time and then gradually expand the set of design variables.

Some optimization algorithms require the user to provide initial design variable values. This initial point is usually based on the best guess the user can produce. This might be an already good design that the optimization refines further by making small changes. Another possibility is that the initial guess is a bad design or a "blank slate" that the optimization changes significantly.

Example 1.1 Design variables for wing design

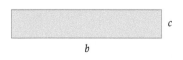

Fig. 1.5 Wingspan (b) and chord (c).

Consider a wing design problem where the wing planform shape is rectangular. The planform could be parametrized by the span (b) and the chord (c), as shown in Fig. 1.5, so that $x = [b, c]$. However, this choice is not unique. Two other variables are often used in aircraft design: wing area (S) and wing aspect ratio (AR), as shown in Fig. 1.6. Because these variables are not independent ($S = bc$ and $AR = b^2/S$), we cannot just add them to the set

1.2 Optimization Problem Formulation

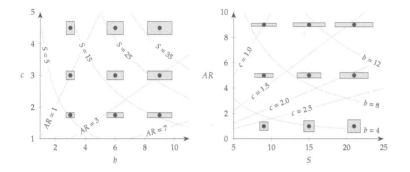

Fig. 1.6 Wing design space for two different sets of design variables, $x = [b, c]$ and $x = [S, AR]$.

of design variables. Instead, we must pick any two variables out of the four to parametrize the design because we have four possible variables and two dependency relationships.

For this wing, the variables must be positive to be physically meaningful, so we must remember to explicitly bound these variables to be greater than zero in an optimization. The variables should be bound from below by small positive values because numerical models are probably not prepared to take zero values. No upper bound is needed unless the optimization algorithm requires it.

Tip 1.1 Use splines to parameterize curves

Many problems that involve shapes, functional distributions, and paths are sometimes implemented with a large number of discrete points. However, these can be represented more compactly with splines. This is a commonly used technique in optimization because reducing the number of design variables often speeds up an optimization with little if any loss in the model parameterization fidelity. Figure 1.7 shows an example spline describing the shape of a turbine blade. In this example, only four design variables are used to represent the curved shape.

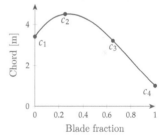

Fig. 1.7 Parameterizing the chord distribution of a wing or turbine blade using a spline reduces the number of design variables while still allowing for a wide range of shape changes.

1.2.2 Objective Function

To find the best design, we need an *objective function*, which is a quantity that determines if one design is better than another. This function must be a scalar that is computable for a given design variable vector x. The objective function can be minimized or maximized, depending on the problem. For example, a designer might want to minimize the weight or cost of a given structure. An example of a function to be maximized could be the range of a vehicle.

The convention adopted in this book is that the objective function, f, is to be *minimized*. This convention does not prevent us from maximizing a function because we can reformulate it as a minimization problem by finding the minimum of the negative of f and then changing the sign, as follows:

$$\max[f(x)] = -\min[-f(x)]. \tag{1.3}$$

This transformation is illustrated in Fig. 1.8.‡

‡Inverting the function $(1/f)$ is another way to turn a maximization problem into a minimization problem, but it is generally less desirable because it alters the scale of the problem and could introduce a divide-by-zero problem.

The objective function is computed through a numerical model whose complexity can range from a simple explicit equation to a system of coupled implicit models (more on this in Chapter 3).

The choice of objective function is crucial for successful design optimization. If the function does not represent the true intent of the designer, it does not matter how precisely the function and its optimum point are computed—the mathematical optimum will be non-optimal from the engineering point of view. A bad choice for the objective function is a common mistake in design optimization.

The choice of objective function is not always obvious. For example, minimizing the weight of a vehicle might sound like a good idea, but this might result in a vehicle that is too expensive to manufacture. In this case, manufacturing cost would probably be a better objective. However, there is a trade-off between manufacturing cost and the performance of the vehicle. It might not be obvious which of these objectives is the most appropriate one because this trade-off depends on customer preferences. This issue motivates *multiobjective optimization*, which is the subject of Chapter 9. Multiobjective optimization does not yield a single design but rather a range of designs that settle for different trade-offs between the objectives.

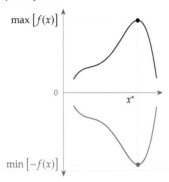

Fig. 1.8 A maximization problem can be transformed into an equivalent minimization problem.

Experimenting with different objectives should be part of the design exploration process (this is represented by the outer loop in the design optimization process in Fig. 1.2). Results from optimizing the "wrong" objective can still yield insights into the design trade-offs and trends for the system at hand.

In Ex. 1.1, we have the luxury of being able to visualize the design space because we have only two variables. For more than three variables, it becomes impossible to visualize the design space. We can also visualize the objective function for two variables, as shown in Fig. 1.9. In this figure, we plot the function values using the vertical axis, which results in a three-dimensional surface. Although plotting the surface might provide intuition about the function, it is not possible to locate the points accurately when drawing on a two-dimensional surface.

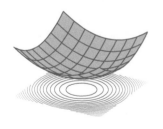

Fig. 1.9 A function of two variables ($f = x_1^2 + x_2^2$ in this case) can be visualized by plotting a three-dimensional surface or contour plot.

Another possibility is to plot the contours of the function, which are lines of constant value, as shown in Fig. 1.10. We prefer this type of plot and use it extensively throughout this book because we can locate points accurately and get the correct proportions in the axes (in Fig. 1.10, the contours are perfect circles, and the location of the minimum is clear). Our convention is to represent lower function values with darker lines and higher values with lighter ones. Unless otherwise stated, the function variation between two adjacent lines is constant, and therefore, the closer together the contour lines are, the faster the function is changing. The equivalent of a contour line in n-dimensional space is a hypersurface of constant value with dimensions of $n-1$, called an *isosurface*.

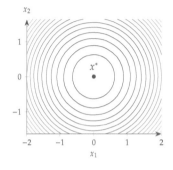

Fig. 1.10 Contour plot of $f = x_1^2 + x_2^2$.

Example 1.2 Objective function for wing design

Let us discuss the appropriate objective function for Ex. 1.1 for a small airplane. A common objective for a wing is to minimize drag. However, this does not take into account the propulsive efficiency, which is strongly affected by speed. A better objective might be to minimize the required power, which balances drag and propulsive efficiency.§

§The simple models used in this example are described in Appendix D.1.6.

The contours for the required power are shown in Fig. 1.11 for the two choices of design variable sets discussed in Ex. 1.1. We can locate the minimum graphically (denoted by the dot). Although the two optimum solutions are the same, the shapes of the objective function contours are different. In this case, using the aspect ratio and wing area simplifies the relationship between the design variables and the objective by aligning the two main curvature trends with each design variable. Thus, the parameterization can change the effectiveness of the optimization.

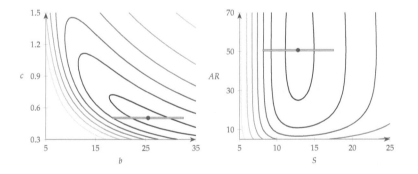

Fig. 1.11 Required power contours for two different choices of design variable sets. The optimal wing is the same for both cases, but the functional form of the objective is simplified in the one on the right.

The optimal wing for this problem has an aspect ratio that is much higher than that typically seen in airplanes or birds. Although the high aspect ratio increases aerodynamic efficiency, it adversely affects the structural strength, which we did not consider here. Thus, as in most engineering problems, we need to add constraints and consider multiple disciplines.

We use mostly two-dimensional examples throughout the book because we can visualize them conveniently. Such visualizations should give you an intuition about the methods and problems. However, keep in mind that general problems have many more dimensions, and only mathematics can help you in such cases.

Although we can sometimes visualize the variation of the objective function in a contour plot as in Ex. 1.2, this is not possible for problems with more design variables or more computationally demanding function evaluations. This motivates numerical optimization algorithms, which aim to find the minimum in a multidimensional design space using as few function evaluations as possible.

1.2.3 Constraints

The vast majority of practical design optimization problems require the enforcement of constraints. These are functions of the design variables that we want to restrict in some way. Like the objective function, constraints are computed through a model whose complexity can vary widely. The *feasible region* is the set of points that satisfy all constraints. We seek to minimize the objective function within this feasible design space.

When we restrict a function to being equal to a fixed value, we call this an *equality constraint*, denoted by $h(x) = 0$. When the function is required to be less than or equal to a certain value, we have an *inequality constraint*, denoted by $g(x) \leq 0$.¶ Although we use the "less or equal" convention, some texts and software programs use "greater or equal" instead. There is no loss of generality with either convention because we can always multiply the constraint by -1 to convert between the two.

¶A strict inequality, $g(x) < 0$, is never used because then x could be arbitrarily close to the equality. Because the optimum is at $g = 0$ for an active constraint, the exact solution would then be ill-defined from a mathematical perspective. Also, the difference is not meaningful when using finite-precision arithmetic (which is always the case when using a computer).

Tip 1.2 Check the inequality convention

When using optimization software, do not forget to check the convention for the inequality constraints (i.e., determine whether it is "less than", "greater than", or "allow two-sided constraints") and convert your constraints as needed.

Some texts and papers omit the equality constraints without loss of generality because an equality constraint can be replaced by two inequality constraints. More specifically, an equality constraint, $h(x) = 0$, is equivalent to enforcing two inequality constraints, $h(x) \geq 0$ and $h(x) \leq 0$.

1.2 Optimization Problem Formulation

Inequality constraints can be *active* or *inactive* at the optimum point. An active inequality constraint means that $g(x^*) = 0$, whereas for an inactive one, $g(x^*) < 0$. If a constraint is inactive at the optimum, this constraint could have been removed from the problem with no change in its solution, as illustrated in Fig. 1.12. In this case, constraints g_2 and g_3 can be removed without affecting the solution of the problem. Furthermore, active constraints (g_1 in this case) can equivalently be replaced by equality constraints. However, it is difficult to know in advance which constraints are active or not at the optimum for a general problem. Constrained optimization is the subject of Chapter 5.

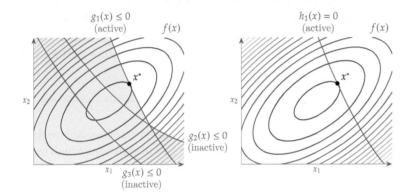

Fig. 1.12 Two-dimensional problem with one active and two inactive inequality constraints (left). The shaded area indicates regions that are *infeasible* (i.e., the constraints are violated). If we only had the active single equality constraint in the formulation, we would obtain the same result (right).

It is possible to overconstrain the problem such that there is no solution. This can happen as a result of a programming error but can also occur at the problem formulation stage. For more complicated design problems, it might not be possible to satisfy all the specified constraints, even if they seem to make sense. When this happens, constraints have to be relaxed or removed.

The problem must not be overconstrained, or else there is no feasible region in the design space over which the function can be minimized. Thus, the number of independent equality constraints must be less than or equal to the number of design variables ($n_h \leq n_x$). There is no limit on the number of inequality constraints. However, they must be such that there is a feasible region, and the number of active constraints plus the equality constraints must still be less than or equal to the number of design variables.

The feasible region grows when constraints are removed and shrinks when constraints are added (unless these constraints are redundant). As the feasible region grows, the optimum objective function usually improves or at least stays the same. Conversely, the optimum worsens or stays the same when the feasible region shrinks.

One common issue in optimization problem formulation is distinguishing objectives from constraints. For example, we might be tempted to minimize the stress in a structure, but this would inevitably result in an overdesigned, heavy structure. Instead, we might want minimum weight (or cost) with sufficient safety factors on stress, which can be enforced by an inequality constraint.

Most engineering problems require constraints—often a large number of them. Although constraints may at first appear limiting, they enable the optimizer to find useful solutions.

As previously mentioned, some algorithms require the user to provide an initial guess for the design variable values. Although it is easy to assign values within the bounds, it might not be as easy to ensure that the initial design satisfies the constraints. This is not an issue for most optimization algorithms, but some require starting with a feasible design.

Example 1.3 Constraints for wing design

We now add a design constraint for the power minimization problem of Ex. 1.2. The unconstrained optimal wing had unrealistically high aspect ratios because we did not include structural considerations. If we add an inequality constraint on the bending stress at the root of the wing for a fixed amount of material, we get the curve and feasible region shown in Fig. 1.13. The unconstrained optimum violates this constraint. The constrained optimum results in a lower span and higher chord, and the constraint is active.

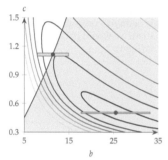

Fig. 1.13 Minimum-power wing with a constraint on bending stress compared with the unconstrained case.

As previously mentioned, it is generally not possible to visualize the design space as shown in Ex. 1.2 and obtain the solution graphically. In addition to the possibility of a large number of design variables and computationally expensive objective function evaluations, we now add the possibility of a large number of constraints, which might also be expensive to evaluate. Again, this is further motivation for the optimization techniques covered in this book.

1.2.4 Optimization Problem Statement

Now that we have discussed the design variables, the objective function, and constraints, we can put them all together in an optimization problem statement. In words, this statement is as follows: *minimize the objective function by varying the design variables within their bounds subject to the constraints.*

1.2 Optimization Problem Formulation

Mathematically, we write this statement as follows:

$$\begin{aligned}
\text{minimize} \quad & f(x) \\
\text{by varying} \quad & \underline{x}_i \leq x_i \leq \overline{x}_i \quad i = 1, \ldots, n_x \\
\text{subject to} \quad & g_j(x) \leq 0 \quad j = 1, \ldots, n_g \\
& h_l(x) = 0 \quad l = 1, \ldots, n_h \, .
\end{aligned} \quad (1.4)$$

This is the standard formulation used in this book; however, other books and software manuals might differ from this.∥ For example, they might use different symbols, use "greater than or equal to" for the inequality constraint, or maximize instead of minimizing. In any case, it is possible to convert between standard formulations to get equivalent problems.

∥ Instead of "by varying", some textbooks use "with respect to" or "w.r.t." as shorthand.

All continuous optimization problems with a single-objective can be written in the standard form shown in Eq. 1.4. Although our target applications are engineering design problems, many other problems can be stated in this form, and thus, the methods covered in this book can be used to solve those problems.

The values of the objective and constraint functions for a given set of design variables are computed through the analysis, which consists of one or more numerical models. The analysis must be fully automatic so that multiple optimization cycles can be completed without human intervention, as shown in Fig. 1.14. The optimizer usually requires an initial design x_0 and then queries the analysis for a sequence of designs until it finds the optimum design, x^*.

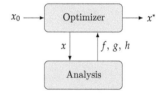

Fig. 1.14 The analysis computes the objective (f) and constraint values (g, h) for a given set of design variables (x).

Tip 1.3 Using an optimization software package

The setup of an optimization problem varies depending on the particular software package, so read the documentation carefully. Most optimization software requires you to define the objective and constraints as *callback functions*. These are passed to the optimizer, which calls them back as needed during the optimization process. The functions take the design variable values as inputs and output the function values, as shown in Fig. 1.14. Study the software documentation for the details on how to use it.** To make sure you understand how to use a given optimization package, test it on simple problems for which you know the solution first (see Prob. 1.4).

**Optimization software resources include the optimization toolboxes in MATLAB, scipy.optimize.minimize in Python, Optim.jl or Ipopt.jl in Julia, NLopt for multiple languages, and the Solver add-in in Microsoft Excel. The pyOptSparse framework provides a common Python wrapper for many existing optimization codes and facilitates the testing of different methods.[1] SNOW.jl wraps a few optimizers and multiple derivative computation methods in Julia.

1. Wu et al., *pyOptSparse: A Python framework for large-scale constrained nonlinear optimization of sparse systems*, 2020.

When the optimizer queries the analysis for a given x, for most methods, the constraints do not have to be feasible. The optimizer is responsible for changing x so that the constraints are satisfied.

The objective and constraint functions must depend on the design variables; if a function does not depend on any variable in the whole domain, it can be ignored and should not appear in the problem statement.

Ideally, f, g, and h should be computable for all values of x that make physical sense. Lower and upper design variable bounds should be set to avoid nonphysical designs as much as possible. Even after taking this precaution, models in the analysis sometimes fail to provide a solution. A good optimizer can handle such eventualities gracefully.

There are some mathematical transformations that do not change the solution of the optimization problem (Eq. 1.4). Multiplying either the objective or the constraints by a constant does not change the optimal design; it only changes the optimum objective value. Adding a constant to the objective does not change the solution, but adding a constant to any constraint changes the feasible space and can change the optimal design.

Determining an appropriate set of design variables, objective, and constraints is a crucial aspect of the outer loop shown in Fig. 1.2, which requires human expertise in engineering design and numerical optimization.

Tip 1.4 Ease into the problem

It is tempting to set up the full problem and attempt to solve it right away. This rarely works, especially for a new problem. Before attempting any optimization, you should run the analysis models and explore the solution space manually. Particularly if using gradient-based methods, it helps to plot the output functions across multiple input sweeps to assess if the numerical outputs display the expected behavior and smoothness.

Instead of solving the full problem, ease into it by setting up the simplest subproblem possible. If the function evaluations are costly, consider using computational models that are less costly (but still representative). It is advisable to start by solving a subproblem with a small set of variables and then gradually expand it. The removal of some constraints has to be done more carefully because it might result in an ill-defined problem. For multidisciplinary problems, you should run optimizations with each component separately before attempting to solve the coupled problem.

Solving simple problems for which you know the answer (or at least problems for which you know the trends) helps identify any issues with the models and problem formulation. Solving a sequence of increasingly complicated problems gradually builds an understanding of how to solve the optimization problem and interpret its results.

1.3 Optimization Problem Classification

To choose the most appropriate optimization algorithm for solving a given optimization problem, we must classify the optimization problem and know how its attributes affect the efficacy and suitability of the available optimization algorithms. This is important because no optimization algorithm is efficient or even appropriate for all types of problems.

We classify optimization problems based on two main aspects: the problem formulation and the characteristics of the objective and constraint functions, as shown in Fig. 1.15.

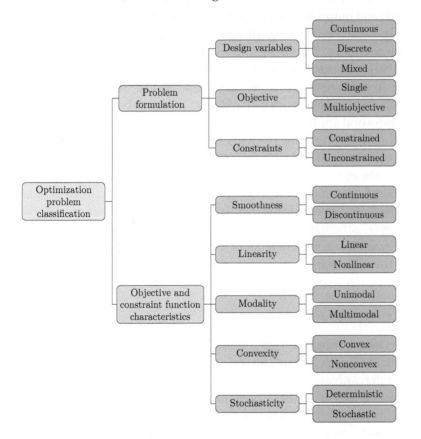

Fig. 1.15 Optimization problems can be classified by attributes associated with the different aspects of the problem. The two main aspects are the problem formulation and the objective and constraint function characteristics.

In the problem formulation, the design variables can be either discrete or continuous. Most of this book assumes continuous design variables, but Chapter 8 provides an introduction to discrete optimization. When the design variables include both discrete and continuous variables, the problem is said to be *mixed*. Most of the book assumes a single objective function, but we explain how to solve multiobjective problems in Chapter 9. Finally, as previously mentioned, unconstrained problems are rare in engineering design optimization. However, we explain unconstrained optimization algorithms (Chapter 4) because they provide the foundation for constrained optimization algorithms (Chapter 5).

The characteristics of the objective and constraint functions also determine the type of optimization problem at hand and ultimately limit the type of optimization algorithm that is appropriate for solving the optimization problem.

In this section, we will view the function as a "black box", that is, a computation for which we only see inputs (including the design variables) and outputs (including objective and constraints), as illustrated in Fig. 1.16. When dealing with black-box models, there is limited or no understanding of the modeling and numerical solution process used to obtain the function values. We discuss these types of models and how to solve them in Chapter 3, but here, we can still characterize the functions based purely on their outputs. The black-box view is common in real-world applications. This might be because the source code is not provided, the modeling methods are not described, or simply because the user does not bother to understand them.

Fig. 1.16 A model is considered a black box when we only see its inputs and outputs.

In the remainder of this section, we discuss the attributes of objectives and constraints shown in Fig. 1.15. Strictly speaking, many of these attributes cannot typically be identified from a black-box model. For example, although the model may appear smooth, we cannot know that it is smooth everywhere without a more detailed inspection. However, for this discussion, we assume that the black box's outputs can be exhaustively explored so that these characteristics can be identified.

1.3.1 Smoothness

The degree of function smoothness with respect to variations in the design variables depends on the continuity of the function values and their derivatives. When the value of the function varies continuously, the function is said to be C^0 continuous. If the first derivatives also vary continuously, then the function is C^1 continuous, and so on. A function is smooth when the derivatives of all orders vary continuously everywhere in its domain. Function smoothness with respect to continuous design variables affects what type of optimization algorithm can be used. Figure 1.17 shows one-dimensional examples for a discontinuous function, a C^0 function, and a C^1 function.

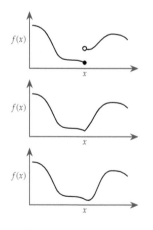

Fig. 1.17 Discontinuous function (top), C^0 continuous function (middle), and C^1 continuous function (bottom).

As we will see later, discontinuities in the function value or derivatives limit the type of optimization algorithm that can be used because some algorithms assume C^0, C^1, and even C^2 continuity. In practice, these algorithms usually still work with functions that have only a few discontinuities that are located away from the optimum.

1.3.2 Linearity

The functions of interest could be linear or nonlinear. When both the objective and constraint functions are linear, the optimization problem is known as a *linear optimization problem*. These problems are easier to solve than general nonlinear ones, and there are entire books and courses dedicated to the subject. The first numerical optimization algorithms were developed to solve linear optimization problems, and there are many applications in operations research (see Chapter 2). An example of a linear optimization problem is shown in Fig. 1.18.

When the objective function is quadratic and the constraints are linear, we have a quadratic optimization problem, which is another type of problem for which specialized solution methods exist.* Linear optimization and quadratic optimization are covered in Sections 5.5, 11.2, and 11.3.

Although many problems can be formulated as linear or quadratic problems, most engineering design problems are nonlinear. However, it is common to have at least a subset of constraints that are linear, and some general nonlinear optimization algorithms take advantage of the techniques used to solve linear and quadratic problems.

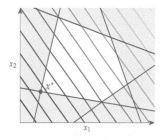

Fig. 1.18 Example of a linear optimization problem in two dimensions.

*Historically, optimization problems were referred to as *programming* problems, so much of the existing literature refers to linear optimization as *linear programming* and similarly for other types of optimization.

1.3.3 Multimodality and Convexity

Functions can be either unimodal or multimodal. Unimodal functions have a single minimum, whereas multimodal functions have multiple minima. When we find a minimum without knowledge of whether the function is unimodal or not, we can only say that it is a *local minimum*; that is, this point is better than any point within a small neighborhood. When we know that a local minimum is the best in the whole domain (because we somehow know that the function is unimodal), then this is also the *global minimum*, as illustrated in Fig. 1.19. Sometimes, the function might be flat around the minimum, in which case we have a *weak minimum*.

For functions involving more complicated numerical models, it is usually impossible to prove that the function is unimodal. Proving that such a function is unimodal would require evaluating the function at every point in the domain, which is computationally prohibitive. However, it much easier to prove multimodality—all we need to do is find two distinct local minima.

Fig. 1.19 Types of minima.

Just because a function is complicated or the design space has many dimensions, it does not mean that the function is multimodal. By default, we should not assume that a given function is either unimodal or multimodal. As we explore the problem and solve it starting from different points or using different optimizers, there are two main possibilities.

One possibility is that we find more than one minimum, thus proving that the function is multimodal. To prove this conclusively, we must make sure that the minima do indeed satisfy the mathematical optimality conditions with good enough precision.

The other possibility is that the optimization consistently converges to the same optimum. In this case, we can become increasingly confident that the function is unimodal with every new optimization that converges to the same optimum.[†]

[†] For example, Lyu et al.[2] and He et al.[3] show consistent convergence to the same optimum in an aerodynamic shape optimization problem.

2. Lyu et al., *Aerodynamic Shape Optimization Investigations of the Common Research Model Wing Benchmark*, 2015.

3. He et al., *Robust aerodynamic shape optimization—From a circle to an airfoil*, 2019.

Often, we need not be too concerned about the possibility of multiple local minima. From an engineering design point of view, achieving a local optimum that is better than the initial design is already a useful result.

Convexity is a concept related to multimodality. A function is convex if all line segments connecting any two points in the function lie above the function and never intersect it. Convex functions are always unimodal. Also, all multimodal functions are nonconvex, but not all unimodal functions are convex (see Fig. 1.20).

Convex optimization seeks to minimize convex functions over convex sets. Like linear optimization, convex optimization is another subfield of numerical optimization with many applications. When the objective and constraints are convex functions, we can use specialized formulations and algorithms that are much more efficient than general nonlinear algorithms to find the global optimum, as explained in Chapter 11.

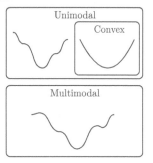

Fig. 1.20 Multimodal functions have multiple minima, whereas unimodal functions have only one minimum. All multimodal functions are nonconvex, but not all unimodal functions are convex.

1.3.4 Deterministic versus Stochastic

Some functions are inherently stochastic. A stochastic model yields different function values for repeated evaluations with the same input (Fig. 1.21). For example, the numerical value from a roll of dice is a stochastic function.

Stochasticity can also arise from deterministic models when the inputs are subject to *uncertainty*. The input variables are then described as probability distributions, and their uncertainties need to be propagated through the model. For example, the bending stress in a beam may follow a deterministic model, but the beam's geometric properties may be subject to uncertainty because of manufacturing deviations. For most of this text, we assume that functions are deterministic, except in Chapter 12, where we explain how to perform optimization when the model inputs are uncertain.

1.4 Optimization Algorithms

No single optimization algorithm is effective or even appropriate for all possible optimization problems. This is why it is important to understand the problem before deciding which optimization algorithm to use. By "effective" algorithm, we mean that the algorithm can solve the problem, and secondly, it does so reliably and efficiently. Figure 1.22 lists the attributes for the classification of optimization algorithms, which we cover in more detail in the following discussion. These attributes are often amalgamated, but they are independent, and any combination is possible. In this text, we cover a wide variety of optimization algorithms corresponding to several of these combinations. However, this overview still does not cover a wide variety of specialized algorithms designed to solve specific problems where a particular structure can be exploited.

When multiple models are involved, we also need to consider how the models are coupled, solved, and integrated with the optimizer. These considerations lead to different MDO *architectures*, which may

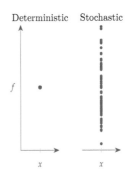

Fig. 1.21 Deterministic functions yield the same output when evaluated repeatedly for the same input, whereas stochastic functions do not.

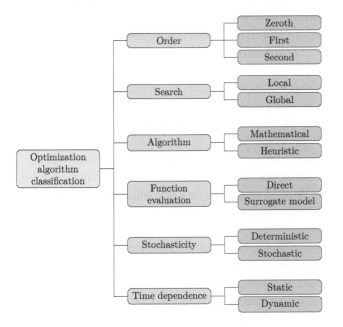

Fig. 1.22 Optimization algorithms can be classified by using the attributes in the rightmost column. As in the problem classification step, these attributes are independent, and any combination is possible.

involve multiple levels of optimization problems. Coupled models and MDO architectures are covered in Chapter 13.

1.4.1 Order of Information

At the minimum, an optimization algorithm requires users to provide the models that compute the objective and constraint values—zeroth-order information—for any given set of allowed design variables. We call algorithms that use only these function values *gradient-free* algorithms (also known as *derivative-free* or *zeroth-order* algorithms). We cover a selection of these algorithms in Chapters 7 and 8. The advantage of gradient-free algorithms is that the optimization is easier to set up because they do not need additional computations other than what the models for the objective and constraints already provide.

Gradient-based algorithms use gradients of both the objective and constraint functions with respect to the design variables—first-order information. The gradients provide much richer information about the function behavior, which the optimizer can use to converge to the optimum more efficiently. Figure 1.23 shows how the cost of gradient-based versus gradient-free optimization algorithms typically scales when the number of design variables increases. The number of function evaluations required by gradient-free methods increases dramatically, whereas the number of evaluations required by gradient-based methods does not increase as much and is many orders of magnitude lower for the larger numbers of design variables.

In addition, gradient-based methods use more rigorous criteria for optimality. The gradients are used to establish whether the optimizer converged to a point that satisfies mathematical optimality conditions, something that is difficult to verify in a rigorous way without gradients.

We first cover gradient-based algorithms for unconstrained problems in Chapter 4 and then extend them to constrained problems in Chapter 5. Gradient-based algorithms also include algorithms that use curvature—second-order information. Curvature is even richer information that tells us the rate of the change in the gradient, which provides an idea of where the function might flatten out.

There is a distinction between the order of information provided by the user and the order of information actually used in the algorithm. For example, a user might only provide function values to a gradient-based algorithm and rely on the algorithm to internally estimate gradients. Optimization algorithms estimate the gradients by requesting additional function evaluations for finite difference approximations (see Section 6.4). Gradient-based algorithms can also internally estimate

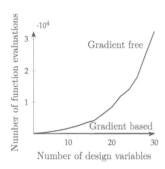

Fig. 1.23 Gradient-based algorithms scale much better with the number of design variables. In this example, the gradient-based curve (with exact derivatives) grows from 67 to 206 function calls, but it is overwhelmed by the gradient-free curve, which grows from 103 function calls to over 32,000.

curvature based on gradient values (see Section 4.4.4).

In theory, gradient-based algorithms require the functions to be sufficiently smooth (at least C^1 continuous). However, in practice, they can tolerate the occasional discontinuity, as long as this discontinuity is not near the optimum.

We devote a considerable portion of this book to gradient-based algorithms because they scale favorably with the number of design variables, and they have rigorous mathematical criteria for optimality. We also cover the various approaches for computing gradients in detail because the accurate and efficient computation of these gradients is crucial for the efficacy and efficiency of these methods (see Chapter 6).

Current state-of-the-art optimization algorithms also use second-order information to implement Newton-type methods for second-order convergence. However, these algorithms tend to build second-order information based on the provided gradients, as opposed to requiring users to provide the second-order information directly (see Section 4.4.4).

Because gradient-based methods require accurate gradients and smooth enough functions, they require more knowledge about the models and optimization algorithm than gradient-free methods. Chapters 3 through 6 are devoted to making the power of gradient-based methods more accessible by providing the necessary theoretical and practical knowledge.

1.4.2 Local versus Global Search

The many ways to search the design space can be classified as being local or global. A local search takes a series of steps starting from a single point to form a sequence of points that hopefully converges to a local optimum. In spite of the name, local methods can traverse large portions of the design space and can even step between convex regions (although this happens by chance). A global search tries to span the whole design space in the hope of finding the global optimum. As previously mentioned when discussing multimodality, even when using a global method, we cannot prove that any optimum found is a global one except for particular cases.

The classification of local versus global searches often gets conflated with the gradient-based versus gradient-free attributes because gradient-based methods usually perform a local search. However, these should be viewed as independent attributes because it is possible to use a global search strategy to provide starting points for a gradient-based algorithm. Similarly, some gradient-free algorithms are based on local search strategies.

The choice of search type is intrinsically linked to the modality of the design space. If the design space is unimodal, then a local search is sufficient because it converges to the global optimum. If the design space is multimodal, a local search converges to an optimum that might be local (or global if we are lucky enough). A global search increases the likelihood that we converge to a global optimum, but this is by no means guaranteed.

1.4.3 Mathematical versus Heuristic

There is a big divide regarding the extent to which an algorithm is based on provable mathematical principles versus heuristics. Optimization algorithms require an iterative process, which determines the sequence of points evaluated when searching for an optimum, and optimality criteria, which determine when the iterative process ends. Heuristics are rules of thumb or commonsense arguments that are not based on a strict mathematical rationale.

Gradient-based algorithms are usually based on mathematical principles, both for the iterative process and for the optimality criteria. Gradient-free algorithms are more evenly split between the mathematical and heuristic for both the optimality criteria and the iterative procedure. The mathematical gradient-free algorithms are often called *derivative-free optimization* algorithms. Heuristic gradient-free algorithms include a wide variety of nature-inspired algorithms (see Section 7.2).

Heuristic optimality criteria are an issue because, strictly speaking, they do not prove that a given point is a local (let alone global) optimum; they are only expected to find a point that is "close enough". This contrasts with mathematical optimality criteria, which are unambiguous about (local) optimality and converge to the optimum within the limits of the working precision. This is not to suggest that heuristic methods are not useful. Finding a better solution is often desirable regardless of whether or not it is strictly optimal. Not converging tightly to optimality criteria does, however, make it harder to compare results from different methods.

Iterative processes based on mathematical principles tend to be more efficient than those based on heuristics. However, some heuristic methods are more robust because they tend to make fewer assumptions about the modality and smoothness of the functions and handle noisy functions more effectively.

Most algorithms mix mathematical arguments and heuristics to some degree. Mathematical algorithms often include constants whose values end up being tuned based on experience. Conversely, algorithms primarily based on heuristics sometimes include steps with mathematical justification.

1.4.4 Function Evaluation

The optimization problem setup that we described previously assumes that the function evaluations are obtained by solving numerical models of the system. We call these *direct* function evaluations. However, it is possible to create *surrogate models* (also known as *metamodels*) of these models and use them in the optimization process. These surrogates can be interpolation-based or projection-based models. Surrogate-based optimization is discussed in Chapter 10.

1.4.5 Stochasticity

This attribute is independent of the stochasticity of the model that we mentioned previously, and it is strictly related to whether the optimization algorithm itself contains steps that are determined at random or not.

A deterministic optimization algorithm always evaluates the same points and converges to the same result, given the same initial conditions. In contrast, a stochastic optimization algorithm evaluates a different set of points if run multiple times from the same initial conditions, even if the models for the objective and constraints are deterministic. For example, most evolutionary algorithms include steps determined by generating random numbers. Gradient-based algorithms are usually deterministic, but some exceptions exist, such as stochastic gradient descent (see Section 10.5).

1.4.6 Time Dependence

In this book, we assume that the optimization problem is *static*. This means that we formulate the problem as a single optimization and solve the complete numerical model at each optimization iteration. In contrast, *dynamic optimization problems* solve a sequence of optimization problems to make decisions at different time instances based on information that becomes available as time progresses.

For some problems that involve time dependence, we can perform time integration to solve for the entire time history of the states and then compute the objective and constraint function values for an optimization iteration. This means that every optimization iteration requires solving for the entire time history. An example of this type of problem is a trajectory optimization problem where the design variables are the coordinates representing the path, and the objective is to minimize the total energy expended to get to a given destination.[4] Although such a problem involves a time dependence, we still classify it as static because we solve a single optimization problem. As a more specific example, consider a car going around a racetrack. We could optimize the time history of the throttle, braking, and steering of a car to get a trajectory that minimizes the total time in a known racetrack for fixed conditions. This is an open-loop optimal control problem because the car control is predetermined and does not react to any disturbances.

For dynamic optimization problems (also known as *dynamic programming*), the design variables are decisions made in a sequence of time steps.[5,6] The decision at a given time step is influenced by the decisions and system states from previous steps. Sometimes, the decision at a given time step also depends on a prediction of the states a few steps into the future.

4. Betts, *Survey of numerical methods for trajectory optimization*, 1998.

5. Bryson and Ho, *Applied Optimal Control; Optimization, Estimation, and Control*, 1969.

6. Bertsekas, *Dynamic Programming and Optimal Control*, 1995.

The car example that we previously mentioned could also be a dynamic optimization problem if we optimized the throttle, braking, or steering of a car at each time instance in response to some measured output. We could, for example, maximize the instantaneous acceleration based on real-time acceleration sensor information and thus react to varying conditions, such as surface traction. This is an example of a closed-loop (or feedback) optimal control problem, a type of dynamic optimization problem where a control law is optimized for a dynamical system over a period of time.

Dynamic optimization is not covered in this book, except in the context of discrete optimization (see Section 8.5). Different approaches are used in general, but many of the concepts covered here are instrumental in the numerical solution of dynamic optimization and optimal control problems.

1.5 Selecting an Optimization Approach

This section provides guidance on how to select an appropriate approach for solving a given optimization problem. This process cannot always be distilled to a simple decision tree; however, it is still helpful to have a framework as a first guide. Many of these decisions will become more apparent as you progress through the book and gain experience, so you may want to revisit this section periodically. Eventually, selecting an appropriate methodology will become second nature.

1.5 Selecting an Optimization Approach

Figure 1.24 outlines one approach to algorithm selection and also serves as an overview of the chapters in this book. The first two characteristics in the decision tree (convex problem and discrete variables) are not the most common within the broad spectrum of engineering optimization problems, but we list them first because they are the more restrictive in terms of usable optimization algorithms.

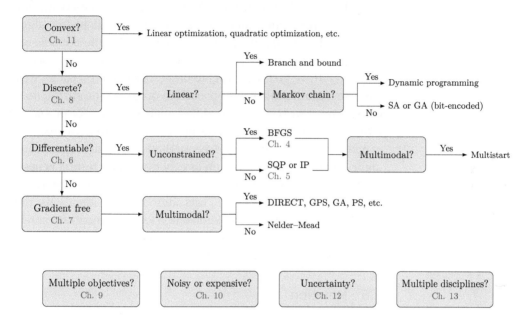

Fig. 1.24 Decision tree for selecting an optimization algorithm.

The first node asks about convexity. Although it is often not immediately apparent if the problem is convex, with some experience, we can usually discern whether we should attempt to reformulate it as a convex problem. In most instances, convexity occurs for problems with simple objectives and constraints (e.g., linear or quadratic), such as in control applications where the optimization is performed repeatedly. A convex problem can be solved with general gradient-based or gradient-free algorithms, but it would be inefficient not to take advantage of the convex formulation structure if we can do so.

The next node asks about discrete variables. Problems with discrete design variables are generally much harder to solve, so we might consider alternatives that avoid using discrete variables when possible. For example, a wind turbine's position in a field could be posed as a discrete variable within a discrete set of options. Alternatively, we could represent the wind turbine's position as a continuous variable with two continuous coordinate variables. That level of flexibility may or may not be desirable but generally leads to better solutions. Many problems are fundamentally discrete, and there is a wide variety of available methods.

Next, we consider whether the model is continuous and differentiable or can be made smooth through model improvements. If the problem is high dimensional (more than a few tens of variables as a rule of thumb), gradient-free algorithms are generally intractable and gradient-based algorithms are preferable. We would either need to make the model smooth enough to use a gradient-based algorithm or reduce the problem dimensionality to use a gradient-free algorithm. Another alternative if the problem is not readily differentiable is to use surrogate-based optimization (the box labeled "Noisy or expensive" in Fig. 1.24). If we go the surrogate-based optimization route, we could still use a gradient-based approach to optimize the surrogate model because most such models are differentiable. Finally, for problems with a relatively small number of design variables, gradient-free methods can be a good fit. Gradient-free methods have the largest variety of algorithms, and a combination of experience and testing is needed to determine an appropriate algorithm for the problem at hand.

The bottom row in Fig. 1.24 lists additional considerations: multiple objectives, surrogate-based optimization for noisy (nondifferentiable) or computationally expensive functions, optimization under uncertainty in the design variables and other model parameters, and MDO.

1.6 Notation

We do not use bold font to represent vectors or matrices. Instead, we follow the convention of many optimization and numerical linear algebra books, which try to use Greek letters (e.g., α and β) for scalars, lowercase roman letters (e.g., x and u) for vectors, and capitalized roman letters (e.g., A and H) for matrices. There are exceptions to this notation because of the wide variety of topics covered in this book and a desire not to deviate from the standard conventions used in each field. We explicitly note these exceptions as needed. For example, the objective function f is a scalar function and the Lagrange multipliers (λ and σ) are vectors.

By default, a vector x is a column vector, and thus x^T is a row vector. We denote the ith element of the vector as x_i, as shown in Fig. 1.25. For more compact notation, we may write a column vector horizontally, with its components separated by commas, for example, $x = [x_1, x_2, \ldots, x_n]$. We refer to a vector with n components as an n-vector, which is equivalent to writing $x \in \mathbb{R}^n$.

An ($n \times m$) matrix has n rows and m columns, which is equivalent to defining $A \in \mathbb{R}^{n \times m}$. The matrix element A_{ij} is the element in the ith

Fig. 1.25 An n-vector, x.

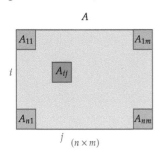

Fig. 1.26 An ($n \times m$) matrix, A.

row of the *j*the column, as shown in Fig. 1.26. Occasionally, additional letters beyond i and j are needed for indices, but those are explicitly noted when used.

The subscript k usually refers to iteration number. Thus, x_k is the complete vector x at iteration k. The subscript zero is used for the same purpose, so x_0 would be the complete vector x at the initial iteration. Other subscripts besides those listed are used for naming. A superscript star (x^*) refers to a quantity at the optimum.

Tip 1.5 Work out the dimensions of the vectors and matrices

As you read this book, we encourage you to work out the dimensions of the vectors and matrices in the operations within each equation and verify the dimensions of the result for consistency. This will enhance your understanding of the equations.

1.7 Summary

Optimization is compelling, and there are opportunities to apply it everywhere. Numerical optimization fully automates the design process but requires expertise in the problem formulation, optimization algorithm selection, and the use of that algorithm. Finally, design expertise is also required to interpret and critically evaluate the results given by the optimization.

There is no single optimization algorithm that is effective in the solution of all types of problems. It is crucial to classify the optimization problem and understand the optimization algorithms' characteristics to select the appropriate algorithm to solve the problem.

In seeking a more automated design process, we must not dismiss the value of engineering intuition, which is often difficult (if not impossible) to convert into a rigid problem formulation and algorithm.

Problems

1.1 Answer *true* or *false* and justify your answer.

   a. MDO arose from the need to consider multiple design objectives.

   b. The preliminary design phase takes place after the conceptual design phase.

   c. Design optimization is a completely automated process from which designers can expect to get their final design.

   d. The design variables for a problem consist of all the inputs needed to compute the objective and constraint functions.

   e. The design variables must always be independent of each other.

   f. An optimization algorithm designed for minimization can be used to maximize an objective function without modifying the algorithm.

   g. Compared with the global optimum of a given problem, adding more design variables to that problem results in a global optimum that is no worse than that of the original problem.

   h. Compared with the global optimum objective value of a given problem, adding more constraints sometimes results in a better global optimum.

   i. A function is C^1 continuous if its derivative varies continuously.

   j. All unimodal functions are convex.

   k. Global search algorithms always converge to the global optimum.

   l. Gradient-based methods are largely based on mathematical principles as opposed to heuristics.

   m. Solving a problem that involves a stochastic model requires a stochastic optimization algorithm.

   n. If a problem is multimodal, it requires a gradient-free optimization algorithm.

1.2 *Plotting a two-dimensional function.* Consider the two-dimensional function
$$f(x_1, x_2) = x_1^3 + 2x_1 x_2^2 - x_2^3 - 20x_1.$$

Plot the function contours and find the approximate location of the minimum point(s). Is there a global minimum? *Exploration*: Plot other functions to get an intuition about their trends and minima. You can start with simple low-order polynomials and then add higher-order terms, trying different coefficients. Then you can also try nonalgebraic functions. This will give you an intuition about the function trends and minima.

1.3 *Standard form.* Convert the following problem to the standard formulation (Eq. 1.4):

$$\text{maximize} \quad 2x_1^2 - x_1^4 x_2^2 - e^{x_3} + e^{-x_3} + 12$$
$$\text{by varying} \quad x_1, x_2, x_3$$
$$\text{subject to} \quad x_1 \geq 1 \tag{1.5}$$
$$x_2 + x_3 \geq 10$$
$$x_1^2 + 3x_2^2 \leq 4.$$

1.4 *Using an unconstrained optimizer.* Consider the two-dimensional function

$$f(x_1, x_2) = (1 - x_1)^2 + (1 - x_2)^2 + \frac{1}{2}(2x_2 - x_1^2)^2.$$

Plot the contours of this function and find the minimum graphically. Then, use optimization software to find the minimum (see Tip 1.3). Verify that the optimizer converges to the minimum you found graphically. *Exploration*: (1) Try minimizing the function in Prob. 1.2 starting from different points. (2) Minimize other functions of your choosing. (3) Study the options provided by the optimization software and explore different settings.

1.5 *Using a constrained optimizer.* Now we add constraints to Prob. 1.4. The objective is the same, but we now have two inequality constraints:

$$x_1^2 + x_2^2 \leq 1$$
$$x_1 - 3x_2 + \frac{1}{2} \geq 0,$$

and bound constraints:

$$x_1 \geq 0, \quad x_2 \geq 0.$$

Plot the constraints and identify the feasible region. Find the constrained minimum graphically. Use optimization software to solve the constrained minimization problem. Which of the inequality constraints and bounds are active at the solution?

1.6 *Paper review.* Select a paper on design optimization that seems interesting to you, preferably from a peer-reviewed journal. Write the full optimization problem statement in the standard form (Eq. 1.4) for the problem solved in the paper. Classify the problem according to Fig. 1.15 and the optimization algorithm according to Fig. 1.22. Use the decision tree in Fig. 1.24 to determine if the optimization algorithm was chosen appropriately. Write a critique of the paper, highlighting its strengths and weaknesses.

1.7 *Problem formulation.* Choose an engineering system that you are familiar with, and use the process outlined in Fig. 1.4 to formulate a problem for the design optimization of that system. Write the statement in the standard form (Eq. 1.4). Critique your statement by asking the following: Does the objective function truly capture the design intent? Are there other objectives that could be considered? Do the design variables provide enough freedom? Are the design variables bounded such that nonphysical designs are prevented? Are you sure you have included all the constraints needed to get a practical design? Can you think of any loophole that the optimizer can exploit in your statement?

A Short History of Optimization   2

This chapter provides helpful historical context for the methods discussed in this book. Nothing else in the book depends on familiarity with the material in this chapter, so it can be skipped. However, this history makes connections between the various topics that will enrich the big picture of optimization as you become familiar with the material in the rest of the book, so you might want to revisit this chapter.

Optimization has a long history that started with geometry problems solved by ancient Greek mathematicians. The invention of algebra and calculus opened the door to many more types of problems, and the advent of numerical computing increased the range of problems that could be solved in terms of type and scale.

> By the end of this chapter you should be able to:
>
> 1. Appreciate a range of historical advances in optimization.
>
> 2. Describe current frontiers in optimization.

2.1 The First Problems: Optimizing Length and Area

Ancient Greek and Egyptian mathematicians made numerous contributions to geometry, including solving optimization problems that involved length and area. They adopted a geometric approach to solving problems that are now more easily solved using calculus.

Archimedes of Syracuse (287–212 BCE) showed that of all possible spherical caps of a given surface area, hemispherical caps have the largest volume. Euclid of Alexandria (325–265 BCE) showed that the shortest distance between a point and a line is the segment perpendicular to that line. He also proved that among rectangles of a given perimeter, the square has the largest area.

Geometric problems involving perimeter and area had practical value. The classic example of such practicality is Dido's problem. According to the legend, Queen Dido, who had fled to Tunis, purchased

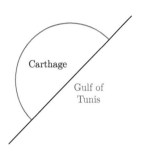

Fig. 2.1 Queen Dido intuitively maximized the area for a given perimeter, thus acquiring enough land to found the city of Carthage.

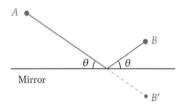

Fig. 2.2 The law of reflection can be derived by minimizing the length of the light beam.

7. Kepler, *Nova stereometria doliorum vinariorum (New Solid Geometry of Wine Barrels)*, 1615.

Fig. 2.3 Wine barrels were measured by inserting a ruler in the tap hole until it hit the corner.

8. Ferguson, *Who solved the secretary problem?* 1989.

from a local leader as much land as could be enclosed by an ox's hide. The leader agreed because this seemed like a modest amount of land. To maximize her land area, Queen Dido had the hide cut into narrow strips to make the longest possible string. Then, she intuitively found the curve that maximized the area enclosed by the string: a semicircle with the diameter segment set along the sea coast (see Fig. 2.1). As a result of the maximization, she acquired enough land to found the ancient city of Carthage. Later, Zenodorus (200–140 BCE) proved this optimal solution using geometrical arguments. A rigorous solution to this problem requires using calculus of variations, which was invented much later.

Geometric optimization problems are also applicable to the laws of physics. Hero of Alexandria (10–70 CE) derived the law of reflection by finding the shortest path for light reflecting from a mirror, which results in an angle of reflection equal to the angle of incidence (Fig. 2.2).

2.2 Optimization Revolution: Derivatives and Calculus

The scientific revolution generated significant optimization developments in the seventeenth and eighteenth centuries that intertwined with other developments in mathematics and physics.

In the early seventeenth century, Johannes Kepler published a book in which he derived the optimal dimensions of a wine barrel.[7] He became interested in this problem when he bought a barrel of wine, and the merchant charged him based on a diagonal length (see Fig. 2.3). This outraged Kepler because he realized that the amount of wine could vary for the same diagonal length, depending on the barrel proportions.

Incidentally, Kepler also formulated an optimization problem when looking for his second wife, seeking to maximize the likelihood of satisfaction. This "marriage problem" later became known as the "secretary problem", which is an optimal-stopping problem that has since been solved using dynamic optimization (mentioned in Section 1.4.6 and discussed in Section 8.5).[8]

Willebrord Snell discovered the law of refraction in 1621, a formula that describes the relationship between the angles of incidence and refraction when light passes through a boundary between two different media, such as air, glass, or water. Whereas Hero minimized the length to derive the law of reflection, Snell minimized time. These laws were generalized by Fermat in the *principle of least time* (or Fermat's principle), which states that a ray of light going from one point to another follows the path that takes the least time.

Pierre de Fermat derived Snell's law by applying the principle of least time, and in the process, he devised a mathematical technique for finding maxima and minima using what amounted to derivatives (he missed the opportunity to generalize the notion of derivative, which came later in the development of calculus).[9] Today, we learn about derivatives before learning about optimization, but Fermat did the reverse.

9. Fermat, *Methodus ad disquirendam maximam et minimam (Method for the Study of Maxima and Minima)*, 1636.

During this period, optimization was not yet considered an important area of mathematics, and contributions to optimization were scattered among other areas. Therefore, most mathematicians did not appreciate seminal contributions in optimization at the time.

In 1669, Isaac Newton wrote about a numerical technique to find the roots of polynomials by successively linearizing them, achieving quadratic convergence. In 1687, he used this technique to find the roots of a nonpolynomial equation (Kepler's equation),* but only after using polynomial expansions. In 1690, Joseph Raphson improved on Newton's method by keeping all the decimals in each linearization and making it a fully iterative scheme. The multivariable "Newton's method" that is widely used today was actually introduced in 1740 by Thomas Simpson. He generalized the method by using the derivatives (which allowed for solving nonpolynomial equations without expansions) and by extending it to a system of two equations and two unknowns.†

*Kepler's equation describes orbits by $E - e \sin(E) = M$, where M is the mean anomaly, e is the eccentricity, and E is the eccentric anomaly. This equation does not have a closed-form solution for E.

In 1685, Newton studied a shape optimization problem where he sought the shape of a body of revolution that minimizes fluid drag and even mentioned a possible application to ship design. Although he used the wrong model for computing the drag, he correctly solved what amounted to a calculus of variations problem.

†For this reason, Kollerstrom[10] argues that the method should be called neither Newton nor Newton–Raphson.

10. Kollerstrom, *Thomas Simpson and 'Newton's method of approximation': an enduring myth*, 1992.

In 1696, Johann Bernoulli challenged other mathematicians to find the path of a body subject to gravity that minimizes the travel time between two points of different heights. This is now a classic calculus of variations problem called the *brachistochrone problem* (Fig. 2.4). Bernoulli already had a solution that he kept secret. Five mathematicians respond with solutions: Newton, Jakob Bernoulli (Johann's brother), Gottfried Wilhelm von Leibniz, Ehrenfried Walther von Tschirnhaus, and Guillaume de l'Hôpital. Newton reportedly started working on the problem as soon as he received it and stayed up all night before sending the solution anonymously to Bernoulli the next day.

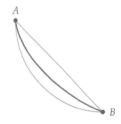

Fig. 2.4 Suppose that you have a bead on a wire that goes from A to B. The brachistochrone curve is the shape of the wire that minimizes the time for the bead to slide between the two points under gravity alone. It is faster than a straight-line trajectory or a circular arc.

Starting in 1736, Leonhard Euler derived the general optimality conditions for solving calculus of variations problems, but the derivation included geometric arguments. In 1755, Joseph-Louis Lagrange used a purely analytic approach to derive the same optimality conditions (he was 19 years old at the time!). Euler recognized Lagrange's derivation, which uses variations of a function, as a superior approach and adopted it, calling it "calculus of variations". This is a second-order partial differential equation that has become known as the *Euler–Lagrange equation*. Lagrange used this equation to develop a reformulation of classical mechanics in 1788, which became known as *Lagrangian mechanics*. When deriving the general equations of equilibrium for problems with constraints, Lagrange introduced the "method of the multipliers".[11] Lagrange multipliers eventually became a fundamental concept in constrained optimization (see Section 5.3).

11. Lagrange, *Mécanique analytique*, 1788.

In 1784, Gaspard Monge developed a geometric method to solve a transportation problem. Although the method was not entirely correct, it established *combinatorial optimization*, a branch of discrete optimization (Chapter 8).

2.3 The Birth of Optimization Algorithms

Several more theoretical contributions related to optimization occurred in the nineteenth century and the early twentieth century. However, it was not until the 1940s that optimization started to gain traction with the development of algorithms and their use in practical applications, thanks to the advent of computer hardware.

In 1805, Adrien-Marie Legendre described the method of least squares, which was used to predict asteroid orbits and for curve fitting. Friedrich Gauss published a rigorous mathematical foundation for the method of least squares and claimed he used it to predict the orbit of the asteroid Ceres in 1801. Legendre and Gauss engaged in a bitter dispute on who first developed the method.

In one of his 789 papers, Augustin-Louis Cauchy proposed the steepest-descent method for solving systems of nonlinear equations.[12] He did not seem to put much thought into it and promised a "paper to follow" on the subject, which never happened. He proposed this method for solving systems of nonlinear equations, but it is directly applicable to unconstrained optimization (see Section 4.4.1).

12. Cauchy, *Méthode générale pour la résolution des systèmes d'équations simultanées*, 1847.

In 1902, Gyula Farkas proved a theorem on the solvability of a system of linear inequalities. This became known as *Farkas' lemma*, which is crucial in the derivation of the optimality conditions for constrained problems (see Section 5.3.2). In 1917, Harris Hancock published the first textbook on optimization, which included the optimality conditions for multivariable unconstrained and constrained problems.[13]

13. Hancock, *Theory of Minima and Maxima*, 1917.

14. Menger, *Das botenproblem*, 1932.

In 1932, Karl Menger presented "the messenger problem",[14] an optimization problem that seeks to minimize the shortest travel path that connects a set of destinations, observing that going to the closest point each time does not, in general, result in the shortest overall path. This is a combinatorial optimization problem that later became known as the *traveling salesperson problem*, one of the most intensively studied problems in optimization (Chapter 8).

In 1939, William Karush derived the necessary conditions for inequality constrained problems in his master's thesis. His approach generalized the method of Lagrange multipliers, which only allowed for equality constraints. Harold Kuhn and Albert Tucker independently rediscovered these conditions and published their seminal paper in 1951.[15] These became known as the *Karush–Kuhn–Tucker (KKT) conditions*, which constitute the foundation of gradient-based constrained optimization algorithms (see Section 5.3).

Leonid Kantorovich developed a technique to solve linear programming problems in 1939 after having been given the task of optimizing production in the Soviet government's plywood industry. However, his contribution was neglected for ideological reasons. In the United States, Tjalling Koopmans rediscovered linear programming in the early 1940s when working on ship-transportation problems. In 1947, George Dantzig published the first complete algorithm for solving linear programming problems—the simplex algorithm.[16] In the same year, von Neumann developed the theory of duality for linear programming problems. Kantorovich and Koopmans later shared the 1975 Nobel Memorial Prize in Economic Sciences "for their contributions to the theory of optimum allocation of resources". Dantzig was not included, presumably because his work was more theoretical. The development of the simplex algorithm and the widespread practical applications of linear programming sparked a revolution in optimization. The first international conference on optimization, the International Symposium on Mathematical Programming, was held in Chicago in 1949.

In 1951, George Box and Kenneth Wilson developed the response-surface methodology (surrogate modeling), which enables optimization of systems based on experimental data (as opposed to a physics-based model). They developed a method to build a quadratic model where the number of data points scales linearly with the number of inputs instead of exponentially, striking a balance between accuracy and ease of application. In the same year, Danie Krige developed a surrogate model based on a stochastic process, which is now known as the *kriging model*. He developed this model in his master's thesis to estimate the most likely distribution of gold based on a limited number of borehole samples.[17] These approaches are foundational in surrogate-based optimization (Chapter 10).

In 1952, Harry Markowitz published a paper on portfolio theory

15. Karush, *Minima of functions of several variables with inequalities as side constraints*, 1939.

16. Dantzig, *Linear programming and extensions*, 1998.

17. Krige, *A statistical approach to some mine valuation and allied problems on the Witwatersrand*, 1951.

that formalized the idea of investment diversification, marking the birth of modern financial economics.[18] The theory is based on a quadratic optimization problem. He received the 1990 Nobel Memorial Prize in Economic Sciences for developing this theory.

In 1955, Lester Ford and Delbert Fulkerson created the first known algorithm to solve the maximum-flow problem, which has applications in transportation, electrical circuits, and data transmission. Although the problem could already be solved with the simplex algorithm, they proposed a more efficient algorithm for this specialized problem.

In 1957, Richard Bellman derived the necessary optimality conditions for dynamic programming problems.[19] These are expressed in what became known as the *Bellman equation* (Section 8.5), which was first applied to engineering control theory and subsequently became a core principle in economic theory.

In 1959, William Davidon developed the first quasi-Newton method for solving nonlinear optimization problems that rely on approximations of the curvature based on gradient information. He was motivated by his work at Argonne National Laboratory, where he used a coordinate-descent method to perform an optimization that kept crashing the computer before converging. Although Davidon's approach was a breakthrough in nonlinear optimization, his original paper was rejected. It was eventually published more than 30 years later in the first issue of the *SIAM Journal on Optimization*.[20] Fortunately, his valuable insight had been recognized well before that by Roger Fletcher and Michael Powell, who further developed the method.[21] The method became known as the *Davidon–Fletcher–Powell (DFP) method* (Section 4.4.4).

Another quasi-Newton approximation method was independently proposed in 1970 by Charles Broyden, Roger Fletcher, Donald Goldfarb, and David Shanno, now called the *Broyden–Fletcher–Goldfarb–Shanno (BFGS) method*. Larry Armijo, A. Goldstein, and Philip Wolfe developed the conditions for the line search that ensure convergence in gradient-based methods (see Section 4.3.2).[22]

Leveraging the developments in unconstrained optimization, researchers sought methods for solving constrained problems. Penalty and barrier methods were developed but fell out of favor because of numerical issues (see Section 5.4). In another effort to solve nonlinear constrained problems, Robert Wilson proposed the sequential quadratic programming (SQP) method in his PhD thesis.[23] SQP consists of applying the Newton method to solve the KKT conditions (see Section 5.5). Shih-Ping Han reinvented SQP in 1976,[24] and Michael Powell popularized this method in a series of papers starting from 1977.[25]

---

18. Markowitz, *Portfolio selection*, 1952.

19. Bellman, *Dynamic Programming*, 1957.

20. Davidon, *Variable metric method for minimization*, 1991.

21. Fletcher and Powell, *A rapidly convergent descent method for minimization*, 1963.

22. Wolfe, *Convergence conditions for ascent methods*, 1969.

23. Wilson, *A simplicial algorithm for concave programming*, 1963.

24. Han, *Superlinearly convergent variable metric algorithms for general nonlinear programming problems*, 1976.

25. Powell, *Algorithms for nonlinear constraints that use Lagrangian functions*, 1978.

There were attempts to model the natural process of evolution starting in the 1950s. In 1975, John Holland proposed genetic algorithms (GAs) to solve optimization problems (Section 7.6).[26] Research in GAs increased dramatically after that, thanks in part to the exponential increase in computing power.

Hooke and Jeeves[27] proposed a gradient-free method called *coordinate search*. In 1965, Nelder and Mead[28] developed the nonlinear simplex method, another gradient-free nonlinear optimization based on heuristics (Section 7.3).*

The Mathematical Programming Society was founded in 1973, an international association for researchers active in optimization. It was renamed the Mathematical Optimization Society in 2010 to reflect the more modern name for the field.

Narendra Karmarkar presented a revolutionary new method in 1984 to solve large-scale linear optimization problems as much as a hundred times faster than the simplex method.[29] The *New York Times* published a related news item on the front page with the headline "Breakthrough in Problem Solving". This heralded the age of interior-point methods, which are related to the barrier methods dismissed in the 1960s. Interior-point methods were eventually adapted to solve nonlinear problems (see Section 5.6) and contributed to the unification of linear and nonlinear optimization.

26. Holland, *Adaptation in Natural and Artificial Systems*, 1975.

27. Hooke and Jeeves, 'Direct search' solution of numerical and statistical problems, 1961.

28. Nelder and Mead, *A simplex method for function minimization*, 1965.

*The Nelder–Mead algorithm has no connection to the simplex algorithm for linear programming problems mentioned earlier.

29. Karmarkar, *A new polynomial-time algorithm for linear programming*, 1984.

2.4 The Last Decades

The relentless exponential increase in computer power throughout the 1980s and beyond has made it possible to perform engineering design optimization with increasingly sophisticated models, including multidisciplinary models. The increased computer power has also been contributing to the gain in popularity of heuristic optimization algorithms. Computer power has also enabled large-scale deep neural networks (see Section 10.5), which have been instrumental in the explosive rise of artificial intelligence (AI).

The field of optimal control flourished after Bellman's contribution to dynamic programming. Another important optimality principle for control, the maximum principle, was derived by Pontryagin et al.[30] This principle makes it possible to transform a calculus of variations problem into a nonlinear optimization problem. Gradient-based nonlinear optimization algorithms were then used to numerically solve for the optimal trajectories of rockets and aircraft, with an adjoint method to compute the gradients of the objective with respect to the control histories.[31] The adjoint method efficiently computes gradients with

30. Pontryagin et al., *The Mathematical Theory of Optimal Processes*, 1961.

31. Bryson Jr, *Optimal control—1950 to 1985*, 1996.

respect to large numbers of variables and has proven to be useful in other disciplines. Optimal control then expanded to include the optimization of feedback control laws that guarantee closed-loop stability. Optimal control approaches include model predictive control, which is widely used today.

In 1960, Schmit[32] proposed coupling numerical optimization with structural computational models to perform structural design, establishing the field of structural optimization. Five years later, he presented applications, including aerodynamics and structures, representing the first known multidisciplinary design optimization (MDO) application.[33] The direct method for computing gradients for structural computational models was developed shortly after that,[34] eventually followed by the adjoint method (Section 6.7).[35] In this early work, the design variables were the cross-sectional areas of the members of a truss structure. Researchers then added joint positions to the set of design variables. Structural optimization was generalized further with shape optimization, which optimizes the shape of arbitrary three-dimensional structural parts.[36] Another significant development was topology optimization, where a structural layout emerges from a solid block of material.[37] It took many years of further development in algorithms and computer hardware for structural optimization to be widely adopted by industry, but this capability has now made its way to commercial software.

Aerodynamic shape optimization began when Pironneau[38] used optimal control techniques to minimize the drag of a body by varying its shape (the "control" variables). Jameson[39] extended the adjoint method with more sophisticated computational fluid dynamics (CFD) models and applied it to aircraft wing design. CFD-based optimization applications spread beyond aircraft wing design to the shape optimization of wind turbines, hydrofoils, ship hulls, and automobiles. The adjoint method was then generalized for any discretized system of equations (see Section 6.7).

MDO developed rapidly in the 1980s following the application of numerical optimization techniques to structural design. The first conference in MDO, the Multidisciplinary Analysis and Optimization Conference, took place in 1985. The earliest MDO applications focused on coupling the aerodynamics and structures in wing design, and other early applications integrated structures and controls.[40] The development of MDO methods included efforts toward decomposing the problem into optimization subproblems, leading to distributed MDO architectures.[41] Sobieszczanski–Sobieski[42] proposed a formulation for computing the derivatives for coupled systems, which is necessary

32. Schmit, *Structural design by systematic synthesis*, 1960.

33. Schmit and Thornton, *Synthesis of an airfoil at supersonic Mach number*, 1965.

34. Fox, *Constraint surface normals for structural synthesis techniques*, 1965.

35. Arora and Haug, *Methods of design sensitivity analysis in structural optimization*, 1979.

36. Haftka and Grandhi, *Structural shape optimization—A survey*, 1986.

37. Eschenauer and Olhoff, *Topology optimization of continuum structures: A review*, 2001.

38. Pironneau, *On optimum design in fluid mechanics*, 1974.

39. Jameson, *Aerodynamic design via control theory*, 1988.

40. Sobieszczanski–Sobieski and Haftka, *Multidisciplinary aerospace design optimization: Survey of recent developments*, 1997.

41. Martins and Lambe, *Multidisciplinary design optimization: A survey of architectures*, 2013.

42. Sobieszczanski–Sobieski, *Sensitivity of complex, internally coupled systems*, 1990.

when performing MDO with gradient-based optimizers. This concept was later combined with the adjoint method for the efficient computation of coupled derivatives.[43] More recently, efficient computation of coupled derivatives and hierarchical solvers have made it possible to solve large-scale MDO problems[44] (Chapter 13). Engineering design has been focusing on achieving improvements made possible by considering the interaction of all relevant disciplines. MDO applications have extended beyond aircraft to the design of bridges, buildings, automobiles, ships, wind turbines, and spacecraft.

In continuous nonlinear optimization, SQP has remained the state-of-the-art approach since its popularization in the late 1970s. However, the interior-point approach, which, as mentioned previously, revolutionized linear optimization, was successfully adapted for the solution of nonlinear problems and has made great strides since the 1990s.[45] Today, both SQP and interior-point methods are considered to be state of the art.

Interior-point methods have contributed to the connection between linear and nonlinear optimization, which were treated as entirely separate fields before 1984. Today, state-of-the-art linear optimization software packages have options for both the simplex and interior-point approaches because the best approach depends on the problem.

Convex optimization emerged as a generalization of linear optimization (Chapter 11). Like linear optimization, it was initially mostly used in operations research applications,* such as transportation, manufacturing, supply-chain management, and revenue management, and there were only a few applications in engineering. Since the 1990s, convex optimization has increasingly been used in engineering applications, including optimal control, signal processing, communications, and circuit design. A disciplined convex programming methodology facilitated this expansion to construct convex problems and convert them to a solvable form.[46] New classes of convex optimization problems have also been developed, such as geometric programming (see Section 11.6), semidefinite programming, and second-order cone programming.

As mathematical models became increasingly complex computer programs, and given the need to differentiate those models when performing gradient-based optimization, new methods were developed to compute derivatives. Wengert[47] was among the first to propose the automatic differentiation of computer programs (or algorithmic differentiation). The reverse mode of algorithmic differentiation, which is equivalent to the adjoint method, was proposed later (see Section 6.6).[48] This field has evolved immensely since then, with techniques to handle more functions and increase efficiency. Algorithmic differentiation tools

43. Martins et al., *A coupled-adjoint sensitivity analysis method for high-fidelity aero-structural design*, 2005.

44. Hwang and Martins, *A computational architecture for coupling heterogeneous numerical models and computing coupled derivatives*, 2018.

45. Wright, *The interior-point revolution in optimization: history, recent developments, and lasting consequences*, 2005.

*The field of operations research was established in World War II to aid in making better strategical decisions.

46. Grant et al., *Disciplined convex programming*, 2006.

47. Wengert, *A simple automatic derivative evaluation program*, 1964.

48. Speelpenning, *Compiling fast partial derivatives of functions given by algorithms*, 1980.

have been developed for an increasing number of programming languages. One of the more recently developed programming languages, Julia, features prominent support for algorithmic differentiation. At the same time, algorithmic differentiation has spread to a wide range of applications.

Another technique to compute derivatives numerically, the complex-step derivative approximation, was proposed by Squire and Trapp.[49] Soon after, this technique was generalized to computer programs, applied to CFD, and found to be related to algorithmic differentiation (see Section 6.5).[50]

The pattern-search algorithms that Hooke and Jeeves and Nelder and Meade developed were disparaged by applied mathematicians, who preferred the rigor and efficiency of the gradient-based methods developed soon after that. Nevertheless, they were further developed and remain popular with engineering practitioners because of their simplicity. Pattern-search methods experienced a renaissance in the 1990s with the development of convergence proofs that added mathematical rigor and the availability of more powerful parallel computers.[51] Today, pattern-search methods (Section 7.4) remain a useful option, sometimes one of the only options, for certain types of optimization problems.

Global optimization algorithms also experienced further developments. Jones et al.[52] developed the DIRECT algorithm, which uses a rigorous approach to find the global optimum (Section 7.5). This seminal development was followed by various extensions and improvements.[53]

The first genetic algorithms started the development of the broader class of evolutionary optimization algorithms inspired by natural and societal processes. Optimization by simulated annealing (Section 8.6) represents one of the early examples of this broader perspective.[54] Another example is particle swarm optimization (Section 7.7).[55] Since then, there has been an explosion in the number of evolutionary algorithms, inspired by any process imaginable (see the sidenote at the end of Section 7.2 for a partial list). Evolutionary algorithms have remained heuristic and have not experienced the mathematical treatment applied to pattern-search methods.

There has been a sustained interest in surrogate models (also known as *metamodels*) since the seminal contributions in the 1950s. Kriging surrogate models are still used and have been the focus of many improvements, but new techniques, such as radial-basis functions, have also emerged.[56] Surrogate-based optimization is now an area of active research (Chapter 10).

AI has experienced a revolution in the last decade and is connected

49. Squire and Trapp, *Using complex variables to estimate derivatives of real functions*, 1998.

50. Martins et al., *The complex-step derivative approximation*, 2003.

51. Torczon, *On the convergence of pattern search algorithms*, 1997.

52. Jones et al., *Lipschitzian optimization without the Lipschitz constant*, 1993.

53. Jones and Martins, *The DIRECT algorithm—25 years later*, 2021.

54. Kirkpatrick et al., *Optimization by simulated annealing*, 1983.

55. Kennedy and Eberhart, *Particle swarm optimization*, 1995.

56. Forrester and Keane, *Recent advances in surrogate-based optimization*, 2009.

to optimization in several ways. The early AI efforts focused on solving problems that could be described formally using logic and decision trees. A design optimization problem statement can be viewed as an example of a formal logic description. Since the 1980s, AI has focused on machine learning, which uses algorithms and statistics to learn from data. In the 2010s, machine learning rose explosively owing to the development of deep learning neural networks, the availability of large data sets for training the neural networks, and increased computer power. Today, machine learning solves problems that are difficult to describe formally, such as face and speech recognition. Deep learning neural networks learn to map a set of inputs to a set of outputs based on training data and can be viewed as a type of surrogate model (see Section 10.5). These networks are trained using optimization algorithms that minimize the loss function (analogous to model error), but they require specialized optimization algorithms such as stochastic gradient descent. [57] The gradients for such problems are efficiently computed with backpropagation, a specialization of the reverse mode of algorithmic differentiation (AD) (see Section 6.6).[58]

57. Bottou et al., *Optimization methods for large-scale machine learning*, 2018.

58. Baydin et al., *Automatic differentiation in machine learning: A survey*, 2018.

2.5 Toward a Diverse Future

In the history of optimization, there is a glaring lack of diversity in geography, culture, gender, and race. Many contributions to mathematics have more diverse origins. This section is just a brief comment on this diversity and is not meant to be comprehensive. For a deeper analysis of the topics mentioned here, please see the cited references and other specialized bibliographies.

One of the oldest known mathematical objects is the Ishango bone, which originates from Africa and shows the construction of a numeral system.[59] Ancient Egyptians and Babylonians had a profound influence on ancient Greek mathematics. The Mayan civilization developed a sophisticated counting system that included zero and made precise astronomical observations to measure the solar year's length accurately.[60] In China, a textbook called *Nine Chapters on the Mathematical Art*, the compilation of which started in 200 BCE, includes a guide on solving equations using a matrix-based method. [61] The word *algebra* derives from a book entitled *Al-jabr wa'l muqabalah* by the Persian mathematician al-Khwarizmi in the ninth century, the title of which originated the term *algorithm*.[62] Finally, some of the basic components of calculus were discovered in India 250 years before Newton's breakthroughs.[63]

We also must recognize that there has been, and still is, a gender gap in science, engineering, and mathematics that has prevented

59. Gerdes, *On mathematics in the history of sub-Saharan Africa*, 1994.

60. Closs, *Native American Mathematics*, 1986.

61. Shen et al., *The Nine Chapters on the Mathematical Art: Companion and Commentary*, 1999.

62. Hodgkin, *A History of Mathematics: From Mesopotamia to Modernity*, 2005.

63. Joseph, *The Crest of the Peacock: Non-European Roots of Mathematics*, 2010.

women from having the same opportunities as men. The first known female mathematician, Hypatia, lived in Alexandria (Egypt) in the fourth century and was brutally murdered for political motives. In the eighteenth century, Sophie Germain corresponded with famous mathematicians under a male pseudonym to avoid gender bias. She could not get a university degree because she was female but was nevertheless a pioneer in elasticity theory. Ada Lovelace famously wrote the first computer program in the nineteenth century.[64] In the late nineteenth century, Sofia Kovalevskaya became the first woman to obtain a doctorate in mathematics but had to be tutored privately because she was not allowed to attend lectures. Similarly, Emmy Noether, who made many fundamental contributions to abstract algebra in the early twentieth century, had to overcome rules that prevented women from enrolling in universities and being employed as faculty.[65]

In more recent history, many women made crucial contributions in computer science. Grace Hopper invented the first compiler and influenced the development of the first high-level programming language (COBOL). Lois Haibt was part of a small team at IBM that developed Fortran, an extremely successful programming language that is still used today. Frances Allen was a pioneer in optimizing compilers (an altogether different type of optimization from the topic in this book) and was the first woman to win the Turing Award. Finally, Margaret Hamilton was the director of a laboratory that developed the flight software for NASA's Apollo program and coined the term *software engineering*.

Many other researchers have made key contributions despite facing discrimination. One of the most famous examples is that of mathematician and computer scientist Alan Turing, who was prosecuted in 1952 for having a relationship with another man. His punishment was chemical castration, which he endured for a time but ultimately led him to commit suicide at the age of 41.[66]

Some races and ethnicities have been historically underrepresented in science, engineering, and mathematics. One of the most apparent disparities has been the lack of representation of African Americans in the United States in these fields. This underrepresentation is a direct result of slavery and, among other factors, segregation, redlining, and anti-black racism.[67,68] In the eighteenth-century United States, Benjamin Banneker, a free African American who was a self-taught mathematician and astronomer, corresponded directly with Thomas Jefferson and successfully challenged the morality of the U.S. government's views on race and humanity.[69] Historically black colleges and universities were established in the United States after the American Civil War because

64. Hollings et al., *Ada Lovelace: The Making of a Computer Scientist*, 2014.

65. Osen, *Women in Mathematics*, 1974.

66. Hodges, *Alan Turing: The Enigma*, 2014.

67. Lipsitz, *How Racism Takes Place*, 2011.

68. Rothstein, *The Color of Law: A Forgotten History of How Our Government Segregated America*, 2017.

69. King, *More than slaves: Black founders, Benjamin Banneker, and critical intellectual agency*, 2014.

African Americans were denied admission to traditional institutions. In 1925, Elbert Frank Cox was the first black man to get a PhD in mathematics, and he then became a professor at Howard University. Katherine Johnson and fellow female African American mathematicians Dorothy Vaughan and Mary Jackson played a crucial role in the U.S. space program despite the open prejudice they had to overcome.[70]

"Talent is equally distributed, opportunity is not."* The arc of recent history has bent toward more diversity and equity,† but it takes concerted action to bend it. We have much more work to do before everyone has the same opportunity to contribute to our scientific progress. Only when that is achieved can we unleash the true potential of humankind.

70. Shetterly, *Hidden Figures: The American Dream and the Untold Story of the Black Women Who Helped Win the Space Race*, 2016.

*Variations of this quote abound; this one is attributed to social entrepreneur Leila Janah.

†A rephrasing of Martin Luther King Jr.'s quote: "The arc of the moral universe is long, but it bends toward justice."

2.6 Summary

The history of optimization is as old as human civilization and has had many twists and turns. Ancient geometric optimization problems that were correctly solved by intuition required mathematical developments that were only realized much later. The discovery of calculus laid the foundations for optimization. Computer hardware and algorithms then enabled the development and deployment of numerical optimization.

Numerical optimization was first motivated by operations research problems but eventually made its way into engineering design. Soon after numerical models were developed to simulate engineering systems, the idea arose to couple those models to optimization algorithms in an automated cycle to optimize the design of such systems. The first application was in structural design, but many other engineering design applications followed, including applications coupling multiple disciplines, establishing MDO. Whenever a new numerical model becomes fast enough and sufficiently robust, there is an opportunity to integrate it with numerical optimization to go beyond simulation and perform design optimization.

Many insightful connections have been made in the history of optimization, and the trend has been to unify the theory and methods. There are no doubt more connections and contributions to be made—hopefully from a more diverse research community.

Numerical Models and Solvers 3

In the introductory chapter, we discussed function characteristics from the point of view of the function's output—the black-box view shown in Fig. 1.16. Here, we discuss *how* the function is modeled and computed. The better your understanding of the model and the more access you have to its details, the more effectively you can solve the optimization problem. We explain the errors involved in the modeling process so that we can interpret optimization results correctly.

> By the end of this chapter you should be able to:
>
> 1. Identify different types of numerical errors and understand the limitations of finite-precision arithmetic.
> 2. Estimate an algorithm's rate of convergence.
> 3. Use Newton's method to solve systems of equations.

3.1 Model Development for Analysis versus Optimization

A good understanding of numerical models and solvers is essential because numerical optimization demands more from the models and solvers than does pure analysis. In an analysis or a parametric study, we may cycle through a range of plausible designs. However, optimization algorithms seek to explore the design space, and therefore, intermediate evaluations may use atypical design variables combinations. The mathematical model, numerical model, and solver must be robust enough to handle these design variable combinations.

A related issue is that an optimizer exploits errors in ways an engineer would not do in analysis. For example, consider the aerodynamic analysis of a car. In a parametric study, we might try a dozen designs, compare the drag, and choose the best one. If we passed this procedure to an optimizer, it might flatten the car to zero height (the minimum drag solution) if there are no explicit constraints on interior volume or structural integrity. Thus, we often need to develop additional models for optimization. A designer often considers some of these requirements implicitly and approximately, but we need to model these requirements explicitly and pose them as constraints in optimization.

Another consideration that affects both the mathematical and the numerical model is the overall computational cost of optimization. An analysis might only be run dozens of times, whereas an optimization often runs the analysis thousands of times. This computational cost can affect the level of fidelity or discretization we can afford to use.

The level of precision desirable for analysis is often insufficient for optimization. In an analysis, a few digits of precision may be sufficient. However, using fewer significant digits limits the types of optimization algorithms we can employ effectively. Convergence failures can cause premature termination of algorithms. Noisy outputs can mislead or terminate an optimization prematurely. A common source of these errors involves programs that work through input and output files (see Tip 6.1). Even though the underlying code may use double-precision arithmetic, output files rarely include all the significant digits (another separate issue is that reading and writing files at every iteration considerably slows down the analysis).

Another common source of errors involves converging systems of equations, as discussed later in this chapter. Optimization generally requires tighter tolerances than are used for analysis. Sometimes this is as easy as changing a default tolerance, but other times we need to rethink the solvers.

3.2 Modeling Process and Types of Errors

Design optimization problems usually involve modeling a physical system to compute the objective and constraint function values for a given design. Figure 3.1 shows the steps in the modeling process. Each of these steps in the modeling process introduces errors.

The physical system represents the reality that we want to *model*. The mathematical model can range from simple mathematical expressions to continuous differential or integral equations for which closed-form solutions over an arbitrary domain are not possible. *Modeling errors* are introduced in the idealizations and approximations performed in the derivation of the mathematical model. The errors involved in the rest of the process are *numerical errors*, which we detail in Section 3.5. In Section 3.3, we discuss mathematical models in more detail and establish the notation for representing them.

When a mathematical model is given by differential or integral equations, we must *discretize* the continuous equations to obtain the

3.2 Modeling Process and Types of Errors

numerical model. Section 3.4 provides a brief overview of the discretization process, and Section 3.5.2 defines the associated errors.

The numerical model must then be *programmed* using a computer language to develop a numerical solver. Because this process is susceptible to human error, we discuss strategies for addressing such errors in Section 3.5.4.

Finally, the solver *computes* the system state variables using finite-precision arithmetic, which introduces roundoff errors (see Section 3.5.1). Section 3.6 includes a brief overview of solvers, and we dedicate a separate section to Newton-based solvers in Section 3.8 because they are used later in this book.

The total error in the modeling process is the sum of the modeling errors and numerical errors. Validation and verification processes quantify and reduce these errors. Verification ensures that the model and solver are correctly implemented so that there are no errors in the code. It also ensures that the errors introduced by discretization and numerical computations are acceptable. Validation compares the numerical results with experimental observations of the physical system, which are themselves subject to experimental errors. By making these comparisons, we can validate the modeling step of the process and ensure that the mathematical model idealizations and assumptions are acceptable.

Modeling and numerical errors relate directly to the concepts of precision and accuracy. An *accurate* solution compares well with the actual physical system (validation), whereas a *precise* solution means that the model is programmed and solved correctly (verification).

It is often said that "all models are wrong, but some are useful".[71] Because there are always errors involved, we must prioritize which aspects of a given model should be improved to reduce the overall error. When developing a new model, we should start with the simplest model that includes the system's dominant behavior. Then, we might selectively add more detail as needed. One common pitfall in numerical modeling is to confuse precision with accuracy. Increasing precision by reducing the numerical errors is usually desirable. However, when we look at the bigger picture, the model might have limited utility if the modeling errors are more significant than the numerical errors.

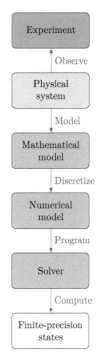

Fig. 3.1 Physical problems are modeled and then solved numerically to produce function values.

71. Box, *Science and statistics*, 1976.

Example 3.1 Modeling a structure

As an example of a physical system, consider the timber roof truss structure shown in Fig. 3.2. A typical mathematical model of such a structure idealizes the wood as a homogeneous material and the joints as pinned. It is also common to assume that the loads are applied only at the joints and that the

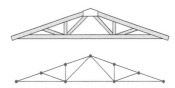

Fig. 3.2 Timber roof truss and idealized model.

structure's weight does not contribute to the loading. Finally, the displacements are assumed to be small relative to the dimensions of the truss members. The structure is discretized by pinned bar elements. The discrete governing equations for any truss structure can be derived using the finite-element method. This leads to the linear system

$$Ku = q,$$

where K is the stiffness matrix, q is the vector of applied loads, and u represents the displacements that we want to compute. At each joint, there are two degrees of freedom (horizontal and vertical) that describe the displacement and applied force. Because there are 9 joints, each with 2 degrees of freedom, the size of this linear system is 18.

3.3 Numerical Models as Residual Equations

Mathematical models vary significantly in complexity and scale. In the simplest case, a model can be represented by one or more explicit functions, which are easily coded and computed. Many examples in this book use explicit functions for simplicity. In practice, however, many numerical models are defined by *implicit* equations.

Implicit equations can be written in the *residual form* as

$$r_i(u_1, \ldots, u_n) = 0, \quad i = 1, \ldots, n, \tag{3.1}$$

where r is a vector of residuals that has the same size as the vector of state variables u. The equations defining the residuals could be any expression that can be coded in a computer program. No matter how complex the mathematical model, it can always be written as a set of equations in this form, which we write more compactly as $r(u) = 0$.

Finding the state variables that satisfy this set of equations requires a *solver*, as illustrated in Fig. 3.3. We review the various types of solvers in Section 3.6. Solving a set of implicit equations is more costly than computing explicit functions, and it is typically the most expensive step in the optimization cycle.

Mathematical models are often referred to as *governing equations*, which determine the *state* (u) of a given physical system at specific conditions. Many governing equations consist of differential equations, which require discretization. The discretization process yields implicit equations that can be solved numerically (see Section 3.4). After discretization, the governing equations can always be written as a set of residuals, $r(u) = 0$.

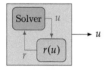

Fig. 3.3 Numerical models use a solver to find the state variables u that satisfy the governing equations, such that $r(u) = 0$.

3.3 Numerical Models as Residual Equations

Example 3.2 Implicit and explicit equations in structural analysis

The linear system from Ex. 3.1 is an example of a system of implicit equations, which we can write as a set of residuals by moving the right-hand-side vector to the left to obtain

$$r(u) = Ku - q = 0,$$

where u represents the state variables. Although the solution for u could be written as an explicit function, $u = K^{-1}f$, this is usually not done because it is computationally inefficient and intractable for large-scale systems. Instead, we use a linear solver that does not explicitly form the inverse of the stiffness matrix (see Appendix B).

In addition to computing the displacements, we might also want to compute the axial stress (σ) in each of the 15 truss members. This is an explicit function of the displacements, which is given by the linear relationship

$$\sigma = Su,$$

where S is a (15 × 18) matrix.

We can still use the residual notation to represent explicit functions to write all the functions in a model (implicit and explicit) as $r(u) = 0$ without loss of generality. Suppose we have an implicit system of equations, $r_r(u_r) = 0$, followed by a set of explicit functions whose output is a vector $u_f = f(u_r)$, as illustrated in Fig. 3.4. We can rewrite the explicit function as a residual by moving all the terms to one side to get $r_f(u_r, u_f) = f(u_r) - u_f = 0$. Then, we can concatenate the residuals and variables for the implicit and explicit equations as

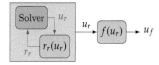

Fig. 3.4 A model with implicit and explicit functions.

$$r(u) \equiv \begin{bmatrix} r_r(u_r) \\ f(u_r) - u_f \end{bmatrix} = 0, \quad \text{where} \quad u \equiv \begin{bmatrix} u_r \\ u_f \end{bmatrix}. \quad (3.2)$$

The solver arrangement would then be as shown in Fig. 3.5.

Even though it is more natural to just evaluate explicit functions instead of adding them to a solver, in some cases, it is helpful to use the residual to represent the entire model with the compact notation, $r(u) = 0$. This will be helpful in later chapters when we compute derivatives (Chapter 6) and solve systems that mix multiple implicit and explicit sets of equations (Chapter 13).

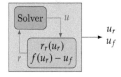

Fig. 3.5 Explicit functions can be written in residual form and added to the solver.

Example 3.3 Expressing an explicit function as an implicit equation

Suppose we have the following mathematical model:

$$u_1^2 + 2u_2 - 1 = 0$$
$$u_1 + \cos(u_1) - u_2 = 0$$
$$f(u_1, u_2) = u_1 + u_2.$$

The first two equations are written in implicit form, and the third equation is given as an explicit function. The first equation could be manipulated to obtain an explicit function of either u_1 or u_2. The second equation does not have a closed-form solution and cannot be written as an explicit function for u_1. The third equation is an explicit function of u_1 and u_2. In this case, we could solve the first two equations for u_1 and u_2 using a nonlinear solver and then evaluate $f(u_1, u_2)$. Alternatively, we can write the whole system as implicit residual equations by defining the value of $f(u_1, u_2)$ as u_3,

$$\begin{aligned} r_1(u_1, u_2) &= u_1^2 + 2u_2 - 1 &= 0 \\ r_2(u_1, u_2) &= u_1 + \cos(u_1) - u_2 &= 0 \\ r_3(u_1, u_2, u_3) &= u_1 + u_2 - u_3 &= 0. \end{aligned}$$

Then we can use the same nonlinear solver to solve for all three equations simultaneously.

3.4 Discretization of Differential Equations

Many physical systems are modeled by differential equations defined over a domain. The domain can be spatial (one or more dimensions), temporal, or both. When time is considered, then we have a dynamic model. When a differential equation is defined in a domain with one degree of freedom (one-dimensional in space or time), then we have an ordinary differential equation (ODE), whereas any domain defined by more than one variable results in a partial differential equation (PDE).

Differential equations need to be discretized over the domain to be solved numerically. There are three main methods for the discretization of differential equations: the finite-difference method, the finite-volume method, and the finite-element method. The finite-difference method approximates the derivatives in the differential equations by the value of the relevant quantities at a discrete number of points in a mesh (see Fig. 3.6). The finite-volume method is based on the integral form of the PDEs. It divides the domain into control volumes called *cells* (which also form a mesh), and the integral is evaluated for each cell. The values of the relevant quantities can be defined either at the centroids of the cells or at the cell vertices. The finite-element model divides the domain into elements (which are similar to cells) over which the quantities are interpolated using predefined shape functions. The states are computed at specific points in the element that are not necessarily at the element boundaries. Governing equations can also include integrals, which can be discretized with quadrature rules.

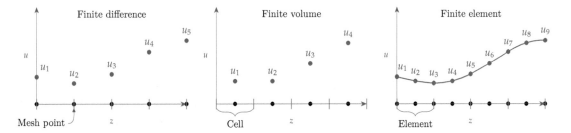

With any of these discretization methods, the final result is a set of algebraic equations that we can write as $r(u) = 0$ and solve for the state variables u. This is a potentially large set of equations depending on the domain and discretization (e.g., it is common to have millions of equations in three-dimensional computational fluid dynamics problems). The number of state variables of the discretized model is equal to the number of equations for a complete and well-defined model. In the most general case, the set of equations could be implicit and nonlinear.

When a problem involves both space and time, the prevailing approach is to decouple the discretization in space from the discretization in time—called the *method of lines* (see Fig. 3.7). The discretization in space is performed first, yielding an ODE in time. The time derivative can then be approximated as a finite difference, leading to a time-integration scheme.

The discretization process usually yields implicit algebraic equations that require a solver to obtain the solution. However, discretization in some cases yields explicit equations, in which case a solver is not required.

Fig. 3.6 Discretization methods in one spatial dimension.

3.5 Numerical Errors

Numerical errors (or computation errors) can be categorized into three main types: roundoff errors, truncation errors, and errors due to coding. Numerical errors are involved with each of the modeling steps between the mathematical model and the states (see Fig. 3.1). The error involved in the discretization step is a type of truncation error. The errors introduced in the coding step are not usually discussed as numerical errors, but we include them here because they are a likely source of error in practice. The errors in the computation step involve both roundoff and truncation errors. The following subsections describe each of these error sources.

An *absolute error* is the magnitude of the difference between the exact value (x^*) and the computed value (x), which we can write as $|x - x^*|$.

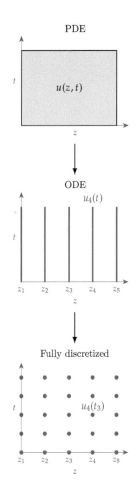

Fig. 3.7 PDEs in space and time are often discretized in space first to yield an ODE in time.

The *relative error* is a more intrinsic error measure and is defined as

$$\varepsilon = \frac{|x - x^*|}{|x^*|}. \tag{3.3}$$

This is the more useful error measure in most cases. When the exact value x^* is close to zero, however, this definition breaks down. To address this, we avoid the division by zero by using

$$\varepsilon = \frac{|x - x^*|}{1 + |x^*|}. \tag{3.4}$$

This error metric combines the properties of absolute and relative errors. When $|x^*| \gg 1$, this metric is similar to the relative error, but when $|x^*| \ll 1$, it becomes similar to the absolute error.

3.5.1 Roundoff Errors

Roundoff errors stem from the fact that a computer cannot represent all real numbers with exact precision. Errors in the representation of each number lead to errors in each arithmetic operation, which in turn might accumulate throughout a program.

There is an infinite number of real numbers, but not all numbers can be represented in a computer. When a number cannot be represented exactly, it is rounded. In addition, a number might be too small or too large to be represented.

Computers use bits to represent numbers, where each bit is either 0 or 1. Most computers use the Institute of Electrical and Electronics Engineers (IEEE) standard for representing numbers and performing finite-precision arithmetic. A typical representation uses 32 bits for integers and 64 bits for real numbers.

Basic operations that only involve integers and whose result is an integer do not incur numerical errors. However, there is a limit on the range of integers that can be represented. When using 32-bit integers, 1 bit is used for the sign, and the remaining 31 bits can be used for the digits, which results in a range from $-2^{31} = -2,147,483,648$ to $2^{31} - 1 = 2,147,483,647$. Any operation outside this range would result in *integer overflow*.*

*Some programming languages, such as Python, have arbitrary precision integers and are not subject to this issue, albeit with some performance trade-offs.

Real numbers are represented using scientific notation in base 2:

$$x = \text{significand} \times 2^{\text{exponent}}. \tag{3.5}$$

The 64-bit representation is known as the *double-precision floating-point format*, where some digits store the significand and others store the exponent. The greatest positive and negative real numbers that can

3.5 Numerical Errors

be represented using the IEEE double-precision representation are approximately 10^{308} and -10^{308}. Operations that result in numbers outside this range result in *overflow*, which is a fatal error in most computers and interrupts the program execution.

There is also a limit on how close a number can come to zero, approximately 10^{-324} when using double precision. Numbers smaller than this result in *underflow*. The computer sets such numbers to zero by default, and the program usually proceeds with no harmful consequences.

One important number to consider in roundoff error analysis is the *machine precision*, ε_M, which represents the precision of the computations. This is the smallest positive number ε such that

$$1 + \varepsilon > 1 \tag{3.6}$$

when calculated using a computer. This number is also known as *machine zero*. Typically, the double precision 64-bit representation uses 1 bit for the sign, 11 bits for the exponent, and 52 bits for the significand. Thus, when using double precision, $\varepsilon_M = 2^{-52} \approx 2.2 \times 10^{-16}$. A double-precision number has about 16 digits of precision, and a relative representation error of up to ε_M may occur.

Example 3.4 Machine precision

Suppose that three decimal digits are available to represent a number (and that we use base 10 for simplicity). Then, $\varepsilon_M = 0.005$ because any number smaller than this results in $1 + \varepsilon = 1$ when rounded to three digits. For example, $1.00 + 0.00499 = 1.00499$, which rounds to 1.00. On the other hand, $1.00 + 0.005 = 1.005$, which rounds to 1.01 and satisfies Eq. 3.6.

Example 3.5 Relative representation error

If we try to store 24.11 using three digits, we get 24.1. The relative error is

$$\frac{24.11 - 24.1}{24.11} \approx 0.0004,$$

which is lower than the maximum possible representation error of $\varepsilon_M = 0.005$ established in Ex. 3.4.

When operating with numbers that contain errors, the result is subject to a *propagated error*. For multiplication and division, the relative propagated error is approximately the sum of the relative errors of the respective two operands.

For addition and subtraction, an error can occur even when the two operands are represented exactly. Before addition and subtraction, the computer must convert the two numbers to the same exponent. When adding numbers with different exponents, several digits from the small number vanish (see Fig. 3.8). If the difference in the two exponents is greater than the magnitude of the exponent of ε_M, the small number vanishes completely—a consequence of Eq. 3.6. The relative error incurred in addition is still ε_M.

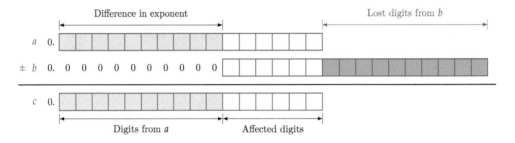

Fig. 3.8 Adding or subtracting numbers of differing exponents results in a loss in the number of digits corresponding to the difference in the exponents. The gray boxes indicate digits that are identical between the two numbers.

On the other hand, subtraction can incur much greater relative errors when subtracting two numbers that have the same exponent and are close to each other. In this case, the digits that match between the two numbers cancel each other and reduce the number of significant digits. When the relative difference between two numbers is less than machine precision, all digits match, and the subtraction result is zero (see Fig. 3.9). This is called *subtractive cancellation* and is a serious issue when approximating derivatives via finite differences (see Section 6.4).

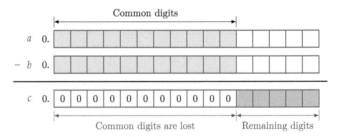

Fig. 3.9 Subtracting two numbers that are close to each other results in a loss of the digits that match.

Sometimes, minor roundoff errors can propagate and result in much more significant errors. This can happen when a problem is ill-conditioned or when the algorithm used to solve the problem is unstable. In both cases, small changes in the inputs cause large changes in the output. Ill-conditioning is not a consequence of the finite-precision computations but is a characteristic of the model itself. Stability is a property of the algorithm used to solve the problem. When a problem

3.5 Numerical Errors

is ill-conditioned, it is challenging to solve it in a stable way. When a problem is well conditioned, there is a stable algorithm to solve it.

Example 3.6 Effect of roundoff error on function representation

Let us examine the function $f(x) = x^2 - 4x + 4$ near its minimum, at $x = 2$. If we use double precision and plot many points in a small interval, we can see that the function exhibits the step pattern shown in Fig. 3.10. The numerical minimum of this function is anywhere in the interval around $x = 2$ where the numerical value is zero. This interval is much larger than the machine precision ($\varepsilon_M = 2.2 \times 10^{-16}$). An additional error is incurred in the function computation around $x = 2$ as a result of subtractive cancellation. This illustrates the fact that all functions are discontinuous when using finite-precision arithmetic.

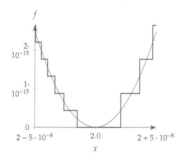

Fig. 3.10 With double precision, the minimum of this quadratic function is in an interval much larger than machine zero.

3.5.2 Truncation Errors

In the most general sense, truncation errors arise from performing a finite number of operations where an infinite number of operations would be required to get an exact result.[†] Truncation errors would arise even if we could do the arithmetic with infinite precision.

When discretizing a mathematical model with partial derivatives as described in Section 3.4, these are approximated by truncated Taylor series expansions that ignore higher-order terms. When the model includes integrals, they are approximated as finite sums. In either case, a mesh of points where the relevant states and functions are evaluated is required. Discretization errors generally decrease as the spacing between the points decreases.

[†]Roundoff error, discussed in the previous section, is sometimes also referred to as *truncation error* because digits are *truncated*. However, we avoid this confusing naming and only use *truncation error* to refer to a truncation in the number of operations.

Tip 3.1 Perform a mesh refinement study

When using a model that depends on a mesh, perform a mesh refinement study. This involves solving the model for increasingly finer meshes to check if the metrics of interest converge in a stable way and verify that the convergence rate is as expected for the chosen numerical discretization scheme. A mesh refinement study is also useful for finding the mesh that provides the best compromise between computational time and accuracy. This is especially important in optimization because the model is solved many times.

3.5.3 Iterative Solver Tolerance Error

Many methods for solving numerical models involve an iterative procedure that starts with a guess for the states u and then improves that guess at each iteration until reaching a specified convergence tolerance. The convergence is usually measured by a norm of the residuals, $\|r(u)\|$, which we want to drive to zero. Iterative linear solvers and Newton-type solvers are examples of iterative methods (see Section 3.6).

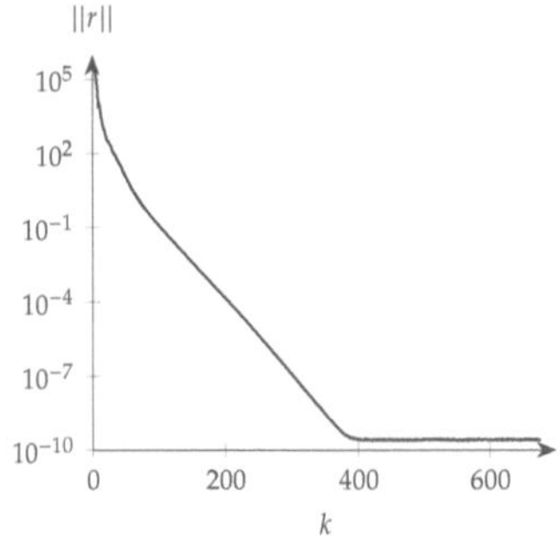

A typical convergence history for an iterative solver is shown in Fig. 3.11. The norm of the residuals decreases gradually until a limit is reached (near 10^{-10} in this case). This limit represents the lowest error achieved with the iterative solver and is determined by other sources of error, such as roundoff and truncation errors. If we terminate before reaching the limit (either by setting a convergence tolerance to a value higher than 10^{-10} or setting an iteration limit to lower than 400 iterations), we incur an additional error. However, it might be desirable to trade off a less precise solution for a lower computational effort.

Fig. 3.11 Norm of residuals versus the number of iterations for an iterative solver.

Tip 3.2 Find the level of the numerical noise in your model

It is crucial to know the error level in your model because this limits the type of optimizer you can use and how well you can optimize. In Ex. 3.6, we saw that if we plot a function at a small enough scale, we can see discrete steps in the function due to roundoff errors. When accumulating all sources of error in a more elaborate model (roundoff, truncation, and iterative), we no longer have a neat step pattern. Instead, we get *numerical noise*, as shown in Fig. 3.12. The noise level can be estimated by the amplitude of the oscillations and gives us the order of magnitude of the total numerical error.

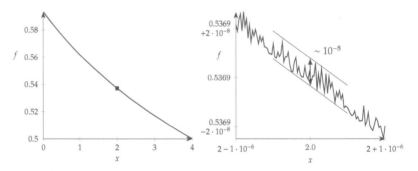

Fig. 3.12 To find the level of numerical noise of a function of interest with respect to an input parameter (left), we magnify both axes by several orders of magnitude and evaluate the function at points that are closely spaced (right).

3.5.4 Programming Errors

Most of the literature on numerical methods is too optimistic and does not explicitly discuss programming errors, commonly known as *bugs*. Most programmers, especially beginners, underestimate the likelihood that their code has bugs.

It is helpful to adopt sound programming practices, such as writing clear, modular code. Clear code has consistent formatting, meaningful naming of variable functions, and helpful comments. Modular code reuses and generalizes functions as much as possible and avoids copying and pasting sections of code.[72] Modular code allows for more flexible usage. Breaking up programs into smaller functions with well-defined inputs and outputs makes debugging much more manageable.‡

There are different types of bugs relevant to numerical models: generic programming errors, incorrect memory handling, and algorithmic or logical errors. Programming errors are the most frequent and include typos, type errors, copy-and-paste errors, faulty initializations, missing logic, and default values. In theory, careful programming and code inspection can avoid these errors, but you must always test your code in practice. This testing involves comparing your result with a case where you know the solution—the reference result. You should start with the simplest representative problem and then build up from that. Interactive debuggers are helpful because let you step through the code and check intermediate variable values.

72. Wilson et al., *Best practices for scientific computing*, 2014.

‡The term *debugging* was used in engineering prior to computers, but Grace Hopper popularized this term in the programming context after a glitch in the Harvard Mark II computer was found to be caused by a moth.

Tip 3.3 Debugging is a skill that takes practice

The overall attitude toward programming should be that all code has bugs until it is verified through testing. Programmers who are skilled at debugging are not necessarily any better at spotting errors by reading code or by stepping through a debugger than average programmers. Instead, effective programmers use a systematic approach to narrow down where the problem is occurring.

Beginners often try to debug by running the entire program. Even experienced programmers have a hard time debugging at that level. One primary strategy discussed in this section is to write modular code. It is much easier to test and debug small functions. Reliable complex programs are built up through a series of well-tested modular functions. Sometimes we need to simplify or break up functions even further to narrow down the problem. We might need to streamline and remove pieces, make sure a simple case works, then slowly rebuild the complexity.

You should also become comfortable reading and understanding the error messages and stack traces produced by the program. These messages seem obscure at first, but through practice and researching what the error messages mean, they become valuable information sources.

Of course, you should carefully reread the code, looking for errors, but reading through it again and again is unlikely to yield new insights. Instead, it can be helpful to step away from the code and hypothesize the most likely ways the function could fail. You can then test and eliminate hypotheses to narrow down the problem.

Memory handling issues are much less frequent than programming errors, but they are usually more challenging to track. These issues include memory leaks (a failure to free unused memory), incorrect use of memory management, buffer overruns (e.g., array bound violations), and reading uninitialized memory. Memory issues are challenging to track because they can result in strange behavior in parts of the code that are far from the source of the error. In addition, they might manifest themselves in specific conditions that are hard to reproduce consistently. Memory debuggers are essential tools for addressing memory issues. They perform detailed bookkeeping of all allocation, deallocation, and memory access to detect and locate any irregularities.§

§See Grotker et al.[73] for more details on how to debug and profile code.

73. Grotker et al., *The Developer's Guide to Debugging*, 2012.

Whereas programming errors are due to a mismatch between the programmer's intent and what is coded, the root cause of algorithmic or logical errors is in the programmer's intent itself. Again, testing is the key to finding these errors, but you must be sure that the reference result is correct.

Tip 3.4 Use sound code testing practices

Automated testing takes effort to implement but ultimately saves time, especially for larger, long-term projects. *Unit tests* check for the internal consistency of a small piece (a "unit") of code and should be implemented as each piece of code is developed. *Integration tests* are designed to demonstrate that different code components work together as expected. *Regression testing* consists of running all the tests (usually automatically) anytime the code has changed to ensure that the changes have not introduced bugs. It is usually impossible to test for all potential issues, but the more you can test, the more *coverage* you have. Whenever a bug has been found, a test should be developed to catch that same type of bug in the future.

Running the analysis within an optimization loop can reveal bugs that do not manifest themselves in a single analysis. Therefore, you should only run an optimization test case after you have tested the analysis code in isolation.

As previously mentioned, there is a higher incentive to reduce the computational cost of an analysis when it runs in an optimization loop because it will run many times. When you first write your code, you should prioritize clarity and correctness as opposed to speed. Once the code is verified through testing, you should identify any bottlenecks using a performance profiling tool. Memory performance issues can also arise from running the analysis in an optimization loop instead of running a single case. In addition to running a memory debugger, you can also run a memory profiling tool to identify opportunities to reduce memory usage.

3.6 Overview of Solvers

There are several methods available for solving the discretized governing equations (Eq. 3.1). We want to solve the governing equations for a fixed set of design variables, so x will not appear in the solution algorithms. Our objective is to find the state variables u such that $r(u) = 0$.

This is not a book about solvers, but it is essential to understand the characteristics of these solvers because they affect the cost and precision of the function evaluations in the overall optimization process. Thus, we provide an overview and some of the most relevant details in this section.* In addition, the solution of coupled systems builds on these solvers, as we will see in Section 13.2. Finally, some of the optimization algorithms detailed in later chapters use these solvers.

There are two main types of solvers, depending on whether the equations to be solved are linear or nonlinear (Fig. 3.13). Linear solution methods solve systems of the form $r(u) = Au - b = 0$, where the matrix A and vector b are not dependent on u. Nonlinear methods can handle any algebraic system of equations that can be written as $r(u) = 0$.

*Ascher and Greif[74] provide a more detailed introduction to the numerical methods mentioned in this chapter.

74. Ascher and Greif, *A First Course in Numerical Methods*, 2011.

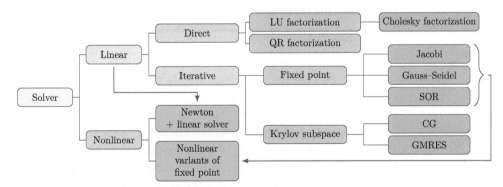

Fig. 3.13 Overview of solution methods for linear and nonlinear systems.

Linear systems can be solved directly or iteratively. Direct methods are based on the concept of Gaussian elimination, which can be expressed in matrix form as a factorization into lower and upper triangular matrices that are easier to solve (LU factorization). Cholesky factorization is a more efficient variant of LU factorization that applies only to symmetric positive-definite matrices.

Whereas direct solvers obtain the solution u at the end of a process, iterative solvers start with a guess for u and successively improve it with each iteration, as illustrated in Fig. 3.14. Iterative methods can be fixed-point iterations, such as Jacobi, Gauss–Seidel, and successive over-relaxation (SOR), or Krylov subspace methods. Krylov subspace methods include the conjugate gradient (CG) and generalized minimum residual (GMRES) methods.[†] Direct solvers are well established and are included in the standard libraries for most programming languages. Iterative solvers are less widespread in standard libraries, but they are becoming more commonplace. Appendix B describes linear solvers in more detail.

Direct methods are the right choice for many problems because they are generally robust. Also, the solution is guaranteed for a fixed number of operations, $O(n^3)$ in this case. However, for large systems where A is sparse, the cost of direct methods can become prohibitive, whereas iterative methods remain viable. Iterative methods have other advantages, such as being able to trade between computational cost and precision. They can also be restarted from a good guess (see Appendix B.4).

[†]See Saad[75] for more details on iterative methods in the context of large-scale numerical models.

75. Saad, *Iterative Methods for Sparse Linear Systems*, 2003.

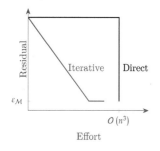

Fig. 3.14 Whereas direct methods only yield the solution at the end of the process, iterative methods produce approximate intermediate results.

| Tip 3.5 | Do not compute the inverse of A |

Because some numerical libraries have functions to compute A^{-1}, you might be tempted to do this and then multiply by a vector to compute $u = A^{-1}b$. This is a bad idea because finding the inverse is computationally expensive. Instead, use LU factorization or another method from Fig. 3.13.

When it comes to nonlinear solvers, the most efficient methods are based on Newton's method, which we explain later in this chapter (Section 3.8). Newton's method solves a sequence of problems that are linearizations of the nonlinear problem about the current iterate. The linear problem at each Newton iteration can be solved using any linear solver, as indicated by the incoming arrow in Fig. 3.13. Although efficient, Newton's method is not robust in that it does not always converge. Therefore, it requires modifications so that it can converge reliably.

Finally, it is possible to adapt linear fixed-point iteration methods to solve nonlinear equations as well. However, unlike the linear case, it might not be possible to derive explicit expressions for the iterations in the nonlinear case. For this reason, fixed-point iteration methods are often not the best choice for solving a system of nonlinear equations. However, as we will see in Section 13.2.5, these methods are useful for solving systems of coupled nonlinear equations.

For time-dependent problems, we require a way to solve for the time history of the states, $u(t)$. As mentioned in Section 3.3, the most popular approach is to decouple the temporal discretization from the spatial one. By discretizing a PDE in space first, this method formulates an ODE in time of the following form:

$$\frac{du}{dt} = -r(u,t), \qquad (3.7)$$

which is called the *semi-discrete* form. A time-integration scheme is then used to solve for the time history. The integration scheme can be either explicit or implicit, depending on whether it involves evaluating explicit expressions or solving implicit equations. If a system under a certain condition has a steady state, these techniques can be used to solve the steady state ($du/dt = 0$).

3.7 Rate of Convergence

Iterative solvers compute a *sequence* of approximate solutions that hopefully converge to the exact solution. When characterizing convergence, we need to first establish if the algorithm converges and, if so, how fast it converges. The first characteristic relates to the stability of the algorithm. Here, we focus on the second characteristic quantified through the *rate of convergence*.

The cost of iterative algorithms is often measured by counting the number of iterations required to achieve the solution. Iterative algorithms often require an infinite number of iterations to converge to the exact solution. In practice, we want to converge to an approximate solution close enough to the exact one. Determining the rate of convergence arises from the need to quantify how fast the approximate solution is approaching the exact one.

In the following, we assume that we have a sequence of points, $x_0, x_1, \ldots, x_k, \ldots$, that represent approximate solutions in the form of vectors in any dimension and converge to a solution x^*. Then,

$$\lim_{k \to \infty} \|x_k - x^*\| = 0, \qquad (3.8)$$

which means that the norm of the error tends to zero as the number of iterations tends to infinity.

The rate of convergence of a sequence is of *order p* with *asymptotic error constant* γ when p is the largest number that satisfies*

$$0 \leq \lim_{k \to \infty} \frac{\|x_{k+1} - x^*\|}{\|x_k - x^*\|^p} = \gamma < \infty. \qquad (3.9)$$

*Some authors refer to p as the rate of convergence. Here, we characterize the rate of convergence by two metrics: *order* and *error constant*.

Asymptotic here refers to the fact that this is the behavior in the limit, when we are close to the solution. There is no guarantee that the initial and intermediate iterations satisfy this condition.

To avoid dealing with limits, let us consider the condition expressed in Eq. 3.9 at all iterations. We can relate the error from one iteration to the next by

$$\|x_{k+1} - x^*\| = \gamma_k \|x_k - x^*\|^p . \quad (3.10)$$

When $p = 1$, we have *linear* order of convergence; when $p = 2$, we have *quadratic* order of convergence. Quadratic convergence is a highly valued characteristic for an iterative algorithm, and in practice, orders of convergence greater than $p = 2$ are usually not worthwhile to consider.

When we have linear convergence, then

$$\|x_{k+1} - x^*\| = \gamma_k \|x_k - x^*\| , \quad (3.11)$$

where γ_k converges to a constant but varies from iteration to iteration. In this case, the convergence is highly dependent on the value of the asymptotic error constant γ. If $\gamma_k > 1$, then the sequence diverges—a situation to be avoided. If $0 < \gamma_k < 1$ for every k, then the norm of the error decreases by a constant factor for every iteration. Suppose that $\gamma = 0.1$ for all iterations. Starting with an initial error norm of 0.1, we get the sequence

$$10^{-1}, 10^{-2}, 10^{-3}, 10^{-4}, 10^{-5}, 10^{-6}, 10^{-7}, \ldots . \quad (3.12)$$

Thus, after six iterations, we get six-digit precision. Now suppose that $\gamma = 0.9$. Then we would have

$$10^{-1}, 9.0 \times 10^{-2}, 8.1 \times 10^{-2}, 7.3 \times 10^{-2}, 6.6 \times 10^{-2},$$
$$5.9 \times 10^{-2}, 5.3 \times 10^{-2}, \ldots . \quad (3.13)$$

This corresponds to only one-digit precision after six iterations. It would take 131 iterations to achieve six-digit precision.

When we have quadratic convergence, then

$$\|x_{k+1} - x^*\| = \gamma_k \|x_k - x^*\|^2 . \quad (3.14)$$

If $\gamma = 1$, then the error norm sequence with a starting error norm of 0.1 would be

$$10^{-1}, 10^{-2}, 10^{-4}, 10^{-8}, \ldots . \quad (3.15)$$

This yields more than six digits of precision in just three iterations! In this case, the number of correct digits doubles at every iteration. When $\gamma > 1$, the convergence will not be as fast, but the series will still converge.

3.7 Rate of Convergence

If $p \geq 1$ and $\lim_{k \to \infty} \gamma_k = 0$, we have *superlinear* convergence, which includes quadratic and higher rates of convergence. There is a special case of superlinear convergence that is relevant for optimization algorithms, which is when $p = 1$ and $\gamma \to 0$. This case is desirable because in practice, it behaves similarly to quadratic convergence and can be achieved by gradient-based algorithms that use first derivatives (as opposed to second derivatives). In this case, we can write

$$\|x_{k+1} - x^*\| = \gamma_k \|x_k - x^*\|, \quad (3.16)$$

where $\lim_{k \to \infty} \gamma_k = 0$. Now we need to consider a sequence of values for γ that tends to zero. For example, if $\gamma_k = 1/(k+1)$, starting with an error norm of 0.1, we get

$$10^{-1}, 5 \times 10^{-1}, 1.7 \times 10^{-1}, 4.2 \times 10^{-2}, 8.3 \times 10^{-4},$$
$$1.4 \times 10^{-4}, 2.0 \times 10^{-5}, \ldots. \quad (3.17)$$

Thus, we achieve four-digit precision after six iterations. This special case of superlinear convergence is not quite as good as quadratic convergence, but it is much better than either of the previous linear convergence examples.

We plot these sequences in Fig. 3.15. Because the points are just scalars and the exact solution is zero, the error norm is just x_k. The first plot uses a linear scale, so we cannot see any differences beyond two digits. To examine the differences more carefully, we need to use a logarithmic axis for the sequence values, as shown in the plot on the right. In this scale, each decrease in order of magnitude represents one more digit of precision.

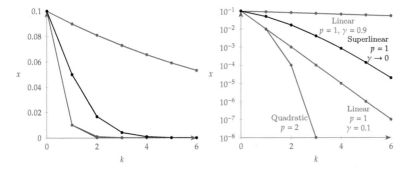

Fig. 3.15 Sample sequences for linear, superlinear, and quadratic cases plotted on a linear scale (left) and a logarithmic scale (right).

The linear convergence sequences show up as straight lines in Fig. 3.15 (right), but the slope of the lines varies widely, depending on the value of the asymptotic error constant. Quadratic convergence exhibits an increasing slope, reflecting the doubling of digits for each iteration. The superlinear sequence exhibits poorer convergence than the best linear one, but we can see that the slope of the superlinear curve is increasing, which means that for a large enough k, it will converge at a higher rate than the linear one.

> **Tip 3.6** Use a logarithmic scale when plotting convergence
>
> When using a linear scale plot, you can only see differences in two significant digits. To reveal changes beyond three digits, you should use a logarithmic scale. This need frequently occurs in plotting the convergence behavior of optimization algorithms.

When solving numerical models iteratively, we can monitor the norm of the residual. Because we know that the residuals should be zero for an exact solution, we have

$$\|r_{k+1}\| = \gamma_k \|r_k\|^p . \tag{3.18}$$

If we monitor another quantity, we do not usually know the exact solution. In these cases, we can use the ratio of the step lengths of each iteration:

$$\frac{\|x_{k+1} - x^*\|}{\|x_k - x^*\|} \approx \frac{\|x_{k+1} - x_k\|}{\|x_k - x_{k-1}\|}. \tag{3.19}$$

The order of convergence can be estimated numerically with the values of the last available four iterates using

$$p \approx \frac{\log_{10} \frac{\|x_{k+1}-x_k\|}{\|x_k-x_{k-1}\|}}{\log_{10} \frac{\|x_k - x_{k-1}\|}{\|x_{k-1} - x_{k-2}\|}}. \tag{3.20}$$

Finally, we can also monitor any quantity (function values, state variables, or design variables) by normalizing the step length in the same way as Eq. 3.4,

$$\frac{\|x_{k+1} - x_k\|}{1 + \|x_k\|}. \tag{3.21}$$

3.8 Newton-Based Solvers

As mentioned in Section 3.6, Newton's method is the basis for many nonlinear equation solvers. Newton's method is also at the core of the most efficient gradient-based optimization algorithms, so we explain it here in more detail. We start with the single-variable case for simplicity and then generalize it to the n-dimensional case.

3.8 Newton-Based Solvers

We want to find u^* such that $r(u^*) = 0$, where, for now, r and u are scalars. Newton's method for root finding estimates a solution at each iteration k by approximating $r(u_k)$ to be a linear function. The linearization is done by taking a Taylor series of r about u_k and truncating it to obtain the approximation

$$r(u_k + \Delta u) \approx r(u_k) + \Delta u\, r'(u_k), \tag{3.22}$$

where $r' = dr/du$. For conciseness, we define $r_k = r(u_k)$. Now we can find the step Δu that makes this approximate residual zero,

$$r_k + \Delta u\, r'_k = 0 \quad \Rightarrow \quad \Delta u = -\frac{r_k}{r'_k}, \tag{3.23}$$

where we need to assume that $r'_k \neq 0$.

Thus, the update for each step in Newton's algorithm is

$$u_{k+1} = u_k - \frac{r_k}{r'_k}. \tag{3.24}$$

If $r'_k = 0$, the algorithm will not converge because it yields a step to infinity. Small enough values of r'_k also cause an issue with large steps, but the algorithm might still converge.

One useful modification of Newton's method is to replace the derivative with a forward finite-difference approximation (see Section 6.4) based on the residual values of the current and last iterations,

$$r'_k \approx \frac{r_{k+1} - r_k}{u_{k+1} - u_k}. \tag{3.25}$$

Then, the update is given by

$$u_{k+1} = u_k - r_k \left(\frac{u_{k+1} - u_k}{r_{k+1} - r_k} \right). \tag{3.26}$$

This is the *secant method*, which is useful when the derivative is not available. The convergence rate is not quadratic like Newton's method, but it is superlinear.

Example 3.7 Newton's method and the secant method for a single variable

Suppose we want to solve the equation $r(u) = 2u^3 + 4u^2 + u - 2 = 0$. Because $r'(u) = 6u^2 + 8u + 1$, the Newton iteration is

$$u_{k+1} = u_k - \frac{2u_k^3 + 4u_k^2 + u_k - 2}{6u_k^2 + 8u_k + 1}.$$

When we start with the guess $u_0 = 1.5$ (left plot in Fig. 3.16), the iterations are well behaved, and the method converges quadratically. We can see the

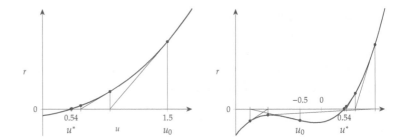

Fig. 3.16 Newton iterations starting from different starting points.

geometric interpretation of Newton's method: For each iteration, it takes the tangent to the curve and finds the intersection with $r = 0$.

When we start with $u_0 = -0.5$ (right plot in Fig. 3.16), the first step goes in the wrong direction but recovers in the second iteration. The third iteration is close to the point with the zero derivative and takes a large step. In this case, the iterations recover and then converge normally. However, we can easily envision a case where an iteration is much closer to the point with the zero derivative, causing an arbitrarily long step.

We can also use the secant method (Eq. 3.26) for this problem, which gives the following update:

$$u_{k+1} = u_k - \frac{\left(2u_k^3 + 4u_k^2 + u_k - 2\right)(u_{k+1} - u_k)}{2u_{k+1}^3 + 4u_{k+1}^2 + u_{k+1} - 2u_k^3 - 4u_k^2 - u_k}.$$

The iterations for the secant method are shown in Fig. 3.17, where we can see the successive secant lines replacing the exact tangent lines used in Newton's method.

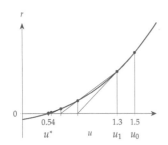

Fig. 3.17 Secant method applied to a one-dimensional function.

Newton's method converges quadratically as it approaches the solution with a convergence constant of

$$\gamma = \left|\frac{r''(u^*)}{2r'(u^*)}\right|. \qquad (3.27)$$

This means that if the derivative is close to zero or the curvature tends to a large number at the solution, Newton's method will not converge as well or not at all.

Now we consider the general case where we have n nonlinear equations of n unknowns, expressed as $r(u) = 0$. Similar to the single-variable case, we derive the Newton step from a truncated Taylor series. However, the Taylor series needs to be multidimensional in both the independent variable and the function. Consider first the multidimensionality of the independent variable, u, for a component of the residuals, $r_i(u)$. The first two terms of the Taylor series about u_k for a step Δu (which is now a vector with arbitrary direction and

3.8 Newton-Based Solvers

magnitude) are

$$r_i(u_k + \Delta u) \approx r_i(u_k) + \sum_{j=1}^{n} \Delta u_j \frac{\partial r_i}{\partial u_j}\bigg|_{u=u_k}. \quad (3.28)$$

Because we have n residuals, $i = 1, \ldots, n$, we can write the second term in matrix form as $J\Delta u$, where J is an $(n \times n)$ square matrix whose elements are

$$J_{ij} = \frac{\partial r_i}{\partial u_j}. \quad (3.29)$$

This matrix is called the *Jacobian*.

Similar to the single-variable case, we want to find the step that makes the two terms zero, which yields the linear system

$$J_k \Delta u_k = -r_k. \quad (3.30)$$

After solving this linear system, we can update the solution to

$$u_{k+1} = u_k + \Delta u_k. \quad (3.31)$$

Thus, Newton's method involves solving a sequence of linear systems given by Eq. 3.30. The linear system can be solved using any of the linear solvers mentioned in Section 3.6. One popular option for solving the Newton step is the Krylov method, which results in the Newton–Krylov method for solving nonlinear systems. Because the Krylov method only requires matrix-vector products of the form Jv, we can avoid computing and storing the Jacobian by computing this product directly (using finite differences or other methods from Chapter 6). In Section 4.4.3 we adapt Newton's method to perform function minimization instead of solving nonlinear equations.

The multivariable version of Newton's method is subject to the same issues we uncovered for the single-variable case: it only converges if the starting point is within a specific region, and it can be subject to ill-conditioning. To increase the likelihood of convergence from any starting point, Newton's method requires a *globalization strategy* (see Section 4.2). The ill-conditioning issue has to do with the linear system (Eq. 3.30) and can be quantified by the condition number of the Jacobian matrix. Ill-conditioning can be addressed by scaling and preconditioning.

There is an analog of the secant method for n dimensions, which is called *Broyden's method*. This method is much more involved than its one-dimensional counterpart because it needs to create an approximate Jacobian based on directional finite-difference derivatives. Broyden's method is described in Appendix C.1 and is related to the quasi-Newton optimization methods of Section 4.4.4.

Example 3.8 Newton's method applied to two nonlinear equations

Suppose we have the following nonlinear system of two equations:

$$u_2 = \frac{1}{u_1}, \quad u_2 = \sqrt{u_1}.$$

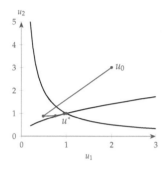

Fig. 3.18 Newton iterations.

This corresponds to the two lines shown in Fig. 3.18, where the solution is at their intersection, $u = (1, 1)$. (In this example, the two equations are explicit, and we could solve them by substitution, but they could have been implicit.)

To solve this using Newton's method, we need to write these as residuals:

$$r_1 = u_2 - \frac{1}{u_1} = 0$$

$$r_2 = u_2 - \sqrt{u_1} = 0.$$

The Jacobian can be derived analytically, and the Newton step is given by the linear system

$$\begin{bmatrix} \frac{1}{u_1^2} & 1 \\ -\frac{1}{2\sqrt{u_1}} & 1 \end{bmatrix} \begin{bmatrix} \Delta u_1 \\ \Delta u_2 \end{bmatrix} = - \begin{bmatrix} u_2 - \frac{1}{u_1} \\ u_2 - \sqrt{u_1} \end{bmatrix}.$$

Starting from $u = (2, 3)$ yields the iterations shown in the following table, with the quadratic convergence shown in Fig. 3.19.

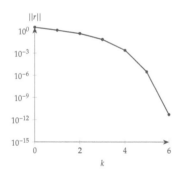

Fig. 3.19 The norm of the residual exhibits quadratic convergence.

u_1	u_2	$\|u - u^*\|$	$\|r\|$
2.000 000	3.000 000	2.24	2.96
0.485 281	0.878 680	5.29×10^{-1}	1.20
0.760 064	0.893 846	2.62×10^{-1}	4.22×10^{-1}
0.952 668	0.982 278	5.05×10^{-2}	6.77×10^{-2}
0.998 289	0.999 417	1.81×10^{-3}	2.31×10^{-3}
0.999 998	0.999 999	2.32×10^{-6}	2.95×10^{-6}
1.000 000	1.000 000	3.82×10^{-12}	4.87×10^{-12}
1.000 000	1.000 000	0.0	0.0

3.9 Models and the Optimization Problem

When performing design optimization, we must compute the values of the objective and constraint functions in the optimization problem (Eq. 1.4). Computing these functions usually requires solving a model for the given design x at one or more specific conditions.* The model often includes governing equations that define the state variables u as

---

*As previously mentioned, the process of solving a model is also known as the *analysis* or *simulation*.

3.9 Models and the Optimization Problem

an implicit function of x. In other words, for a given x, there is a u that solves $r(u; x) = 0$, as illustrated in Fig. 3.20. Here, the semicolon in $r(u; x)$ indicates that x is fixed when the governing equations are solved for u.

The objective and constraints are typically explicit functions of the state and design variables, as illustrated in Fig. 3.21 (this is a more detailed version of Fig. 1.14). There is also an implicit dependence of the objective and constraint functions on x through u. Therefore, the objective and constraint functions are ultimately fully determined by the design variables. In design optimization applications, solving the governing equations is usually the most computationally intensive part of the overall optimization process.

When we first introduced the general optimization problem (Eq. 1.4), the governing equations were not included because they were assumed to be part of the computation of the objective and constraints for a given x. However, we can include them in the problem statement for completeness as follows:

$$\begin{aligned}
\text{minimize} \quad & f(x; u) \\
\text{by varying} \quad & x_i & i = 1, \ldots, n_x \\
\text{subject to} \quad & g_j(x; u) \leq 0 & j = 1, \ldots, n_g \\
& h_k(x; u) = 0 & k = 1, \ldots, n_h \\
& \underline{x}_i \leq x_i \leq \overline{x}_i & i = 1, \ldots, n_x \\
\text{while solving} \quad & r_l(u; x) = 0 & l = 1, \ldots, n_u \\
\text{by varying} \quad & u_l & l = 1, \ldots, n_u.
\end{aligned} \quad (3.32)$$

Here, "while solving" means that the governing equations are solved at each optimization iteration to find a valid u for each value of x. The semicolon in $f(x; u)$ indicates that u is fixed while the optimizer determines the next value of x.

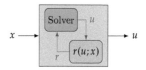

Fig. 3.20 For a general model, the state variables u are implicit functions of the design variables x through the solution of the governing equations.

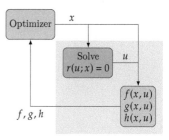

Fig. 3.21 Computing the objective (f) and constraint functions (g, h) for a given set of design variables (x) usually involves the solution of a numerical model ($r = 0$) by varying the state variables (u).

Example 3.9 Structural sizing optimization

Recalling the truss problem of Ex. 3.2, suppose we want to minimize the mass of the structure (m) by varying the cross-sectional areas of the truss members (a), subject to stress constraints.

The structural mass is an explicit function that can be written as

$$m = \sum_{i=1}^{15} \rho a_i l_i,$$

where ρ is the material density, a_i is the cross-sectional area of each member i, and l_i is the member length. This function depends on the design variables directly and does not depend on the displacements.

We can write the optimization problem statement as follows:

$$
\begin{aligned}
&\text{minimize} && m(a) \\
&\text{by varying} && a_i \geq a_{\min} && i = 1, \ldots, 15 \\
&\text{subject to} && |\sigma_j(a, u)| - \sigma_{\max} \leq 0 && j = 1, \ldots, 15 \\
&\text{while solving} && Ku - q = 0 && \text{(system of 18 equations)} \\
&\text{by varying} && u_l && l = 1, \ldots, 18.
\end{aligned}
$$

The governing equations are a linear set of equations whose solution determines the displacements (u) of a given design (a) for a load condition (q). We mentioned previously that the objective and constraint functions are usually explicit functions of the state variables, design variables, or both. As we saw in Ex. 3.2, the mass is an explicit function of the cross-sectional areas. In this case, it does not even depend on the state variables. The constraint function is also explicit, but in this case, it is just a function of the state variables. This example illustrates a common situation where the solution of the state variables requires the solution of implicit equations (structural solver), whereas the constraints (stresses) and objective (weight) are explicit functions of the states and design variables.

From a mathematical perspective, the model governing equations $r(x, u) = 0$ can be considered equality constraints in an optimization problem. Some specialized optimization approaches add these equations to the optimization problem and let the optimization algorithm solve both the governing equations and optimization simultaneously. This is called a *full-space* approach and is also known as *simultaneous analysis and design* (SAND) or *one-shot optimization*. The approach is illustrated in Fig. 3.22 and stated as follows:

$$
\begin{aligned}
&\text{minimize} && f(x, u) \\
&\text{by varying} && x_i && i = 1, \ldots, n_x \\
& && u_l && l = 1, \ldots, n_u \\
&\text{subject to} && g_j(x, u) \leq 0 && j = 1, \ldots, n_g \\
& && h_k(x, u) = 0 && k = 1, \ldots, n_h \\
& && \underline{x}_i \leq x_i \leq \overline{x}_i && i = 1, \ldots, n_x \\
& && r_l(x, u) = 0 && l = 1, \ldots, n_u.
\end{aligned}
\qquad (3.33)
$$

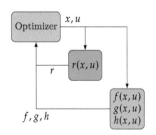

Fig. 3.22 In the full-space approach, the governing equations are solved by the optimizer by varying the state variables.

This approach is described in more detail in Section 13.4.3.

More generally, the optimization constraints and equations in a model are interchangeable. Suppose a set of equations in a model can be satisfied by varying a corresponding set of state variables. In that case, these equations and variables can be moved to the optimization problem statement as equality constraints and design variables, respectively.

Unless otherwise stated, we assume that the optimization model governing equations are solved by a dedicated solver for each optimization iteration, as stated in Eq. 3.32.

Example 3.10 Structural sizing optimization using a full-space approach

To solve the structural sizing problem (Ex. 3.9) using a full-space approach, we forgo the linear solver by adding u to the set of design variables and letting the optimizer enforce the governing equations. This results in the following problem:

$$\begin{aligned}
\text{minimize} \quad & m(a) \\
\text{by varying} \quad & a_i \geq a_{\min} & i = 1, \ldots, 15 \\
& u_l & l = 1, \ldots, 18 \\
\text{subject to} \quad & |\sigma_j(a, u)| - \sigma_{\max} \leq 0 & j = 1, \ldots, 15 \\
& Ku - q = 0 & \text{(system of 18 equations)}.
\end{aligned}$$

Tip 3.7 Test your analysis before you attempt optimization

Before you optimize, you should be familiar with the analysis (model and solver) that computes the objective and constraints. If possible, make several parameter sweeps to see what the functions look like—whether they are smooth, whether they seem unimodal or not, what the trends are, and the range of values. You should also get an idea of the computational effort required and if that varies significantly. Finally, you should test the robustness of the analysis to different inputs because the optimization is likely to ask for extreme values.

3.10 Summary

It is essential to understand the models that compute the objective and constraint functions because they directly affect the performance and effectiveness of the optimization process.

The modeling process introduces several types of numerical errors associated with each step of the process (discretization, programming, computation), limiting the achievable precision of the optimization. Knowing the level of numerical error is necessary to establish what precision can be achieved in the optimization. Understanding the types of errors involved helps us find ways to reduce those errors. Programming errors—"bugs"—are often underestimated; thorough testing is required to verify that the numerical model is coded correctly.

A lack of understanding of a given model's numerical errors is often the cause of a failure in optimization, especially when using gradient-based algorithms.

Modeling errors arise from discrepancies between the mathematical model and the actual physical system. Although they do not affect the optimization process's performance and precision, modeling errors affect the accuracy and determine how valid the result is in the real world. Therefore, model validation and an understanding of modeling error are also critical.

In engineering design optimization problems, the models usually involve solving large sets of nonlinear implicit equations. The computational time required to solve these equations dominates the overall optimization time, and therefore, solver efficiency is crucial. Solver robustness is also vital because optimization often asks for designs that are very different from what a human designer would ask for, which tests the limits of the model and the solver.

We presented an overview of the various types of solvers available for linear and nonlinear equations. Newton-type methods are highly desirable for solving nonlinear equations because they exhibit second-order convergence. Because Newton-type methods involve solving a linear system at each iteration, a linear solver is always required. These solvers are also at the core of several of the optimization algorithms in later chapters.

Problems

3.1 Answer *true* or *false* and justify your answer.

   a. A model developed to perform well for analysis will always do well in a numerical optimization process.
   b. Modeling errors have nothing to do with computations.
   c. Explicit and implicit equations can always be written in residual form.
   d. Subtractive cancellation is a type of roundoff error.
   e. Programming errors can always be eliminated by carefully reading the code.
   f. Quadratic convergence is only better than linear convergence if the asymptotic convergence error constant is less than or equal to one.
   g. Logarithmic scales are desirable when plotting convergence because they show errors of all magnitudes.
   h. Newton solvers always require a linear solver.
   i. Some linear iterative solvers can be used to solve nonlinear problems.
   j. Direct methods allow us to trade between computational cost and precision, whereas iterative methods do not.
   k. Newton's method requires the derivatives of all the state variables with respect to the residuals.
   l. In the full-space optimization approach, the state variables become design variables, and the governing equations become constraints.

3.2 Choose an engineering system that you are familiar with and describe each of the components illustrated in Fig. 3.1 for that system. List all the options for the mathematical and numerical models that you can think of, and describe the assumptions for each model. What type of solver is usually used for each model (see Section 3.6)? What are the state variables for each model?

3.3 Consider the following mathematical model:

$$\frac{u_1^2}{4} + u_2^2 = 1$$
$$4u_1 u_2 = \pi$$
$$f = 4(u_1 + u_2).$$

Solve this model by hand. Write these equations in residual form and use a numerical solver to obtain the same solution.

3.4 Reproduce a plot similar to the one shown in Fig. 3.10 for

$$f(x) = \cos(x) + 1$$

in the neighborhood of $x = \pi$.

3.5 Consider the residual equation

$$r(u) = u^3 - 6u^2 + 12u - 8 = 0.$$

   a. Find the solution using your own implementation of Newton's method.
   b. Tabulate the residual for each iteration number.
   c. What is the lowest error you can achieve?
   d. Plot the residual versus the iteration number using a linear axis. How many digits can you discern in this plot?
   e. Make the same plot using a logarithmic axis for the residual and estimate the rate of convergence. Discuss whether the rate is as expected or not.
   f. *Exploration*: Try different starting points. Can you find a predictable trend and explain it?

3.6 Kepler's equation, which we mentioned in Section 2.2, defines the relationship between a planet's polar coordinates and the time elapsed from a given initial point and is stated as follows:

$$E - e \sin(E) = M,$$

where M is the mean anomaly (a parameterization of time), E is the eccentric anomaly (a parameterization of polar angle), and e is the eccentricity of the elliptical orbit.

   a. Use Newton's method to find E when $e = 0.7$ and $M = \pi/2$.
   b. Devise a fixed-point iteration to solve the same problem.
   c. Compare the number of iterations and rate of convergence.
   d. *Exploration*: Plot E versus M in the interval $[0, 2\pi]$ for $e = [0, 0.1, 0.5, 0.9]$ and interpret your results physically.

3.7 Consider the equation from Prob. 3.5 where we replace one of the coefficients with a parameter a as follows:

$$r(u) = au^3 - 6u^2 + 12u - 8 = 0.$$

a. Produce a plot similar to Fig. 3.12 by perturbing a in the neighborhood of $a = 1.2$ using a solver convergence tolerance of $|r| \leq 10^{-6}$.

b. *Exploration*: Try smaller tolerances and see how much you can decrease the numerical noise.

3.8 Reproduce the solution of Ex. 3.8 and then try different initial guesses. Can you define a distinct region from where Newton's method converges?

3.9 Choose a problem that you are familiar with and find the magnitude of numerical noise in one or more outputs of interest with respect to one or more inputs of interest. What means do you have to decrease the numerical noise? What is the lowest possible level of noise you can achieve?

4 Unconstrained Gradient-Based Optimization

In this chapter we focus on unconstrained optimization problems with continuous design variables, which we can write as

$$\underset{x}{\text{minimize}}\; f(x), \tag{4.1}$$

where $x = [x_1, \ldots, x_n]$ is composed of the design variables that the optimization algorithm can change.

We solve these problems using gradient information to determine a series of steps from a starting guess (or initial design) to the optimum, as shown in Fig. 4.1. We assume the objective function to be nonlinear, C^2 continuous, and deterministic. We do not assume unimodality or multimodality, and there is no guarantee that the algorithm finds the global optimum. Referring to the attributes that classify an optimization problem (Fig. 1.22), the optimization algorithms discussed in this chapter range from first to second order, perform a local search, and evaluate the function directly. The algorithms are based on mathematical principles rather than heuristics.

Although most engineering design problems are constrained, the constrained optimization algorithms in Chapter 5 build on the methods explained in the current chapter.

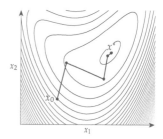

Fig. 4.1 Gradient-based optimization starts with a guess, x_0, and takes a sequence of steps in n-dimensional space that converge to an optimum, x^*.

> By the end of this chapter you should be able to:
>
> 1. Understand the significance of gradients, Hessians, and directional derivatives.
>
> 2. Mathematically define the optimality conditions for an unconstrained problem.
>
> 3. Describe, implement, and use line-search-based methods.
>
> 4. Explain the pros and cons of the various search direction methods.
>
> 5. Understand the trust-region approach and how it contrasts with the line search approach.

4.1 Fundamentals

To determine the step directions shown in Fig. 4.1, gradient-based methods need the gradient (first-order information). Some methods also use curvature (second-order information). Gradients and curvature are required to build a second-order Taylor series, a fundamental building block in establishing optimality and developing gradient-based optimization algorithms.

4.1.1 Derivatives and Gradients

Recall that we are considering a scalar objective function $f(x)$, where x is the vector of design variables, $x = [x_1, x_2, \ldots, x_n]$. The *gradient* of this function, $\nabla f(x)$, is a column vector of first-order partial derivatives of the function with respect to each design variable:

$$\nabla f(x) = \left[\frac{\partial f}{\partial x_1}, \frac{\partial f}{\partial x_2}, \ldots, \frac{\partial f}{\partial x_n} \right], \quad (4.2)$$

where each partial derivative is defined as the following limit:

$$\frac{\partial f}{\partial x_i} = \lim_{\varepsilon \to 0} \frac{f(x_1, \ldots, x_i + \varepsilon, \ldots, x_n) - f(x_1, \ldots, x_i, \ldots, x_n)}{\varepsilon}. \quad (4.3)$$

Each component in the gradient vector quantifies the function's local rate of change with respect to the corresponding design variable, as shown in Fig. 4.2 for the two-dimensional case. In other words, these components represent the slope of the function along each coordinate direction. The gradient is a vector pointing in the direction of the greatest function increase from the current point.

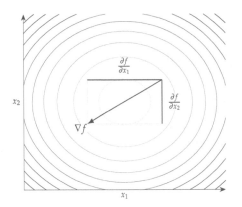

Fig. 4.2 Components of the gradient vector in the two-dimensional case.

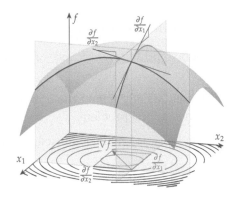

The gradient vectors are normal to the surfaces of constant f in n-dimensional space (*isosurfaces*). In the two-dimensional case, gradient

4.1 Fundamentals

vectors are perpendicular to the function contour lines, as shown in Fig. 4.2.*

*In this book, most of the illustrations and examples are based on two-dimensional problems because they are easy to visualize. However, the principles and methods apply to n dimensions.

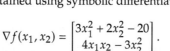

Example 4.1 Gradient of a polynomial function

Consider the following function of two variables:

$$f(x_1, x_2) = x_1^3 + 2x_1 x_2^2 - x_2^3 - 20x_1.$$

The gradient can be obtained using symbolic differentiation, yielding

$$\nabla f(x_1, x_2) = \begin{bmatrix} 3x_1^2 + 2x_2^2 - 20 \\ 4x_1 x_2 - 3x_2^2 \end{bmatrix}.$$

This defines the vector field plotted in Fig. 4.3, where each vector points in the direction of the steepest local increase.

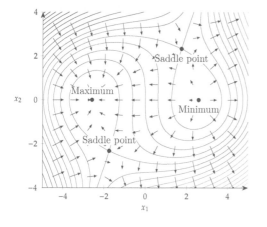
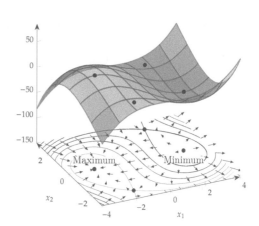

Fig. 4.3 Gradient vector field shows how gradients point toward maxima and away from minima.

If a function is simple, we can use symbolic differentiation as we did in Ex. 4.1. However, symbolic differentiation has limited utility for general engineering models because most models are far more complicated; they may include loops, conditionals, nested functions, and implicit equations. Fortunately, there are several methods for computing derivatives numerically; we cover these methods in Chapter 6.

Each gradient component has units that correspond to the units of the function divided by the units of the corresponding variable. Because the variables might represent different physical quantities, each gradient component might have different units.

From an engineering design perspective, it might be helpful to think about *relative* changes, where the derivative is given as the percentage change in the function for a 1 percent increase in the variable. This relative derivative can be computed by nondimensionalizing both the function and the variable, that is,

$$\frac{\partial f}{\partial x}\frac{x}{f},\tag{4.4}$$

where f and x are the values of the function and variable, respectively, at the point where the derivative is computed.

Example 4.2 Interpretation of derivatives for wing design problem

Consider the wing design problem from Ex. 1.1, where the objective function is the required power (P). For the derivative of power with respect to span ($\partial P/\partial b$), the units are watts per meter (W/m). For a wing with $c = 1$ m and $b = 12$ m, we have $P = 1087.85$ W and $\partial P/\partial b = -41.65$ W/m. This means that for an increase in span of 1 m, the linear approximation predicts a decrease in power of 41.65 W (to $P = 1046.20$). However, the actual power at $b = 13\ m$ is 1059.77 W because the function is nonlinear (see Fig. 4.4). The relative derivative for this same design can be computed as $(\partial P/\partial b)(b/P) = -0.459$, which means that for a 1 percent increase in span, the linear approximation predicts a 0.459 percent decrease in power. The actual decrease is 0.310 percent.

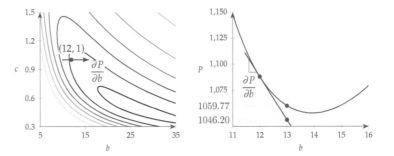

Fig. 4.4 Power versus span and the corresponding derivative.

The gradient components quantify the function's rate of change in each coordinate direction (x_i), but sometimes we are interested in the rate of change in a direction that is not a coordinate direction. The rate of change in a direction p is quantified by a *directional derivative*, defined as

$$\nabla_p f(x) = \lim_{\varepsilon \to 0} \frac{f(x + \varepsilon p) - f(x)}{\varepsilon}.\tag{4.5}$$

We can find this derivative by projecting the gradient onto the desired direction p using the dot product

$$\nabla_p f(x) = \nabla f^\mathsf{T} p.\tag{4.6}$$

4.1 Fundamentals

When p is a unit vector aligned with one of the Cartesian coordinates i, this dot product yields the corresponding partial derivative $\partial f / \partial x_i$. A two-dimensional example of this projection is shown in Fig. 4.5.

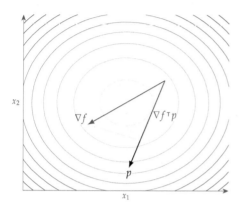
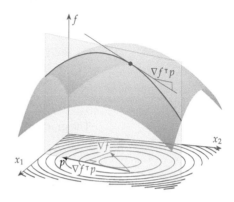

Fig. 4.5 Projection of the gradient in an arbitrary unit direction p.

From the gradient projection, we can see why the gradient is the direction of the steepest increase. If we use this definition of the dot product,

$$\nabla_p f(x) = \nabla f^\mathsf{T} p = \|\nabla f\| \|p\| \cos \theta , \qquad (4.7)$$

where θ is the angle between the two vectors, we can see that this is maximized when $\theta = 0°$. That is, the directional derivative is largest when p points in the same direction as ∇f.

If θ is in the interval $(-90, 90)°$, the directional derivative is positive and is thus in a direction of increase, as shown in Fig. 4.6. If θ is in the interval $(90, 180]°$, the directional derivative is negative, and p points in a descent direction. Finally, if $\theta = \pm 90°$, the directional derivative is 0, and thus the function value does not change for small steps; it is locally flat in that direction. This condition occurs when ∇f and p are orthogonal; therefore, the gradient is orthogonal to the function isosurfaces.

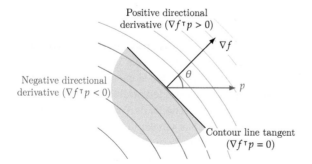

Fig. 4.6 The gradient ∇f is always orthogonal to contour lines (surfaces), and the directional derivative in the direction p is given by $\nabla f^\mathsf{T} p$.

To get the correct slope in the original units of x, the direction should be normalized as $\hat{p} = p/\|p\|$. However, in some of the gradient-based algorithms of this chapter, p is not normalized because the length contains useful information. If p is not normalized, the slopes and variable axis are scaled by a constant.

Example 4.3 Directional derivative of a quadratic function

Consider the following function of two variables:
$$f(x_1, x_2) = x_1^2 + 2x_2^2 - x_1 x_2.$$

The gradient can be obtained using symbolic differentiation, yielding
$$\nabla f(x_1, x_2) = \begin{bmatrix} 2x_1 - x_2 \\ 4x_2 - x_1 \end{bmatrix}.$$

At point $x = [-1, 1]$, the gradient is
$$\nabla f(-1, 1) = \begin{bmatrix} -3 \\ 5 \end{bmatrix}.$$

Taking the derivative in the normalized direction $p = [2/\sqrt{5}, -1/\sqrt{5}]$, we obtain
$$\nabla f^\mathsf{T} p = [-3, 5] \begin{bmatrix} 2/\sqrt{5} \\ -1/\sqrt{5} \end{bmatrix} = -\frac{11}{\sqrt{5}},$$

which we show in Fig. 4.7 (left). We use a p with unit length to get the slope of the function in the original units.

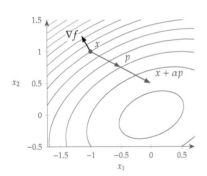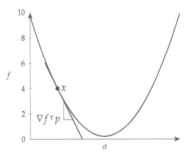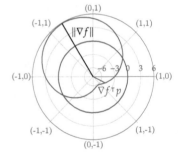

Fig. 4.7 Function contours and direction p (left), one-dimensional slice along p (middle), directional derivative for all directions on polar plot (right).

A projection of the function in the p direction can be obtained by plotting f along the line defined by $x + \alpha p$, where α is the independent variable, as shown in Fig. 4.7 (middle). The projected slope of the function in that direction corresponds to the slope of this single-variable function. The polar plot in Fig. 4.7 (right) shows how the directional derivative changes with the direction of p. The directional derivative has a maximum in the direction of the gradient, has the largest negative magnitude in the opposite direction, and has zero values in the directions orthogonal to the gradient.

4.1.2 Curvature and Hessians

The rate of change of the gradient—the *curvature*—is also useful information because it tells us if a function's slope is increasing (positive curvature), decreasing (negative curvature), or stationary (zero curvature).

In one dimension, the gradient reduces to a scalar (the slope), and the curvature is also a scalar that can be calculated by taking the second derivative of the function. To quantify curvature in n dimensions, we need to take the partial derivative of each gradient component j with respect to each coordinate direction i, yielding

$$\frac{\partial^2 f}{\partial x_i \partial x_j}. \tag{4.8}$$

If the function f has continuous second partial derivatives, the order of differentiation does not matter, and the mixed partial derivatives are equal; thus

$$\frac{\partial^2 f}{\partial x_i \partial x_j} = \frac{\partial^2 f}{\partial x_j \partial x_i}. \tag{4.9}$$

This property is known as the *symmetry of second derivatives* or *equality of mixed partials*.[†]

Considering all gradient components and their derivatives with respect to all coordinate directions results in a second-order tensor. This tensor can be represented as a square ($n \times n$) matrix of second-order partial derivatives called the *Hessian*:

$$H_f(x) = \begin{bmatrix} \frac{\partial^2 f}{\partial x_1^2} & \frac{\partial^2 f}{\partial x_1 \partial x_2} & \cdots & \frac{\partial^2 f}{\partial x_1 \partial x_n} \\ \frac{\partial^2 f}{\partial x_2 \partial x_1} & \frac{\partial^2 f}{\partial x_2^2} & \cdots & \frac{\partial^2 f}{\partial x_2 \partial x_n} \\ \vdots & \vdots & \ddots & \vdots \\ \frac{\partial^2 f}{\partial x_n \partial x_1} & \frac{\partial^2 f}{\partial x_n \partial x_2} & \cdots & \frac{\partial^2 f}{\partial x_n^2} \end{bmatrix}. \tag{4.10}$$

The Hessian is expressed in index notation as:

$$H_{f_{ij}} = \frac{\partial^2 f}{\partial x_i \partial x_j}. \tag{4.11}$$

Because of the symmetry of second derivatives, the Hessian is a symmetric matrix with $n(n + 1)/2$ independent elements.

[†] The discovery and proof of the symmetry of second derivatives property has a long history.[76]

76. Higgins, *A note on the history of mixed partial derivatives*, 1940.

Each row i of the Hessian is a vector that quantifies the rate of change of all components j of the gradient vector with respect to the direction i. On the other hand, each column j of the matrix quantifies the rate of change of component j of the gradient vector with respect to all coordinate directions i. Because the Hessian is symmetric, the rows and columns are transposes of each other, and these two interpretations are equivalent.

We can find the rate of change of the gradient in an arbitrary normalized direction p by taking the product Hp. This yields an n-vector that quantifies the rate of change of the gradient in the direction p, where each component of the vector is the rate of the change of the corresponding partial derivative with respect to a movement along p. Therefore, this product is defined as follows:

$$Hp = \nabla_p \left(\nabla f(x) \right) = \lim_{\varepsilon \to 0} \frac{\nabla f(x + \varepsilon p) - \nabla f(x)}{\varepsilon}. \tag{4.12}$$

Because of the symmetry of second derivatives, we can also interpret this as the rate of change in the directional derivative of the function along p with respect to each of the components of p.

To find the curvature of the one-dimensional function along a direction p, we need to project Hp onto direction p as

$$\nabla_p \left(\nabla_p f(x) \right) = p^\mathsf{T} H p, \tag{4.13}$$

which yields a scalar quantity. Again, if we want to get the curvature in the original units of x, p should be normalized.

For an n-dimensional Hessian, it is possible to find directions v_i (where $i = 1, \ldots, n$) along which the projected curvature aligns with that direction, that is,

$$Hv = \kappa v. \tag{4.14}$$

This is an eigenvalue problem whose eigenvectors represent the *principal curvature* directions, and the eigenvalues κ quantify the corresponding curvatures. If each eigenvector is normalized as $\hat{v} = v/\|v\|$, then the corresponding κ is the curvature in the original units.

Example 4.4 Hessian and principal curvature directions of a quadratic

Consider the following quadratic function of two variables:

$$f(x_1, x_2) = x_1^2 + 2x_2^2 - x_1 x_2,$$

whose contours are shown in Fig. 4.8 (left). These contours are ellipses that have the same center. The Hessian of this quadratic is

$$H = \begin{bmatrix} 2 & -1 \\ -1 & 4 \end{bmatrix},$$

4.1 Fundamentals

which is constant. To find the curvature in the direction $p = [-1/2, -\sqrt{3}/2]$, we compute

$$p^\mathsf{T} H p = \begin{bmatrix} \frac{-1}{2} & \frac{-\sqrt{3}}{2} \end{bmatrix} \begin{bmatrix} 2 & -1 \\ -1 & 4 \end{bmatrix} \begin{bmatrix} \frac{-1}{2} \\ \frac{-\sqrt{3}}{2} \end{bmatrix} = \frac{7 - \sqrt{3}}{2}.$$

The principal curvature directions can be computed by solving the eigenvalue problem (Eq. 4.14). This yields two eigenvalues and two corresponding eigenvectors,

$$\kappa_A = 3 + \sqrt{2}, \quad v_A = \begin{bmatrix} 1 - \sqrt{2} \\ 1 \end{bmatrix}, \quad \text{and} \quad \kappa_B = 3 - \sqrt{2}, \quad v_B = \begin{bmatrix} 1 + \sqrt{2} \\ 1 \end{bmatrix}.$$

By plotting the principal curvature directions superimposed on the function contours (Fig. 4.8, left), we can see that they are aligned with the ellipses' major and minor axes. To see how the curvature varies as a function of the direction, we make a polar plot of the curvature $p^\mathsf{T} H p$, where p is normalized (Fig. 4.8, right). The maximum curvature aligns with the first principal curvature direction, as expected, and the minimum curvature corresponds to the second principal curvature direction.

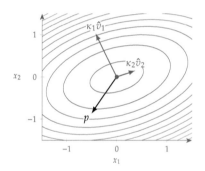
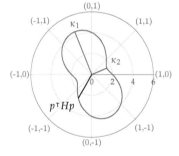

Fig. 4.8 Contours of f for Ex. 4.4 and the two principal curvature directions in red. The polar plot shows the curvature, with the eigenvectors pointing at the directions of principal curvature; all other directions have curvature values in between.

Example 4.5 Hessian of two-variable polynomial

Consider the same polynomial from Ex. 4.1. Differentiating the gradient we obtained previously yields the Hessian:

$$H(x_1, x_2) = \begin{bmatrix} 6x_1 & 4x_2 \\ 4x_2 & 4x_1 - 6x_2 \end{bmatrix}.$$

We can visualize the variation of the Hessian by plotting the principal curvatures at different points (Fig. 4.9).

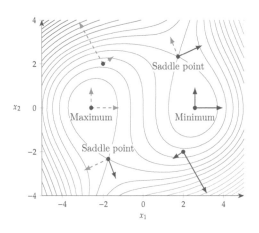
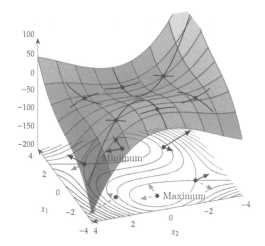

Fig. 4.9 Principal curvature direction and magnitude variation. Solid lines correspond to positive curvature, whereas dashed lines are for negative curvature.

4.1.3 Taylor Series

The Taylor series provides a local approximation to a function and is the foundation for gradient-based optimization algorithms.

For an n-dimensional function, the Taylor series can predict the function along any direction p. This is done by projecting the gradient and Hessian onto the desired direction p to get an approximation of the function at any nearby point $x + p$:[‡]

‡For a more extensive introduction to the Taylor series, see Appendix A.1.

$$f(x + p) = f(x) + \nabla f(x)^\mathsf{T} p + \frac{1}{2} p^\mathsf{T} H(x) p + \mathcal{O}\left(\|p\|^3\right). \quad (4.15)$$

We use a second-order Taylor series (ignoring the cubic term) because it results in a quadratic, the lowest-order Taylor series that can have a minimum. For a function that is C^2 continuous, this approximation can be made arbitrarily accurate by making $\|p\|$ small enough.

Example 4.6 Second-order Taylor series expansion of two-variable function

Using the gradient and Hessian of the two-variable polynomial from Ex. 4.1 and Ex. 4.5, we can use Eq. 4.15 to construct a second-order Taylor expansion about x_0,

$$\tilde{f}(p) = f(x_0) + \begin{bmatrix} 3x_1^2 + 2x_2^2 - 20 \\ 4x_1 x_2 - 3x_2^2 \end{bmatrix}^\mathsf{T} p + p^\mathsf{T} \begin{bmatrix} 6x_1 & 4x_2 \\ 4x_2 & 4x_1 - 6x_2 \end{bmatrix} p.$$

Figure 4.10 shows the resulting Taylor series expansions about different points. We perform three expansions, each about three critical points: the minimum (left), the maximum (middle), and the saddle point (right). The expansion about the minimum yields a convex quadratic that is a good approximation of the original function near the minimum but becomes worse as we step farther

4.1 Fundamentals

away. The expansion about the maximum shows a similar trend except that the approximation is a concave quadratic. Finally, the expansion about the saddle point yields a saddle function.

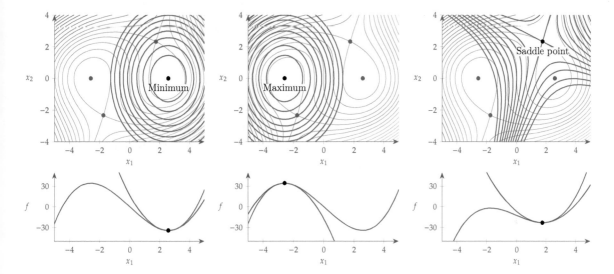

Fig. 4.10 The second-order Taylor series expansion uses the function value, gradient, and Hessian at a point to construct a quadratic model about that point. The model can vary drastically, depending on the function and the point location. The one-dimensional slices are in the x_1 direction and at x_2 values corresponding to the critical points.

4.1.4 Optimality Conditions

To find the minimum of a function, we must determine the mathematical conditions that identify a given point x as a minimum. There is only a limited set of problems for which we can prove global optimality, so in general, we are only interested in local optimality. A point x^* is a local minimum if $f(x^*) \leq f(x)$ for all x in the neighborhood of x^*. In other words, there must be no descent direction starting from the local minimum.

A second-order Taylor series expansion about x^* for small steps of size p yields

$$f(x^* + p) = f(x^*) + \nabla f(x^*)^\mathsf{T} p + \frac{1}{2} p^\mathsf{T} H(x^*) p + \ldots . \qquad (4.16)$$

For x^* to be an optimal point, we must have $f(x^* + p) \geq f(x^*)$ for all p. This implies that the first- and second-order terms in the Taylor series have to be nonnegative, that is,

$$\nabla f(x^*)^\mathsf{T} p + \frac{1}{2} p^\mathsf{T} H(x^*) p \geq 0 \quad \text{for all} \quad p . \qquad (4.17)$$

Because the magnitude of p is small, we can always find a p such that the first term dominates. Therefore, we require that

$$\nabla f(x^*)^\mathsf{T} p \geq 0 \quad \text{for all} \quad p. \tag{4.18}$$

Because p can be in any arbitrary direction, the only way this inequality can be satisfied is if all the elements of the gradient are zero (refer to Fig. 4.6),

$$\nabla f(x^*) = 0. \tag{4.19}$$

This is the *first-order necessary optimality condition* for an unconstrained problem. This is necessary because if any element of p is nonzero, there are descent directions (e.g., $p = -\nabla f$) for which the inequality would not be satisfied.

Because the gradient term has to be zero, we must now satisfy the remaining term in the inequality (Eq. 4.17), that is,

$$p^\mathsf{T} H(x^*) p \geq 0 \quad \text{for all} \quad p. \tag{4.20}$$

From Eq. 4.13, we know that this term represents the curvature in direction p, so this means that the function curvature must be positive or zero when projected in any direction. You may recognize this inequality as the definition of a *positive-semidefinite* matrix. In other words, the Hessian $H(x^*)$ must be positive semidefinite.

For a matrix to be positive semidefinite, its eigenvalues must all be greater than or equal to zero. Recall that the eigenvalues of the Hessian quantify the principal curvatures, so as long as all the principal curvatures are greater than or equal to zero, the curvature along an arbitrary direction is also greater than or equal to zero.

These conditions on the gradient and curvature are *necessary conditions* for a local minimum but are not sufficient. They are not sufficient because if the curvature is zero in some direction p (i.e., $p^\mathsf{T} H(x^*) p = 0$), we have no way of knowing if it is a minimum unless we check the third-order term. In that case, even if it is a minimum, it is a weak minimum.

The *sufficient conditions* for optimality require the curvature to be positive in any direction, in which case we have a *strong minimum*. Mathematically, this means that $p^\mathsf{T} H(x^*) p > 0$ for all nonzero p, which is the definition of a *positive-definite* matrix. If H is a positive-definite matrix, every eigenvalue of H is positive.[§]

Figure 4.11 shows some examples of quadratic functions that are positive definite (all positive eigenvalues), positive semidefinite (nonnegative eigenvalues), indefinite (mixed eigenvalues), and negative definite (all negative eigenvalues).

[§]For other approaches to determine if a matrix is positive definite, see Appendix A.6.

4.1 Fundamentals

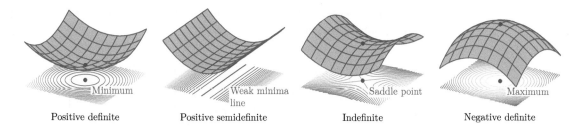

Positive definite — Minimum
Positive semidefinite — Weak minima line
Indefinite — Saddle point
Negative definite — Maximum

Fig. 4.11 Quadratic functions with different types of Hessians.

In summary, the *necessary optimality conditions* for an unconstrained optimization problem are

$$\nabla f(x^*) = 0 \\ H(x^*) \text{ is positive semidefinite}. \tag{4.21}$$

The *sufficient optimality conditions* are

$$\nabla f(x^*) = 0 \\ H(x^*) \text{ is positive definite}. \tag{4.22}$$

Example 4.7 Finding minima analytically

Consider the following function of two variables:

$$f = 0.5x_1^4 + 2x_1^3 + 1.5x_1^2 + x_2^2 - 2x_1 x_2.$$

We can find the minima of this function by solving for the optimality conditions analytically.

To find the critical points of this function, we solve for the points at which the gradient is equal to zero,

$$\nabla f = \begin{bmatrix} \dfrac{\partial f}{\partial x_1} \\ \dfrac{\partial f}{\partial x_2} \end{bmatrix} = \begin{bmatrix} 2x_1^3 + 6x_1^2 + 3x_1 - 2x_2 \\ 2x_2 - 2x_1 \end{bmatrix} = \begin{bmatrix} 0 \\ 0 \end{bmatrix}.$$

From the second equation, we have that $x_2 = x_1$. Substituting this into the first equation yields

$$x_1 \left(2x_1^2 + 6x_1 + 1\right) = 0.$$

The solution of this equation yields three points:

$$x_A = \begin{bmatrix} 0 \\ 0 \end{bmatrix}, \quad x_B = \begin{bmatrix} -\dfrac{3}{2} - \dfrac{\sqrt{7}}{2} \\ -\dfrac{3}{2} - \dfrac{\sqrt{7}}{2} \end{bmatrix}, \quad x_C = \begin{bmatrix} \dfrac{\sqrt{7}}{2} - \dfrac{3}{2} \\ \dfrac{\sqrt{7}}{2} - \dfrac{3}{2} \end{bmatrix}.$$

To classify these points, we need to compute the Hessian matrix. Differentiating the gradient, we get

$$H(x_1, x_2) = \begin{bmatrix} \dfrac{\partial^2 f}{\partial x_1^2} & \dfrac{\partial^2 f}{\partial x_1 \partial x_2} \\ \dfrac{\partial^2 f}{\partial x_2 \partial x_1} & \dfrac{\partial^2 f}{\partial x_2^2} \end{bmatrix} = \begin{bmatrix} 6x_1^2 + 12x_1 + 3 & -2 \\ -2 & 2 \end{bmatrix}.$$

The Hessian at the first point is

$$H(x_A) = \begin{bmatrix} 3 & -2 \\ -2 & 2 \end{bmatrix},$$

whose eigenvalues are $\kappa_1 \approx 0.438$ and $\kappa_2 \approx 4.561$. Because both eigenvalues are positive, this point is a local minimum. For the second point,

$$H(x_B) = \begin{bmatrix} 3\left(3 + \sqrt{7}\right) & -2 \\ -2 & 2 \end{bmatrix}.$$

The eigenvalues are $\kappa_1 \approx 1.737$ and $\kappa_2 \approx 17.200$, so this point is another local minimum. For the third point,

$$H(x_C) = \begin{bmatrix} 9 - 3\sqrt{7} & -2 \\ -2 & 2 \end{bmatrix}.$$

The eigenvalues for this Hessian are $\kappa_1 \approx -0.523$ and $\kappa_2 \approx 3.586$, so this point is a saddle point.

Figure 4.12 shows these three critical points. To find out which of the two local minima is the global one, we evaluate the function at each of these points. Because $f(x_B) < f(x_A)$, x_B is the global minimum.

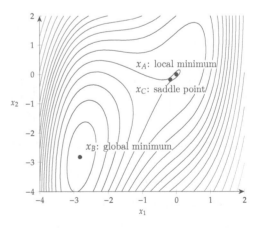

Fig. 4.12 Minima and saddle point locations.

We may be able to solve the optimality conditions analytically for simple problems, as we did in Ex. 4.7. However, this is not possible

4.1 Fundamentals

in general because the resulting equations might not be solvable in closed form. Therefore, we need numerical methods that solve for these conditions.

When using a numerical approach, we seek points where $\nabla f(x^*) = 0$, but the entries in ∇f do not converge to exactly zero because of finite-precision arithmetic. Instead, we define convergence for the first criterion based on the maximum component of the gradient, such that

$$\|\nabla f\|_\infty \leq \tau, \tag{4.23}$$

where τ is some tolerance. A typical absolute tolerance is $\tau = 10^{-6}$ or a six-order magnitude reduction in gradient when using a relative tolerance. Absolute and relative criteria are often combined in a metric such as the following:

$$\|\nabla f\|_\infty \leq \tau \left(1 + \|\nabla f_0\|_\infty\right), \tag{4.24}$$

where ∇f_0 is the gradient at the starting point.

The second optimality condition (that H must be positive semidefinite) is not usually checked explicitly. If we satisfy the first condition, then all we know is that we have reached a stationary point, which could be a maximum, a minimum, or a saddle point. However, as shown in Section 4.4, the search directions for the algorithms of this chapter are always descent directions, and therefore in practice, they should converge to a local minimum.

For a practical algorithm, other exit conditions are often used besides the reduction in the norm of the gradient. A function might be poorly scaled, be noisy, or have other numerical issues that prevent it from ever satisfying this optimality condition (Eq. 4.24). To prevent the algorithm from running indefinitely, it is common to set a limit on the computational budget, such as the number of function calls, the number of major iterations, or the clock time. Additionally, to detect a case where the optimizer is not making significant progress and not likely to improve much further, we might set criteria on the minimum step size and the minimum change in the objective. Similar to the conditions on the gradient, the minimum change in step size could be limited as follows:

$$\|x_k - x_{k-1}\|_\infty < \tau_x \left(1 + \|x_{k-1}\|_\infty\right). \tag{4.25}$$

The absolute and relative conditions on the objective are of the same form, although they only use an absolute value rather than a norm because the objective is scalar.

> **Tip 4.1** Check the exit message when using an optimizer

Optimizers usually include an exit message when returning a result. Inexperienced users often take whatever solution the optimizer returns without checking this message. However, as discussed previously, the optimizer may terminate without satisfying first-order optimality (Eq. 4.24). Check the exit message and study the optimizer's documentation to make sure you understand the result. If the message indicates that this is not a definite optimum, you should investigate further.

You might have to increase the limit on the number of iterations if the optimization reached this limit. When terminating due to small step sizes or function changes, you might need to improve your numerical model by reducing the noise (see Tip 3.2) or by smoothing it (Tip 4.7). Another likely culprit is scaling (Tip 4.4). Finally, you might want to explore the design space around the point where the optimizer is stuck (Tip 4.2) and more specifically, see what is happening with the line search (Tip 4.3).

4.2 Two Overall Approaches to Finding an Optimum

Although the optimality conditions derived in the previous section can be solved analytically to find the function minima, this analytic approach is not possible for functions that result from numerical models. Instead, we need iterative numerical methods to find minima based only on the function values and gradients.

In Chapter 3, we reviewed methods for solving simultaneous systems of nonlinear equations, which we wrote as $r(u) = 0$. Because the first-order optimality condition ($\nabla f = 0$) can be written in this residual form (where $r = \nabla f$ and $u = x$), we could try to use the solvers from Chapter 3 directly to solve unconstrained optimization problems. Although several components of general solvers for $r(u) = 0$ are used in optimization algorithms, these solvers are not the most effective approaches in their original form. Furthermore, solving $\nabla f = 0$ is not necessarily sufficient—it finds a stationary point but not necessarily a minimum. Optimization algorithms require additional considerations to ensure convergence to a minimum.

Like the iterative solvers from Chapter 3, gradient-based algorithms start with a guess, x_0, and generate a series of points, $x_1, x_2, \ldots, x_k, \ldots$ that converge to a local optimum, x^*, as previously illustrated in Fig. 4.1. At each iteration, some form of the Taylor series about the current point is used to find the next point.

A truncated Taylor series is, in general, only a good model within a small neighborhood, as shown in Fig. 4.13, which shows three quadratic

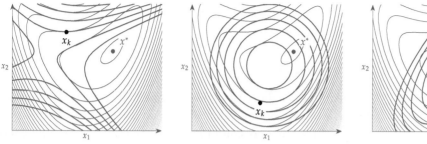
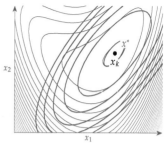

models of the same function based on three different points. All quadratic approximations match the local gradient and curvature at the respective points. However, the Taylor series quadratic about the first point (left plot) yields a quadratic without a minimum (the only critical point is a saddle point). The second point (middle plot) yields a quadratic whose minimum is closer to the true minimum. Finally, the Taylor series about the actual minimum point (right plot) yields a quadratic with the same minimum, as expected, but we can see how the quadratic model worsens the farther we are from the point.

Fig. 4.13 Taylor series quadratic models are only guaranteed to be accurate near the point about which the series is expanded (x_k).

Because the Taylor series is only guaranteed to be a good model locally, we need a *globalization strategy* to ensure convergence to an optimum. *Globalization* here means making the algorithm robust enough that it can converge to a local minimum when starting from any point in the domain. This should not be confused with finding the global minimum, which is a separate issue (see Tip 4.8). There are two main globalization strategies: line search and trust region.

The line search approach consists of three main steps for every iteration (Fig. 4.14):

1. Choose a suitable search direction from the current point. The choice of search direction is based on a Taylor series approximation.
2. Determine how far to move in that direction by performing a *line search*.
3. Move to the new point and update all values.

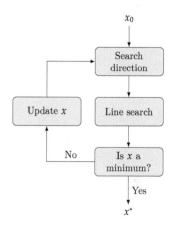

Fig. 4.14 Line search approach.

The two first steps can be seen as two separate subproblems. We address the line search subproblem in Section 4.3 and the search direction subproblem in Section 4.4.

Trust-region methods also consist of three steps (Fig. 4.15):

1. Create a model about the current point. This model can be based on a Taylor series approximation or another type of surrogate model.

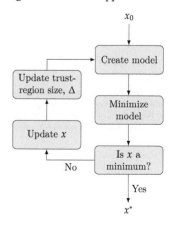

Fig. 4.15 Trust-region approach.

2. Minimize the model within a *trust region* around the current point to find the step.
3. Move to the new point, update values, and adapt the size of the trust region.

We introduce the trust-region approach in Section 4.5. However, we devote more attention to algorithms that use the line search approach because they are more common in general nonlinear optimization.

Both line search and trust-region approaches use iterative processes that must be repeated until some convergence criterion is satisfied. The first step in both approaches is usually referred to as a *major* iteration, whereas the second step might require more function evaluations corresponding to *minor* iterations.

Tip 4.2 Before optimizing, explore the design space

Before coupling your model solver with an optimizer, it is a good idea to explore the design space. Ensure that the solver is robust and can handle a wide variety of inputs within your provided bounds without errors. Plotting the multidimensional design space is generally impossible, but you can perform a series of one-dimensional sweeps. From the starting point, plot the objective with all design variables fixed except one. Vary that design variable across a range, and repeat that process for several design variables. These one-dimensional plots can identify issues such as analysis failures, noisy outputs, and discontinuous outputs, which you can then fix. These issues should be addressed before attempting to optimize. This same technique can be helpful when an optimizer becomes stuck; you can plot the behavior in a small neighborhood around the point of failure (see Tip 4.3).

4.3 Line Search

Gradient-based unconstrained optimization algorithms that use a line search follow Alg. 4.1. We start with a guess x_0 and provide a convergence tolerance τ for the optimality condition.* The final output is an optimal point x^* and the corresponding function value $f(x^*)$. As mentioned in the previous section, there are two main subproblems in line search gradient-based optimization algorithms: choosing the search direction and determining how far to step in that direction. In the next section, we introduce several methods for choosing the search direction. The line search method determines how far to step in the chosen direction and is usually independent of the method for choosing the search direction. Therefore, line search methods can be combined

*This algorithm, and others in this section, use a basic convergence check for simplicity. See the end of Section 4.1.4 for better alternatives and additional exit criteria.

4.3 Line Search

with any method for finding the search direction. However, the search direction method determines the name of the overall optimization algorithm, as we will see in the next section.

Algorithm 4.1 Gradient-based unconstrained optimization using a line search

Inputs:
 x_0: Starting point
 τ: Convergence tolerance

Outputs:
 x^*: Optimal point
 $f(x^*)$: Minimum function value

$k = 0$ Initialize iteration counter
while $\|\nabla f\|_\infty > \tau$ **do**   Optimality condition
    Determine search direction, p_k Use any of the methods from Section 4.4
    Determine step length, α_k Use a line search algorithm
    $x_{k+1} = x_k + \alpha_k p_k$ Update design variables
    $k = k + 1$ Increment iteration index
end while

For the line search subproblem, we assume that we are given a starting location at x_k and a suitable search direction p_k along which to search (Fig. 4.16). The line search then operates solely on points along direction p_k starting from x_k, which can be written as

$$x_{k+1} = x_k + \alpha p_k, \tag{4.26}$$

where the scalar α is always positive and represents how far we go in the direction p_k. This equation produces a one-dimensional slice of n-dimensional space, as illustrated in Fig. 4.17.

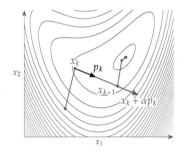
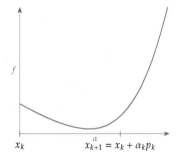

Fig. 4.16 The line search starts from a given point x_k and searches solely along direction p_k.

Fig. 4.17 The line search projects the n-dimensional problem onto one dimension, where the independent variable is α.

The line search determines the magnitude of the scalar α_k, which in turn determines the next point in the iteration sequence. Even though

x_k and p_k are n-dimensional, the line search is a one-dimensional problem with the goal of selecting α_k.

Line search methods require that the search direction p_k be a *descent direction* so that $\nabla f_k^\mathsf{T} p_k < 0$ (see Fig. 4.18). This guarantees that f can be reduced by stepping some distance along this direction with a positive α.

The goal of the line search is *not* to find the value of α that minimizes $f(x_k + \alpha p_k)$ but to find a point that is "good enough" using as few function evaluations as possible. This is because finding the exact minimum along the line would require too many evaluations of the objective function and possibly its gradient. Because the overall optimization needs to find a point in n-dimensional space, the search direction might change drastically between line searches, so spending too many iterations on each line search is generally not worthwhile.

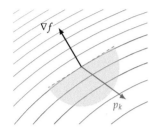

Fig. 4.18 The line search direction must be a descent direction.

Consider the bean function whose contours are shown in Fig. 4.19. At point x_k, the direction p_k is a descent direction. However, it would be wasteful to spend a lot of effort determining the exact minimum in the p_k direction because it would not take us any closer to the minimum of the overall function (the dot on the right side of the plot). Instead, we should find a point that is good enough and then update the search direction.

To simplify the notation for the line search, we define the single-variable function

$$\phi(\alpha) = f(x_k + \alpha p_k), \quad (4.27)$$

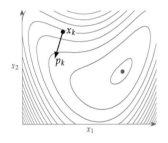

Fig. 4.19 The descent direction does not necessarily point toward the minimum, in which case it would be wasteful to do an exact line search.

where $\alpha = 0$ corresponds to the start of the line search (x_k in Fig. 4.17), and thus $\phi(0) = f(x_k)$. Then, using $x = x_k + \alpha p_k$, the slope of the single-variable function is

$$\phi'(\alpha) = \frac{\partial(f(x))}{\partial \alpha} = \sum_{i=1}^{n} \frac{\partial f}{\partial x_i} \frac{\partial x_i}{\partial \alpha}. \quad (4.28)$$

Substituting into the derivatives results in

$$\phi'(\alpha) = \nabla f(x_k + \alpha p_k)^\mathsf{T} p_k, \quad (4.29)$$

which is the directional derivative along the search direction. The slope at the start of a given line search is

$$\phi'(0) = \nabla f_k^\mathsf{T} p_k. \quad (4.30)$$

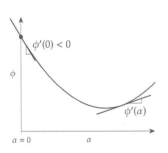

Fig. 4.20 For the line search, we denote the function as $\phi(\alpha)$ with the same value as f. The slope $\phi'(\alpha)$ is the gradient of f projected onto the search direction.

Because p_k must be a descent direction, $\phi'(0)$ is always negative. Figure 4.20 is a version of the one-dimensional slice from Fig. 4.17 in this notation. The α axis and the slopes scale with the magnitude of p_k.

4.3.1 Sufficient Decrease and Backtracking

The simplest line search algorithm to find a "good enough" point relies on the *sufficient decrease condition* combined with a *backtracking algorithm*. The sufficient decrease condition, also known as the *Armijo condition*, is given by the inequality

$$\phi(\alpha) \leq \phi(0) + \mu_1 \alpha \phi'(0), \quad (4.31)$$

where μ_1 is a constant such that $0 < \mu_1 \leq 1$.[†] The quantity $\alpha \phi'(0)$ represents the expected decrease of the function, assuming the function continued at the same slope. The multiplier μ_1 states that Eq. 4.31 will be satisfied as long we achieve even a small fraction of the expected decrease, as shown in Fig. 4.21. In practice, this constant is several orders of magnitude smaller than 1, typically $\mu_1 = 10^{-4}$. Because p_k is a descent direction, and thus $\phi'(0) = \nabla f_k^\mathsf{T} p_k < 0$, there is always a positive α that satisfies this condition for a smooth function.

The concept is illustrated in Fig. 4.22, which shows a function with a negative slope at $\alpha = 0$ and a sufficient decrease line whose slope is a fraction of that initial slope. When starting a line search, we know the function value and slope at $\alpha = 0$, so we do not really know how the function varies until we evaluate it. Because we do not want to evaluate the function too many times, the first point whose value is below the sufficient decrease line is deemed acceptable. The sufficient decrease line slope in Fig. 4.22 is exaggerated for illustration purposes; for typical values of μ_1, the line is indistinguishable from horizontal when plotted.

[†]This condition can be problematic near a local minimum because $\phi(0)$ and $\phi(\alpha)$ are so similar that their subtraction is inaccurate. Hager and Zhang[77] introduced a condition with improved accuracy, along with an efficient line search based on a secant method.

[77]. Hager and Zhang, *A new conjugate gradient method with guaranteed descent and an efficient line search*, 2005.

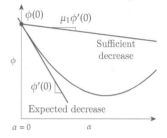

Fig. 4.21 The sufficient decrease line has a slope that is a small fraction of the slope at the start of the line search.

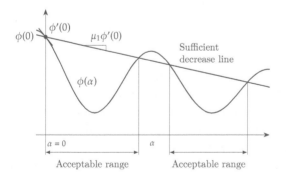

Fig. 4.22 Sufficient decrease condition.

Line search algorithms require a first guess for α. As we will see later, some methods for finding the search direction also provide good guesses for the step length. However, in many cases, we have no idea of the scale of function, so our initial guess may not be suitable. Even if we do have an educated guess for α, it is only a guess, and the first step might not satisfy the sufficient decrease condition.

A straightforward algorithm that is guaranteed to find a step that satisfies the sufficient decrease condition is backtracking (Alg. 4.2). This algorithm starts with a maximum step and successively reduces the step by a constant ratio ρ until it satisfies the sufficient decrease condition (a typical value is $\rho = 0.5$). Because the search direction is a descent direction, we know that we will achieve an acceptable decrease in function value if we backtrack enough.

Algorithm 4.2 Backtracking line search algorithm

Inputs:
   $\alpha_{\text{init}} > 0$: Initial step length
   $0 < \mu_1 < 1$: Sufficient decrease factor (typically small, e.g., $\mu_1 = 10^{-4}$)
   $0 < \rho < 1$: Backtracking factor (e.g., $\rho = 0.5$)

Outputs:
   α^*: Step size satisfying sufficient decrease condition

$\alpha = \alpha_{\text{init}}$
while $\phi(\alpha) > \phi(0) + \mu_1 \alpha \phi'(0)$ **do**   Function value is above sufficient decrease line
   $\alpha = \rho \alpha$ Backtrack
end while

Although backtracking is guaranteed to find a point that satisfies sufficient decrease, there are two undesirable scenarios where this algorithm performs poorly. The first scenario is that the guess for the initial step is far too large, and the step sizes that satisfy sufficient decrease are smaller than the starting step by several orders of magnitude. Depending on the value of ρ, this scenario requires a large number of backtracking evaluations.

The other undesirable scenario is where our initial guess immediately satisfies sufficient decrease. However, the function's slope is still highly negative, and we could have decreased the function value by much more if we had taken a larger step. In this case, our guess for the initial step is far too small.

Even if our original step size is not too far from an acceptable step size, the basic backtracking algorithm ignores any information we have about the function values and gradients. It blindly takes a reduced step based on a preselected ratio ρ. We can make more intelligent estimates of where an acceptable step is based on the evaluated function values (and gradients, if available). The next section introduces a more

4.3 Line Search

sophisticated line search algorithm that deals with these scenarios much more efficiently.

Example 4.8 Backtracking line search

Consider the following function:

$$f(x_1, x_2) = 0.1x_1^6 - 1.5x_1^4 + 5x_1^2 + 0.1x_2^4 + 3x_2^2 - 9x_2 + 0.5x_1x_2.$$

Suppose we do a line search starting from $x = [-1.25, 1.25]$ in the direction $p = [4, 0.75]$, as shown in Fig. 4.23. Applying the backtracking algorithm with $\mu_1 = 10^{-4}$ and $\rho = 0.7$ produces the iterations shown in Fig. 4.24. The sufficient decrease line appears to be horizontal, but that is because the small negative slope cannot be discerned in a plot for typical values of μ_1. Using a large initial step of $\alpha_{\text{init}} = 1.2$ (Fig. 4.24, left), several iterations are required. For a small initial step of $\alpha_{\text{init}} = 0.05$ (Fig. 4.24, right), the algorithm satisfies sufficient decrease at the first iteration but misses the opportunity for further reductions.

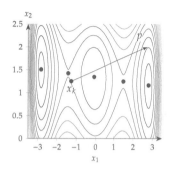

Fig. 4.23 A line search direction for an example problem.

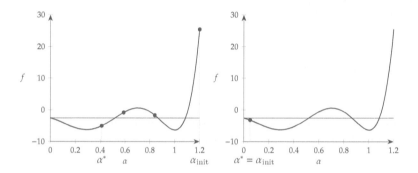

Fig. 4.24 Backtracking using different initial steps.

4.3.2 Strong Wolfe Conditions

One major weakness of the sufficient decrease condition is that it accepts small steps that marginally decrease the objective function because μ_1 in Eq. 4.31 is typically small. We could increase μ_1 (i.e., tilt the red line downward in Fig. 4.22) to prevent these small steps; however, that would prevent us from taking large steps that still result in a reasonable decrease. A large step that provides a reasonable decrease is desirable because large steps generally lead to faster convergence. Therefore, we want to prevent overly small steps while not making it more difficult to accept reasonable large steps. We can accomplish this by adding a second condition to construct a more efficient line search algorithm.

Just like guessing the step size, it is difficult to know in advance how much of a function value decrease to expect. However, if we compare the slope of the function at the candidate point with the slope at the start of the line search, we can get an idea if the function is "bottoming out", or flattening, using the *sufficient curvature condition*:

$$|\phi'(\alpha)| \leq \mu_2 |\phi'(0)|. \tag{4.32}$$

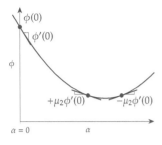

Fig. 4.25 The sufficient curvature condition requires the function slope magnitude to be a fraction of the initial slope.

This condition requires that the magnitude of the slope at the new point be lower than the magnitude of the slope at the start of the line search by a factor of μ_2, as shown in Fig. 4.25. This requirement is called the *sufficient curvature condition* because by comparing the two slopes, we quantify the function's rate of change in the slope—the curvature. Typical values of μ_2 range from 0.1 to 0.9, and the best value depends on the method for determining the search direction and is also problem dependent. As μ_2 tends to zero, enforcing the sufficient curvature condition tends toward a point where $\phi'(\alpha) = 0$, which would yield an exact line search.

The sign of the slope at a point satisfying this condition is not relevant; all that matters is that the function slope be shallow enough. The idea is that if the slope $\phi'(\alpha)$ is still negative with a magnitude similar to the slope at the start of the line search, then the step is too small, and we expect the function to decrease even further by taking a larger step. If the slope $\phi'(\alpha)$ is positive with a magnitude similar to that at the start of the line search, then the step is too large, and we expect to decrease the function further by taking a smaller step. On the other hand, when the slope is shallow enough (either positive or negative), we assume that the candidate point is near a local minimum, and additional effort yields only incremental benefits that are wasteful in the context of the larger problem.

The sufficient decrease and sufficient curvature conditions are collectively known as the *strong Wolfe conditions*. Figure 4.26 shows acceptable intervals that satisfy the strong Wolfe conditions, which are more restrictive than the sufficient decrease condition (Fig. 4.22).

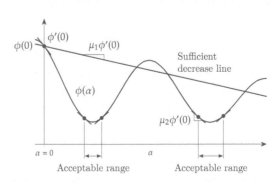

Fig. 4.26 Steps that satisfy the strong Wolfe conditions.

The sufficient decrease slope must be shallower than the sufficient curvature slope, that is, $0 < \mu_1 < \mu_2 < 1$. This is to guarantee that there are steps that satisfy both the sufficient decrease and sufficient curvature conditions. Otherwise, the situation illustrated in Fig. 4.27 could take place.

We now present a line search algorithm that finds a step satisfying the strong Wolfe conditions. Enforcing the sufficient curvature condition means we require derivative information (ϕ'), at least using the derivative at the beginning of the line search that we already computed from the gradient. There are various line search algorithms in the literature, including some that are derivative-free. Here, we detail a line search algorithm based on the one developed by Moré and Thuente.[78][‡] The algorithm has two phases:

1. The *bracketing* phase finds an interval within which we are certain to find a point that satisfies the strong Wolfe conditions.
2. The *pinpointing* phase finds a point that satisfies the strong Wolfe conditions within the interval provided by the bracketing phase.

The bracketing phase is given by Alg. 4.3 and illustrated in Fig. 4.28. For brevity, we use a notation in the following algorithms where, for example, $\phi_0 \equiv \phi(0)$ and $\phi_{\text{low}} \equiv \phi(\alpha_{\text{low}})$. Overall, the bracketing algorithm increases the step size until it either finds an interval that must contain a point satisfying the strong Wolfe conditions or a point that already meets those conditions.

We start the line search with a guess for the step size, which defines the first interval. For a smooth continuous function, we are guaranteed to have a minimum within an interval if either of the following hold:

1. The function value at the candidate step is higher than the value at the start of the line search.
2. The step satisfies sufficient decrease, and the slope is positive.

These two scenarios are illustrated in the top two rows of Fig. 4.28. In either case, we have an interval within which we can find a point that satisfies the strong Wolfe conditions using the pinpointing algorithm. The order in arguments to the `pinpoint` function in Alg. 4.3 is significant because this function assumes that the function value corresponding to the first α is the lower one. The third row in Fig. 4.28 illustrates the scenario where the point satisfies the strong Wolfe conditions, in which case the line search is finished.

If the point satisfies sufficient decrease and the slope at that point is negative, we assume that there are better points farther along the line, and the algorithm increases the step size. This larger step and the

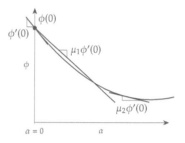

Fig. 4.27 If $\mu_2 < \mu_1$, there might be no point that satisfies the strong Wolfe conditions.

78. Moré and Thuente, *Line search algorithms with guaranteed sufficient decrease*, 1994.

[‡]A similar algorithm is detailed in Chapter 3 of Nocedal and Wright.[79]

79. Nocedal and Wright, *Numerical Optimization*, 2006.

previous one define a new interval that has moved away from the line search starting point. We repeat the procedure and check the scenarios for this new interval. To save function calls, bracketing should return not just α^* but also the corresponding function value and gradient to the outer function.

Algorithm 4.3 Bracketing phase for the line search algorithm

Inputs:
- $\alpha_{init} > 0$: Initial step size
- ϕ_0, ϕ_0': computed in outer routine, pass in to save function call
- $0 < \mu_1 < 1$: Sufficient decrease factor
- $\mu_1 < \mu_2 < 1$: Sufficient curvature factor
- $\sigma > 1$: Step size increase factor (e.g., $\sigma = 2$)

Outputs:
- α^*: Acceptable step size (satisfies the strong Wolfe conditions)

$\alpha_1 = 0$ *Define initial bracket*
$\alpha_2 = \alpha_{init}$
$\phi_1 = \phi_0$
$\phi_1' = \phi_0'$ *Used in pinpoint*
first = true
while true **do**
    $\phi_2 = \phi(\alpha_2)$ *Compute ϕ_2' on this line if user provides derivatives*
    if $\left[\phi_2 > \phi_0 + \mu_1 \alpha_2 \phi_0'\right]$ or $\left[\text{not first and } \phi_2 > \phi_1\right]$ **then**
       $\alpha^* = \text{pinpoint}(\alpha_1, \alpha_2, \ldots)$ $1 \Rightarrow \text{low}, 2 \Rightarrow \text{high}$
       return α^*
    end if
    $\phi_2' = \phi'(\alpha_2)$ *If not computed previously*
    if $|\phi_2'| \leq -\mu_2 \phi_0'$ **then**   *Step is acceptable; exit line search*
       return $\alpha^* = \alpha_2$
    else if $\phi_2' \geq 0$ **then**   *Bracketed minimum*
       $\alpha^* = \text{pinpoint}(\alpha_2, \alpha_1, \ldots)$ *Find acceptable step, $2 \Rightarrow \text{low}, 1 \Rightarrow \text{high}$*
       return α^*
    else   *Slope is negative*
       $\alpha_1 = \alpha_2$
       $\alpha_2 = \sigma \alpha_2$ *Increase step*
    end if
    first = false
end while

If the bracketing phase does not find a point that satisfies the strong Wolfe conditions, it finds an interval where we are guaranteed to find such a point in the pinpointing phase described in Alg. 4.4

4.3 Line Search

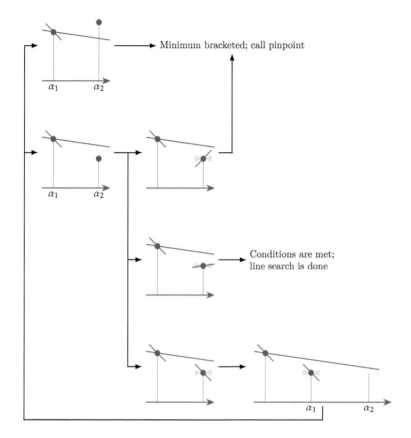

Fig. 4.28 Visual representation of the bracketing algorithm. The sufficient decrease line is drawn as if α_1 were the starting point for the line search, which is the case for the first line search iteration but not necessarily the case for later iterations.

and illustrated in Fig. 4.29. The intervals generated by this algorithm, bounded by α_{low} and α_{high}, always have the following properties:

1. The interval has one or more points that satisfy the strong Wolfe conditions.
2. Among all the points generated so far that satisfy the sufficient decrease condition, α_{low} has the lowest function value.
3. The slope at α_{low} decreases toward α_{high}.

The first step of pinpointing is to find a new point within the given interval. Various techniques can be used to find such a point. The simplest one is to select the midpoint of the interval (bisection), but this method is limited to a linear convergence rate. It is more efficient to perform interpolation and select the point that minimizes the interpolation function, which can be done analytically (see Section 4.3.3). Using this approach, we can achieve quadratic convergence.

Once we have a new point within the interval, four scenarios are possible, as shown in Fig. 4.29. The first scenario is that $\phi\left(\alpha_p\right)$ is above the sufficient decrease line or greater than or equal to $\phi(\alpha_{\text{low}})$. In that scenario, α_p becomes the new α_{high}, and we have a new smaller interval.

In the second, third, and fourth scenarios, $\phi(\alpha_p)$ is below the sufficient decrease line, and $\phi(\alpha_p) < \phi(\alpha_{\text{low}})$. In those scenarios, we check the value of the slope $\phi'(\alpha_p)$. In the second and third scenarios, we choose the new interval based on the direction in which the slope predicts a local decrease. If the slope is shallow enough (fourth scenario), we have found a point that satisfies the strong Wolfe conditions.

Algorithm 4.4 Pinpoint function for the line search algorithm

Inputs:
α_{low}: Interval endpoint with lower function value
α_{high}: Interval endpoint with higher function value
$\phi_0, \phi_{\text{low}}, \phi_{\text{high}}, \phi'_0$: Computed in outer routine
$\phi'_{\text{low}}, \phi'_{\text{high}}$: One, if not both, computed previously
$0 < \mu_1 < 1$: Sufficient decrease factor
$\mu_1 < \mu_2 < 1$: Sufficient curvature factor

Outputs:
α^*: Step size satisfying strong Wolfe conditions

$k = 0$
while true **do**
    Find α_p in interval $(\alpha_{\text{low}}, \alpha_{\text{high}})$ Use interpolation (see Section 4.3.3)
                                         Uses $\phi_{\text{low}}, \phi_{\text{high}}$, and ϕ' from at least one endpoint
    $\phi_p = \phi(\alpha_p)$ Also evaluate ϕ'_p if derivatives available
    if $\phi_p > \phi_0 + \mu_1 \alpha_p \phi'_0$ or $\phi_p > \phi_{\text{low}}$ **then**
        $\alpha_{\text{high}} = \alpha_p$ Also update $\phi_{\text{high}} = \phi_p$, and if cubic interpolation $\phi'_{\text{high}} = \phi'_p$
    else
        $\phi'_p = \phi'(\alpha_p)$ If not already computed
        if $|\phi'_p| \leq -\mu_2 \phi'_0$ **then**
            $\alpha^* = \alpha_p$
            return α_p
        else if $\phi'_p(\alpha_{\text{high}} - \alpha_{\text{low}}) \geq 0$ **then**
            $\alpha_{\text{high}} = \alpha_{\text{low}}$
        end if
        $\alpha_{\text{low}} = \alpha_p$
    end if
    $k = k + 1$
end while

In theory, the line search given in Alg. 4.3 followed by Alg. 4.4 is guaranteed to find a step length satisfying the strong Wolfe conditions. In practice, some additional considerations are needed for improved

4.3 Line Search

robustness. One of these criteria is to ensure that the new point in the pinpoint algorithm is not so close to an endpoint as to cause the interpolation to be ill-conditioned. A fallback option in case the interpolation fails could be a simpler algorithm, such as bisection. Another criterion is to ensure that the loop does not continue indefinitely in case finite-precision arithmetic leads to indistinguishable function value changes. A limit on the number of iterations might be necessary.

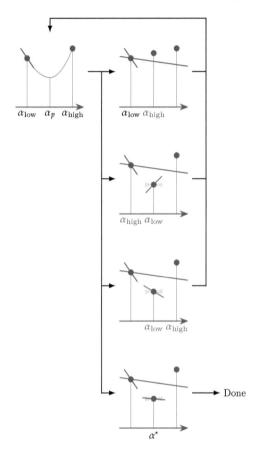

Fig. 4.29 Visual representation of the pinpointing algorithm. The labels in red indicate the new interval endpoints.

Example 4.9 Line search with bracketing and pinpointing

Let us perform the same line search as in Alg. 4.2 but using bracketing and pinpointing instead of backtracking. In this example, we use quadratic interpolation, the pinpointing phase uses a step size increase factor of $\sigma = 2$, and the sufficient curvature factor is $\mu_2 = 0.9$. Bracketing is achieved in the first iteration by using a large initial step of $\alpha_{\text{init}} = 1.2$ (Fig. 4.30, left). Then pinpointing finds an improved point through interpolation. The small initial step of $\alpha_{\text{init}} = 0.05$ (Fig. 4.30, right) does not satisfy the strong Wolfe conditions,

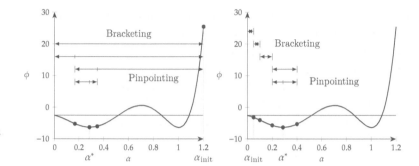

Fig. 4.30 Example of a line search iteration with different initial steps.

and the bracketing phase moves forward toward a flatter part of the function. The result is a point that is much better than the one obtained with backtracking.

Tip 4.3 When stuck, plot the line search

When gradient-based optimizers cannot move away from a non-optimal point, it usually happens during the line search. To understand why the optimizer is stuck, plot the iterations along the line search, add more points, or plot the whole line if you can afford to. Even if you have a high-dimensional problem, you can always plot the line search, which will be understandable because it is one-dimensional.

4.3.3 Interpolation for Pinpointing

Interpolation is recommended to find a new point within each interval at the pinpointing phase. Once we have an interpolation function, we find the new point by determining the analytic minimum of that function. This accelerates the convergence compared with bisection. We consider two options: quadratic interpolation and cubic interpolation.

Because we have the function value and derivative at one endpoint of the interval and at least the function value at the other endpoint, one option is to perform *quadratic interpolation* to estimate the minimum within the interval.

The quadratic can be written as

$$\tilde{\phi}(\alpha) = c_0 + c_1 \alpha + c_2 \alpha^2, \qquad (4.33)$$

where c_0, c_1, and c_2 are constants to be determined by interpolation. Suppose that we have the function value and the derivative at α_1 and the function value at α_2, as illustrated in Fig. 4.31. These values correspond to α_{low} and α_{high} in the pinpointing algorithm, but we use

4.3 Line Search

the more generic indices 1 and 2 because the formulas of this section are not dependent on which one is lower or higher. Then, the boundary conditions at the endpoints are

$$\phi(\alpha_1) = c_0 + c_1\alpha_1 + c_2\alpha_1^2$$
$$\phi(\alpha_2) = c_0 + c_2\alpha_2 + c_2\alpha_2^2 \quad (4.34)$$
$$\phi'(\alpha_1) = c_1 + 2c_2\alpha_1.$$

We can use these three equations to find the three coefficients based on function and derivative values. Once we have the coefficients for the quadratic, we can find the minimum of the quadratic analytically by finding the point α^* such that $\tilde{\phi}'(\alpha^*) = 0$, which is $\alpha^* = -c_1/2c_2$. Substituting the analytic solution for the coefficients as a function of the given values into this expression yields the final expression for the minimizer of the quadratic:

$$\alpha^* = \frac{2\alpha_1\left[\phi(\alpha_2) - \phi(\alpha_1)\right] + \phi'(\alpha_1)\left(\alpha_1^2 - \alpha_2^2\right)}{2\left[\phi(\alpha_2) - \phi(\alpha_1) + \phi'(\alpha_1)(\alpha_1 - \alpha_2)\right]}. \quad (4.35)$$

Performing this quadratic interpolation for successive intervals is similar to the Newton method and also converges quadratically. The pure Newton method also models a quadratic, but it is based on the information at a single point (function value, derivative, and curvature), as opposed to information at two points.

If computing additional derivatives is inexpensive, or we already evaluated $\phi'(\alpha_i)$ (either as part of Alg. 4.3 or as part of checking the strong Wolfe conditions in Alg. 4.4), then we have the function values and derivatives at both points. With these four pieces of information, we can perform a *cubic interpolation*,

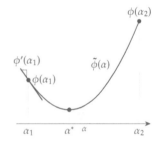

Fig. 4.31 Quadratic interpolation with two function values and one derivative.

$$\tilde{\phi}(\alpha) = c_0 + c_1\alpha + c_2\alpha^2 + c_3\alpha^3, \quad (4.36)$$

as shown in Fig. 4.32. To determine the four coefficients, we apply the boundary conditions:

$$\phi(\alpha_1) = c_0 + c_1\alpha_1 + c_2\alpha_1^2 + c_3\alpha_1^3$$
$$\phi(\alpha_2) = c_0 + c_2\alpha_2 + c_2\alpha_2^2 + c_3\alpha_2^3$$
$$\phi'(\alpha_1) = c_1 + 2c_2\alpha_1 + 3c_3\alpha_1^2 \quad (4.37)$$
$$\phi'(\alpha_2) = c_1 + 2c_2\alpha_2 + 3c_3\alpha_2^2.$$

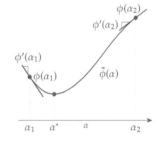

Fig. 4.32 Cubic interpolation with function values and derivatives at endpoints.

Using these four equations, we can find expressions for the four coefficients as a function of the four pieces of information. Similar to the quadratic interpolation function, we can find the solution for $\tilde{\phi}'(\alpha^*) = c_1 + 2c_2\alpha^* + 3c_3\alpha^{*2} = 0$ as a function of the coefficients. There could be two valid solutions, but we are only interested in the minimum, for which the curvature is positive; that is, $\tilde{\phi}''(\alpha^*) = 2c_2 + 6c_3\alpha^* > 0$. Substituting the coefficients with the expressions obtained from solving the boundary condition equations and selecting the minimum solution yields

$$\alpha^* = \alpha_2 - (\alpha_2 - \alpha_1)\frac{\phi'(\alpha_2) + \beta_2 - \beta_1}{\phi'(\alpha_2) - \phi'(\alpha_1) + 2\beta_2}, \tag{4.38}$$

where

$$\beta_1 = \phi'(\alpha_1) + \phi'(\alpha_2) - 3\frac{\phi(\alpha_1) - \phi(\alpha_2)}{\alpha_1 - \alpha_2}$$

$$\beta_2 = \text{sign}(\alpha_2 - \alpha_1)\sqrt{\beta_1^2 - \phi'(\alpha_1)\phi'(\alpha_2)}. \tag{4.39}$$

These interpolations become ill-conditioned if the interval becomes too small. The interpolation may also lead to points outside the bracket. In such cases, we can switch to bisection for the problematic iterations.

4.4 Search Direction

As stated at the beginning of this chapter, each iteration of an unconstrained gradient-based algorithm consists of two main steps: determining the search direction and performing the line search (Alg. 4.1). The optimization algorithms are named after the method used to find the search direction, p_k, and can use any suitable line search. We start by introducing two first-order methods that only require the gradient and then explain two second-order methods that require the Hessian, or at least an approximation of the Hessian.

4.4.1 Steepest Descent

The steepest-descent method (also called *gradient descent*) is a simple and intuitive method for determining the search direction. As discussed in Section 4.1.1, the gradient points in the direction of steepest increase, so $-\nabla f$ points in the direction of steepest descent, as shown in Fig. 4.33. Thus, our search direction at iteration k is simply

$$p = -\nabla f. \tag{4.40}$$

One major issue with the steepest descent is that, in general, the entries in the gradient and its overall scale can vary greatly depending on the magnitudes of the objective function and design variables. The gradient itself contains no information about an appropriate step length, and therefore the search direction is often better posed as a normalized

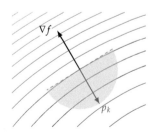

Fig. 4.33 The steepest-descent direction points in the opposite direction of the gradient.

direction,
$$p_k = -\frac{\nabla f_k}{\|\nabla f_k\|}. \quad (4.41)$$

Algorithm 4.5 provides the complete steepest descent procedure.

Algorithm 4.5 Steepest descent

Inputs:
 x_0: Starting point
 τ: Convergence tolerance
Outputs:
 x^*: Optimal point
 $f(x^*)$: Minimum function value

$k = 0$	Initialize iteration counter
while $\|\nabla f\|_\infty > \tau$ do	Optimality condition
$\quad p_k = -\frac{\nabla f_k}{\|\nabla f_k\|}$	Normalized steepest descent direction
$\quad$ Estimate α_{init} from Eq. 4.43	
$\quad \alpha_k = \texttt{linesearch}(p_k, \alpha_{\text{init}})$	Perform a line search
$\quad x_{k+1} = x_k + \alpha_k p_k$	Update design variables
$\quad k = k + 1$	Increment iteration index
end while	

Regardless of whether we choose to normalize the search direction or not, the gradient does not provide enough information to inform a good guess of the initial step size for the line search. As we saw in Section 4.3, this initial choice has a large impact on the efficiency of the line search because the first guess could be orders of magnitude too small or too large. The second-order methods described later in this section are better in this respect. In the meantime, we can make a guess of the step size for a given line search based on the result of the previous one. Assuming that we will obtain a decrease in objective function at the current line search that is comparable to the previous one, we can write

$$\alpha_k \nabla f_k^\mathsf{T} p_k \approx \alpha_{k-1} \nabla f_{k-1}^\mathsf{T} p_{k-1}. \quad (4.42)$$

Solving for the step length, we obtain the guess

$$\alpha_k = \alpha_{k-1} \frac{\nabla f_{k-1}^\mathsf{T} p_{k-1}}{\nabla f_k^\mathsf{T} p_k}. \quad (4.43)$$

Although this expression could be simplified for the steepest descent, we leave it as is so that it is applicable to other methods. If the slope of the function increases in magnitude relative to the previous line search, this guess decreases relative to the previous line search step length, and vice versa. This is just the first step length in the new line search, after which we proceed as usual.

Although steepest descent sounds like the best possible search direction for decreasing a function, it generally is not. The reason is that when a function curvature varies significantly with direction, the gradient alone is a poor representation of function behavior beyond a small neighborhood, as illustrated previously in Fig. 4.19.

Example 4.10 Steepest descent with varying amount of curvature

Consider the following quadratic function:

$$f(x_1, x_2) = x_1^2 + \beta x_2^2,$$

where β can be set to adjust the curvature in the x_2 direction. In Fig. 4.34, we show this function for $\beta = 1, 5, 15$. The starting point is $x_0 = (10, 1)$. When $\beta = 1$ (left), this quadratic has the same curvature in all directions, and the steepest-descent direction points directly to the minimum. When $\beta > 1$ (middle and right), this is no longer the case, and steepest descent shows abrupt changes in the subsequent search directions. This zigzagging is an inefficient way to approach the minimum. The higher the difference in curvature, the more iterations it takes.

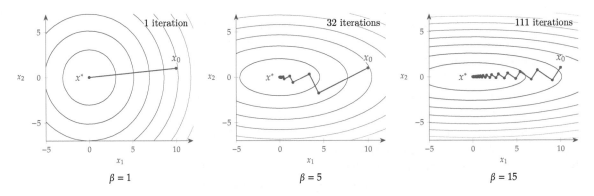

Fig. 4.34 Iteration history for a quadratic function, with three different curvatures, using the steepest-descent method with an exact line search (small enough μ_2).

The behavior shown in Ex. 4.10 is expected, and we can show it mathematically. Assuming we perform an exact line search at each iteration, this means selecting the optimal value for α along the line

4.4 Search Direction

search:

$$\frac{\partial f(x_k + \alpha p_k)}{\partial \alpha} = 0 \Rightarrow$$
$$\frac{\partial f(x_{k+1})}{\partial \alpha} = 0 \Rightarrow$$
$$\frac{\partial f(x_{k+1})}{\partial x_{k+1}} \frac{\partial (x_k + \alpha p_k)}{\partial \alpha} = 0 \Rightarrow \quad (4.44)$$
$$\nabla f_{k+1}{}^\mathsf{T} p_k = 0 \Rightarrow$$
$$-p_{k+1}{}^\mathsf{T} p_k = 0.$$

Hence, each search direction is orthogonal to the previous one. When performing an exact line search, the gradient projection in the line search direction vanishes at the minimum, which means that the gradient is orthogonal to the search direction, as shown in Fig. 4.35.

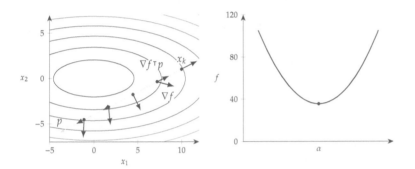

Fig. 4.35 The gradient projection in the line search direction vanishes at the line search minimum.

As discussed in the last section, exact line searches are not desirable, so the search directions are not orthogonal. However, the overall zigzagging behavior still exists.

Example 4.11 Steepest descent applied to the bean function

We now find the minimum of the bean function,

$$f(x_1, x_2) = (1 - x_1)^2 + (1 - x_2)^2 + \frac{1}{2}\left(2x_2 - x_1^2\right)^2,$$

using the steepest-descent algorithm with an exact line search, and a convergence tolerance of $\|\nabla f\|_\infty \leq 10^{-6}$. The optimization path is shown in Fig. 4.36. Although it takes only a few iterations to get close to the minimum, it takes many more to satisfy the specified convergence tolerance.

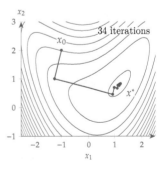

Fig. 4.36 Steepest-descent optimization path.

> **Tip 4.4** Scale the design variables and the objective function

Problem scaling is one of the most crucial considerations in practical optimization. Steepest descent is susceptible to scaling, as demonstrated in Ex. 4.10. Even though we will learn about less sensitive methods, poor scaling can decrease the effectiveness of any method for general nonlinear functions.

A common cause of poor scaling is unit choice. For example, consider a problem with two types of design variables, where one type is the material thickness, on the order of 10^{-6} m, and the other type is the length of the structure, on the order of 1 m. If both distances are measured in meters, the derivative in the thickness direction is much larger than the derivative in the length direction. In other words, the design space would have a valley that is steep in one direction and shallow in the other. The optimizer would have great difficulty in navigating this type of design space.

Similarly, if the objective is power and a typical value is 10^6 W, the gradients would likely be relatively small, and satisfying convergence tolerances may be challenging.

A good rule of thumb is to scale the objective function and every design variable to be around unity. The scaling of the objective is only needed after the model analysis computes the function and can be written as

$$\bar{f} = f/s_f, \tag{4.45}$$

where s_f is the scaling factor, which could be the value of the objective at the starting point, $f(x_0)$, or another typical value. Multiplying the functions by a scalar does not change the optimal solution but can significantly improve the ability of the optimizer to find the optimum.

Scaling the design variables is more involved because scaling them changes the value that the optimizer would pass to the model and thus changes their meaning. In general, we might use different scaling factors for different types of variables, so we represent these as an n-vector, s_x. Starting with the physical design variables, x_0, we obtain the scaled variables by dividing them by the scaling factors:

$$\bar{x}_0 = x_0 \oslash s_x, \tag{4.46}$$

where $\oslash$ denotes element-wise division. Then, because the optimizer works with the scaled variables, we need to convert them back to physical variables by multiplying them by the scaling factors:

$$x = \bar{x} \odot s_x, \tag{4.47}$$

where $\odot$ denotes element-wise multiplication. Finally, we must also convert the scaled variables to their physical values after the optimization is completed. The complete process is shown in Fig. 4.37.

It is not necessary that the objective and all variables be precisely 1—which is impossible to maintain as the optimization progresses. Instead, this heuristic suggests that the objective and all variables should have an order of magnitude of 1. If one of the variables or functions is expected to vary across multiple orders of magnitude during an optimization, one effective way to scale is to

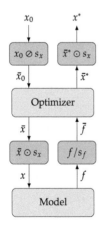

Fig. 4.37 Scaling works by providing a scaled version of the design variables and objective function to the optimizer. However, the model analysis still needs to work with the original variables and function.

take the logarithm. For example, suppose the objective was expected to vary across multiple orders of magnitude. In that case, we could minimize $\log(f)$ instead of minimizing f.*

*If f can be negative, a transformation is required to ensure that the logarithm argument is always positive.

This heuristic still does not guarantee that the derivatives are well scaled, but it provides a reasonable starting point for further fine-tuning of the problem scaling. A scaling example is discussed in Ex. 4.19.

Sometimes, additional adjustment is needed if the objective is far less sensitive to some of the design variables than others (i.e., the entries in the gradient span various orders of magnitude). A more appropriate but more involved approach is to scale the variables and objective function such that the gradient elements have a similar magnitude (ideally of order 1). Achieving a well-scaled gradient sometimes requires adjusting inputs and outputs away from the earlier heuristic. Sometimes this occurs because the objective is much less sensitive to a particular variable.

4.4.2 Conjugate Gradient

Steepest descent generally performs poorly, especially if the problem is not well scaled, like the quadratic example in Fig. 4.34. The conjugate gradient method updates the search directions such that they do not zigzag as much. This method is based on the linear conjugate gradient method, which was designed to solve linear equations. We first introduce the linear conjugate gradient method and then adapt it to the nonlinear case.

For the moment, let us assume that we have the following quadratic objective function:

$$f(x) = \frac{1}{2}x^\mathsf{T} A x - b^\mathsf{T} x, \tag{4.48}$$

where A is a positive definite Hessian, and b is the gradient at $x = 0$. The constant term is omitted with no loss of generality because it does not change the location of the minimum. To find the minimum of this quadratic, we require

$$\nabla f(x^*) = Ax^* - b = 0. \tag{4.49}$$

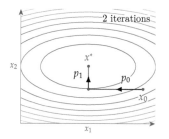

Fig. 4.38 For a quadratic function with elliptical contours and the principal axis aligned with the coordinate axis, we can find the minimum in n steps, where n is the number of dimensions, by using a coordinate search.

Thus, finding the minimum of a quadratic amounts to solving the linear system $Ax = b$, and the residual vector is the gradient of the quadratic.

If A were a positive-definite diagonal matrix, the contours would be elliptical, as shown in Fig. 4.38 (or hyper-ellipsoids in the n-dimensional case), and the axes of the ellipses would align with the coordinate directions. In that case, we could converge to the minimum by successively performing an exact line search in each coordinate direction for a total of n line searches.

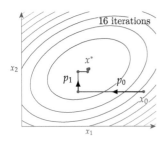

Fig. 4.39 For a quadratic function with the elliptical principal axis not aligned with the coordinate axis, more iterations are needed to find the minimum using a coordinate search.

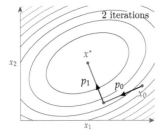

Fig. 4.40 We can converge to the minimum of a quadratic function by minimizing along each Hessian eigenvector.

In the more general case (but still assuming A to be positive definite), the axes of the ellipses form an orthogonal coordinate system in some other orientation. A coordinate search would no longer work as well in this case, as illustrated in Fig. 4.39.

Recall from Section 4.1.2 that the eigenvectors of the Hessian represent the directions of principal curvature, which correspond to the axes of the ellipses. Therefore, we could successively perform a line search along the direction defined by each eigenvector and again converge to the minimum with n line searches, as illustrated in Fig. 4.40. The problem with this approach is that we would have to compute the eigenvectors of A, a computation whose cost is $O(n^3)$.

Fortunately, the eigenvector directions are not the only set of directions that can minimize the quadratic function in n line searches. To find out which directions can achieve this, let us express the path from the origin to the minimum of the quadratic as a sequence of n steps with directions p_i and length α_i,

$$x^* = \sum_{i=0}^{n-1} \alpha_i p_i. \tag{4.50}$$

Thus, we have represented the solution as a linear combination of n vectors. Substituting this into the quadratic (Eq. 4.48), we get

$$\begin{aligned}
f(x^*) &= f\left(\sum_{i=0}^{n-1} \alpha_i p_i\right) \\
&= \frac{1}{2}\left(\sum_{i=0}^{n-1} \alpha_i p_i\right)^\mathsf{T} A \left(\sum_{j=0}^{n-1} \alpha_j p_j\right) - b^\mathsf{T}\left(\sum_{i=0}^{n-1} \alpha_i p_i\right) \\
&= \frac{1}{2} \sum_{i=0}^{n-1} \sum_{j=0}^{n-1} \alpha_i \alpha_j p_i^\mathsf{T} A p_j - \sum_{i=0}^{n-1} \alpha_i b^\mathsf{T} p_i.
\end{aligned} \tag{4.51}$$

Suppose that the vectors $p_0, p_1, \ldots, p_{n-1}$ are *conjugate* with respect to A; that is, they have the following property:

$$p_i^\mathsf{T} A p_j = 0, \quad \text{for all} \quad i \neq j. \tag{4.52}$$

Then, the double-sum term in Eq. 4.51 can be simplified to a single sum and we can write

$$f(x^*) = \sum_{i=0}^{n-1} \left(\frac{1}{2}\alpha_i^2 p_i^\mathsf{T} A p_i - \alpha_i b^\mathsf{T} p_i\right). \tag{4.53}$$

Because each term in this sum involves only one direction p_i, we have reduced the original problem to a series of one-dimensional quadratic

4.4 Search Direction

functions that can be minimized one at a time. Two possible conjugate directions are shown for the two-dimensional case in Fig. 4.41.

Each one-dimensional problem corresponds to minimizing the quadratic with respect to the step length α_i. Differentiating each term and setting it to zero yields

$$\alpha_i p_i^\mathsf{T} A p_i - b^\mathsf{T} p_i = 0 \Rightarrow \alpha_i = \frac{b^\mathsf{T} p_i}{p_i^\mathsf{T} A p_i}, \quad (4.54)$$

which corresponds to the result of an exact line search in direction p_i.

There are many possible sets of vectors that are conjugate with respect to A, including the eigenvectors. The conjugate gradient method finds these directions starting with the steepest-descent direction,

$$p_0 = -\nabla f(x_0), \quad (4.55)$$

and then finds each subsequent direction using the update,

$$p_k = -\nabla f_k + \beta_{k-1} p_{k-1}. \quad (4.56)$$

For a positive β, the result is a new direction somewhere between the current steepest descent and the previous search direction, as shown in Fig. 4.42. The factor β is set such that p_k and p_{k-1} are conjugate with respect to A. One option to compute a β that achieves conjugacy is given by the *Fletcher–Reeves formula*,

$$\beta_k = \frac{\nabla f_k^\mathsf{T} \nabla f_k}{\nabla f_{k-1}^\mathsf{T} \nabla f_{k-1}}. \quad (4.57)$$

This formula is derived in Appendix B.4 as Eq. B.40 in the context of linear solvers. Here, we replace the residual of the linear system with the gradient of the quadratic because they are equivalent. Using the directions given by Eq. 4.56 and the step lengths given by Eq. 4.54, we can minimize a quadratic in n steps, where n is the size of x. The minimization shown in Fig. 4.41 starts with the steepest-descent direction and then computes one update to converge to the minimum in two iterations using exact line searches. The linear conjugate gradient method is detailed in Alg. B.2.

However, we are interested in minimizing general nonlinear functions. We can adapt the linear conjugate gradient method to the nonlinear case by doing the following:

1. Use the gradient of the nonlinear function in the search direction update (Eq. 4.56) and the expression for β (Eq. 4.57). This gradient can be computed using any of the methods in Chapter 6.

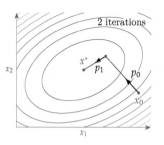

Fig. 4.41 By minimizing along a sequence of conjugate directions in turn, we can find the minimum of a quadratic in n steps, where n is the number of dimensions.

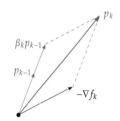

Fig. 4.42 The conjugate gradient search direction update combines the steepest-descent direction with the previous conjugate gradient direction.

2. Perform an inexact line search instead of doing the exact line search. This frees us from providing the Hessian vector products required for an exact line search (see Eq. 4.54). A line search that satisfies the strong Wolfe conditions is a good choice, but we need a stricter range in the sufficient decrease and sufficient curvature parameters ($0 < \mu_1 < \mu_2 < 1/2$).[†] This stricter requirement on μ_2 is necessary with the Fletcher–Reeves formula (Eq. 4.57) to ensure descent directions. As a first guess for α in the line search, we can use the same estimate proposed for steepest descent (Eq. 4.43).

3. Reset the search direction periodically back to the steepest-descent direction. In practice, resetting is often helpful to remove old information that is no longer useful. Some methods reset every n iterations, motivated by the fact that the linear case only generates n conjugate vectors. A more mathematical approach resets the direction when

$$\frac{|\nabla f_k^T \nabla f_{k-1}|}{|\nabla f_k^T \nabla f_k|} \geq 0.1. \qquad (4.58)$$

[†] For more details on the line search requirements, see Sec. 5.2 in Nocedal and Wright.[79]

79. Nocedal and Wright, *Numerical Optimization*, 2006.

The full procedure is given in Alg. 4.6. As with steepest descent, we may use normalized search directions.

The nonlinear conjugate gradient method is no longer guaranteed to converge in n steps like its linear counterpart, but it significantly outperforms the steepest-descent method. The change required relative to steepest descent is minimal: save information on the search direction and gradient from the previous iteration, and add the β term to the search direction update. Therefore, there is rarely a reason to prefer steepest descent. The parameter β can be interpreted as a "damping parameter" that prevents each search direction from varying too much relative to the previous one. When the function steepens, the damping becomes larger, and vice versa.

The formula for β in Eq. 4.57 is only one of several options. Another well-known option is the *Polak–Ribière formula*, which is given by

$$\beta_k = \frac{\nabla f_k^T \left(\nabla f_k - \nabla f_{k-1} \right)}{\nabla f_{k-1}^T \nabla f_{k-1}}. \qquad (4.59)$$

The conjugate gradient method with the Polak–Ribière formula tends to converge more quickly than with the Fletcher–Reeves formula, and this method does not require the more stringent range for μ_2. However, regardless of the value of μ_2, the strong Wolfe conditions still do not guarantee that p_k is a descent direction (β might become negative). This issue can be addressed by forcing β to remain nonnegative:

$$\beta \leftarrow \max(0, \beta). \qquad (4.60)$$

4.4 Search Direction

This equation automatically triggers a reset whenever $\beta = 0$ (see Eq. 4.56), so in this approach, other checks on resetting can be removed from Alg. 4.6.

Algorithm 4.6 Nonlinear conjugate gradient

Inputs:
  x_0: Starting point
  τ: Convergence tolerance
Outputs:
  x^*: Optimal point
  $f(x^*)$: Minimum function value

$k = 0$ Initialize iteration counter
while $\|\nabla f_k\|_\infty > \tau$ **do**   Optimality condition
     if $k = 0$ or reset = **true then**   first direction, and at resets
         $p_k = -\dfrac{\nabla f_k}{\|\nabla f_k\|}$
     else
         $\beta_k = \dfrac{\nabla f_k^\mathsf{T} \nabla f_k}{\nabla f_{k-1}^\mathsf{T} \nabla f_{k-1}}$
         $p_k = -\dfrac{\nabla f_k}{\|\nabla f_k\|} + \beta_k p_{k-1}$ Conjugate gradient direction update
     end if
     $\alpha_k = \text{linesearch}(p_k, \alpha_{\text{init}})$ Perform a line search
     $x_{k+1} = x_k + \alpha_k p_k$ Update design variables
     $k = k + 1$ Increment iteration index
end while

Example 4.12 Conjugate gradient applied to the bean function

Minimizing the same bean function from Ex. 4.11 and the same line search algorithm and settings, we get the optimization path shown in Fig. 4.43. The changes in direction for the conjugate gradient method are smaller than for steepest descent, and it takes fewer iterations to achieve the same convergence tolerance.

4.4.3 Newton's Method

The steepest-descent and conjugate gradient methods use only first-order information (the gradient). Newton's method uses second-order (curvature) information to get better estimates for search directions. The main advantage of Newton's method is that, unlike first-order methods,

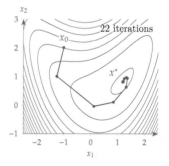

Fig. 4.43 Conjugate gradient optimization path.

it provides an estimate of the step length because the curvature predicts where the function derivative is zero.

In Section 3.8, we presented Newton's method for solving nonlinear equations. Newton's method for minimizing functions is based on the same principle, but instead of solving $r(u) = 0$, we solve for $\nabla f(x) = 0$.

As in Section 3.8, we can derive Newton's method for one-dimensional function minimization from the Taylor series approximation,

$$f(x_k + s) \approx f(x_k) + sf'(x_k) + \frac{1}{2}s^2 f''(x_k). \tag{4.61}$$

We now include a second-order term to get a quadratic that we can minimize. We minimize this quadratic approximation by differentiating with respect to the step s and setting the derivative to zero, which yields

$$f'(x_k) + sf''(x_k) = 0 \quad \Rightarrow \quad s = -\frac{f'(x_k)}{f''(x_k)}. \tag{4.62}$$

Thus, the Newton update is

$$x_{k+1} = x_k - \frac{f'_k}{f''_k}. \tag{4.63}$$

We could also derive this equation by taking Newton's method for root finding (Eq. 3.24) and replacing $r(u)$ with $f'(x)$.

Example 4.13 Newton's method for one-dimensional minimization

Suppose we want to minimize the following single-variable function:

$$f(x) = (x-2)^4 + 2x^2 - 4x + 4.$$

The first derivative is
$$f'(x) = 4(x-2)^3 + 4x - 4,$$

and the second derivative is
$$f''(x) = 12(x-2)^2 + 4.$$

Starting from $x_0 = 3$, we can form the quadratic (Eq. 4.61) using the function value and the first and second derivatives evaluated at that point, as shown in the top plot in Fig. 4.44. Then, the minimum of the quadratic is given analytically by the Newton update (Eq. 4.63). We successively form quadratics at each iteration and minimize them to find the next iteration. This is equivalent to finding the zero of the function's first derivative, as shown in the bottom plot in Eq. 4.63.

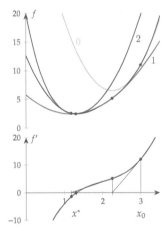

Fig. 4.44 Newton's method for finding roots can be adapted for function minimization by formulating it to find a zero of the derivative. We step to the minimum of a quadratic at each iteration (top) or equivalently find the root of the function's first derivative (bottom).

4.4 Search Direction

Like the one-dimensional case, we can build an n-dimensional Taylor series expansion about the current design point:

$$f(x_k + s) \approx f_k + \nabla f_k{}^\mathsf{T} s + \frac{1}{2} s^\mathsf{T} H_k s, \qquad (4.64)$$

where s is a vector centered at x_k. Similar to the one-dimensional case, we can find the step s_k that minimizes this quadratic model by taking the derivative with respect to s and setting that equal to zero:

$$\frac{df(x_k + s)}{ds} = \nabla f_k + H_k s = 0. \qquad (4.65)$$

Thus, each Newton step is the solution of a linear system where the matrix is the Hessian,

$$H_k s_k = -\nabla f_k. \qquad (4.66)$$

This linear system is analogous to the one used for solving nonlinear systems with Newton's method (Eq. 3.30), except that the Jacobian becomes the Hessian, the residual is the gradient, and the design variables replace the states. We can use any of the linear solvers mentioned in Section 3.6 and Appendix B to solve this system.

When minimizing the quadratic function from Ex. 4.10, Newton's method converges in one step for any value of β, as shown in Fig. 4.45. Thus, Newton's method is *scale invariant*

Because the function is quadratic, the quadratic "approximation" from the Taylor series is exact, so we can find the minimum in one step. It will take more iterations for a general nonlinear function, but using curvature information generally yields a better search direction than first-order methods. In addition, Newton's method provides a step length embedded in s_k because the quadratic model estimates the stationary point location. Furthermore, Newton's method exhibits quadratic convergence.

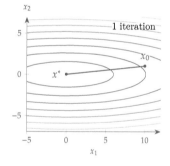

Fig. 4.45 Iteration history for a quadratic function using an exact line search and Newton's method. Unsurprisingly, only one iteration is required.

Although Newton's method is powerful, it suffers from a few issues in practice. One issue is that the Newton step does not necessarily result in a function decrease. This issue can occur if the Hessian is not positive definite or if the quadratic predictions overshoot because the actual function has more curvature than predicted by the quadratic approximation. Both of these possibilities are illustrated in Fig. 4.46.

If the Hessian is not positive definite, the step might not even be in a descent direction. Replacing the real Hessian with a positive-definite Hessian can mitigate this issue. The quasi-Newton methods in the next section force a positive-definite Hessian by construction.

To fix the overshooting issue, we can use a line search instead of blindly accepting the Newton step length. We would set $p_k = s_k$,

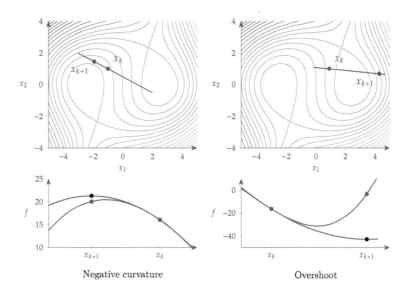

Fig. 4.46 Newton's method in its pure form is vulnerable to negative curvature (in which case it might step away from the minimum) and overshooting (which might result in a function increase).

with $\alpha_{\text{init}} = 1$ as the first guess for the step length. In this case, we have a much better guess for α compared with the steepest-descent or conjugate gradient cases because this guess is based on the local curvature. Even if the first step length given by the Newton step overshoots, the line search would find a point with a lower function value.

The trust-region methods in Section 4.5 address both of these issues by minimizing the function approximation within a specified region around the current iteration.

Another major issue with Newton's method is that the Hessian can be difficult or costly to compute. Even if available, the solution of the linear system in Eq. 4.65 can be expensive. Both of these considerations motivate the quasi-Newton methods, which we explain next.

Example 4.14 Newton method applied to the bean function

Minimizing the same bean function from Exs. 4.11 and 4.12, we get the optimization path shown in Fig. 4.47. Newton's method takes fewer iterations than steepest descent (Ex. 4.11) or conjugate gradient (Ex. 4.12) to achieve the same convergence tolerance. The first quadratic approximation is a saddle function that steps to the saddle point, away from the minimum of the function. However, in subsequent iterations, the quadratic approximation becomes convex, and the steps take us along the valley of the bean function toward the minimum.

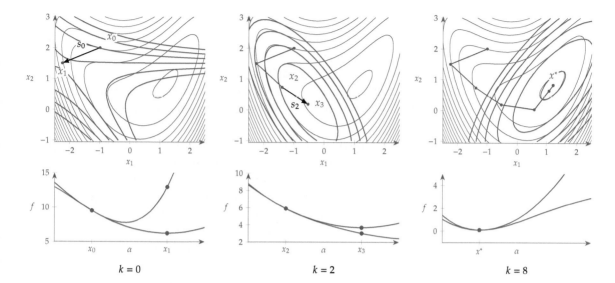

Fig. 4.47 Newton's method minimizes a sequence of quadratic approximations of the function at each iteration. In this case, it converges in 8 major iterations.

4.4.4 Quasi-Newton Methods

As mentioned in Section 4.4.3, Newton's method is efficient because the second-order information results in better search directions, but it has the significant shortcoming of requiring the Hessian. Quasi-Newton methods are designed to address this issue. The basic idea is that we can use first-order information (gradients) along each step in the iteration path to build an approximation of the Hessian.

In one dimension, we can adapt the secant method (see Eq. 3.26) for function minimization. Instead of estimating the first derivative, we now estimate the second derivative (curvature) using two successive first derivatives, as follows:

$$f''_{k+1} = \frac{f'_{k+1} - f'_k}{x_{k+1} - x_k}. \tag{4.67}$$

Then we can use this approximation in the Newton step (Eq. 4.63) to obtain an iterative procedure that requires only first derivatives instead of first and second derivatives.

The quadratic approximation based on this approximation of the second derivative is

$$\tilde{f}_{k+1}(x_{k+1} + s) = f_{k+1} + s f'_{k+1} + \frac{s^2}{2} \left(\frac{f'_{k+1} - f'_k}{x_{k+1} - x_k} \right). \tag{4.68}$$

Taking the derivative of this approximation with respect to s, we get

$$\tilde{f}'_{k+1}(x_{k+1} + s) = f'_{k+1} + s \left(\frac{f'_{k+1} - f'_k}{x_{k+1} - x_k} \right). \tag{4.69}$$

For $s = 0$, which corresponds to x_{k+1}, we get $\tilde{f}'_{k+1}(x_{k+1}) = f'_{k+1}$, which tells us that the slope of the approximation matches the slope of the actual function at x_{k+1}, as expected.

Also, by stepping backward to x_k by setting $s = -(x_{k+1} - x_k)$, we find that $\tilde{f}'_{k+1}(x_k) = f'_k$. Thus, the nature of this approximation is such that it matches the slope of the actual function at the last two points, as shown in Fig. 4.48.

In n dimensions, things are more involved, but the principle is the same: use first-derivative information from the last two points to approximate second-derivative information. Instead of iterating along the x-axis as we would in one dimension, the optimization in n dimensions follows a sequence of steps (as shown in Fig. 4.1) for the separate line searches. We have gradients at the endpoints of each step, so we can take the difference between the gradients at those points to get the curvature along that direction. The question is: How do we update the Hessian, which is expressed in the coordinate system of x, based on directional curvatures in directions that are not necessarily aligned with the coordinate system?

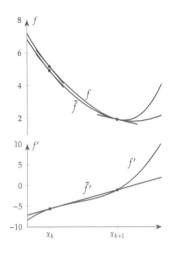

Fig. 4.48 The quadratic approximation based on the secant method matches the slopes at the two last points and the function value at the last point.

Quasi-Newton methods use the quadratic approximation of the objective function,

$$\tilde{f}(x_k + p) = f_k + \nabla f_k^\mathsf{T} p + \frac{1}{2} p^\mathsf{T} \tilde{H}_k p, \tag{4.70}$$

where $\tilde{H}$ is an approximation of the Hessian. Similar to Newton's method, we minimize this quadratic with respect to p, which yields the linear system

$$\tilde{H}_k p_k = -\nabla f_k. \tag{4.71}$$

We solve this linear system for p_k, but instead of accepting it as the final step, we perform a line search in the p_k direction. Only after finding a step size α_k that satisfies the strong Wolfe conditions do we update the point using

$$x_{k+1} = x_k + \alpha_k p_k. \tag{4.72}$$

Quasi-Newton methods update the approximate Hessian at every iteration based on the latest information using an update of the form

$$\tilde{H}_{k+1} = \tilde{H}_k + \Delta \tilde{H}_k, \tag{4.73}$$

where the update $\Delta \tilde{H}_k$ is a function of the last two gradients. The first Hessian approximation is usually set to the identity matrix (or a scaled version of it), which yields a steepest-descent direction for the first line search (set $\tilde{H} = I$ in Eq. 4.71 to verify this).

4.4 Search Direction

We now develop the requirements for the approximate Hessian update. Suppose we just obtained the new point x_{k+1} after a line search starting from x_k in the direction p_k. We can write the new quadratic based on an updated Hessian as follows:

$$\tilde{f}(x_{k+1} + p) = f_{k+1} + \nabla f_{k+1}^\mathsf{T} p + \frac{1}{2} p^\mathsf{T} \tilde{H}_{k+1} p. \quad (4.74)$$

We can assume that the new point's function value and gradient are given, but we do not have the new approximate Hessian yet. Taking the gradient of this quadratic with respect to p, we obtain

$$\nabla \tilde{f}(x_{k+1} + p) = \nabla f_{k+1} + \tilde{H}_{k+1} p. \quad (4.75)$$

In the single-variable case, we observed that the quadratic approximation based on the secant method matched the slope of the actual function at the last two points. Therefore, it is logical to require the n-dimensional quadratic based on the approximate Hessian to match the gradient of the actual function at the last two points.

The gradient of the new approximation (Eq. 4.75) matches the gradient at the new point x_{k+1} by construction (just set $p = 0$). To find the gradient predicted by the new approximation (Eq. 4.75) at the previous point x_k, we set $p = x_k - x_{k+1} = -\alpha_k p_k$ (which is a backward step from the end of the last line search to the start of the line search) to get

$$\nabla \tilde{f}(x_{k+1} - \alpha_k p_k) = \nabla f_{k+1} - \alpha_k \tilde{H}_{k+1} p_k. \quad (4.76)$$

Now, we enforce that this must be equal to the actual gradient at that point,

$$\begin{aligned} \nabla f_{k+1} - \alpha_k \tilde{H}_{k+1} p_k &= \nabla f_k \Rightarrow \\ \alpha_k \tilde{H}_{k+1} p_k &= \nabla f_{k+1} - \nabla f_k. \end{aligned} \quad (4.77)$$

To simplify the notation, we define the step as

$$s_k = x_{k+1} - x_k = \alpha_k p_k, \quad (4.78)$$

and the difference in the gradient as

$$y_k = \nabla f_{k+1} - \nabla f_k. \quad (4.79)$$

Figure 4.49 shows the step and the corresponding gradients.

Rewriting Eq. 4.77 using this notation, we get

$$\tilde{H}_{k+1} s_k = y_k. \quad (4.80)$$

This is called the *secant equation* and is a fundamental requirement for quasi-Newton methods. The result is intuitive when we recall the

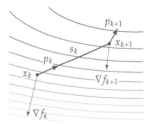

Fig. 4.49 Quasi-Newton methods use the gradient at the endpoint of each step to estimate the curvature in the step direction and update an approximation of the Hessian.

meaning of the product of the Hessian with a vector (Eq. 4.12): it is the rate of change of the Hessian in the direction defined by that vector. Thus, it makes sense that the rate of change of the curvature predicted by the approximate Hessian should match the difference between the gradients.[‡]

‡The secant equation is also known as the *quasi-Newton condition*.

We need $\tilde{H}$ to be positive definite. Using the secant equation (Eq. 4.80) and the definition of positive definiteness ($s^\mathsf{T} H s > 0$), we see that this requirement implies that the predicted curvature is positive along the step; that is,

$$s_k{}^\mathsf{T} y_k > 0. \qquad (4.81)$$

This is called the *curvature condition*, and it is automatically satisfied if the line search finds a step that satisfies the strong Wolfe conditions.

The secant equation (Eq. 4.80) is a linear system of n equations where the step and the gradients are known. However, there are $n(n+1)/2$ unknowns in the approximate Hessian matrix (recall that it is symmetric), so this equation is not sufficient to determine the elements of $\tilde{H}$. The requirement of positive definiteness adds one more equation, but those are not enough to determine all the unknowns, leaving us with an infinite number of possibilities for $\tilde{H}$.

To find a unique $\tilde{H}_{k+1}$, we rationalize that among all the matrices that satisfy the secant equation (Eq. 4.80), $\tilde{H}_{k+1}$ should be the one closest to the previous approximate Hessian, $\tilde{H}_k$. This makes sense intuitively because the curvature information gathered in one step is limited (because it is along a single direction) and should not change the Hessian approximation more than necessary to satisfy the requirements.

The original quasi-Newton update, known as *DFP*, was first proposed by Davidon and then refined by Fletcher and also Powell (see historical note in Section 2.3).[20,21] The DFP update formula has been superseded by the BFGS formula, which was independently developed by Broyden, Fletcher, Goldfarb, and Shanno.[80–83] BFGS is currently considered the most effective quasi-Newton update, so we focus on this update. However, Appendix C.2.1 has more details on DFP.

The formal derivation of the BFGS update formula is rather involved, so we do not include it here. Instead, we work through an informal derivation that provides intuition about this update and quasi-Newton methods in general. We also include more details in Appendix C.2.2.

Recall that quasi-Newton methods add an update to the previous Hessian approximation (Eq. 4.73). One way to think about an update that yields a matrix close to the previous one is to consider the rank of the update, $\Delta \tilde{H}$. The lower the rank of the update, the closer the updated matrix is to the previous one. Also, the curvature information contained in this update is minimal because we are only gathering

20. Davidon, *Variable metric method for minimization*, 1991.
21. Fletcher and Powell, *A rapidly convergent descent method for minimization*, 1963.
80. Broyden, *The convergence of a class of double-rank minimization algorithms 1. General considerations*, 1970.
81. Fletcher, *A new approach to variable metric algorithms*, 1970.
82. Goldfarb, *A family of variable-metric methods derived by variational means*, 1970.
83. Shanno, *Conditioning of quasi-Newton methods for function minimization*, 1970.

information in one direction for each update. Therefore, we can reason that the rank of the update matrix should be the lowest possible rank that satisfies the secant equation (Eq. 4.80).

The update must be symmetric and positive definite to ensure a symmetric positive-definite Hessian approximation. If we start with a symmetric positive-definite approximation, then all subsequent approximations remain symmetric and positive definite. As it turns out, it is possible to derive a rank 1 update matrix that satisfies the secant equation, but this update is not guaranteed to be positive definite. However, we can get positive definiteness with a rank 2 update.

We can obtain a symmetric rank 2 update by adding two symmetric rank 1 matrices. One convenient way to obtain a symmetric rank 1 matrix is to perform a *self outer product* of a vector, which takes a vector of size n and multiplies it with its transpose to obtain an $(n \times n)$ matrix, as shown in Fig. 4.50. Matrices resulting from vector outer products have rank 1 because all the columns are linearly dependent.

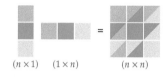

Fig. 4.50 The self outer product of a vector produces a symmetric $(n \times n)$ matrix of rank 1.

With two linearly independent vectors (u and v), we can get a rank 2 update using

$$\tilde{H}_{k+1} = \tilde{H}_k + \alpha u u^\mathsf{T} + \beta v v^\mathsf{T}, \qquad (4.82)$$

where α and β are scalar coefficients. Substituting this into the secant equation (Eq. 4.80), we have

$$\tilde{H}_k s_k + \alpha u u^\mathsf{T} s_k + \beta v v^\mathsf{T} s_k = y_k. \qquad (4.83)$$

Because the new information about the function is encapsulated in the vectors y and s, we can reason that u and v should be based on these vectors. It turns out that using s on its own does not yield a useful update (one term cancels out), but $\tilde{H}s$ does. Setting $u = y$ and $v = \tilde{H}s$ in Eq. 4.83 yields

$$\tilde{H}_k s_k + \alpha y_k y_k^\mathsf{T} s_k + \beta \tilde{H}_k s_k \left(\tilde{H}_k s_k\right)^\mathsf{T} s_k = y_k. \qquad (4.84)$$

Rearranging this equation, we have

$$y_k \left(1 - \alpha y_k^\mathsf{T} s_k\right) = \tilde{H}_k s_k \left(1 + \beta s_k^\mathsf{T} \tilde{H}_k s_k\right). \qquad (4.85)$$

Because the vectors y_k and $\tilde{H}_k s_k$ are not parallel in general (because the secant equation applies to $\tilde{H}_{k+1}$, not to $\tilde{H}_k$), the only way to guarantee this equality is to set the terms in parentheses to zero. Thus, the scalar coefficients are

$$\alpha = \frac{1}{y_k^\mathsf{T} s_k}, \quad \beta = -\frac{1}{s_k^\mathsf{T} \tilde{H}_k s_k}. \qquad (4.86)$$

Substituting these coefficients and the chosen vectors back into Eq. 4.82, we get the BFGS update,

$$\tilde{H}_{k+1} = \tilde{H}_k + \frac{y_k y_k^\mathsf{T}}{y_k^\mathsf{T} s_k} - \frac{\tilde{H}_k s_k s_k^\mathsf{T} \tilde{H}_k}{s_k^\mathsf{T} \tilde{H}_k s_k}. \qquad (4.87)$$

Although we did not explicitly enforce positive definiteness, the rank 2 update is positive definite, and therefore, all the Hessian approximations are positive definite, as long as we start with a positive-definite approximation.

Now recall that we want to solve the linear system that involves this matrix (Eq. 4.71), so it would be more efficient to approximate the inverse of the Hessian directly instead. The inverse can be found analytically from the update (Eq. 4.87) using the Sherman–Morrison–Woodbury formula.[§] Defining $\tilde{V}$ as the approximation of the inverse of the Hessian, the final result is

[§]This formula is also known as the *Woodbury matrix identity*. Given a matrix and an update to that matrix, it yields an explicit expression for the inverse of the updated matrix in terms of the inverses of the matrix and the update (see Appendix C.3).

$$\tilde{V}_{k+1} = \left(I - \sigma_k s_k y_k^\mathsf{T}\right) \tilde{V}_k \left(I - \sigma_k y_k s_k^\mathsf{T}\right) + \sigma_k s_k s_k^\mathsf{T}, \qquad (4.88)$$

where

$$\sigma_k = \frac{1}{y_k^\mathsf{T} s_k}. \qquad (4.89)$$

Figure 4.51 shows the sizes of the vectors and matrices involved in this equation.

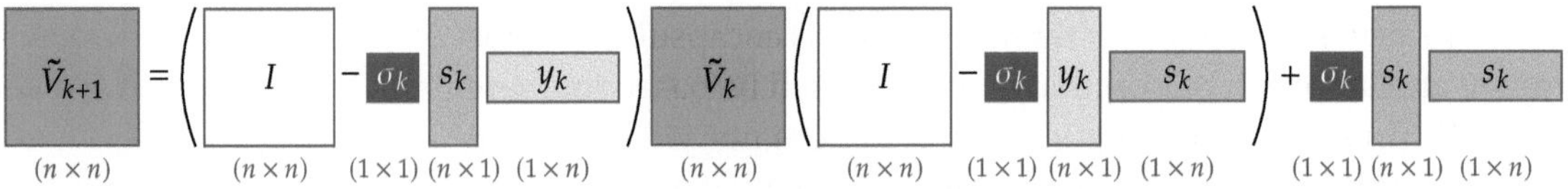

Fig. 4.51 Sizes of each term of the BFGS update (Eq. 4.88).

Now we can replace the potentially costly solution of the linear system (Eq. 4.71) with the much cheaper matrix-vector product,

$$p_k = -\tilde{V}_k \nabla f_k, \qquad (4.90)$$

where $\tilde{V}$ is the estimate for the inverse of the Hessian.

Algorithm 4.7 details the steps for the BFGS algorithm. Unlike first-order methods, we should not normalize the direction vector p_k because the length of the vector is meaningful. Once we have curvature information, the quasi-Newton step should give a reasonable estimate of where the function slope flattens. Thus, as advised for Newton's method, we set $\alpha_{\text{init}} = 1$. Alternatively, this would be equivalent to using a normalized direction vector and then setting α_{init} to the initial magnitude of p. However, optimization algorithms in practice use

4.4 Search Direction

$\alpha_{\text{init}} = 1$ to signify that a full (quasi-) Newton step was accepted (see Tip 4.5).

As discussed previously, we need to start with a positive-definite estimate to maintain a positive-definite inverse Hessian. Typically, this is the identity matrix or a weighted identity matrix, for example:

$$\tilde{V}_0 = \frac{1}{\|\nabla f_0\|} I. \quad (4.91)$$

This makes the first step a normalized steepest-descent direction:

$$p_0 = -\tilde{V}_0 \nabla f_0 = -\frac{\nabla f_0}{\|\nabla f_0\|}. \quad (4.92)$$

Algorithm 4.7 BFGS

Inputs:
  x_0: Starting point
  τ: Convergence tolerance

Outputs:
  x^*: Optimal point
  $f(x^*)$: Minimum function value

$k = 0$ — Initialize iteration counter
$\alpha_{\text{init}} = 1$ — Initial step length for line search
while $\|\nabla f_k\|_\infty > \tau$ **do** — Optimality condition
  if $k = 0$ or reset = true **then**
    $\tilde{V}_k = \frac{1}{\|\nabla f\|} I$
  else
    $s = x_k - x_{k-1}$ — Last step
    $y = \nabla f_k - \nabla f_{k-1}$ — Curvature along last step
    $\sigma = \frac{1}{s^T y}$
    $\tilde{V}_k = (I - \sigma s y^T) \tilde{V}_{k-1} (I - \sigma y s^T) + \sigma s s^T$ — Quasi-Newton update
  end if
  $p = -\tilde{V}_k \nabla f_k$ — Compute quasi-Newton step
  $\alpha = \text{linesearch}(p, \alpha_{\text{init}})$ — Should satisfy the strong Wolfe conditions
  $x_{k+1} = x_k + \alpha p$ — Update design variables
  $k = k + 1$ — Increment iteration index
end while

In a practical algorithm, $\tilde{V}$ might require occasional resets to the scaled identity matrix. This is because as we iterate in the design space, curvature information gathered far from the current point might become irrelevant and even counterproductive. The trigger for this reset could occur when the directional derivative $\nabla f^\mathsf{T} p$ is greater than some threshold. That would mean that the slope along the search direction is shallow; in other words, the search direction is close to orthogonal to the steepest-descent direction.

Another well-known quasi-Newton update is the symmetric rank 1 (SR1) update, which we derive in Appendix C.2.3. Because the update is rank 1, it does not guarantee positive definiteness. Why would we be interested in a Hessian approximation that is potentially indefinite? In practice, the matrices produced by SR1 have been found to approximate the true Hessian matrix well, often better than BFGS. This alternative is more common in trust-region methods (see Section 4.5), which depend more strongly on an accurate Hessian and do not require positive definiteness. It is also sometimes used for constrained optimization problems where the Hessian of the Lagrangian is often indefinite, even at the minimizer.

Example 4.15 BFGS applied to the bean function

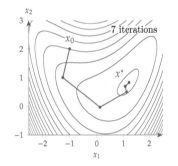

Fig. 4.52 BFGS optimization path.

Minimizing the same bean function from previous examples using BFGS, we get the optimization path shown in Fig. 4.52. We also show the corresponding quadratic approximations for a few selected steps of this minimization in Fig. 4.53. Because we generate approximations to the inverse, we invert those approximations to get the Hessian approximation for the purpose of illustration.

We initialize the inverse Hessian to the identity matrix, which results in a quadratic with circular contours and a steepest-descent step (Fig. 4.53, left). Using the BFGS update procedure, after two iterations,

$$x_2 = (0.1197030, -0.043079),$$

and the inverse Hessian approximation is

$$\tilde{V}_2 = \begin{bmatrix} 0.435747 & -0.202020 \\ -0.202020 & 0.222556 \end{bmatrix}.$$

The exact inverse Hessian at the same point is

$$H^{-1}(x_2) = \begin{bmatrix} 0.450435 & 0.035946 \\ 0.035946 & 0.169535 \end{bmatrix}.$$

The predicted curvature improves, and it results in a good step toward the minimum, as shown in the middle plot of Fig. 4.53. The one-dimensional slice reveals how the approximation curvature in the line search direction is higher than the actual; however, the line search moves past the approximation minimum toward the true minimum.

By the end of the optimization, at $x^* = (1.213412, 0.824123)$, the BFGS estimate is

$$\tilde{V}^* = \begin{bmatrix} 0.276946 & 0.224010 \\ 0.224010 & 0.347847 \end{bmatrix},$$

4.4 Search Direction

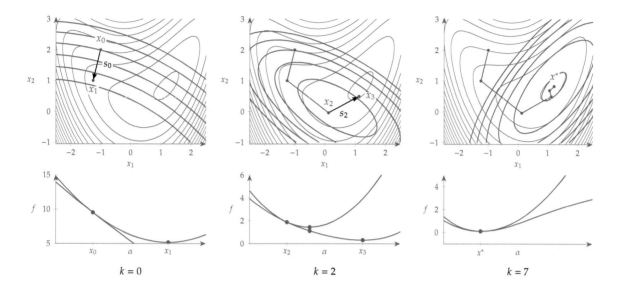

whereas the exact one is

$$H^{-1}(x^*) = \begin{bmatrix} 0.276901 & 0.223996 \\ 0.223996 & 0.347867 \end{bmatrix}.$$

Now the estimate is much more accurate. In the right plot of Fig. 4.53, we can see that the minimum of the approximation coincides with the actual minimum. The approximation is only accurate locally, worsening away from the minimum.

Fig. 4.53 Minimization of the bean function using BFGS. The first quadratic approximation has circular contours (left). After two iterations, the quadratic approximation improves, and the step approaches the minimum (middle). Once converged, the minimum of the quadratic approximation coincides with the bean function minimum (right).

4.4.5 Limited-Memory Quasi-Newton Methods

When the number of design variables is large (millions or billions), it might not be possible to store the Hessian inverse approximation matrix in memory. This motivates limited-memory quasi-Newton methods, which make it possible to handle such problems. In addition, these methods also improve the computational efficiency of medium-sized problems (hundreds or thousands of design variables) with minimal sacrifice in accuracy.

Recall that we are only interested in the matrix-vector product $\tilde{V}\nabla f$ to find each search direction using Eq. 4.90. As we will see in this section, we can compute this product without ever actually forming the matrix $\tilde{V}$. We focus on doing this for the BFGS update because this is the most popular approach (known as *L-BFGS*), although similar techniques apply to other quasi-Newton update formulas.

The BFGS update (Eq. 4.88) is a recursive sequence:

$$\tilde{V}_k = \left[(I - \sigma s y^\mathsf{T})\tilde{V}(I - \sigma y s^\mathsf{T}) + \sigma s s^\mathsf{T}\right]_{k-1}, \qquad (4.93)$$

where
$$\sigma = \frac{1}{s^\mathsf{T} y}. \tag{4.94}$$

If we save the sequence of s and y vectors and specify a starting value for $\tilde{V}_0$, we can compute any subsequent $\tilde{V}_k$. Of course, what we want is $\tilde{V}_k \nabla f_k$, which we can also compute using an algorithm with the recurrence relationship. However, such an algorithm would not be advantageous from the memory-usage perspective because we would have to store a long sequence of vectors and a starting matrix.

To reduce the memory usage, we do not store the entire history of vectors. Instead, we limit the storage to the last m vectors for s and y. In practice, m is usually between 5 and 20. Next, we make the starting Hessian diagonal such that we only require vector storage (or scalar storage if we make all entries in the diagonal equal). A common choice is to use a scaled identity matrix, which just requires storing one number,
$$\tilde{V}_0 = \frac{s^\mathsf{T} y}{y^\mathsf{T} y} I, \tag{4.95}$$

where the s and y correspond to the previous iteration. Algorithm 4.8 details the procedure.

Algorithm 4.8 Compute search direction using L-BFGS

Inputs:
  ∇f_k: Gradient at point x_k
  $s_{k-1,\ldots,k-m}$: History of steps $x_k - x_{k-1}$
  $y_{k-1,\ldots,k-m}$: History of gradient differences $\nabla f_k - \nabla f_{k-1}$

Outputs:
  p: Search direction $-\tilde{V}_k \nabla f_k$

$d = \nabla f_k$
for $i = k-1$ **to** $k-m$ **by** -1 **do**
  $\alpha_i = \sigma_i s_i^\mathsf{T} d$
  $d = d - \alpha_i y_i$
end for
$\tilde{V}_0 = \left(\frac{s_{k-1}^\mathsf{T} y_{k-1}}{y_{k-1}^\mathsf{T} y_{k-1}} \right) I$ Initialize Hessian inverse approximation as a scaled identity matrix
$d = \tilde{V}_0 d$
for $i = k-m$ **to** $k-1$ **do**
  $\beta_i = \sigma_i y_i^\mathsf{T} d$
  $d = d + (\alpha_i - \beta_i) s_i$
end for
$p = -d$

4.4 SEARCH DIRECTION

Using this technique, we no longer need to bear the memory cost of storing a large matrix or incur the computational cost of a large matrix-vector product. Instead, we store a small number of vectors and require fewer vector-vector products (a cost that scales linearly with n rather than quadratically).

Example 4.16 L-BFGS compared with BFGS for the bean function

Minimizing the same bean function from the previous examples, the optimization iterations using BFGS and L-BFGS are the same, as shown in Fig. 4.54. The L-BFGS method is applied to the same sequence using the last five iterations. The number of variables is too small to benefit from the limited-memory approach, but we show it in this small problem as an example. At the same x^* as in Ex. 4.15, the product $\tilde{V}\nabla f$ is estimated using Alg. 4.8 as

$$d^* = \begin{bmatrix} -7.38683 \times 10^{-5} \\ 5.75370 \times 10^{-5} \end{bmatrix},$$

whereas the exact value is:

$$\tilde{V}^*\nabla f^* = \begin{bmatrix} -7.49228 \times 10^{-5} \\ 5.90441 \times 10^{-5} \end{bmatrix}.$$

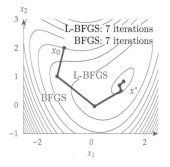

Fig. 4.54 Optimization paths using BFGS and L-BFGS.

Example 4.17 Minimizing the total potential energy for a spring system

Many structural mechanics models involve solving an unconstrained energy minimization problem. Consider a mass supported by two springs, as shown in Fig. 4.55. Formulating the total potential energy for the system as a function of the mass position yields the following problem:[¶]

[¶]Appendix D.1.8 has details on this problem.

$$\underset{x_1,x_2}{\text{minimize}} \quad \frac{1}{2}k_1\left(\sqrt{(\ell_1+x_1)^2+x_2^2}-\ell_1\right)^2 + \frac{1}{2}k_2\left(\sqrt{(\ell_2-x_1)^2+x_2^2}-\ell_2\right)^2 - mgx_2.$$

The contours of this function are shown in Fig. 4.56 for the case where $l_1 = 12, l_2 = 8, k_1 = 1, k_2 = 10, mg = 7$. There is a minimum and a maximum. The minimum represents the position of the mass at the stable equilibrium condition. The maximum also represents an equilibrium point, but it is unstable. All methods converge to the minimum when starting near the maximum. All four methods use the same parameters, convergence tolerance, and starting point. Depending on the starting point, Newton's method can become stuck at the saddle point, and if a line search is not added to safeguard it, it could have terminated at the maximum instead.

As expected, steepest descent is the least efficient, and the second-order methods are the most efficient. The number of iterations and the relative

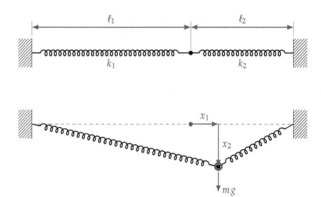

Fig. 4.55 Two-spring system with no applied force (top) and with applied force (bottom).

performance are problem dependent and sensitive to the optimization algorithm parameters, so we should not analyze the number of iterations too closely. However, these results show the expected trends for most problems.

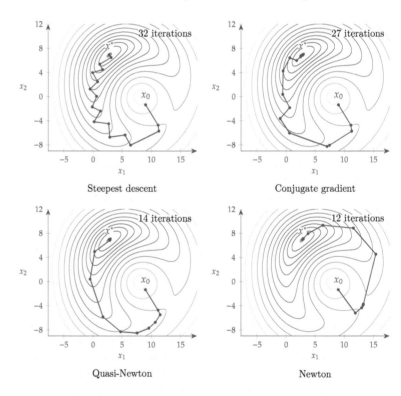

Fig. 4.56 Minimizing the total potential for two-spring system.

4.4 Search Direction

Example 4.18 Comparing methods for the Rosenbrock function

We now test the methods on the following more challenging function:

$$f(x_1, x_2) = (1 - x_1)^2 + 100\left(x_2 - x_1^2\right)^2,$$

which is known as the *Rosenbrock function*. This is a well-known optimization problem because a narrow, highly curved valley makes it challenging to minimize.∥ The optimization path and the convergence history for four methods starting from $x = (-1.2, 1.0)$ are shown in Figs. 4.57 and 4.58, respectively. All four methods use an inexact line search with the same parameters and a convergence tolerance of $\|\nabla f\|_\infty \leq 10^{-6}$. Compared with the previous two examples, the difference between the steepest-descent and second-order methods is much more dramatic (two orders of magnitude more iterations!), owing to the more challenging variation in the curvature (recall Ex. 4.10).

∥ The "bean" function we used in previous examples is a milder version of the Rosenbrock function.

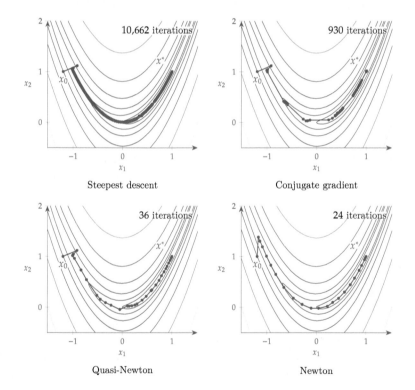

Fig. 4.57 Optimization paths for the Rosenbrock function using steepest descent, conjugate gradient, BFGS, and Newton.

The steepest-descent method converges, but it takes many iterations because it bounces between the steep walls of the valley while making little progress along the bottom of the valley. The conjugate gradient method is much more efficient because it damps the steepest-descent oscillations. Eventually, the conjugate gradient method achieves superlinear convergence near the optimum, saving many iterations over the last several orders of magnitude in

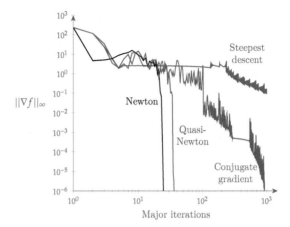

Fig. 4.58 Convergence of the four methods shows the dramatic difference between the linear convergence of steepest descent, the superlinear convergence of the conjugate gradient method, and the quadratic convergence of the methods that use second-order information.

the convergence criterion. The methods that use second-order information are even more efficient, exhibiting quadratic convergence in the last few iterations.

The number of major iterations is not always an effective way to compare performance. For example, Newton's method takes fewer major iterations, but each iteration in Newton's method is more expensive than each iteration in the quasi-Newton method. This is because Newton's method requires a linear solution, which is an $O(n^3)$ operation, as opposed to a matrix-vector multiplication, which is an $O(n^2)$ operation. For a small problem like the two-dimensional Rosenbrock function, this is an insignificant difference, but this is a significant difference in computational effort for large problems. Additionally, each major iteration includes a line search, and depending on the quality of the search direction, the number of function calls contained in each iteration will differ.

Tip 4.5 Unit steps indicate good progress

When performing a line search within a quasi-Newton algorithm, we pick $\alpha_{\text{init}} = 1$ (a unit step) because this corresponds to the minimum if the quadratic approximation were perfect. When the quadratic approximation matches the actual function well enough, the line search should exit after the first evaluation. On the other hand, if the line search takes many iterations, this indicates a poor match or other numerical difficulties. If difficulties persist over many major iterations, plot the line search (Tip 4.3).

4.4 SEARCH DIRECTION

Example 4.19 Problem scaling

In Tip 4.4, we discussed the importance of scaling. Let us illustrate this with an example. Consider a stretched version of the Rosenbrock function from Ex. 4.18:

$$f(x_1, x_2) = \left(1 - \frac{x_1}{10^4}\right)^2 + 100\left(x_2 - \left(\frac{x_1}{10^4}\right)^2\right)^2. \tag{4.96}$$

The contours of this function have the same characteristics as those of the original Rosenbrock function shown in Fig. 4.57, but the x_1 axis is stretched, as shown in Fig. 4.59. Because x_1 is scaled by such a large number (10^4), we cannot show it using the same scale as the x_2 axis, otherwise the x_2 axis would disappear. The minimum of this function is at $x^* = [10^4, 1]$, where $f^* = 0$.

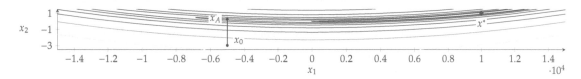

Fig. 4.59 The contours the scaled Rosenbrock function (Eq. 4.96) are highly stretched in the x_1 direction, by orders of magnitude more than what we can show here.

Let us attempt to minimize this function starting from $x_0 = [-5000, -3]$. The gradient at this starting point is $\nabla f(x_0) = [-0.0653, -650.0]$, so the slope in the x_2 direction is four orders of magnitude times larger than the slope in the x_1 direction! Therefore, there is a significant bias toward moving along the x_2 direction but little incentive to move in the x_1 direction. After an exact line search in the steepest descent direction, we obtain the step to $x_A = [-5000, 0.25]$ as shown in Fig. 4.59. The optimization stops at this point, even though it is not a minimum. This premature convergence is because $\partial f / \partial x_1$ is orders of magnitude smaller, so both components of the gradient satisfy the optimality conditions when using a standard relative tolerance.

To address this issue, we scale the design variables as explained in Tip 4.4. Using the scaling $s_x = [10^4, 1]$, the scaled starting point becomes $\tilde{x}_0 = [-5000, -3] \oslash [10^4, 1] = [-0.5, -3]$. Before evaluating the function, we need to convert the design variables back to their unscaled values, that is, $f(x) = f(\tilde{x} \odot s_x)$.

This scaling of the design variables alone is sufficient to improve the optimization convergence. Still, let us also scale the objective because it is large at our starting point (around 900). Dividing the objective by $s_f = 1000$, the initial gradient becomes $\nabla f(x_0) = [-0.00206, -0.6]$. This is still not ideally scaled, but it has much less variation in orders of magnitude—more than sufficient to solve the problem successfully. The optimizer returns $\tilde{x}^* = [1, 1]$, where $\tilde{f}^* = 1.57 \times 10^{-12}$. When rescaled back to the problem coordinates, $x^* = [10^4, 1], f^* = 1.57 \times 10^{-9}$.

In this example, the function derivatives span many orders of magnitude, so dividing the function by a scalar does not have much effect. Instead, we could minimize $\log(f)$, which allows us to solve the problem even without scaling x. If we also scale x, the number of required iterations for convergence decreases. Using $\log(f)$ as the objective and scaling the design variables as before yields $\tilde{x}^* = [1, 1]$, where $\tilde{f}^* = -25.28$, which in the original problem space corresponds to $x^* = [10^4, 1]$, where $f^* = 1.05 \times 10^{-11}$.

Although this example does not correspond to a physical problem, such differences in scaling occur frequently in engineering analysis. For example, optimizing the operating point of a propeller might involve two variables: the pitch angle and the rotation rate. The angle would typically be specified in radians (a quantity of order 1) and the rotation rate in rotations per minute (typically tens of thousands).

Poor scaling causes premature convergence for various reasons. In Ex. 4.19, it was because convergence was based on a tolerance relative to the starting gradient, and some gradient components were much smaller than others. When using an absolute tolerance, premature convergence can occur when the gradients are small to begin with (because of the scale of the problem, not because they are near an optimum). When the scaling is poor, the optimizer is even more dependent on accurate gradients to navigate the narrow regions of function improvement.

Larger engineering simulations are usually more susceptible to numerical noise due to iteration loops, solver convergence tolerances, and longer computational procedures. Another issue arises when the derivatives are not computed accurately. In these cases, poorly scaled problems struggle because the line search directions are not accurate enough to yield a decrease, except for tiny step sizes.

Most practical optimization algorithms terminate early when this occurs, not because the optimality conditions are satisfied but because the step sizes or function changes are too small, and progress is stalled (see Tip 4.1). A lack of attention to scaling is one of the most frequent causes of poor solutions in engineering optimization problems.

Tip 4.6 Accurate derivatives matter

The effectiveness of gradient-based methods depends strongly on providing accurate gradients. Convergence difficulties, or apparent multimodal behavior, are often mistakenly identified as optimization algorithm difficulties or fundamental modeling issues when in reality, the numerical issues are caused by inaccurate gradients. Chapter 6 is devoted to computing accurate derivatives.

4.5 Trust-Region Methods

In Section 4.2, we mentioned two main approaches for unconstrained gradient-based optimization: line search and trust region. We described the line search in Section 4.3 and the associated methods for computing search directions in Section 4.4. Now we describe trust-region methods, also known as *restricted-step methods*. The main motivation for trust-region methods is to address the issues with Newton's method (see Section 4.4.3) and quasi-Newton updates that do not guarantee a positive definite-Hessian approximation (e.g., SR1, which we briefly described in Section 4.4.4).

The trust-region approach is fundamentally different from the line search approach because it finds the direction and distance of the step simultaneously instead of finding the direction first and then the distance. The trust-region approach builds a model of the function to be minimized and then minimizes the model within a *trust region*, within which we trust the model to be good enough for our purposes.

The most common model is a local quadratic function, but other models may be used. When using a quadratic model based on the function value, gradient, and Hessian at the current iteration, the method is similar to Newton's method.

The trust region is centered about the current iteration point and can be defined as an n-dimensional box, sphere, or ellipsoid of a given size. Each trust-region iteration consists of the following main steps:

1. Create or update the model about the current point.
2. Minimize the model within the trust region.
3. Move to the new point, update values, and adapt the size of the trust region.

These steps are illustrated in Fig. 4.60, and they are repeated until convergence. Figure 4.61 shows the steps to minimize the bean function, where the circles show the trust regions for each iteration.

The trust-region subproblem solved at each iteration is

$$\begin{aligned} \underset{s}{\text{minimize}} \quad & \tilde{f}(s) \\ \text{subject to} \quad & \|s\| \leq \Delta, \end{aligned} \quad (4.97)$$

where $\tilde{f}(s)$ is the local trust-region model, s is the step from the current iteration point, and Δ is the size of the trust region. We use s instead of p to indicate that this is a step vector and not simply the direction vector used in methods based on a line search.

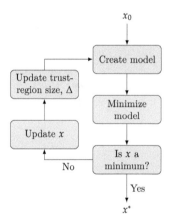

Fig. 4.60 Trust-region methods minimize a model within a trust region for each iteration, and then they update the trust-region size and the model before the next iteration.

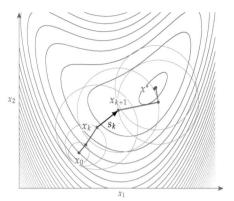

Fig. 4.61 Path for the trust-region approach showing the circular trust regions at each step.

84. Conn et al., *Trust Region Methods*, 2000.

The subproblem (Eq. 4.97) defines the trust region as a norm. The Euclidean norm, $\|s\|_2$, defines a spherical trust region and is the most common type of trust region. Sometimes ∞-norms are used instead because they are easy to apply, but 1-norms are rarely used because they are just as complex as 2-norms but introduce sharp corners that can be problematic.[84] The shape of the trust region is dictated by the norm (see Fig. A.8) and can significantly affect the convergence rate. The ideal trust-region shape depends on the local function space, and some algorithms allow for the trust-region shape to change throughout the optimization.

4.5.1 Quadratic Model with Spherical Trust Region

Using a quadratic trust-region model and the Euclidean norm, we can define the more specific subproblem:

$$\begin{aligned} \underset{s}{\text{minimize}} \quad & \tilde{f}(s) = f_k + \nabla f_k^\mathsf{T} s + \frac{1}{2} s^\mathsf{T} \tilde{H}_k s \\ \text{subject to} \quad & \|s\|_2 \leq \Delta_k, \end{aligned} \quad (4.98)$$

where $\tilde{H}_k$ is the approximate (or true) Hessian at our current iterate. This problem has a quadratic objective and quadratic constraints and is called a *quadratically constrained quadratic program* (QCQP). If the problem is unconstrained and $\tilde{H}$ is positive definite, we can get to the solution using a single step, $s = -\tilde{H}(k)^{-1} \nabla f_k$. However, because of the constraints, there is no analytic solution for the QCQP. Although the problem is still straightforward to solve numerically (because it is a convex problem; see Section 11.4), it requires an iterative solution approach with multiple factorizations.

Similar to the line search, where we only obtain a sufficiently good point instead of finding the exact minimum, in the trust-region

4.5 Trust-Region Methods

subproblem, we seek an approximate solution to the QCQP. Including the trust-region constraint allows us to omit the requirement that $\tilde{H}_k$ be positive definite, which is used in most quasi-Newton methods. We do not detail approximate solution methods to the QCQP, but there are various options.[79,84,85]

Figure 4.62 compares the bean function with a local quadratic model, which is built using information about the point where the arrow originates. The trust-region step seeks the minimum of the local quadratic model within the circular trust region. Unlike line search methods, as the size of the trust region changes, the direction of the step (the solution to Eq. 4.98) might also change, as shown on the right panel of Fig. 4.62.

79. Nocedal and Wright, *Numerical Optimization*, 2006.

84. Conn et al., *Trust Region Methods*, 2000.

85. Steihaug, *The conjugate gradient method and trust regions in large scale optimization*, 1983.

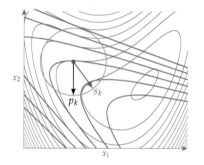
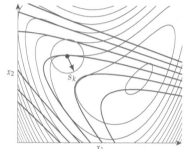

Fig. 4.62 Quadratic model (gray contours) compared to the actual function (blue contours), and two different different trust region sizes (red circles). The trust-region step s_k finds the minimum of the quadratic model while remaining within the trust region. The steepest-descent direction p is shown for comparison.

4.5.2 Trust-Region Sizing Strategy

This section presents an algorithm for updating the size of the trust region at each iteration. The trust region can grow, shrink, or remain the same, depending on how well the model predicts the actual function decrease. The metric we use to assess the model is the actual function decrease divided by the expected decrease:

$$r = \frac{f(x) - f(x+s)}{\tilde{f}(0) - \tilde{f}(s)}. \tag{4.99}$$

The denominator in this definition is the expected decrease, which is always positive. The numerator is the actual change in the function, which could be a reduction or an increase. An r value close to unity means that the model agrees well with the actual function. An r value larger than 1 is fortuitous and means that the actual decrease was even greater than expected. A negative value of r means that the function actually increased at the expected minimum, and therefore the model is not suitable.

The trust-region sizing strategy in Alg. 4.9 determines the size of the trust region at each major iteration k based on the value of r_k. The parameters in this algorithm are not derived from any theory; instead, they are empirical. This example uses the basic procedure from Nocedal and Wright[79] with the parameters recommended by Conn et al.[84]

79. Nocedal and Wright, *Numerical Optimization*, 2006.

84. Conn et al., *Trust Region Methods*, 2000.

Algorithm 4.9 Trust-region algorithm

Inputs:
  x_0: Starting point
  Δ_0: Initial size of the trust region
Outputs:
  x^*: Optimal point

while not converged **do**
  Compute or estimate the Hessian
  Solve (approximately) for s_k — Use Eq. 4.97
  Compute r_k — Use Eq. 4.99
  ▷ Resize trust region
  if $r_k \leq 0.05$ **then** — Poor model
    $\Delta_{k+1} = \Delta_k/4$ — Shrink trust region
    $s_k = 0$ — Reject step
  else if $r_k \geq 0.9$ and $\|s_k\| = \Delta_k$ **then** — Good model and step to edge
    $\Delta_{k+1} = \min(2\Delta_k, \Delta_{\max})$ — Expand trust region
  else — Reasonable model and step within trust region
    $\Delta_{k+1} = \Delta_k$ — Maintain trust region size
  end if
  $x_{k+1} = x_k + s_k$ — Update location of trust region
  $k = k + 1$ — Update iteration count
end while

The initial value of Δ is usually 1, assuming the problem is already well scaled. One way to rationalize the trust-region method is that the quadratic approximation of a nonlinear function is guaranteed to be reasonable only within a limited region around the current point x_k. Thus, we minimize the quadratic function within a region around x_k within which we *trust* the quadratic model.

When our model performs well, we expand the trust region. When it performs poorly, we shrink the trust region. If we shrink the trust region sufficiently, our local model will eventually be a good approximation of the actual function, as dictated by the Taylor series expansion.

We should also set a maximum trust-region size ($\Delta_{\max}$) to prevent the trust region from expanding too much. Otherwise, it may take

4.5 Trust-Region Methods

too long to reduce the trust region to an acceptable size over other portions of the design space where a smaller trust region is needed. The same convergence criteria used in other gradient-based methods are applicable.*

*Conn et al.[84] provide more detail on trust-region problems, including trust-region norms and scaling, approaches to solving the trust-region subproblem, extensions to the model, and other important practical considerations.

Example 4.20 Trust-region method applied to the total potential energy of spring system

Minimizing the total potential energy function from Ex. 4.17 using a trust-region method starting from the same points as before yields the optimization path shown in Fig. 4.63. The initial trust region size is $\Delta = 0.3$, and the maximum allowable is $\Delta_{\max} = 1.5$.

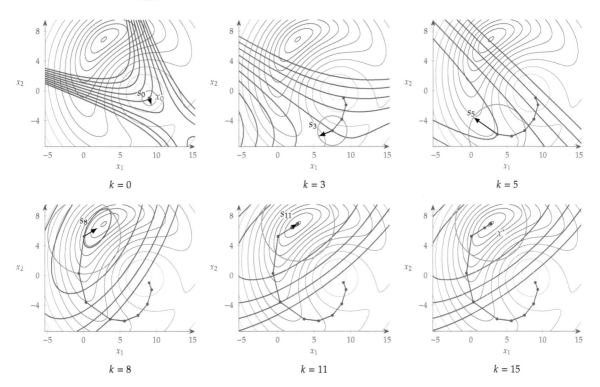

The first few quadratic approximations do not have a minimum because the function has negative curvature around the starting point, but the trust region prevents steps that are too large. When it gets close enough to the bowl containing the minimum, the quadratic approximation has a minimum, and the trust-region subproblem yields a minimum within the trust region. In the last few iterations, the quadratic is a good model, and therefore the region remains large.

Fig. 4.63 Minimizing the total potential for two-spring system using a trust-region method shown at different iterations. The local quadratic approximation is overlaid on the function contours and the trust region is shown as a red circle.

> **Example 4.21** Trust-region method applied to the Rosenbrock function

We now test the trust-region method on the Rosenbrock function. The overall path is similar to the other second-order methods, as shown in Fig. 4.64. The initial trust region size is $\Delta = 1$, and the maximum allowable is $\Delta_{\max} = 5$. At any given point, the direction of maximum curvature of the quadratic approximation matches the maximum curvature across the valley and rotates as we track the bottom of the valley toward the minimum.

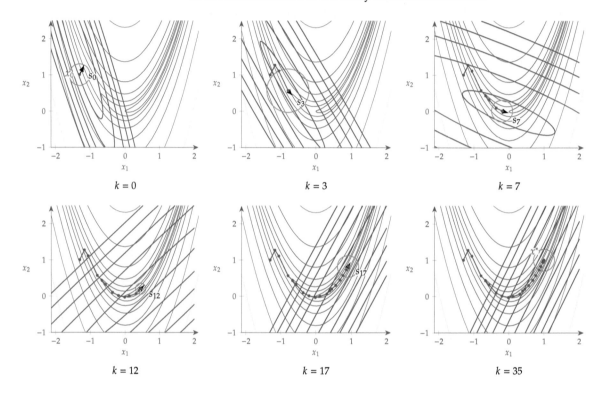

Fig. 4.64 Minimization of the Rosenbrock function using a trust-region method.

4.5.3 Comparison with Line Search Methods

Trust-region methods are typically more strongly dependent on accurate Hessians than are line search methods. For this reason, they are usually only effective when exact gradients (or better yet, an exact Hessian) can be supplied. Many optimization packages require the user to provide the full Hessian, or at least the gradients, to use a trust-region approach. Trust-region methods usually require fewer iterations than quasi-Newton methods with a line search, but each iteration is more computationally expensive because they require at least one matrix factorization.

Scaling can also be more challenging with trust-region approaches. Newton's method is invariant with scaling, but a Euclidean trust-region constraint implicitly assumes that the function changes in each direction at a similar rate. Some enhancements try to address this issue through elliptical trust regions rather than spherical ones.

Tip 4.7 Smooth model discontinuities

Many models are defined in a piecewise manner, resulting in a discontinuous function value, discontinuous derivative, or both. This can happen even if the underlying physical behavior is continuous, such as fitting experimental data using a non-smooth interpolation. The solution is to modify the implementation so that it is continuous while remaining consistent with the physics. If the physics is truly discontinuous, it might still be advisable to artificially smooth the function, as long as there is no significant increase in the modeling error. Even if the smoothed version is highly nonlinear, having a continuous first derivative helps the derivative computation and gradient-based optimization. Some techniques are specific to the problem, but we discuss some examples here.

The absolute value function can often be tolerated as the outermost level of the optimization. However, if propagated through subsequent functions, it can introduce numerical issues from rapid changes in the function. One possibility to smooth this function is to round off the vertex with a quadratic function, as shown in Fig. 4.65. If we force continuity in the function and the first derivative, then the equation of a smooth absolute value is

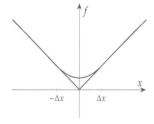

Fig. 4.65 Smoothed absolute value function.

$$f(x) = \begin{cases} |x| & \text{if } |x| > \Delta x \\ \dfrac{x^2}{2\Delta x} + \dfrac{\Delta x}{2} & \text{otherwise,} \end{cases} \quad (4.100)$$

where Δx is a user-adjustable parameter representing the half-width of the transition.

Piecewise functions are often used in fits to empirical data. Cubic splines or a sigmoid function can blend the transition between two functions smoothly. We can also use the same technique to blend discrete steps (where the two functions are constant values) or implement smooth max or min functions.[†] For example, a sigmoid can be used to blend two functions ($f_1(x)$ and $f_2(x)$) together at a transition point x_t using

[†]Another option to smooth the max of multiple functions is aggregation, which is detailed in Section 5.7.

$$f(x) = f_1(x) + (f_2(x) - f_1(x)) \left(\frac{1}{1 + e^{-h(x-x_t)}} \right), \quad (4.101)$$

where h is a user-selected parameter that controls how sharply the transition occurs. The left side of Fig. 4.66 shows an example transitioning x and x^2 with $x_t = 0$ and $h = 50$.

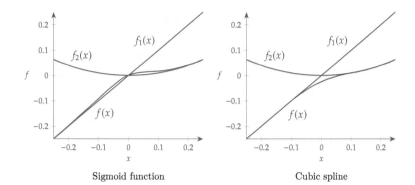

Sigmoid function   Cubic spline

Fig. 4.66 Smoothly blending two functions.

Another approach is to use a cubic spline for the blending. Given a transition point x_t and a half-width Δx, we can compute a cubic spline transition as

$$f(x) = \begin{cases} f_1(x) & \text{if } x < x_1 \\ f_2(x) & \text{if } x > x_2 \\ c_1 x^3 + c_2 x^2 + c_3 x + c_4 & \text{otherwise}, \end{cases} \quad (4.102)$$

where we define $x_1 = x_t - \Delta x$ and $x_2 = x_t + \Delta x$, and the coefficients c are found by solving the following linear system:

$$\begin{bmatrix} x_1^3 & x_1^2 & x_1 & 1 \\ x_2^3 & x_2^2 & x_2 & 1 \\ 3x_1^2 & 2x_1 & 1 & 0 \\ 3x_2^2 & 2x_2 & 1 & 0 \end{bmatrix} \begin{bmatrix} c_1 \\ c_2 \\ c_3 \\ c_4 \end{bmatrix} = \begin{bmatrix} f_1(x_1) \\ f_2(x_2) \\ f_1'(x_1) \\ f_2'(x_2) \end{bmatrix}. \quad (4.103)$$

This ensures continuity in the function and the first derivative. The right side of Fig. 4.66 shows the same two functions and transition location, blended with a cubic spline using a half-width of 0.05.

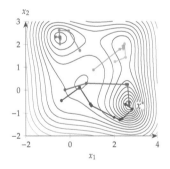

Fig. 4.67 A multistart approach with a gradient-based algorithm finds the global minimum of the Jones function. We successfully apply the same strategy to a discontinuous version of this function in Ex. 7.9.

> **Tip 4.8** Gradient-based optimization can find the global optimum
>
> Gradient-based methods are local search methods. If the design space is fundamentally multimodal, it may be helpful to augment the gradient-based search with a global search. The simplest and most common approach is to use a *multistart* approach, where we run a gradient-based search multiple times, starting from different points, as shown in Fig. 4.67. The starting points might be chosen from engineering intuition, randomly generated points, or sampling methods, such as Latin hypercube sampling (see Section 10.2.1).
>
> Convergence testing is needed to determine a suitable number of starting points. If all points converge to the same optimum and the starting points are well spaced, this suggests that the design space might not be multimodal after all. By using multiple starting points, we increase the likelihood that we find the global optimum, or at least that we find a better optimum than would be

found with a single starting point. One advantage of this approach is that it can easily be run in parallel.

Another approach is to start with a global search strategy (see Chapter 7). After a suitable initial exploration, the design(s) given by the global search become starting points for gradient-based optimization(s). This finds points that satisfy the optimality conditions, which is typically challenging with a pure gradient-free approach. It also improves the convergence rate and finds optima more precisely.

4.6 Summary

Gradient-based optimization is powerful because gradients make it possible to efficiently navigate n-dimensional space in a series of steps converging to an optimum. The gradient also determines when the optimum has been reached, which is when the gradient is zero.

Gradients provide only local information, so an approach that ensures a function decrease when stepping away from the current point is required. There are two approaches to ensure this: line search and trust region. Algorithms based on a line search have two stages: finding an appropriate search direction and determining how far to step in that direction. Trust-region algorithms minimize a surrogate function within a finite region around the current point. The region expands or contracts, depending on how well the optimization within the previous iteration went. Gradient-based optimization algorithms based on a line search are more prevalent than trust-region methods, but trust-region methods can be effective when second derivatives are available.

There are different options for determining the search direction for each line search using gradient information. Although the negative gradient points in the steepest-descent direction, following this direction is not the best approach because it is prone to oscillations. The conjugate gradient method dampens these oscillations and thus converges much faster than steepest descent.

Second-order methods use curvature information, which dramatically improves the rate of convergence. Newton's method converges quadratically but requires the Hessian of the function, which can be prohibitive. Quasi-Newton methods circumvent this requirement by building an approximation of the inverse of the Hessian based on changes in the gradients along the optimization path. Quasi-Newton methods also avoid matrix factorization, requiring matrix-vector multiplication instead. Because they are much less costly while achieving better than linear convergence, quasi-Newton methods are widely used. Limited-memory quasi-Newton methods can be used when the problem is too large to fit in computer memory.

The line search in a given direction does not seek to find a minimum because this is not usually worthwhile. Instead, it seeks to find a "good enough" point that sufficiently decreases the function and the slope. Once such a point is found, we select a new search direction and repeat the process. Second-order methods provide a guess for the first step length in the line search that further improves overall convergence.

This chapter provides the building blocks for the gradient-based constrained optimization covered in the next chapter.

Problems

4.1 Answer *true* or *false* and justify your answer.

  a. Gradient-based optimization requires the function to be continuous and infinitely differentiable.

  b. Gradient-based methods perform a local search.

  c. Gradient-based methods are only effective for problems with one minimum.

  d. The dot product of ∇f with a unit vector p yields the slope of the f along the direction of p.

  e. The Hessian of a unimodal function is positive definite or positive semidefinite everywhere.

  f. Each column j of the Hessian quantifies the rate of change of component j of the gradient vector with respect to all coordinate directions i.

  g. If the function curvature at a point is zero in some direction, that point cannot be a local minimum.

  h. A globalization strategy in a gradient-based algorithm ensures convergence to the global minimum.

  i. The goal of the line search is to find the minimum along a given direction.

  j. For minimization, the line search must always start in a descent direction.

  k. The direction in the steepest-descent algorithm for a given iteration is orthogonal to the direction of the previous iteration.

  l. Newton's method is not affected by problem scaling.

  m. Quasi-Newton methods approximate the function Hessian by using gradients.

  n. Newton's method is a good choice among gradient-based methods because it uses exact second-order information and therefore converges well from any starting point.

  o. The trust-region method does not require a line search.

4.2 Consider the function

$$f(x_1, x_2, x_3) = x_1^2 x_2 + 4x_2^4 - x_2 x_3 + x_3^{-1},$$

and answer the following:

a. Find the gradient of this function. Where is the gradient not defined?

b. Calculate the directional derivative of the function at $x_A = (2, -1, 5)$ in the direction $p = [6, -2, 3]$.

c. Find the Hessian of this function. Is the curvature in the direction p positive or negative?

d. Write the second-order Taylor series expansion of this function. Plot the Taylor series function along the p direction and compare it to the actual function.

4.3 Consider the function from Ex. 4.1,

$$f(x_1, x_2) = x_1^3 + 2x_1 x_2^2 - x_2^3 - 20x_1. \tag{4.104}$$

Find the critical points of this function analytically and classify them. What is the global minimum of this function?

4.4 Review Kepler's wine barrel story from Section 2.2. Approximate the barrel as a cylinder and find the height and diameter of a barrel that maximizes its volume for a diagonal measurement of 1 m.

4.5 Consider the following function:

$$f = x_1^4 + 3x_1^3 + 3x_2^2 - 6x_1 x_2 - 2x_2.$$

Find the critical points analytically and classify them. Where is the global minimum? Plot the function contours to verify your results.

4.6 Consider a slightly modified version of the function from Prob. 4.5, where we add a x_2^4 term to get

$$f = x_1^4 + x_2^4 + 3x_1^3 + 3x_2^2 - 6x_1 x_2 - 2x_2.$$

Can you find the critical points analytically? Plot the function contours. Locate the critical points graphically and classify them.

4.7 Implement the two line search algorithms from Section 4.3, such that they work in n dimensions (x and p can be vectors of any size).

a. As a first test for your code, reproduce the results from the examples in Section 4.3 and plot the function and iterations for both algorithms. For the line search that satisfies the strong Wolfe conditions, reduce the value of μ_2 until you get an exact line search. How much accuracy can you achieve?

b. Test your code on another easy two-dimensional function, such as the bean function from Ex. 4.11, starting from different points and using different directions (but remember that you must always provide a valid descent direction; otherwise, the algorithm might not work!). Does it always find a suitable point? *Exploration*: Try different values of μ_2 and ρ to analyze their effect on the number of iterations.

c. Apply your line search algorithms to the two-dimensional Rosenbrock function and then the n-dimensional variant (see Appendix D.1.2). Again, try different points and search directions to see how robust the algorithm is, and try to tune μ_2 and ρ.

4.8 Consider the one-dimensional function

$$f(x) = -\frac{x}{x^2 + 2}.$$

Solve this problem using your line search implementations from Prob. 4.7. Start from $x_0 = 0$ and with an initial step of $\alpha_0 = -kf'(x_0)$, where $k = 1$.

a. How many function evaluations are required for each of the algorithms? Plot the points where each algorithm terminates on top of the function.

b. Try a different initial step of $k = 20$ from the same starting point. Did your algorithms work as expected? Explain the behaviors.

c. Start from $x_0 = 30$ with $k = 20$ and discuss the results.

4.9 Program the steepest-descent, conjugate gradient, and BFGS algorithms from Section 4.4. You must have a thoroughly tested line search algorithm from the previous exercise first. For the gradients, differentiate the functions analytically and compute them exactly. Solve each problem using your implementations of the various algorithms, as well as off-the-shelf optimization software for comparison.

a. For your first test problem, reproduce the results from the examples in Section 4.4.

b. Minimize the two-dimensional Rosenbrock function (see Appendix D.1.2) using the various algorithms and compare your results starting from $x = (-1, 2)$. Compare the total number of evaluations. Compare the number of minor versus major iterations. Discuss the trends. *Exploration*: Try different starting points and tuning parameters (e.g., ρ and μ_2 in the line search) and compare the number of major and minor iterations.

c. Benchmark your algorithms on the n-dimensional variant of the Rosenbrock function (see Appendix D.1.2). Try $n = 3$ and $n = 4$ first, then $n = 8, 16, 32, \ldots$. What is the highest number of dimensions you can solve? How does the number of function evaluations scale with the number of variables?

d. Optional: Implement L-BFGS and compare it with BFGS.

4.10 Implement a trust-region algorithm and apply it to one or more of the test problems from the previous exercise. Compare the trust-region results with BFGS and the off-the-shelf software.

4.11 Consider the aircraft wing design problem described in Appendix D.1.6. Program the model and solve the problem using an optimizer of your choice. Plot the optimization path and convergence histories. *Exploration:* Change the model to fit an aircraft of your choice by picking the appropriate parameter values and solve the same optimization problem.

4.12 The brachistochrone problem seeks to find the path that minimizes travel time between two points for a particle under the force of gravity.* Solve the discretized version of this problem using an optimizer of your choice (see Appendix D.1.7 for a detailed description).

*This problem was mentioned in Section 2.2 as one of the problems that inspired developments in calculus of variations.

a. Plot the optimal path for the frictionless case with $n = 10$ and compare it to the exact solution (see Appendix D.1.7).

b. Solve the optimal path with friction and plot the resulting path. Report the travel time between the two points and compare it to the frictionless case.

c. Study the effect of increased problem dimensionality. Start with 4 points and double the dimension each time up to 128 points. Plot and discuss the increase in computational expense with problem size. Example metrics include the number of major iterations, function evaluations, and computational time. Hint: When solving the higher-dimensional cases, start with the solution interpolated from a lower-dimensional case—this is called a *warm start*.

Constrained Gradient-Based Optimization 5

Engineering design optimization problems are rarely unconstrained. In this chapter, we explain how to solve constrained problems. The methods in this chapter build on the gradient-based unconstrained methods from Chapter 4 and also assume smooth functions. We first introduce the optimality conditions for a constrained optimization problem and then focus on three main methods for handling constraints: penalty methods, sequential quadratic programming (SQP), and interior-point methods.

Penalty methods are no longer used in constrained gradient-based optimization because they have been replaced by more effective methods. Still, the concept of a penalty is useful when thinking about constraints, partially motivates more sophisticated approaches like interior-point methods, and is often used with gradient-free optimizers.

SQP and interior-point methods represent the state of the art in nonlinear constrained optimization. We introduce the basics for these two optimization methods, but a complete and robust implementation of these methods requires detailed knowledge of a growing body of literature that is not covered here.

> By the end of this chapter you should be able to:
>
> 1. State and understand the optimality conditions for a constrained problem.
>
> 2. Understand the motivation for and the limitations of penalty methods.
>
> 3. Understand the concepts behind state-of-the-art constrained optimization algorithms and use them to solve real engineering problems.

5.1 Constrained Problem Formulation

We can express a general constrained optimization problem as

$$
\begin{aligned}
& \text{minimize} && f(x) \\
& \text{by varying} && x_i && i = 1, \ldots, n_x \\
& \text{subject to} && g_j(x) \le 0 && j = 1, \ldots, n_g \\
& && h_l(x) = 0 && l = 1, \ldots, n_h \\
& && \underline{x}_i \le x_i \le \overline{x}_i && i = 1, \ldots, n_x,
\end{aligned}
\quad (5.1)
$$

where $g(x)$ is the vector of *inequality constraints*, $h(x)$ is the vector of *equality constraints*, and $\underline{x}$ and $\overline{x}$ are lower and upper design variable *bounds* (also known as *bound constraints*). Both objective and constraint functions can be nonlinear, but they should be C^2 continuous to be solved using gradient-based optimization algorithms. The inequality constraints are expressed as "less than" without loss of generality because they can always be converted to "greater than" by putting a negative sign on g. We could also eliminate the equality constraints $h = 0$ without loss of generality by replacing it with two inequality constraints, $h \le \varepsilon$ and $-h \le \varepsilon$, where ε is some small number. In practice, it is desirable to distinguish between equality and inequality constraints because of numerical precision and algorithm implementation.

Example 5.1 Graphical solution of constrained problem

Consider the following two-variable problem with quadratic objective and constraint functions:

$$
\begin{aligned}
& \underset{x_1, x_2}{\text{minimize}} && f(x_1, x_2) = x_1^2 - \tfrac{1}{2} x_1 - x_2 - 2 \\
& \text{subject to} && g_1(x_1, x_2) = x_1^2 - 4x_1 + x_2 + 1 \le 0 \\
& && g_2(x_1, x_2) = \tfrac{1}{2} x_1^2 + x_2^2 - x_1 - 4 \le 0.
\end{aligned}
$$

We can plot the contours of the objective function and the constraint lines ($g_1 = 0$ and $g_2 = 0$), as shown in Fig. 5.1. We can see the feasible region defined by the two constraints. The approximate location of the minimum is evident by inspection. We can visualize the contours for this problem because the functions can be evaluated quickly and because it has only two dimensions. If the functions were more expensive, we would not be able to afford the many evaluations needed to plot the contours. If the problem had more dimensions, it would become difficult or impossible to visualize the functions and feasible space fully.

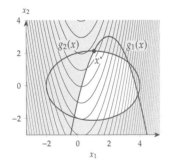

Fig. 5.1 Graphical solution for constrained problem showing contours of the objective, the two constraint curves, and the shaded infeasible region.

5.1 Constrained Problem Formulation

> **Tip 5.1** Do not mistake constraints for objectives
>
> Practitioners sometimes consider metrics to be objectives when it would be more appropriate to pose them as constraints. This can lead to a multiobjective problem, which does not have a single optimum and is costly to solve (more on this in Chapter 9).
>
> A helpful rule of thumb is to ask yourself if improving that metric indefinitely is desirable or whether there is some threshold after which additional improvements do not matter. For example, you might state that you want to maximize the range of an electric car. However, there is probably a threshold beyond which increasing the range does not improve the car's desirability (e.g., if the range is greater than can be driven in one day). In that case, the range should be posed as a constraint, and the objective should be another metric, such as efficiency or profitability.

The constrained problem formulation just described does not distinguish between nonlinear and linear constraints. It is advantageous to make this distinction because some algorithms can take advantage of these differences. However, the methods introduced in this chapter assume general nonlinear functions.

For unconstrained gradient-based optimization (Chapter 4), we only require the gradient of the objective, ∇f. To solve a constrained problem, we also require the gradients of all the constraints. Because the constraints are vectors, their derivatives yield a *Jacobian* matrix. For the equality constraints, the Jacobian is defined as

$$J_h = \frac{\partial h}{\partial x} = \underbrace{\begin{bmatrix} \frac{\partial h_1}{\partial x_1} & \cdots & \frac{\partial h_1}{\partial x_{n_x}} \\ \vdots & \ddots & \vdots \\ \frac{\partial h_{n_h}}{\partial x_1} & \cdots & \frac{\partial h_{n_h}}{\partial x_{n_x}} \end{bmatrix}}_{(n_h \times n_x)} = \begin{bmatrix} \nabla h_1^T \\ \vdots \\ \nabla h_{n_h}^T \end{bmatrix}, \qquad (5.2)$$

which is an $(n_h \times n_x)$ matrix whose rows are the gradients of each constraint. Similarly, the Jacobian of the inequality constraints is an $(n_g \times n_x)$ matrix.

> **Tip 5.2** Do not specify design variable bounds as nonlinear constraints
>
> The design variable bounds in the general nonlinear constrained problem (Eq. 5.1) are expressed as $\underline{x} \leq x \leq \overline{x}$, where $\underline{x}$ is the vector of lower bounds and $\overline{x}$ is the vector of upper bounds. Bounds are treated differently in optimization algorithms, so they should be specified as a bound constraint rather than a

general nonlinear constraint. Some bounds stem from physical limitations on the engineering system. If not otherwise limited, the bounds should be sufficiently wide not to constrain the problem artificially. It is good practice to check your optimal solution against your design variable bounds to ensure that you have not artificially constrained the problem.

5.2 Understanding n-Dimensional Space

Understanding the optimality conditions and optimization algorithms for constrained problems requires basic n-dimensional geometry and linear algebra concepts. Here, we review the concepts in an informal way.* We sketch the concepts for two and three dimensions to provide some geometric intuition but keep in mind that the only way to tackle n dimensions is through mathematics.

There are several essential linear algebra concepts for constrained optimization. The *span* of a set of vectors is the space formed by all the points that can be obtained by a linear combination of those vectors. With one vector, this space is a line, with two linearly independent vectors, this space is a two-dimensional plane (see Fig. 5.2), and so on. With n linearly independent vectors, we can obtain any point in n-dimensional space.

*For a more formal introduction to these concepts, see Chapter 2 in Boyd and Vandenberghe.[86] Strang[87] provides a comprehensive treatment of linear algebra.

86. Boyd and Vandenberghe, *Convex Optimization*, 2004.

87. Strang, *Linear Algebra and its Applications*, 2006.

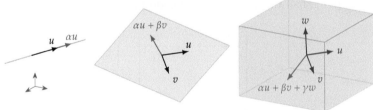

Fig. 5.2 Span of one, two, and three vectors in three-dimensional space.

Because matrices are composed of vectors, we can apply the concept of span to matrices. Suppose we have a rectangular ($m \times n$) matrix A. For our purposes, we are interested in considering the m row vectors in the matrix. The *rank* of A is the number of linearly independent rows of A, and it corresponds to the dimension of the space spanned by the row vectors of A.

The *nullspace* of a matrix A is the set of all n-dimensional vectors p such that $Ap = 0$. This is a subspace of $n - r$ dimensions, where r is the rank of A. One fundamental theorem of linear algebra is that *the nullspace of a matrix contains all the vectors that are perpendicular to the row space of that matrix and vice versa*. This concept is illustrated in Fig. 5.3

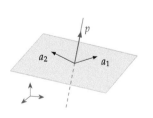

Fig. 5.3 Nullspace of a (2×3) matrix A of rank 2, where a_1 and a_2 are the row vectors of A.

for $n = 3$, where $r = 2$, leaving only one dimension for the nullspace. Any vector v that is perpendicular to p must be a linear combination of the rows of A, so it can be expressed as $v = \alpha a_1 + \beta a_2$.[†]

A *hyperplane* is a generalization of a plane in n-dimensional space and is an essential concept in constrained optimization. In a space of n dimensions, a hyperplane is a subspace with at most $n - 1$ dimensions. In Fig. 5.4, we illustrate hyperplanes in two dimensions (a line) and three dimensions (a two-dimensional plane); higher dimensions cannot be visualized, but the mathematical description that follows holds for any n.

[†]The subspaces spanned by A, A^T, and their respective nullspaces constitute four fundamental subspaces, which we elaborate on in Appendix A.4.

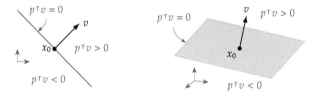

Fig. 5.4 Hyperplanes and half-spaces in two and three dimensions.

To define a hyperplane of $n - 1$ dimensions, we just need a point contained in the hyperplane (x_0) and a vector (v). Then, the hyperplane is defined as the set of all points $x = x_0 + p$ such that $p^\mathsf{T} v = 0$. That is, the hyperplane is defined by all vectors that are perpendicular to v. To define a hyperplane with $n - 2$ dimensions, we would need two vectors, and so on. In n dimensions, a hyperplane of $n - 1$ dimensions divides the space into two *half-spaces*: in one of these, $p^\mathsf{T} v > 0$, and in the other, $p^\mathsf{T} v < 0$. Each half-space is *closed* if it includes the hyperplane ($p^\mathsf{T} v = 0$) and *open* otherwise.

When we have the isosurface of a function f, the function gradient at a point on the isosurface is locally perpendicular to the isosurface. The gradient vector defines the *tangent hyperplane* at that point, which is the set of points such that $p^\mathsf{T} \nabla f = 0$. In two dimensions, the isosurface reduces to a contour and the tangent reduces to a line, as shown in Fig. 5.5 (left). In three dimensions, we have a two-dimensional hyperplane tangent to an isosurface, as shown in Fig. 5.5 (right).

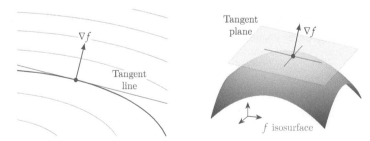

Fig. 5.5 The gradient of a function defines the hyperplane tangent to the function isosurface.

The intersection of multiple half-spaces yields a *polyhedral cone*. A polyhedral cone is the set of all the points that can be obtained by the linear combination of a given set of vectors using nonnegative coefficients. This concept is illustrated in Fig. 5.6 (left) for the two-dimensional case. In this case, only two vectors are required to define a cone uniquely. In three dimensions and higher there could be any number of vectors corresponding to all the possible polyhedral "cross sections", as illustrated in Fig. 5.6 (middle and right).

Fig. 5.6 Polyhedral cones in two and three dimensions.

5.3 Optimality Conditions

The optimality conditions for constrained optimization problems are not as straightforward as those for unconstrained optimization (Section 4.1.4). We begin with equality constraints because the mathematics and intuition are simpler, then add inequality constraints. As in the case of unconstrained optimization, the optimality conditions for constrained problems are used not only for the termination criteria, but they are also used as the basis for optimization algorithms.

5.3.1 Equality Constraints

First, we review the optimality conditions for an unconstrained problem, which we derived in Section 4.1.4. For the unconstrained case, we can take a first-order Taylor series expansion of the objective function with some step p that is small enough that the second-order term is negligible and write

$$f(x + p) \approx f(x) + \nabla f(x)^\mathsf{T} p. \tag{5.3}$$

If x^* is a minimum point, then every point in a small neighborhood must have a greater value,

$$f(x^* + p) \geq f(x^*). \tag{5.4}$$

Given the Taylor series expansion (Eq. 5.3), the only way that this inequality can be satisfied is if

$$\nabla f(x^*)^\mathsf{T} p \geq 0. \tag{5.5}$$

5.3 Optimality Conditions

The condition $\nabla f^\mathsf{T} p = 0$ defines a hyperplane that contains the directions along which the first-order variation of the function is zero. This hyperplane divides the space into an open half-space of directions where the function decreases ($\nabla f^\mathsf{T} p < 0$) and an open half-space where the function increases ($\nabla f^\mathsf{T} p > 0$), as shown in Fig. 5.7. Again, we are considering first-order variations.

If the problem were unconstrained, the only way to satisfy the inequality in Eq. 5.5 would be if $\nabla f(x^*) = 0$. That is because for any nonzero ∇f, there is an open half-space of directions that result in a function decrease (see Fig. 5.7). This is consistent with the first-order unconstrained optimality conditions derived in Section 4.1.4.

However, we now have a constrained problem. The function increase condition (Eq. 5.5) still applies, but p must also be a *feasible* direction. To find the feasible directions, we can write a first-order Taylor series expansion for each equality constraint function as

$$h_j(x + p) \approx h_j(x) + \nabla h_j(x)^\mathsf{T} p, \quad j = 1, \ldots, n_h. \tag{5.6}$$

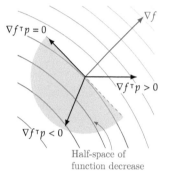

Fig. 5.7 The gradient $f(x)$, which is the direction of steepest function increase, splits the design space into two halves. Here we highlight the open half-space of directions that result in function decrease.

Again, the step size is assumed to be small enough so that the higher-order terms are negligible.

Assuming that x is a feasible point, then $h_j(x) = 0$ for all constraints j, and we are left with the second term in the linearized constraint (Eq. 5.6). To remain feasible a small step away from x, we require that $h_j(x + p) = 0$ for all j. Therefore, first-order feasibility requires that

$$\nabla h_j(x)^\mathsf{T} p = 0, \quad \text{for all} \quad j = 1, \ldots, n_h, \tag{5.7}$$

which means that *a direction is feasible when it is orthogonal to all equality constraint gradients*. We can write this in matrix form as

$$J_h(x) p = 0. \tag{5.8}$$

This equation states that any feasible direction has to lie in the nullspace of the Jacobian of the constraints, J_h.

Assuming that J_h has full row rank (i.e., the constraint gradients are linearly independent), then the feasible space is a subspace of dimension $n_x - n_h$. For optimization to be possible, we require $n_x > n_h$. Figure 5.8 illustrates a case where $n_x = n_h = 2$, where the feasible space reduces to a single point, and there is no freedom for performing optimization.

For one constraint, Eq. 5.8 reduces to a dot product, and the feasible space corresponds to a tangent hyperplane, as illustrated on the left side of Fig. 5.9 for the three-dimensional case. For two or more constraints, the feasible space corresponds to the intersection of all the tangent hyperplanes. On the right side of Fig. 5.9, we show the intersection of two tangent hyperplanes in three-dimensional space (a line).

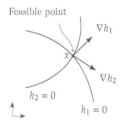

Fig. 5.8 If we have two equality constraints ($n_h = 2$) in two-dimensional space ($n_x = 2$), we are left with no freedom for optimization.

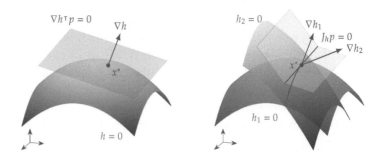

Fig. 5.9 Feasible spaces in three dimensions for one and two constraints.

For constrained optimality, we need to satisfy both $\nabla f(x^*)^\mathsf{T} p \geq 0$ (Eq. 5.5) and $J_h(x)p = 0$ (Eq. 5.8). For equality constraints, if a direction p is feasible, then $-p$ must also be feasible. Therefore, the only way to satisfy $\nabla f(x^*)^\mathsf{T} p \geq 0$ is if $\nabla f(x)^\mathsf{T} p = 0$.

In sum, for x^* to be a constrained optimum, we require

$$\nabla f(x^*)^\mathsf{T} p = 0 \quad \text{for all } p \text{ such that} \quad J_h(x^*)p = 0. \tag{5.9}$$

In other words, *the projection of the objective function gradient onto the feasible space must vanish.* Figure 5.10 illustrates this requirement for a case with two constraints in three dimensions.

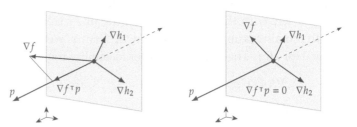

Fig. 5.10 If the projection of ∇f onto the feasible space is nonzero, there is a feasible descent direction (left); if the projection is zero, the point is a constrained optimum (right).

The constrained optimum conditions (Eq. 5.9) require that ∇f be orthogonal to the nullspace of J_h (since p, as defined, is the nullspace of J_h). The row space of a matrix contains all the vectors that are orthogonal to its nullspace.* Because the rows of J_h are the gradients of the constraints, *the objective function gradient must be a linear combination of the gradients of the constraints.* Thus, we can write the requirements defined in Eq. 5.9 as a single vector equation,

*Recall the fundamental theorem of linear algebra illustrated in Fig. 5.3 and the four subspaces reviewed in Appendix A.4.

$$\nabla f(x^*) = -\sum_{j=1}^{n_h} \lambda_j \nabla h_j(x^*), \tag{5.10}$$

where λ_j are called the *Lagrange multipliers*.† There is a multiplier associated with each constraint. The sign of the Lagrange multipliers is arbitrary for equality constraints but will be significant later when dealing with inequality constraints.

†Despite our convention of reserving Greek symbols for scalars, we use λ to represent the n_h-vector of Lagrange multipliers because it is common usage.

5.3 Optimality Conditions

Therefore, the first-order optimality conditions for the equality constrained case are

$$\nabla f(x^*) = -J_h(x)^\mathsf{T} \lambda$$
$$h(x) = 0, \tag{5.11}$$

where we have reexpressed Eq. 5.10 in matrix form and added the constraint satisfaction condition.

In constrained optimization, it is sometimes convenient to use the *Lagrangian function*, which is a scalar function defined as

$$\mathcal{L}(x, \lambda) = f(x) + h(x)^\mathsf{T} \lambda. \tag{5.12}$$

In this function, the Lagrange multipliers are considered to be independent variables. Taking the gradient of $\mathcal{L}$ with respect to both x and λ and setting them to zero yields

$$\nabla_x \mathcal{L} = \nabla f(x) + J_h(x)^\mathsf{T} \lambda = 0$$
$$\nabla_\lambda \mathcal{L} = h(x) = 0, \tag{5.13}$$

which are the first-order conditions derived in Eq. 5.11.

With the Lagrangian function, we have transformed a constrained problem into an unconstrained problem by adding new variables, λ. A constrained problem of n_x design variables and n_h equality constraints was transformed into an unconstrained problem with $n_x + n_h$ variables. Although you might be tempted to simply use the algorithms of Chapter 4 to minimize the Lagrangian function (Eq. 5.12), some modifications are needed in the algorithms to solve these problems effectively (particularly once inequality constraints are introduced).

The derivation of the first-order optimality conditions (Eq. 5.11) assumes that the gradients of the constraints are linearly independent; that is, J_h has full row rank. A point satisfying this condition is called a *regular point* and is said to satisfy *linear independence constraint qualification*. Figure 5.11 illustrates a case where the x^* is not a regular point. A special case that does not satisfy constraint qualification is when one (or more) constraint gradient is zero. In that case, that constraint is not linearly independent, and the point is not regular. Fortunately, these situations are uncommon.

The optimality conditions just described are first-order conditions that are necessary but not sufficient. To make sure that a point is a constrained minimum, we also need to satisfy second-order conditions. For the unconstrained case, the Hessian of the objective function has to be positive definite. In the constrained case, we need to check the Hessian of the Lagrangian with respect to the design variables in the

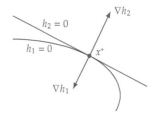

Fig. 5.11 The constraint qualification condition does not hold in this case because the gradients of the two constraints not linearly independent.

space of feasible directions. The Lagrangian Hessian is

$$H_{\mathcal{L}} = H_f + \sum_{j=1}^{n_h} \lambda_j H_{h_j}, \tag{5.14}$$

where H_f is the Hessian of the objective, and H_{h_j} is the Hessian of equality constraint j. The second-order sufficient conditions are as follows:

$$p^\mathsf{T} H_{\mathcal{L}} p > 0 \quad \text{for all } p \text{ such that} \quad J_h p = 0. \tag{5.15}$$

This ensures that the curvature of the Lagrangian is positive when projected onto any feasible direction.

Example 5.2 Equality constrained problem

Consider the following constrained problem featuring a linear objective function and a quadratic equality constraint:

$$\underset{x_1, x_2}{\text{minimize}} \quad f(x_1, x_2) = x_1 + 2x_2$$

$$\text{subject to} \quad h(x_1, x_2) = \frac{1}{4}x_1^2 + x_2^2 - 1 = 0.$$

The Lagrangian for this problem is

$$\mathcal{L}(x_1, x_2, \lambda) = x_1 + 2x_2 + \lambda\left(\frac{1}{4}x_1^2 + x_2^2 - 1\right).$$

Differentiating this to get the first-order optimality conditions,

$$\frac{\partial \mathcal{L}}{\partial x_1} = 1 + \frac{1}{2}\lambda x_1 = 0$$

$$\frac{\partial \mathcal{L}}{\partial x_2} = 2 + 2\lambda x_2 = 0$$

$$\frac{\partial \mathcal{L}}{\partial \lambda} = \frac{1}{4}x_1^2 + x_2^2 - 1 = 0.$$

Solving these three equations for the three unknowns (x_1, x_2, λ), we obtain two possible solutions:

$$x_A = \begin{bmatrix} x_1 \\ x_2 \end{bmatrix} = \begin{bmatrix} -\sqrt{2} \\ -\frac{\sqrt{2}}{2} \end{bmatrix}, \quad \lambda_A = \sqrt{2},$$

$$x_B = \begin{bmatrix} x_1 \\ x_2 \end{bmatrix} = \begin{bmatrix} \sqrt{2} \\ \frac{\sqrt{2}}{2} \end{bmatrix}, \quad \lambda_B = -\sqrt{2}.$$

These two points are shown in Fig. 5.12, together with the objective and constraint gradients. The optimality conditions (Eq. 5.11) state that the gradient must be a linear combination of the gradients of the constraints at the optimum. In the case of one constraint, this means that the two gradients are colinear (which occurs in this example).

5.3 Optimality Conditions

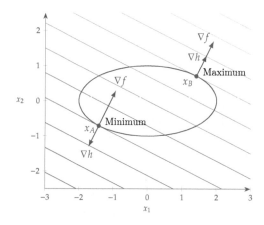

Fig. 5.12 Two points satisfy the first-order optimality conditions; one is a constrained minimum, and the other is a constrained maximum.

To determine if either of these points is a minimum, we check the second-order conditions by evaluating the Hessian of the Lagrangian,

$$H_{\mathcal{L}} = \begin{bmatrix} \tfrac{1}{2}\lambda & 0 \\ 0 & 2\lambda \end{bmatrix}.$$

The Hessian is only positive definite for the case where $\lambda_A = \sqrt{2}$, and therefore x_A is a minimum. Although the Hessian only needs to be positive definite in the feasible directions, in this case, we can show that it is positive or negative definite in all possible directions. The Hessian is negative definite for x_B, so this is not a minimum; instead, it is a maximum.

Figure 5.13 shows the Lagrangian function (with the optimal Lagrange multiplier we solved for) overlaid on top of the original function and constraint. The unconstrained minimum of the Lagrangian corresponds to the constrained minimum of the original function. The Lagrange multiplier can be visualized as a third dimension coming out of the page. Here we show only the slice for the Lagrange multiplier that solves the optimality conditions.

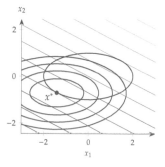

Fig. 5.13 The minimum of the Lagrangian function with the optimum Lagrange multiplier value ($\lambda = \sqrt{2}$) is the constrained minimum of the original problem.

Example 5.3 Second-order conditions for constrained case

Consider the following problem:

$$\underset{x_1, x_2}{\text{minimize}} \quad f(x_1, x_2) = x_1^2 + 3(x_2 - 2)^2$$

$$\text{subject to} \quad h(x_1, x_2) = \beta x_1^2 - x_2 = 0,$$

where β is a parameter that we will vary to change the characteristics of the constraint.

The Lagrangian for this problem is

$$\mathcal{L}(x_1, x_2, \lambda) = x_1^2 + 3(x_2 - 2)^2 + \lambda \left(\beta x_1^2 - x_2\right).$$

Differentiating for the first-order optimality conditions, we get

$$\nabla_x \mathcal{L} = \begin{bmatrix} 2x_1(1 + \lambda\beta) \\ 6(x_2 - 2) - \lambda \end{bmatrix} = 0$$

$$\nabla_\lambda \mathcal{L} = \beta x_1^2 - x_2 = 0.$$

Solving these three equations for the three unknowns (x_1, x_2, λ), the solution is $[x_1, x_2, \lambda] = [0, 0, -12]$, which is independent of β.

To determine if this is a minimum, we must check the second-order conditions by evaluating the Hessian of the Lagrangian,

$$H_\mathcal{L} = \begin{bmatrix} 2(1 - 12\beta) & 0 \\ 0 & 6 \end{bmatrix}.$$

We only need $H_\mathcal{L}$ to be positive definite in the feasible directions. The feasible directions are all p such that $J_h^\mathsf{T} p = 0$. In this case, $J_h = [2\beta x_1, -1]$, yielding $J_h(x^*) = [0, -1]$. Therefore, the feasible directions at the solution can be represented as $p = [\alpha, 0]$, where α is any real number. For positive curvature in the feasible directions, we require that

$$p^\mathsf{T} H_\mathcal{L} p = 2\alpha^2(1 - 12\beta) > 0.$$

Thus, the second-order sufficient condition requires that $\beta < 1/12$.[‡]

We plot the constraint and the Lagrangian for three different values of β in Fig. 5.14. The location of the point satisfying the first-order optimality conditions is the same for all three cases, but the curvature of the constraint changes the Lagrangian significantly.

[‡]This happens to be the same condition for a positive-definite $H_\mathcal{L}$ in this case, but this does not happen in general.

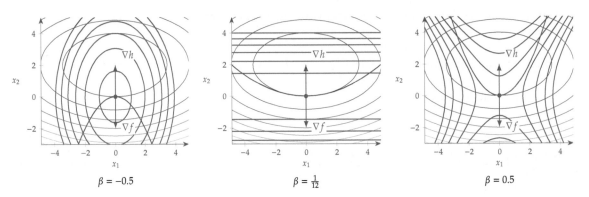

Fig. 5.14 Three different problems illustrating the meaning of the second-order conditions for constrained problems.

For $\beta = -0.5$, the Hessian of the Lagrangian is positive definite, and we have a minimum. For $\beta = 0.5$, the Lagrangian has negative curvature in the feasible directions, so the point is not a minimum; we can reduce the objective by moving along the curved constraint. The first-order conditions alone do not capture this possibility because they linearize the constraint. Finally, in the limiting case ($\beta = 1/12$), the curvature of the constraint matches the curvature of the objective, and the curvature of the Lagrangian is zero in the feasible directions. This point is not a minimum either.

5.3.2 Inequality Constraints

We can reuse some of the concepts from the equality constrained optimality conditions for inequality constrained problems. Recall that an inequality constraint j is feasible when $g_j(x^*) \leq 0$ and it is said to be *active* if $g_j(x^*) = 0$ and *inactive* if $g_j(x^*) < 0$.

As before, if x^* is an optimum, any small enough feasible step p from the optimum must result in a function increase. Based on the Taylor series expansion (Eq. 5.3), we get the condition

$$\nabla f(x^*)^\mathsf{T} p \geq 0, \qquad (5.16)$$

which is the same as for the equality constrained case. We use the arc in Fig. 5.15 to show the descent directions, which are in the open half-space defined by the hyperplane tangent to the gradient of the objective.

To consider inequality constraints, we use the same linearization as the equality constraints (Eq. 5.6), but now we enforce an inequality to get

$$g_j(x+p) \approx g_j(x) + \nabla g_j(x)^\mathsf{T} p \leq 0, \qquad j = 1, \ldots, n_g. \qquad (5.17)$$

For a given candidate point that satisfies all constraints, there are two possibilities to consider for each inequality constraint: whether the constraint is inactive ($g_j(x) < 0$) or active ($g_j(x) = 0$). If a given constraint is inactive, we do not need to add any condition for it because we can take a step p in any direction and remain feasible as long as the step is small enough. Thus, we only need to consider the active constraints for the optimality conditions.

For the equality constraint, we found that all first-order feasible directions are in the nullspace of the Jacobian matrix. Inequality constraints are not as restrictive. From Eq. 5.17, if constraint j is active ($g_j(x) = 0$), then the nearby point $g_j(x+p)$ is only feasible if $\nabla g_j(x)^\mathsf{T} p \leq 0$ for all constraints j that are active. In matrix form, we can write $J_g(x)p \leq 0$, where the Jacobian matrix includes only the gradients of the active constraints. Thus, the feasible directions for inequality constraint j can be any direction in the closed half-space, corresponding to all directions p such that $p^\mathsf{T} g_j \leq 0$, as shown in Fig. 5.16. In this figure, the arc shows the *infeasible* directions.

The set of feasible directions that satisfies all active constraints is the intersection of all the closed half-spaces defined by the inequality constraints, that is, all p such that $J_g(x)p \leq 0$. This intersection of the feasible directions forms a polyhedral cone, as illustrated in Fig. 5.17 for a two-dimensional case with two constraints. To find the cone of

Fig. 5.15 The descent directions are in the open half-space defined by the objective function gradient.

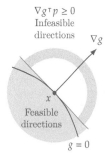

Fig. 5.16 The feasible directions for each constraint are in the closed half-space defined by the inequality constraint gradient.

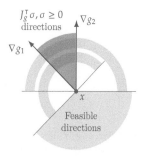

Fig. 5.17 Excluding the infeasible directions with respect to each constraint (red arcs) leaves the cone of feasible directions (blue), which is the polar cone of the active constraint gradients cone (gray).

feasible directions, let us first consider the cone formed by the active inequality constraint gradients (shown in gray in Fig. 5.17). This cone is defined by all vectors d such that

$$d = J_g^T \sigma = \sum_{j=1}^{n_g} \sigma_j \nabla g_j, \quad \text{where} \quad \sigma_j \geq 0. \tag{5.18}$$

A direction p is feasible if $p^T d \leq 0$ for all d in the cone. The set of all feasible directions forms the *polar cone* of the cone defined by Eq. 5.18 and is shown in blue in Fig. 5.17.

Now that we have established some intuition about the feasible directions, we need to establish under which conditions there is no feasible descent direction (i.e., we have reached an optimum). In other words, when is there no intersection between the cone of feasible directions and the open half-space of descent directions? To answer this question, we can use *Farkas' lemma*. This lemma states that given a rectangular matrix (J_g in our case) and a vector with the same size as the rows of the matrix (∇f in our case), one (and only one) of two possibilities occurs:§

§Farkas' lemma has other applications beyond optimization and can be written in various equivalent forms. Using the statement by Dax,[88] we set $A = J_g$, $x = -p$, $c = -\nabla f$, and $y = \sigma$.

88. Dax, *Classroom note: An elementary proof of Farkas' lemma*, 1997.

1. There exists a p such that $J_g p \leq 0$ and $\nabla f^T p < 0$. This means that there is a descent direction that is feasible (Fig. 5.18, left).
2. There exists a σ such that $J_g^T \sigma = -\nabla f$ with $\sigma \geq 0$ (Fig. 5.18, right). This corresponds to optimality because it excludes the first possibility.

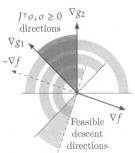
1. Feasible descent direction exists, so point is not an optimum

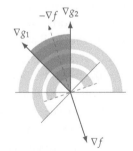
2. No feasible descent direction exists, so point is an optimum

Fig. 5.18 Two possibilities involving active inequality constraints.

The second possibility yields the following optimality criterion for inequality constraints:

$$\nabla f + J_g(x)^T \sigma = 0, \quad \text{with} \quad \sigma \geq 0. \tag{5.19}$$

5.3 Optimality Conditions

Comparing with the corresponding criteria for equality constraints (Eq. 5.13), we see a similar form. However, σ corresponds to the Lagrange multipliers for the inequality constraints and carries the additional restriction that $\sigma \geq 0$.

If equality constraints are present, the conditions for the inequality constraints apply only in the subspace of the directions feasible with respect to the equality constraints.

Similar to the equality constrained case, we can construct a Lagrangian function whose stationary points are candidates for optimal points. We need to include all inequality constraints in the optimality conditions because we do not know in advance which constraints are active. To represent inequality constraints in the Lagrangian, we replace them with the equality constraints defined by

$$g_j + s_j^2 = 0, \quad j = 1, \ldots, n_g, \qquad (5.20)$$

where s_j is a new unknown associated with each inequality constraint called a *slack variable*. The slack variable is squared to ensure it is nonnegative In that way, Eq. 5.20 can only be satisfied when g_j is feasible ($g_j \leq 0$). The significance of the slack variable is that when $s_j = 0$, the corresponding inequality constraint is active ($g_j = 0$), and when $s_j \neq 0$, the corresponding constraint is inactive.

The Lagrangian including both equality and inequality constraints is then

$$\mathcal{L}(x, \lambda, \sigma, s) = f(x) + \lambda^\mathsf{T} h(x) + \sigma^\mathsf{T} \left(g(x) + s \odot s\right), \qquad (5.21)$$

where σ represents the Lagrange multipliers associated with the inequality constraints. Here, we use $\odot$ to represent the element-wise multiplication of s.¶

¶This is a special case of the Hadamard product of two matrices.

Similar to the equality constrained case, we seek a stationary point for the Lagrangian, but now we have additional unknowns: the inequality Lagrange multipliers and the slack variables. Taking partial derivatives of the Lagrangian with respect to each set of unknowns and setting those derivatives to zero yields the first-order optimality conditions:

$$\nabla_x \mathcal{L} = 0 \quad \Rightarrow \quad \frac{\partial \mathcal{L}}{\partial x_i} = \frac{\partial f}{\partial x_i} + \sum_{l=1}^{n_h} \lambda_l \frac{\partial h_l}{\partial x_i} + \sum_{j=1}^{n_g} \sigma_j \frac{\partial g_j}{\partial x_i} = 0$$

$$i = 1, \ldots, n_x. \qquad (5.22)$$

This criterion is the same as before but with additional Lagrange multipliers and constraints. Taking the derivatives with respect to the equality Lagrange multipliers, we have

$$\nabla_\lambda \mathcal{L} = 0 \quad \Rightarrow \quad \frac{\partial \mathcal{L}}{\partial \lambda_l} = h_l = 0, \quad l = 1, \ldots, n_h, \qquad (5.23)$$

which enforces the equality constraints as before. Taking derivatives with respect to the inequality Lagrange multipliers, we get

$$\nabla_\sigma \mathcal{L} = 0 \quad \Rightarrow \quad \frac{\partial \mathcal{L}}{\partial \sigma_j} = g_j + s_j^2 = 0 \quad j = 1, \ldots, n_g, \qquad (5.24)$$

which enforces the inequality constraints. Finally, differentiating the Lagrangian with respect to the slack variables, we obtain

$$\nabla_s \mathcal{L} = 0 \quad \Rightarrow \quad \frac{\partial \mathcal{L}}{\partial s_j} = 2\sigma_j s_j = 0, \quad j = 1, \ldots, n_g, \qquad (5.25)$$

which is called the *complementary slackness condition*. This condition helps us to distinguish the active constraints from the inactive ones. For each inequality constraint, either the Lagrange multiplier is zero (which means that the constraint is inactive), or the slack variable is zero (which means that the constraint is active). Unfortunately, the complementary slackness condition introduces a combinatorial problem. The complexity of this problem grows exponentially with the number of inequality constraints because the number of possible combinations of active versus inactive constraints is 2^{n_g}.

In addition to the conditions for a stationary point of the Lagrangian (Eqs. 5.22 to 5.25), recall that we require the Lagrange multipliers for the active constraints to be nonnegative. Putting all these conditions together in matrix form, the first-order constrained optimality conditions are as follows:

$$\begin{aligned} \nabla f + J_h^T \lambda + J_g^T \sigma &= 0 \\ h &= 0 \\ g + s \odot s &= 0 \\ \sigma \odot s &= 0 \\ \sigma &\geq 0. \end{aligned} \qquad (5.26)$$

These are called the *Karush–Kuhn–Tucker* (KKT) conditions. The equality and inequality constraints are sometimes lumped together using a single Jacobian matrix (and single Lagrange multiplier vector). This can be convenient because the expression for the Lagrangian follows the same form for both cases.

As in the equality constrained case, these first-order conditions are necessary but not sufficient. The second-order sufficient conditions require that the Hessian of the Lagrangian must be positive definite in all feasible directions, that is,

$$\begin{aligned} p^\mathsf{T} H_\mathcal{L} p &> 0 \quad \text{for all } p \text{ such that:} \\ J_h p &= 0 \\ J_g p &\leq 0 \quad \text{for the active constraints.} \end{aligned} \quad (5.27)$$

In other words, we only require positive definiteness in the intersection of the nullspace of the equality constraint Jacobian with the feasibility cone of the active inequality constraints.

Similar to the equality constrained case, the KKT conditions (Eq. 5.26) only apply when a point is regular, that is, when it satisfies linear independence constraint qualification. However, the linear independence applies only to the gradients of the inequality constraints that are active and the equality constraint gradients.

Suppose we have the two constraints shown in the left pane of Fig. 5.19. For the given objective function contours, point x^* is a minimum. At x^*, the gradients of the two constraints are linearly independent, and x^* is thus a regular point. Therefore, we can apply the KKT conditions at this point.

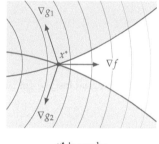
x^* is regular

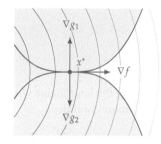
x^* is not regular

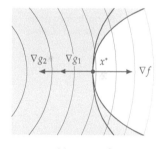
x^* is not regular

The middle and right panes of Fig. 5.19 illustrate cases where x^* is also a constrained minimum. However, x^* is not a regular point in either case because the gradients of the two constraints are not linearly independent. This means that the gradient of the objective cannot be expressed as a unique linear combination of the constraints. Therefore, we cannot use the KKT conditions, even though x^* is a minimum. The problem would be ill-conditioned, and the numerical methods described in this chapter would run into numerical difficulties. Similar to the equality constrained case, this situation is uncommon in practice.

Fig. 5.19 The KKT conditions apply only to regular points. A point x^* is regular when the gradients of the constraints are linearly independent. The middle and right panes illustrate cases where x^* is a constrained minimum but not a regular point.

Example 5.4 Problem with one inequality constraint

Consider a variation of the problem in Ex. 5.2 where the equality is replaced by an inequality, as follows:

$$\underset{x_1,x_2}{\text{minimize}} \quad f(x_1,x_2) = x_1 + 2x_2$$

$$\text{subject to} \quad g(x_1,x_2) = \frac{1}{4}x_1^2 + x_2^2 - 1 \leq 0.$$

The Lagrangian for this problem is

$$\mathcal{L}(x_1,x_2,\sigma,s) = x_1 + 2x_2 + \sigma\left(\frac{1}{4}x_1^2 + x_2^2 - 1 + s^2\right).$$

The objective function and feasible region are shown in Fig. 5.20.

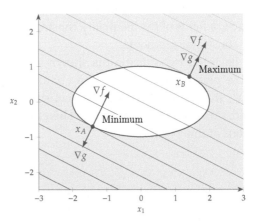

Fig. 5.20 Inequality constrained problem with linear objective and feasible space within an ellipse.

Differentiating the Lagrangian with respect to all the variables, we get the first-order optimality conditions

$$\frac{\partial \mathcal{L}}{\partial x_1} = 1 + \frac{1}{2}\sigma x_1 = 0$$

$$\frac{\partial \mathcal{L}}{\partial x_2} = 2 + 2\sigma x_2 = 0$$

$$\frac{\partial \mathcal{L}}{\partial \sigma} = \frac{1}{4}x_1^2 + x_2^2 - 1 = 0$$

$$\frac{\partial \mathcal{L}}{\partial s} = 2\sigma s = 0.$$

There are two possibilities in the last (complementary slackness) condition: $s = 0$ (meaning the constraint is active) and $\sigma = 0$ (meaning the constraint is not active). However, we can see that setting $\sigma = 0$ in either of the two first equations does not yield a solution. Assuming that $s = 0$ and $\sigma \neq 0$, we can solve the equations to obtain:

$$x_A = \begin{bmatrix} x_1 \\ x_2 \\ \sigma \end{bmatrix} = \begin{bmatrix} -\sqrt{2} \\ -\sqrt{2}/2 \\ \sqrt{2} \end{bmatrix}, \quad x_B = \begin{bmatrix} x_1 \\ x_2 \\ \sigma \end{bmatrix} = \begin{bmatrix} \sqrt{2} \\ \sqrt{2}/2 \\ -\sqrt{2} \end{bmatrix}.$$

5.3 Optimality Conditions

These are the same critical points as in the equality constrained case of Ex. 5.2, as shown in Fig. 5.20. However, now the sign of the Lagrange multiplier is significant.

According to the KKT conditions, the Lagrange multiplier has to be nonnegative. Point x_A satisfies this condition. As a result, there is no feasible descent direction at x_A, as shown in Fig. 5.21 (left). The Hessian of the Lagrangian at this point is the same as in Ex. 5.2, which we have already shown to be positive definite. Therefore, x_A is a minimum.

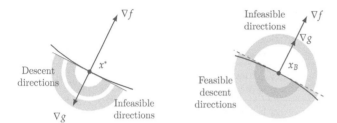

Fig. 5.21 At the minimum (left), the Lagrange multiplier is positive, and there is no feasible descent direction. At the critical point x_B (right), the Lagrange multiplier is negative, and all descent directions are feasible, so this point is not a minimum.

Unlike the equality constrained problem, we do not need to check the Hessian at point x_B because the Lagrange multiplier is negative. As a consequence, there are feasible descent directions, as shown in Fig. 5.21 (right). Therefore, x_B is not a minimum.

Example 5.5 Simple problem with two inequality constraints

Consider a variation of Ex. 5.4 where we add one more inequality constraint, as follows:

$$\underset{x_1, x_2}{\text{minimize}} \quad f(x_1, x_2) = x_1 + 2x_2$$

$$\text{subject to} \quad g_1(x_1, x_2) = \frac{1}{4}x_1^2 + x_2^2 - 1 \le 0$$

$$g_2(x_2) = -x_2 \le 0.$$

The feasible region is the top half of the ellipse, as shown in Fig. 5.22.

The Lagrangian for this problem is

$$\mathcal{L}(x, \sigma, s) = x_1 + 2x_2 + \sigma_1 \left(\frac{1}{4}x_1^2 + x_2^2 - 1 + s_1^2 \right) + \sigma_2 \left(-x_2 + s_2^2 \right).$$

Differentiating the Lagrangian with respect to all the variables, we get the first-order optimality conditions,

$$\frac{\partial \mathcal{L}}{\partial x_1} = 1 + \frac{1}{2}\sigma_1 x_1 = 0$$

$$\frac{\partial \mathcal{L}}{\partial x_2} = 2 + 2\sigma_1 x_2 - \sigma_2 = 0$$

$$\frac{\partial \mathcal{L}}{\partial \sigma_1} = \frac{1}{4}x_1^2 + x_2^2 - 1 + s_1^2 = 0$$

$$\frac{\partial \mathcal{L}}{\partial \sigma_2} = -x_2 + s_2^2 = 0$$

$$\frac{\partial \mathcal{L}}{\partial s_1} = 2\sigma_1 s_1 = 0$$

$$\frac{\partial \mathcal{L}}{\partial s_2} = 2\sigma_2 s_2 = 0.$$

We now have two complementary slackness conditions, which yield the four potential combinations listed in Table 5.1.

Table 5.1 Two inequality constraints yield four potential combinations.

Assumption	Meaning	x_1	x_2	σ_1	σ_2	s_1	s_2	Point
$s_1 = 0$ $s_2 = 0$	g_1 is active g_2 is active	-2 2	0 0	1 -1	2 2	0 0	0 0	x^* x_C
$\sigma_1 = 0$ $\sigma_2 = 0$	g_1 is inactive g_2 is inactive	$-$	$-$	$-$	$-$	$-$	$-$	$-$
$s_1 = 0$ $\sigma_2 = 0$	g_1 is active g_2 is inactive	$\sqrt{2}$	$\frac{\sqrt{2}}{2}$	$-\sqrt{2}$	0	0	$2^{-\frac{1}{4}}$	x_B
$\sigma_1 = 0$ $s_2 = 0$	g_1 is inactive g_2 is active	$-$	$-$	$-$	$-$	$-$	$-$	$-$

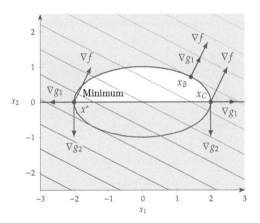

Fig. 5.22 Only one point satisfies the first-order KKT conditions.

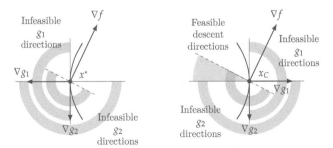

Fig. 5.23 At the minimum (left), the intersection of the feasible directions and descent directions is null, so there is no feasible descent direction. At this point, there is a cone of descent directions that is also feasible, so it is not a minimum.

Assuming that both constraints are active yields two possible solutions (x^* and x_C) corresponding to two different Lagrange multipliers. According to the KKT conditions, the Lagrange multipliers for all active inequality constraints have to be positive, so only the solution with $\sigma_1 = 1$ (x^*) is a candidate for a minimum. This point corresponds to x^* in Fig. 5.22. As shown in Fig. 5.23 (left), there are no feasible descent directions starting from x^*. The Hessian of the Lagrangian at x^* is identical to the previous example and is positive definite when σ_1 is positive. Therefore, x^* is a minimum.

The other solution for which both constraints are active is point x_C in Fig. 5.22. As shown in Fig. 5.23 (right), there is a cone of feasible descent directions, and therefore x_C is not a minimum.

Assuming that neither constraint is active yields 1 = 0 for the first optimality condition, so this situation is not possible. Assuming that g_1 is active yields the solution corresponding to the maximum that we already found in Ex. 5.4, x_B. Finally, assuming that only g_2 is active yields no candidate point.

Although these examples can be solved analytically, they are the exception rather than the rule. The KKT conditions quickly become challenging to solve analytically (try solving Ex. 5.1), and as the number of constraints increases, trying all combinations of active and inactive constraints becomes intractable. Furthermore, engineering problems usually involve functions defined by models with implicit equations, which are impossible to solve analytically. The reason we include these analytic examples is to gain a better understanding of the KKT conditions. For the rest of the chapter, we focus on numerical methods, which are necessary for the vast majority of practical problems.

5.3.3 Meaning of the Lagrange Multipliers

The Lagrange multipliers quantify how much the corresponding constraints drive the design. More specifically, a Lagrange multiplier quantifies the sensitivity of the optimal objective function value $f(x^*)$ to a variation in the value of the corresponding constraint. Here we explain why that is the case. We discuss only inequality constraints, but the same analysis applies to equality constraints.

When a constraint is inactive, the corresponding Lagrange multiplier is zero. This indicates that changing the value of an inactive constraint does not affect the optimum, as expected. This is only valid to the first order because the KKT conditions are based on the linearization of the objective and constraint functions. Because small changes are assumed in the linearization, we do not consider the case where an inactive constraint becomes active after perturbation.

Now let us examine the active constraints. Suppose that we want to quantify the effect of a change in an active (or equality) constraint g_i on the optimal objective function value.∥ The differential of g_i is given by the following dot product:

∥ As an example, we could change the value of the allowable stress constraint in the structural optimization problem of Ex. 3.9.

$$dg_i = \frac{\partial g_i}{\partial x} dx. \tag{5.28}$$

For all the other constraints j that remain unperturbed, which means that

$$\frac{\partial g_j}{\partial x} dx = 0 \quad \text{for all} \quad j \neq i. \tag{5.29}$$

This equation states that any movement dx must be in the nullspace of the remaining constraints to remain feasible with respect to those constraints.** An example with two constraints is illustrated in Fig. 5.24, where g_1 is perturbed and g_2 remains fixed. The objective and constraint functions are linearized because we are considering first-order changes represented by the differentials.

**This condition is similar to Eq. 5.7, but here we apply it to all equality and active constraints except for constraint i.

From the KKT conditions (Eq. 5.22), we know that at the optimum,

$$\frac{\partial f}{\partial x} = -\sigma^\mathsf{T} \frac{\partial g}{\partial x}. \tag{5.30}$$

Using this condition, we can write the differential of the objective, $df = (\partial f/\partial x) dx$, as

$$df = -\sigma^\mathsf{T} \frac{\partial g}{\partial x} dx. \tag{5.31}$$

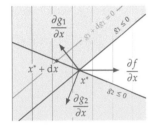

Fig. 5.24 Lagrange multipliers can be interpreted as the change in the optimal objective due a perturbation in the corresponding constraint. In this case, we show the effect of perturbing g_1.

According to Eqs. 5.28 and 5.29, the product with dx is only nonzero for the perturbed constraint i and therefore,

$$df = -\sigma_i \frac{\partial g_i}{\partial x} dx = -\sigma_i \, dg_i. \tag{5.32}$$

This leads to the derivative of the optimal f with respect to a change in the value of constraint i:

$$\sigma_i = -\frac{df}{dg_i}. \tag{5.33}$$

Thus, the Lagrange multipliers can predict how much improvement can be expected if a given constraint is relaxed. For inequality constraints, because the Lagrange multipliers are positive at an optimum, this equation correctly predicts a decrease in the objective function value when the constraint value is increased.

The derivative defined in Eq. 5.33 has practical value because it tells us how much a given constraint drives the design. In this interpretation of the Lagrange multipliers, we need to consider the scaling of the problem and the units. Still, for similar quantities, they quantify the relative importance of the constraints.

5.3.4 Post-Optimality Sensitivities

It is sometimes helpful to find sensitivities of the optimal objective function value with respect to a parameter held fixed during optimization. Suppose that we have found the optimum for a constrained problem. Say we have a scalar parameter ρ held fixed in the optimization, but now want to quantify the effect of a perturbation in that parameter on the optimal objective value. Perturbing ρ changes the objective and the constraint functions, so the optimum point moves, as illustrated in Fig. 5.25. For our current purposes, we use g to represent either active inequality or equality constraints. We assume that the set of active constraints does not change with a perturbation in ρ like we did when perturbing the constraint in Section 5.3.3.

The objective function is affected by ρ through a change in f itself and a change induced by the movement of the constraints. This dependence can be written in the total differential form as:

$$\mathrm{d}f = \frac{\partial f}{\partial \rho} \mathrm{d}\rho + \frac{\partial f}{\partial g}\frac{\partial g}{\partial \rho} \mathrm{d}\rho. \tag{5.34}$$

The derivative $\partial f / \partial g$ corresponds to the derivative of the optimal value of the objective with respect to a perturbation in the constraint, which according to Eq. 5.33, is the negative of the Lagrange multipliers. This means that the post-optimality derivative is

$$\frac{\mathrm{d}f}{\mathrm{d}\rho} = \frac{\partial f}{\partial \rho} - \sigma^\mathsf{T} \frac{\partial g}{\partial \rho}, \tag{5.35}$$

where the partial derivatives with respect to ρ can be computed without re-optimizing.

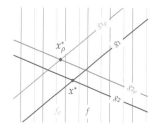

Fig. 5.25 Post-optimality sensitivities quantify the change in the optimal objective due to a perturbation of a parameter that was originally fixed in the optimization. The optimal objective value changes due to changes in the optimum point (which moves to x_ρ^*) and objective function (which becomes f_ρ.)

5.4 Penalty Methods

The concept behind penalty methods is intuitive: to transform a constrained problem into an unconstrained one by adding a penalty to the objective function when constraints are violated or close to being violated. As mentioned in the introduction to this chapter, penalty methods are no longer used directly in gradient-based optimization algorithms because they have difficulty converging to the true solution. However, these methods are still valuable because (1) they are simple and thus ease the transition into understanding constrained optimization; (2) they are useful in some constrained gradient-free methods (Chapter 7); (3) they can be used as merit functions in line search algorithms, as discussed in Section 5.5.3; (4) penalty concepts are used in interior points methods, as discussed in Section 5.6. The penalized function can be written as

$$\hat{f}(x) = f(x) + \mu \pi(x), \quad (5.36)$$

where $\pi(x)$ is a penalty function, and the scalar μ is a penalty parameter. This is similar in form to the Lagrangian, but one difference is that μ is fixed instead of being a variable.

We can use the unconstrained optimization techniques to minimize $\hat{f}(x)$. However, instead of just solving a single optimization problem, penalty methods usually solve a sequence of problems with different values of μ to get closer to the actual constrained minimum. We will see shortly why we need to solve a sequence of problems rather than just one problem.

Various forms for $\pi(x)$ can be used, leading to different penalty methods. There are two main types of penalty functions: *exterior* penalties, which impose a penalty only when constraints are violated, and *interior* penalty functions, which impose a penalty that increases as a constraint is approached.

Figure 5.26 shows both interior and exterior penalties for a two-dimensional function. The exterior penalty leads to slightly infeasible solutions, whereas an interior penalty leads to a feasible solution but underpredicts the objective.

5.4.1 Exterior Penalty Methods

Of the many possible exterior penalty methods, we focus on two of the most popular ones: quadratic penalties and the augmented Lagrangian method. Quadratic penalties are continuously differentiable and straightforward to implement, but they suffer from numerical ill-conditioning. The augmented Lagrangian method is more sophisticated; it is based on the quadratic penalty but adds terms that improve the numerical properties. Many other penalties are possible, such as 1-norms, which are often used when continuous differentiability is unnecessary.

Quadratic Penalty Method

For equality constrained problems, the quadratic penalty method takes the form

$$\hat{f}(x; \mu) = f(x) + \frac{\mu}{2} \sum_i h_i(x)^2, \quad (5.37)$$

where the semicolon denotes that μ is a fixed parameter. The motivation for a quadratic penalty is that it is simple and results in a function that

5.4 Penalty Methods

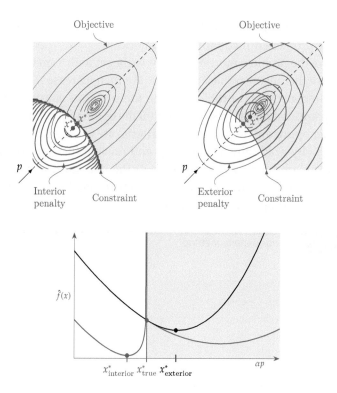

Fig. 5.26 Interior penalties tend to infinity as the constraint is approached from the feasible side of the constraint (left), whereas exterior penalty functions activate when the points are not feasible (right). The minimum for both approaches is different from the true constrained minimum.

is continuously differentiable. The factor of one half is unnecessary but is included by convention because it eliminates the extra factor of two when taking derivatives. The penalty is nonzero unless the constraints are satisfied ($h_i = 0$), as desired.

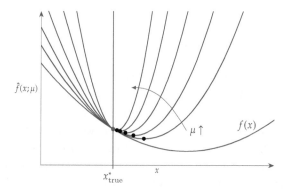

Fig. 5.27 Quadratic penalty for an equality constrained problem. The minimum of the penalized function (black dots) approaches the true constrained minimum (blue circle) as the penalty parameter μ increases.

The value of the penalty parameter μ must be chosen carefully. Mathematically, we recover the exact solution to the constrained problem only as μ tends to infinity (see Fig. 5.27). However, starting with a large value for μ is not practical. This is because the larger the value of μ, the larger the Hessian condition number, which corresponds to the curvature varying greatly with direction (see Ex. 4.10). This behavior makes the problem difficult to solve numerically.

To solve the problem more effectively, we begin with a small value of μ and solve the unconstrained problem. We then increase μ and solve the new unconstrained problem, using the previous solution as the starting point. We repeat this process until the optimality conditions (or some other approximate convergence criteria) are satisfied, as outlined in Alg. 5.1. By gradually increasing μ and reusing the solution from the previous problem, we avoid some of the ill-conditioning issues. Thus, the original constrained problem is transformed into a sequence of unconstrained optimization problems.

Algorithm 5.1 Exterior penalty method

Inputs:
  x_0: Starting point
  $\mu_0 > 0$: Initial penalty parameter
  $\rho > 1$: Penalty increase factor ($\rho \sim 1.2$ is conservative, $\rho \sim 10$ is aggressive)

Outputs:
  x^*: Optimal point
  $f(x^*)$: Corresponding function value

$k = 0$
while not converged **do**
  $x_k^* \leftarrow \underset{x_k}{\text{minimize}}\ \hat{f}(x_k; \mu_k)$
  $\mu_{k+1} = \rho \mu_k$ Increase penalty
  $x_{k+1} = x_k^*$ Update starting point for next optimization
  $k = k + 1$
end while

There are three potential issues with the approach outlined in Alg. 5.1. Suppose the starting value for μ is too low. In that case, the penalty might not be enough to overcome a function that is unbounded from below, and the penalized function has no minimum.

The second issue is that we cannot practically approach $\mu \to \infty$. Hence, the solution to the problem is always slightly infeasible. By comparing the optimality condition of the constrained problem,

$$\nabla_x \mathcal{L} = \nabla f + J_h^T \lambda = 0, \tag{5.38}$$

and the optimality conditional of the penalized function,

$$\nabla_x \hat{f} = \nabla f + \mu J_h^T h = 0, \tag{5.39}$$

5.4 Penalty Methods

we see that for each constraint j,

$$h_j \approx \frac{\lambda_j^*}{\mu}. \tag{5.40}$$

Because $h_j = 0$ at the optimum, μ must be large to satisfy the constraints.

The third issue has to do with the curvature of the penalized function, which is directly proportional to μ. The extra curvature is added in a direction perpendicular to the constraints, making the Hessian of the penalized function increasingly ill-conditioned as μ increases. Thus, the need to increase μ to improve accuracy directly leads to a function space that is increasingly challenging to solve.

Example 5.6 Quadratic penalty for equality constrained problem

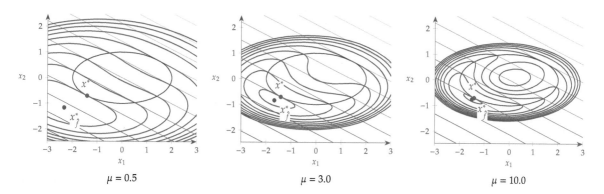

Consider the equality constrained problem from Ex. 5.2. The penalized function for that case is

$$\hat{f}(x;\mu) = x_1 + 2x_2 + \frac{\mu}{2}\left(\frac{1}{4}x_1^2 + x_2^2 - 1\right)^2. \tag{5.41}$$

Figure 5.28 shows this function for different values of the penalty parameter μ. The penalty is active for all points that are infeasible, but the minimum of the penalized function does not coincide with the constrained minimum of the original problem. The penalty parameter needs to be increased for the minimum of the penalized function to approach the correct solution, but this results in a poorly conditioned function.

To show the impact of increasing μ, we solve a sequence of problems starting with a small value of μ and reusing the optimal point for one solution as the starting point for the next. Figure 5.29 shows that large penalty values are required for high accuracy. In this example, even using a penalty parameter of $\mu = 1,000$ (which results in extremely skewed contours), the objective value achieves only three digits of accuracy.

Fig. 5.28 The quadratic penalized function minimum approaches the constrained minimum as the penalty parameter increases.

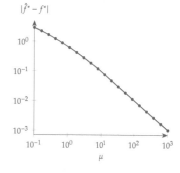

Fig. 5.29 Error in optimal solution for increasing penalty parameter.

The approach discussed so far handles only equality constraints, but we can extend it to handle inequality constraints. Instead of adding a penalty to both sides of the constraints, we add the penalty when the inequality constraint is violated (i.e., when $g_j(x) > 0$). This behavior can be achieved by defining a new penalty function as

$$\hat{f}(x; \mu) = f(x) + \frac{\mu}{2} \sum_{j=1}^{n_g} \max\left(0, g_j(x)\right)^2 . \tag{5.42}$$

The only difference relative to the equality constraint penalty shown in Fig. 5.27 is that the penalty is removed on the feasible side of the inequality constraint, as shown in Fig. 5.30.

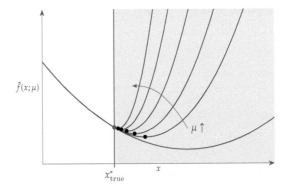

Fig. 5.30 Quadratic penalty for an inequality constrained problem. The minimum of the penalized function approaches the constrained minimum from the infeasible side.

The inequality quadratic penalty can be used together with the quadratic penalty for equality constraints if we need to handle both types of constraints:

$$\hat{f}(x; \mu) = f(x) + \frac{\mu_h}{2} \sum_{l=1}^{n_h} h_l(x)^2 + \frac{\mu_g}{2} \sum_{j=1}^{n_g} \max\left(0, g_j(x)\right)^2 . \tag{5.43}$$

The two penalty parameters can be incremented in lockstep or independently.

Example 5.7 Quadratic penalty for inequality constrained problem

Consider the inequality constrained problem from Ex. 5.4. The penalized function for that case is

$$\hat{f}(x; \mu) = x_1 + 2x_2 + \frac{\mu}{2} \max\left(0, \frac{1}{4}x_1^2 + x_2^2 - 1\right)^2 .$$

This function is shown in Fig. 5.31 for different values of the penalty parameter μ. The contours of the feasible region inside the ellipse coincide with the

5.4 Penalty Methods

original function contours. However, outside the feasible region, the contours change to create a function whose minimum approaches the true constrained minimum as the penalty parameter increases.

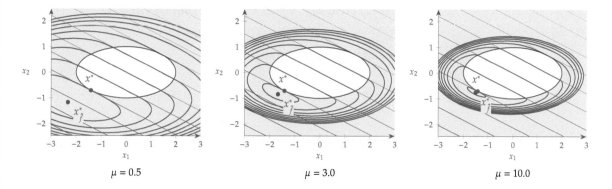

$\mu = 0.5$ $\mu = 3.0$ $\mu = 10.0$

Fig. 5.31 The quadratic penalized function minimum approaches the constrained minimum from the infeasible side.

Tip 5.3 Scaling is also important for constrained problems

The considerations on scaling discussed in Tip 4.4 are just as crucial for constrained problems. Similar to scaling the objective function, a good scaling rule of thumb is to normalize each constraint function such they are of order 1. For constraints, a natural scale is typically already defined by the limits we provide. For example, instead of

$$g_j(x) - g_{\max j} \leq 0, \tag{5.44}$$

we can reexpress a scaled version as

$$\frac{g_j(x)}{g_{\max j}} - 1 \leq 0. \tag{5.45}$$

Augmented Lagrangian

As explained previously, the quadratic penalty method requires a large value of μ for constraint satisfaction, but the large μ degrades the numerical conditioning. The augmented Lagrangian method helps alleviate this dilemma by adding the quadratic penalty to the Lagrangian instead of just adding it to the function. The augmented Lagrangian function for equality constraints is

$$\hat{f}(x; \lambda, \mu) = f(x) + \sum_{j=1}^{n_h} \lambda_j h_j(x) + \frac{\mu}{2} \sum_{j=1}^{n_h} h_j(x)^2. \tag{5.46}$$

To estimate the Lagrange multipliers, we can compare the optimality conditions for the augmented Lagrangian,

$$\nabla_x \hat{f}(x; \lambda, \mu) = \nabla f(x) + \sum_{j=1}^{n_h} \left(\lambda_j + \mu h_j(x) \right) \nabla h_j = 0, \quad (5.47)$$

to those of the actual Lagrangian,

$$\nabla_x \mathcal{L}(x^*, \lambda^*) = \nabla f(x^*) + \sum_{j=1}^{n_h} \lambda_j^* \nabla h_j(x^*) = 0. \quad (5.48)$$

Comparing these two conditions suggests the approximation

$$\lambda_j^* \approx \lambda_j + \mu h_j. \quad (5.49)$$

Therefore, we update the vector of Lagrange multipliers based on the current estimate of the Lagrange multipliers and constraint values using

$$\lambda_{k+1} = \lambda_k + \mu_k h(x_k). \quad (5.50)$$

The complete algorithm is shown in Alg. 5.2.

This approach is an improvement on the plain quadratic penalty because updating the Lagrange multiplier estimates at each iteration allows for more accurate solutions without increasing μ as much. The augmented Lagrangian approximation for each constraint obtained from Eq. 5.49 is

$$h_j \approx \frac{1}{\mu}(\lambda_j^* - \lambda_j). \quad (5.51)$$

The corresponding approximation in the quadratic penalty method is

$$h_j \approx \frac{\lambda_j^*}{\mu}. \quad (5.52)$$

The quadratic penalty relies solely on increasing μ in the denominator to drive the constraints to zero. However, the augmented Lagrangian also controls the numerator through the Lagrange multiplier estimate. If the estimate is reasonably close to the true Lagrange multiplier, then the numerator becomes small for modest values of μ. Thus, the augmented Lagrangian can provide a good solution for x^* while avoiding the ill-conditioning issues of the quadratic penalty.

5.4 Penalty Methods

Algorithm 5.2 Augmented Lagrangian penalty method

Inputs:
- x_0: Starting point
- $\lambda_0 = 0$: Initial Lagrange multiplier
- $\mu_0 > 0$: Initial penalty parameter
- $\rho > 1$: Penalty increase factor

Outputs:
- x^*: Optimal point
- $f(x^*)$: Corresponding function value

$k = 0$
while not converged **do**
$\quad x_k^* \leftarrow \underset{x_k}{\text{minimize}}\, \hat{f}(x_k; \lambda_k, \mu_k)$
$\quad \lambda_{k+1} = \lambda_k + \mu_k h(x_k)$ Update Lagrange multipliers
$\quad \mu_{k+1} = \rho \mu_k$ Increase penalty parameter
$\quad x_{k+1} = x_k^*$ Update starting point for next optimization
$\quad k = k + 1$
end while

So far we have only discussed equality constraints where the definition for the augmented Lagrangian is universal. Example 5.8 included an inequality constraint by assuming it was active and treating it like an equality, but this is not an approach that can be used in general. Several formulations exist for handling inequality constraints using the augmented Lagrangian approach.[89–91] One well-known approach is given by:[92]

$$\hat{f}(x; \mu) = f(x) + \lambda^\mathsf{T} \bar{g}(x) + \frac{1}{2}\mu \left\| \bar{g}(x) \right\|_2^2 . \tag{5.53}$$

where

$$\bar{g}_j(x) \equiv \begin{cases} h_j(x) & \text{for equality constraints} \\ g_j(x) & \text{if } g_j \geq -\lambda_j/\mu \\ -\lambda_j/\mu & \text{otherwise} \end{cases} . \tag{5.54}$$

89. Gill et al., *Some theoretical properties of an augmented Lagrangian merit function*, 1986.

90. Di Pillo and Grippo, *A new augmented Lagrangian function for inequality constraints in nonlinear programming problems*, 1982.

91. Birgin et al., *Numerical comparison of augmented Lagrangian algorithms for nonconvex problems*, 2005.

92. Rockafellar, *The multiplier method of Hestenes and Powell applied to convex programming*, 1973.

Example 5.8 Augmented Lagrangian for inequality constrained problem

Consider the inequality constrained problem from Ex. 5.4. Assuming the inequality constraint is active, the augmented Lagrangian (Eq. 5.46) is

$$\hat{f}(x; \mu) = x_1 + 2x_2 + \lambda \left(\frac{1}{4}x_1^2 + x_2^2 - 1 \right) + \frac{\mu}{2} \left(\frac{1}{4}x_1^2 + x_2^2 - 1 \right)^2 .$$

Applying Alg. 5.2, starting with $\mu = 0.5$ and using $\rho = 1.1$, we get the iterations shown in Fig. 5.32.

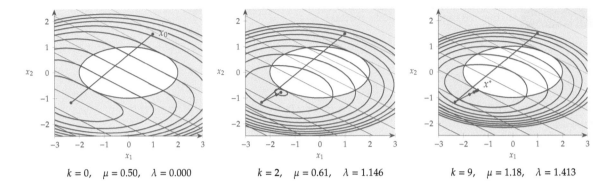

$k = 0$, $\mu = 0.50$, $\lambda = 0.000$ $k = 2$, $\mu = 0.61$, $\lambda = 1.146$ $k = 9$, $\mu = 1.18$, $\lambda = 1.413$

Fig. 5.32 Augmented Lagrangian applied to inequality constrained problem.

Compared with the quadratic penalty in Ex. 5.7, the penalized function is much better conditioned, thanks to the term associated with the Lagrange multiplier. The minimum of the penalized function eventually becomes the minimum of the constrained problem without a large penalty parameter.

As done in Ex. 5.6, we solve a sequence of problems starting with a small value of μ and reusing the optimal point for one solution as the starting point for the next. In this case, we update the Lagrange multiplier estimate between optimizations as well. Figure 5.33 shows that only modest penalty parameters are needed to achieve tight convergence to the true solution, a significant improvement over the regular quadratic penalty.

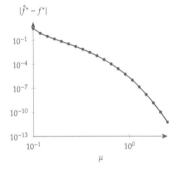

Fig. 5.33 Error in optimal solution as compared with true solution as a function of an increasing penalty parameter.

5.4.2 Interior Penalty Methods

Interior penalty methods work the same way as exterior penalty methods—they transform the constrained problem into a series of unconstrained problems. The main difference with interior penalty methods is that they always seek to maintain feasibility. Instead of adding a penalty only when constraints are violated, they add a penalty as the constraint is approached from the feasible region. This type of penalty is particularly desirable if the objective function is ill-defined outside the feasible region. These methods are called *interior* because the iteration points remain on the interior of the feasible region. They are also referred to as *barrier methods* because the penalty function acts as a barrier preventing iterates from leaving the feasible region.

One possible interior penalty function to enforce $g(x) \leq 0$ is the *inverse barrier*,

$$\pi(x) = \sum_{j=1}^{n_g} -\frac{1}{g_j(x)}, \quad (5.55)$$

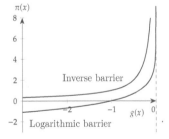

Fig. 5.34 Two different interior penalty functions: inverse barrier and logarithmic barrier.

where $\pi(x) \to \infty$ as $g_j(x) \to 0^-$ (where the superscript "−" indicates a left-sided derivative). A more popular interior penalty function is the

5.4 Penalty Methods

logarithmic barrier,

$$\pi(x) = \sum_{j=1}^{n_g} -\ln\left(-g_j(x)\right), \qquad (5.56)$$

which also approaches infinity as the constraint tends to zero from the feasible side. The penalty function is then

$$\hat{f}(x;\mu) = f(x) - \mu \sum_{j=1}^{n_g} \ln(-g_j(x)). \qquad (5.57)$$

These two penalty functions as illustrated in Fig. 5.34.

Neither of these penalty functions applies when $g > 0$ because they are designed to be evaluated only within the feasible space. Algorithms based on these penalties must be prevented from evaluating infeasible points.

Like exterior penalty methods, interior penalty methods must also solve a sequence of unconstrained problems but with $\mu \to 0$ (see Alg. 5.3). As the penalty parameter decreases, the region across which the penalty acts decreases, as shown in Fig. 5.35.

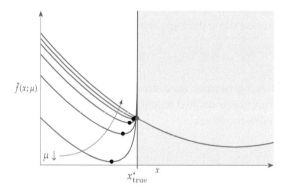

Fig. 5.35 Logarithmic barrier penalty for an inequality constrained problem. The minimum of the penalized function (black circles) approaches the true constrained minimum (blue circle) as the penalty parameter μ decreases.

The methodology is the same as is described in Alg. 5.1 but with a decreasing penalty parameter. One major weakness of the method is that the penalty function is not defined for infeasible points, so a feasible starting point must be provided. For some problems, providing a feasible starting point may be difficult or practically impossible.

The optimization must be safeguarded to prevent the algorithm from becoming infeasible when starting from a feasible point. This can be achieved by checking the constraints values during the line search and backtracking if any of them is greater than or equal to zero. Multiple backtracking iterations might be required.

Algorithm 5.3 Interior penalty method

Inputs:
x_0: Starting point
$\mu_0 > 0$: Initial penalty parameter
$\rho < 1$: Penalty decrease factor

Outputs:
x^*: Optimal point
$f(x^*)$: Corresponding function value

$k = 0$
while not converged **do**
  $x_k^* \leftarrow \underset{x_k}{\text{minimize}}\, \hat{f}(x_k; \mu_k)$
  $\mu_{k+1} = \rho \mu_k$ Decrease penalty parameter
  $x_{k+1} = x_k^*$ Update starting point for next optimization
  $k = k + 1$
end while

Example 5.9 Logarithmic penalty for inequality constrained problem

Consider the equality constrained problem from Ex. 5.4. The penalized function for that case using the logarithmic penalty (Eq. 5.57) is

$$\hat{f}(x; \mu) = x_1 + 2x_2 - \mu \ln\left(-\frac{1}{4}x_1^2 - x_2^2 + 1\right).$$

Figure 5.36 shows this function for different values of the penalty parameter μ. The penalized function is defined only in the feasible space, so we do not plot its contours outside the ellipse.

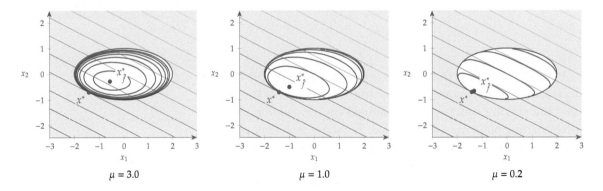

$\mu = 3.0$ $\mu = 1.0$ $\mu = 0.2$

Fig. 5.36 Logarithmic penalty for one inequality constraint. The minimum of the penalized function approaches the constrained minimum from the feasible side.

Like exterior penalty methods, the Hessian for interior penalty methods becomes increasingly ill-conditioned as the penalty parameter

5.5 Sequential Quadratic Programming

tends to zero.[93] There are augmented and modified barrier approaches that can avoid the ill-conditioning issue (and other methods that remain ill-conditioned but can still be solved reliably, albeit inefficiently).[94] However, these methods have been superseded by the modern interior-point methods discussed in Section 5.6, so we do not elaborate on further improvements to classical penalty methods.

[93]. Murray, *Analytical expressions for the eigenvalues and eigenvectors of the Hessian matrices of barrier and penalty functions*, 1971.

[94]. Forsgren et al., *Interior methods for nonlinear optimization*, 2002.

5.5 Sequential Quadratic Programming

SQP is the first of the modern constrained optimization methods we discuss. SQP is not a single algorithm; instead, it is a conceptual method from which various specific algorithms are derived. We present the basic method but mention only a few of the many details needed for robust practical implementations. We begin with equality constrained SQP and then add inequality constraints.

5.5.1 Equality Constrained SQP

To derive the SQP method, we start with the KKT conditions for this problem and treat them as equation residuals that need to be solved. Recall that the Lagrangian (Eq. 5.12) is

$$\mathcal{L}(x, \lambda) = f(x) + h(x)^\mathsf{T} \lambda. \tag{5.58}$$

Differentiating this function with respect to the design variables and Lagrange multipliers and setting the derivatives to zero, we get the KKT conditions,

$$r = \begin{bmatrix} \nabla_x \mathcal{L}(x, \lambda) \\ \nabla_\lambda \mathcal{L}(x, \lambda) \end{bmatrix} = \begin{bmatrix} \nabla f(x) + J_h^\mathsf{T} \lambda \\ h(x) \end{bmatrix} = 0. \tag{5.59}$$

Recall that to solve a system of equations $r(u) = 0$ using Newton's method, we solve a sequence of linear systems,

$$J_r(u_k) p_u = -r(u_k), \tag{5.60}$$

where J_r is the Jacobian of derivatives $\partial r / \partial u$. The step in the variables is $p_u = u_{k+1} - u_k$, where the variables are

$$u \equiv \begin{bmatrix} x \\ \lambda \end{bmatrix}. \tag{5.61}$$

Differentiating the vector of residuals (Eq. 5.59) with respect to the two concatenated vectors in u yields the following block linear system:

$$\begin{bmatrix} H_\mathcal{L} & J_h^\mathsf{T} \\ J_h & 0 \end{bmatrix} \begin{bmatrix} p_x \\ p_\lambda \end{bmatrix} = \begin{bmatrix} -\nabla_x \mathcal{L} \\ -h \end{bmatrix}. \tag{5.62}$$

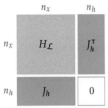

Fig. 5.37 Structure and block shapes for the matrix in the SQP system (Eq. 5.62)

This is a linear system of $n_x + n_h$ equations where the Jacobian matrix is square. The shape of the matrix and its blocks are as shown in Fig. 5.37. We solve a sequence of these problems to converge to the optimal design variables and the corresponding optimal Lagrange multipliers. At each iteration, we update the design variables and Lagrange multipliers as follows:

$$x_{k+1} = x_k + \alpha_k p_x \tag{5.63}$$

$$\lambda_{k+1} = \lambda_k + p_\lambda. \tag{5.64}$$

The inclusion of α_k suggests that we do not automatically accept the Newton step (which corresponds to $\alpha = 1$) but instead perform a line search as previously described in Section 4.3. The function used in the line search needs some modification, as discussed later in this section.

SQP can be derived in an alternative way that leads to different insights. This alternate approach requires an understanding of quadratic programming (QP), which is discussed in more detail in Section 11.3 but briefly described here. A QP problem is an optimization problem with a quadratic objective and linear constraints. In a general form, we can express any equality constrained QP as

$$\begin{aligned} \underset{x}{\text{minimize}} \quad & \tfrac{1}{2} x^\mathsf{T} Q x + q^\mathsf{T} x \\ \text{subject to} \quad & Ax + b = 0. \end{aligned} \tag{5.65}$$

A two-dimensional example with one constraint is illustrated in Fig. 5.38. The constraint is a matrix equation that represents multiple linear equality constraints—one for every row in A. We can solve this optimization problem analytically from the optimality conditions. First, we form the Lagrangian:

$$\mathcal{L}(x, \lambda) = \tfrac{1}{2} x^\mathsf{T} Q x + q^\mathsf{T} x + \lambda^\mathsf{T} (Ax + b). \tag{5.66}$$

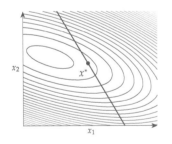

Fig. 5.38 Quadratic problem in two dimensions.

We now take the partial derivatives and set them equal to zero:

$$\begin{aligned} \nabla_x \mathcal{L} &= Qx + q + A^\mathsf{T} \lambda = 0 \\ \nabla_\lambda \mathcal{L} &= Ax + b = 0. \end{aligned} \tag{5.67}$$

We can express those same equations in a block matrix form:

$$\begin{bmatrix} Q & A^\mathsf{T} \\ A & 0 \end{bmatrix} \begin{bmatrix} x \\ \lambda \end{bmatrix} = \begin{bmatrix} -q \\ -b \end{bmatrix}. \tag{5.68}$$

This is like the procedure we used in solving the KKT conditions, except that these are linear equations, so we can solve them directly without any iteration. As in the unconstrained case, finding the minimum of a quadratic objective results in a system of linear equations.

As long as Q is positive definite, then the linear system always has a solution, and it is the global minimum of the QP.* The ease with which a QP can be solved provides a strong motivation for SQP. For a general constrained problem, we can make a local QP approximation of the nonlinear model, solve the QP, and repeat this process until convergence. This method involves iteratively solving a sequence of quadratic programming problems, hence the name *sequential quadratic programming*.

*In other words, this is a *convex* problem. Convex optimization is discussed in Chapter 11.

To form the QP, we use a local quadratic approximation of the Lagrangian (removing the constant term because it does not change the solution) and a linear approximation of the constraints for some step p near our current point. In other words, we locally approximate the problem as the following QP:

$$\underset{p}{\text{minimize}} \quad \frac{1}{2}p^\mathsf{T} H_{\mathcal{L}} p + \nabla_x \mathcal{L}^\mathsf{T} p \qquad (5.69)$$
$$\text{subject to} \quad J_h p + h = 0.$$

We substitute the gradient of the Lagrangian into the objective:

$$\frac{1}{2}p^\mathsf{T} H_{\mathcal{L}} p + \nabla f^\mathsf{T} p + \lambda^\mathsf{T} J_h p. \qquad (5.70)$$

Then, we substitute the constraint $J_h p = -h$ into the objective:

$$\frac{1}{2}p^\mathsf{T} H_{\mathcal{L}} p + \nabla f^\mathsf{T} p - \lambda^\mathsf{T} h. \qquad (5.71)$$

Now, we can remove the last term in the objective because it does not depend on the variable (p), resulting in the following equivalent problem:

$$\underset{p}{\text{minimize}} \quad \frac{1}{2}p^\mathsf{T} H_{\mathcal{L}} p + \nabla f^\mathsf{T} p \qquad (5.72)$$
$$\text{subject to} \quad J_h p + h = 0.$$

Using the QP solution method outlined previously results in the following system of linear equations:

$$\begin{bmatrix} H_{\mathcal{L}} & J_h^\mathsf{T} \\ J_h & 0 \end{bmatrix} \begin{bmatrix} p_x \\ \lambda_{k+1} \end{bmatrix} = \begin{bmatrix} -\nabla f \\ -h \end{bmatrix}. \qquad (5.73)$$

Replacing $\lambda_{k+1} = \lambda_k + p_\lambda$ and multiply through:

$$\begin{bmatrix} H_{\mathcal{L}} & J_h^\mathsf{T} \\ J_h & 0 \end{bmatrix} \begin{bmatrix} p_x \\ p_\lambda \end{bmatrix} + \begin{bmatrix} J_h^\mathsf{T} \lambda_k \\ 0 \end{bmatrix} = \begin{bmatrix} -\nabla f \\ -h \end{bmatrix}. \qquad (5.74)$$

Subtracting the second term on both sides yields

$$\begin{bmatrix} H_{\mathcal{L}} & J_h^\mathsf{T} \\ J_h & 0 \end{bmatrix} \begin{bmatrix} p_x \\ p_\lambda \end{bmatrix} = \begin{bmatrix} -\nabla_x \mathcal{L} \\ -h \end{bmatrix}, \tag{5.75}$$

which is the same linear system we found from applying Newton's method to the KKT conditions (Eq. 5.62).

This derivation relies on the somewhat arbitrary choices of choosing a QP as the subproblem and using an approximation of the Lagrangian with constraints (rather than an approximation of the objective with constraints or an approximation of the Lagrangian with no constraints).[†] Nevertheless, it is helpful to conceptualize the method as solving a sequence of QPs. This concept will motivate the solution process once we add inequality constraints.

[†] The Lagrangian objective can also be considered to be an approximation of the objective along the feasible surface $h(x) = 0$.[95]

95. Gill and Wong, *Sequential quadratic programming methods*, 2012.

5.5.2 Inequality Constraints

Introducing inequality constraints adds complications. For inequality constraints, we cannot solve the KKT conditions directly as we could for equality constraints. This is because the KKT conditions include the complementary slackness conditions $\sigma_j g_j = 0$, which we cannot solve directly. Even though the number of equations in the KKT conditions is equal to the number of unknowns, the complementary conditions do not provide complete information (they just state that each constraint is either active or inactive). Suppose we knew which of the inequality constraints were active ($g_j = 0$) and which were inactive ($\sigma_j = 0$) at the optimum. Then, we could use the same approach outlined in the previous section, treating the active constraints as equality constraints and ignoring the inactive constraints. Unfortunately, we do not know which constraints are active at the optimum beforehand in general. Finding which constraints are active in an iterative way is challenging because we would have to try all possible combinations of active constraints. This is intractable if there are many constraints.

A common approach to handling inequality constraints is to use an *active-set method*. The active set is the set of constraints that are active at the optimum (the only ones we ultimately need to enforce). Although the actual active set is unknown until the solution is found, we can estimate this set at each iteration. This subset of potentially active constraints is called the *working set*. The working set is then updated at each iteration.

Similar to the SQP developed in the previous section for equality constraints, we can create an algorithm based on solving a sequence of QPs that linearize the constraints.[‡] We extend the equality constrained

[‡] Linearizing the constraints can sometimes lead to an infeasible QP subproblem; additional techniques are needed to handle such cases.[79,96]

79. Nocedal and Wright, *Numerical Optimization*, 2006.

96. Gill et al., *SNOPT: An SQP algorithm for large-scale constrained optimization*, 2005.

5.5 Sequential Quadratic Programming

QP (Eq. 5.69) to include the inequality constraints as follows:

$$\begin{aligned}\underset{s}{\text{minimize}} \quad & \tfrac{1}{2}s^\mathsf{T} H_\mathcal{L} s + \nabla_x \mathcal{L}^\mathsf{T} s \\ \text{subject to} \quad & J_h s + h = 0 \\ & J_g s + g \le 0.\end{aligned} \qquad (5.76)$$

The determination of the working set could happen in the inner loop, that is, as part of the inequality constrained QP subproblem (Eq. 5.76). Alternatively, we could choose a working set in the outer loop and then solve the QP subproblem with only equality constraints (Eq. 5.69), where the working-set constraints would be posed as equalities. The former approach is more common and is discussed here. In that case, we need consider only the active-set problem in the context of a QP. Many variations on active-set methods exist; we outline just one such approach based on a *binding-direction method*.

The general QP problem we need to solve is as follows:

$$\begin{aligned}\underset{x}{\text{minimize}} \quad & \tfrac{1}{2}x^\mathsf{T} Q x + q^\mathsf{T} x \\ \text{subject to} \quad & Ax + b = 0 \\ & Cx + d \le 0.\end{aligned} \qquad (5.77)$$

We assume that Q is positive definite so that this problem is convex. Here, Q corresponds to the Lagrangian Hessian. Using an appropriate quasi-Newton approximation (which we will discuss in Section 5.5.4) ensures a positive definite Lagrangian Hessian approximation.

Consider iteration k in an SQP algorithm that handles inequality constraints. At the end of the previous iteration, we have a design point x_k and a working set W_k. The working set in this approach is a set of row indices corresponding to the subset of inequality constraints that are active at x_k.[§] Then, we consider the corresponding inequality constraints to be equalities, and we write:

$$C_w x_k + d_w = 0, \qquad (5.78)$$

where C_w and d_w correspond to the rows of the inequality constraints specified in the working set.

The constraints in the working set, combined with the equality constraints, must be linearly independent. Thus, we cannot include more working-set constraints (plus equality constraints) than design variables. Although the active set is unique, there can be multiple valid choices for the working set.

[§]This is not a universal definition. For example, the constraints in the working set need not be active at x_k in some approaches.

Assume, for the moment, that the working set does not change at nearby points (i.e., we ignore the constraints outside the working set). We seek a step p to update the design variables as follows: $x_{k+1} = x_k + p$. We find p by solving the following simplified QP that considers only the working set:

$$\begin{aligned}
\underset{p}{\text{minimize}} \quad & \frac{1}{2}(x_k + p)^\mathsf{T} Q(x_k + p) + q^\mathsf{T}(x_k + p) \\
\text{subject to} \quad & A(x_k + p) + b = 0 \\
& C_w(x_k + p) + d_w = 0 \,.
\end{aligned} \tag{5.79}$$

We solve this QP by varying p, so after multiplying out the terms in the objective, we can ignore the terms that do not depend on p. We can also simplify the constraints because we know the constraints were satisfied at the previous iteration (i.e., $Ax_k + b = 0$ and $C_w x_k + d_w = 0$). The simplified problem is as follows:

$$\begin{aligned}
\underset{p}{\text{minimize}} \quad & \frac{1}{2} p^\mathsf{T} Q p + (q + Q^\mathsf{T} x_k) p \\
\text{subject to} \quad & A p = 0 \\
& C_w p = 0 \,.
\end{aligned} \tag{5.80}$$

We now have an equality constrained QP that we can solve using the methods from the previous section. Using Eq. 5.68, the KKT solution to this problem is as follows:

$$\begin{bmatrix} Q & A^\mathsf{T} & C_w^\mathsf{T} \\ A & 0 & 0 \\ C_w & 0 & 0 \end{bmatrix} \begin{bmatrix} p \\ \lambda \\ \sigma \end{bmatrix} = \begin{bmatrix} -q - Q^\mathsf{T} x_k \\ 0 \\ 0 \end{bmatrix} . \tag{5.81}$$

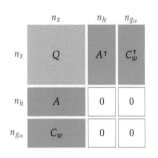

Fig. 5.39 Structure of the QP subproblem within the inequality constrained QP solution process.

Figure 5.39 shows the structure of the matrix in this linear system.

Let us consider the case where the solution of this linear system is nonzero. Solving the KKT conditions in Eq. 5.80 ensures that all the constraints in the working set are still satisfied at $x_k + p$. Still, there is no guarantee that the step does not violate some of the constraints outside of our working set. Suppose that C_n and d_n define the constraints outside of the working set. If

$$C_n(x_k + p) + d_n \leq 0 \tag{5.82}$$

for all rows, all the constraints are still satisfied. In that case, we accept the step p and update the design variables as follows:

$$x_{k+1} = x_k + p \,. \tag{5.83}$$

5.5 Sequential Quadratic Programming

The working set remains unchanged as we proceed to the next iteration.

Otherwise, if some of the constraints are violated, we cannot take the full step p and reduce it the step length by α as follows:

$$x_{k+1} = x_k + \alpha p. \tag{5.84}$$

We cannot take the full step ($\alpha = 1$), but we would like to take as large a step as possible while still keeping all the constraints feasible.

Let us consider how to determine the appropriate step size, α. Substituting the step update (Eq. 5.84) into the equality constraints, we obtain the following:

$$A(x_k + \alpha p) + b = 0. \tag{5.85}$$

We know that $Ax_k + b = 0$ from solving the problem at the previous iteration. Also, we just solved p under the condition that $Ap = 0$. Therefore, the equality constraints (Eq. 5.85) remain satisfied for any choice of α. By the same logic, the constraints in our working set remain satisfied for any choice of α as well.

Now let us consider the constraints that are not in the working set. We denote c_i as row i of the matrix C_n (associated with the inequality constraints outside of the working set). If these constraints are to remain satisfied, we require

$$c_i^T(x_k + \alpha p) + d_i \leq 0. \tag{5.86}$$

After rearranging, this condition becomes

$$\alpha c_i^T p \leq -(c_i^T x_k + d_i). \tag{5.87}$$

We do not divide through by $c_i^T p$ yet because the direction of the inequality would change depending on its sign. We consider the two possibilities separately. Because the QP constraints were satisfied at the previous iteration, we know that $c_i^T x_k + d_i \leq 0$ for all i. Thus, the right-hand side is always positive. If $c_i^T p$ is negative, then the inequality will be satisfied for any choice of α. Alternatively, if $c_i^T p$ is positive, we can rearrange Eq. 5.87 to obtain the following:

$$\alpha_i \leq -\frac{(c_i^T x_k + d_i)}{c_i^T p}. \tag{5.88}$$

This equation determines how large α can be without causing one of the constraints outside of the working set to become active. Because multiple constraints may become active, we have to evaluate α for each one and choose the smallest α among all constraints.

A constraint for which $\alpha < 1$ is said to be *blocking*. In other words, if we had included that constraint in our working set before solving the QP, it would have changed the solution. We add one of the blocking constraints to the working set, and proceed to the next iteration.¶

¶In practice, adding only one constraint to the working set at a time (or removing only one constraint in other steps described later) typically leads to faster convergence.

Now consider the case where the solution to Eq. 5.81 is $p = 0$. If all inequality constraint Lagrange multipliers are positive ($\sigma_i > 0$), the KKT conditions are satisfied and we have solved the original inequality constrained QP. If one or more σ_i values are negative, additional iterations are needed. We find the σ_i value that is most negative, remove constraint i from the working set, and proceed to the next iteration.

As noted previously, all the constraints in the reduced QP (the equality constraints plus all working-set constraints) must be linearly independent and thus $[A\ C_w]^\mathsf{T}$ has full row rank. Otherwise, there would be no solution to Eq. 5.81. Therefore, the starting working set might not include all active constraints at x_0 and must instead contain only a subset, such that linear independence is maintained. Similarly, when adding a blocking constraint to the working set, we must again check for linear independence. At a minimum, we need to ensure that the length of the working set does not exceed n_x. The complete algorithm for solving an inequality constrained QP is shown in Alg. 5.4.

Tip 5.4 Some equality constraints can be posed as inequality constraints

Equality constraints are less common in engineering design problems than inequality constraints. Sometimes we pose a problem as an equality constraint unnecessarily. For example, the simulation of an aircraft in steady-level flight may require the lift to equal the weight. Formally, this is an equality constraint, but it can also be posed as an inequality constraint (lift greater or equal to weight). There is no advantage to having more lift than the required because it increases drag, so the constraint is always active at the optimum. When such a constraint is not active at the solution, it can be a helpful indicator that something is wrong with the formulation, the optimizer, or the assumptions. Although an equality constraint is more natural from the algorithm perspective, the flexibility of the inequality constraint might allow the optimizer to explore the design space more effectively.

Consider another example: a propeller design problem might require a specified thrust. Although an equality constraint would likely work, it is more constraining than necessary. If the optimal design were somehow able to produce excess thrust, we would accept that design. Thus, we should not formulate the constraint in an unnecessarily restrictive way.

5.5 Sequential Quadratic Programming

Algorithm 5.4 Active-set solution method for an inequality constrained QP

Inputs:
Q, q, A, b, C, D: Matrices and vectors defining the QP (Eq. 5.77); Q must be positive definite

ε: Tolerance used for termination and for determining whether constraint is active

Outputs:
x^*: Optimal point

$k = 0$
$x_k = x_0$
$W_k = i$ for all i where $(c_i^\mathsf{T} x_k + d_i) > -\varepsilon$ and $\texttt{length}(W_k) \leq n_x$ One possible initial working set
while true **do**
    set $C_w = C_{i,*}$ and $d_w = d_i$ for all $i \in W_k$ Select rows for working set
    Solve the KKT system (Eq. 5.81)
    if $\|p\| < \varepsilon$ **then**
        if $\sigma \geq 0$ **then**   Satisfied KKT conditions
            $x^* = x_k$
            return
        else
            $i = \mathrm{argmin}\,\sigma$
            $W_{k+1} = W_k \setminus \{i\}$ Remove i from working set
            $x_{k+1} = x_k$
        end if
    else
        $\alpha = 1$ Initialize with optimum step
        $B = \{\}$ Blocking index
        for $i \notin W_k$ **do**   Check constraints outside of working set
            if $c_i^\mathsf{T} p > 0$ **then**   Potential blocking constraint
                $\alpha_b = \dfrac{-(c_i^\mathsf{T} x_k + d_i)}{c_i^\mathsf{T} p}$ c_i is a row of C_n
                if $\alpha_b < \alpha$ **then**
                    $\alpha = \alpha_b$
                    $B = i$ Save or overwrite blocking index
                end if
            end if
        end for
        $W_{k+1} = W_k \cup \{B\}$ Add B to working set (if linearly independent)
        $x_{k+1} = x_k + \alpha p$
    end if
    $k = k + 1$
end while

Example 5.10 Inequality constrained QP

Let us solve the following problem using the active-set QP algorithm:

$$\underset{x_1,x_2}{\text{minimize}} \quad 3x_1^2 + x_2^2 + 2x_1x_2 + x_1 + 6x_2$$
$$\text{subject to} \quad 2x_1 + 3x_2 \geq 4$$
$$x1 \geq 0$$
$$x2 \geq 0.$$

Rewriting in the standard form (Eq. 5.77) yields the following:

$$Q = \begin{bmatrix} 6 & 2 \\ 2 & 2 \end{bmatrix}, \quad q = \begin{bmatrix} 1 \\ 6 \end{bmatrix}, \quad C = \begin{bmatrix} -2 & -3 \\ -1 & 0 \\ 0 & -1 \end{bmatrix}, \quad d = \begin{bmatrix} 4 \\ 0 \\ 0 \end{bmatrix}.$$

We arbitrarily chose $x = [3, 2]$ as a starting point. Because none of the constraints are active, the initial working set is empty, $W = \{\}$. At each iteration, we solve the QP formed by the equality constraints and any constraints in the active set (treated as equality constraints). The sequence of iterations is detailed as follows and is plotted in Fig. 5.40:

$k = 1$ The QP subproblem yields $p = [-1.75, -6.25]$ and $\sigma = [0, 0, 0]$. Next, we check whether any constraints are blocking at the new point $x + p$. Because all three constraints are outside of the working set, we check all three. Constraint 1 is potentially blocking ($c_i^T p > 0$) and leads to $\alpha_b = 0.35955$. Constraint 2 is also potentially blocking and leads to $\alpha_b = 1.71429$. Finally, constraint 3 is also potentially blocking and leads to $\alpha_b = 0.32$. We choose the constraint with the smallest α, which is constraint 3, and add it to our working set. At the end of the iteration, $x = [2.44, 0.0]$ and $W = \{3\}$.

$k = 2$ The new QP subproblem yields $p = [-2.60667, 0.0]$ and $\sigma = [0, 0, 5.6667]$. Constraints 1 and 2 are outside the working set. Constraint 1 is potentially blocking and gives $\alpha_b = 0.1688$; constraint 2 is also potentially blocking and yields $\alpha_b = 0.9361$. Because constraint 1 yields the smaller step, we add it to the working set. At the end of the iteration, $x = [2.0, 0.0]$ and $W = \{1, 3\}$.

$k = 3$ The QP subproblem now yields $p = [0, 0]$ and $\sigma = [6.5, 0, -9.5]$. Because $p = 0$, we check for convergence. One of the Lagrange multipliers is negative, so this cannot be a solution. We remove the constraint associated with the most negative Lagrange multiplier from the working set (constraint 3). At the end of the iteration, x is unchanged at $x = [2.0, 0.0]$, and $W = \{1\}$.

$k = 4$ The QP yields $p = [-1.5, 1.0]$ and $\sigma = [3, 0, 0]$. Constraint 2 is potentially blocking and yields $\alpha_b = 1.333$ (which means it is not blocking because $\alpha_b > 1$). Constraint 3 is also not blocking ($c_i^T p < 0$). None of the α_b values was blocking, so we can take the full step ($\alpha = 1$). The new x point is $x = [0.5, 1.0]$, and the working set is unchanged at $W = \{1\}$.

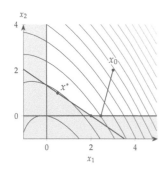

Fig. 5.40 Iteration history for the active-set QP example.

5.5 Sequential Quadratic Programming

$k = 5$ The QP yields $p = [0,0]$, $\sigma = [3,0,0]$. Because $p = 0$, we check for convergence. All Lagrange multipliers are nonnegative, so the problem is solved. The solution to the original inequality constrained QP is then $x^* = [0.5, 1.0]$.

Because SQP solves a sequence of QPs, an effective approach is to use the optimal x and active set from the previous QP as the starting point and working set for the next QP. The algorithm outlined in this section requires both a feasible starting point and a working set of linearly independent constraints. Although the previous starting point and working set usually satisfy these conditions, this is not guaranteed, and adjustments may be necessary.

Algorithms to determine a feasible point are widely used (often by solving a linear programming problem). There are also algorithms to remove or add to the constraint matrix as needed to ensure full rank.[96]

96. Gill et al., *SNOPT: An SQP algorithm for large-scale constrained optimization*, 2005.

Tip 5.5 Consider reformulating your constraints

There are often multiple mathematically equivalent ways to pose the problem constraints. Reformulating can sometimes yield equivalent problems that are significantly easier to solve. In some cases, it can help to add redundant constraints to avoid areas of the design space that are not useful. Similarly, we should consider whether the model that computes the objective and constraint functions should be solved separately or posed as constraints at the optimizer level (as we did in Eq. 3.33).

5.5.3 Merit Functions and Filters

Similar to what we did in unconstrained optimization, we do not directly accept the step p returned from solving the subproblem (Eq. 5.62 or Eq. 5.76). Instead, we use p as the first step length in a line search.

In the line search for unconstrained problems (Section 4.3), determining if a point was good enough to terminate the search was based solely on comparing the objective function value (and the slope when enforcing the strong Wolfe conditions). For constrained optimization, we need to make some modifications to these methods and criteria.

In constrained optimization, objective function decrease and feasibility often compete with each other. During a line search, a new point may decrease the objective but increase the infeasibility, or it may decrease the infeasibility but increase the objective. We need to take these two metrics into account to determine the line search termination criterion.

The Lagrangian is a function that accounts for the two metrics. However, at a given iteration, we only have an *estimate* of the Lagrange multipliers, which can be inaccurate.

One way to combine the objective value with the constraints in a line search is to use *merit functions*, which are similar to the penalty functions introduced in Section 5.4. Common merit functions include functions that use the norm of constraint violations:

$$\hat{f}(x; \mu) = f(x) + \mu \left\| \bar{g}(x) \right\|_p , \quad (5.89)$$

where p is 1 or 2 and $\bar{g}$ are the constraint violations, defined as

$$\bar{g}_j(x) = \begin{cases} h_j(x) & \text{for equality constraints} \\ \max(0, g_j(x)) & \text{for inequality constraints} \end{cases}. \quad (5.90)$$

The augmented Lagrangian from Section 5.4.1 can also be repurposed for a constrained line search (see Eqs. 5.53 and 5.54).

Like penalty functions, one downside of merit functions is that it is challenging to choose a suitable value for the penalty parameter μ. This parameter needs to be large to ensure feasibility. However, if it is too large, a full Newton step might not be permitted. This might slow the convergence unnecessarily. Using the augmented Lagrangian can help, as discussed in Section 5.4.1. However, there are specific techniques used in SQP line searches and various safeguarding techniques needed for robustness.

Filter methods are an alternative to using penalty-based methods in a line search.[97] Filter methods interfere less with the full Newton step and are effective for both SQP and interior-point methods (which are introduced in Section 5.6).[98,99] The approach is based on concepts from multiobjective optimization, which is the subject of Chapter 9. In the filter method, there are two objectives: decrease the objective function and decrease infeasibility. A point is said to *dominate* another if its objective is lower *and* the sum of its constraint violations is lower. The filter consists of all the points that have been found to be non-dominated in the line searches so far. The line search terminates when it finds a point that is not dominated by any point in the current filter. That new point is then added to the filter, and any points that it dominates are removed from the filter.‖

97. Fletcher and Leyffer, *Nonlinear programming without a penalty function*, 2002.

98. Benson et al., *Interior-point methods for nonconvex nonlinear programming: Filter methods and merit functions*, 2002.

99. Fletcher et al., *A brief history of filter methods*, 2006.

‖ See Section 9.2 for more details on the concept of dominance.

This is only the basic concept. Robust implementation of a filter method requires imposing sufficient decrease conditions, not unlike those in the unconstrained case, and several other modifications. Fletcher et al.[99] provide more details on filter methods.

5.5 Sequential Quadratic Programming

Example 5.11 Using a filter

A filter consists of pairs $(f(x), \|\tilde{g}\|_1)$, where $\|\tilde{g}\|_1$ is the sum of the constraint violations (Eq. 5.90). Suppose that the current filter contains the following three points: $\{(2,5), (3,2), (7,1)\}$. None of the points in the filter dominates any other. These points are plotted as the blue dots in Fig. 5.41, where the shaded regions correspond to all the points that are dominated by the points in the filter. During a line search, a new candidate point is evaluated. There are three possible outcomes. Consider the following three points that illustrate these three outcomes (corresponding to the labeled points in Fig. 5.41):

1. $(1,4)$: This point is not dominated by any point in the filter. The step is accepted, the line search ends, and this point is added to the filter. Because this new point dominates one of the points in the filter, $(2,5)$, that dominated point is removed from the filter. The current set in the filter is now $\{(1,4), (3,2), (7,1)\}$.

2. $(1,6)$: This point is not dominated by any point in the filter. The step is accepted, the line search ends, and this new point is added to the filter. Unlike the previous case, none of the points in the filter are dominated. Therefore, no points are removed from the filter set, which becomes $\{(1,6), (2,5), (3,2), (7,1)\}$.

3. $(4,3)$: This point is dominated by a point in the filter, $(3,2)$. The step is rejected, and the line search continues by selecting a new candidate point. The filter is unchanged.

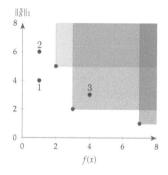

Fig. 5.41 Filter method example showing three points in the filter (blue dots); the shaded regions correspond to all the points that are dominated by the filter. The red dots illustrate three different possible outcomes when new points are considered.

5.5.4 Quasi-Newton SQP

In the discussion of the SQP method so far, we have assumed that we have the Hessian of the Lagrangian $H_{\mathcal{L}}$. Similar to the unconstrained optimization case, the Hessian might not be available or be too expensive to compute. Therefore, it is desirable to use a quasi-Newton approach that approximates the Hessian, as we did in Section 4.4.4.

The difference now is that we need an approximation of the Lagrangian Hessian instead of the objective function Hessian. We denote this approximation at iteration k as $\tilde{H}_{\mathcal{L}_k}$.

Similar to the unconstrained case, we can approximate $\tilde{H}_{\mathcal{L}_k}$ using the gradients of the Lagrangian and a quasi-Newton update, such as the Broyden—Fletcher—Goldfarb—Shanno (BFGS) update. Unlike in unconstrained optimization, we do not want the inverse of the Hessian directly. Therefore, we use the version of the BFGS formula that computes the Hessian (Eq. 4.87):

$$\tilde{H}_{\mathcal{L}_{k+1}} = \tilde{H}_{\mathcal{L}_k} - \frac{\tilde{H}_{\mathcal{L}_k} s_k s_k^T \tilde{H}_{\mathcal{L}_k}}{s_k^T \tilde{H}_{\mathcal{L}_k} s_k} + \frac{y_k y_k^T}{y_k^T s_k}, \quad (5.91)$$

where:
$$s_k = x_{k+1} - x_k$$
$$y_k = \nabla_x \mathcal{L}(x_{k+1}, \lambda_{k+1}) - \nabla_x \mathcal{L}(x_k, \lambda_{k+1}).$$
(5.92)

The step in the design variable space, s_k, is the step that resulted from the latest line search. The Lagrange multiplier is fixed to the latest value when approximating the curvature of the Lagrangian because we only need the curvature in the space of the design variables.

Recall that for the QP problem (Eq. 5.76) to have a solution, $\tilde{H}_{\mathcal{L}_k}$ must be positive definite. To ensure a positive definite approximation, we can use a *damped BFGS update*.[25]** This method replaces y with a new vector r, defined as

$$r_k = \theta_k y_k + (1 - \theta_k) \tilde{H}_{\mathcal{L}_k} s_k,$$
(5.93)

where the scalar θ_k is defined as

$$\theta_k = \begin{cases} 1 & \text{if } s_k^T y_k \geq 0.2 s_k^T \tilde{H}_{\mathcal{L}_k} s_k \\ \dfrac{0.8 s_k^T \tilde{H}_{\mathcal{L}_k} s_k}{s_k^T \tilde{H}_{\mathcal{L}_k} s_k - s_k^T y_k} & \text{if } s_k^T y_k < 0.2 s_k^T \tilde{H}_{\mathcal{L}_k} s_k, \end{cases}$$
(5.94)

which can range from 0 to 1. We then use the same BFGS update formula (Eq. 5.91), except that we replace each y_k with r_k.

To better understand this update, let us consider the two extremes for θ. If $\theta_k = 0$, then Eq. 5.93 in combination with Eq. 5.91 yields $\tilde{H}_{\mathcal{L}_{k+1}} = \tilde{H}_{\mathcal{L}_k}$; that is, the Hessian approximation is unmodified. At the other extreme, $\theta_k = 1$ yields the full BFGS update formula (r_k is set to y_k). Thus, the parameter θ_k provides a linear weighting between keeping the current Hessian approximation and using the full BFGS update.

The definition of θ_k (Eq. 5.94) ensures that $\tilde{H}_{\mathcal{L}_{k+1}}$ stays close enough to $\tilde{H}_{\mathcal{L}_k}$ and remains positive definite. The damping is activated when the predicted curvature in the new latest step is below one-fifth of the curvature predicted by the latest approximate Hessian. This could happen when the function is flattening or when the curvature becomes negative.

5.5.5 Algorithm Overview

We now put together the various pieces in a high-level description of SQP with quasi-Newton approximations in Alg. 5.5.[††] For the convergence criterion, we can use an infinity norm of the KKT system residual vector. For better control over the convergence, we can consider two separate tolerances: one for the norm of the optimality and another

---

25. Powell, *Algorithms for nonlinear constraints that use Lagrangian functions*, 1978.

**The damped BFGS update is not always the best approach. There are approaches built around other approximation methods, such as symmetric rank 1 (SR1).[100] Limited-memory updates similar to L-BFGS (see Section 4.4.5) can be used when storing a dense Hessian for large problems is prohibitive.[101]

100. Fletcher, *Practical Methods of Optimization*, 1987.

101. Liu and Nocedal, *On the limited memory BFGS method for large scale optimization*, 1989.

††A few popular SQP implementations include SNOPT,[96] Knitro,[102] MATLAB's fmincon, and SLSQP.[103] The first three are commercial options, whereas SLSQP is open source. There are interfaces in different programming languages for these optimizers, including pyOptSparse (for SNOPT and SLSQP).[1]

1. Wu et al., *pyOptSparse: A Python framework for large-scale constrained nonlinear optimization of sparse systems*, 2020.

102. Byrd et al., *Knitro: An Integrated Package for Nonlinear Optimization*, 2006.

103. Kraft, *A software package for sequential quadratic programming*, 1988.

5.5 Sequential Quadratic Programming

for the norm of the feasibility. For problems that only have equality constraints, we can solve the corresponding QP (Eq. 5.62) instead.

Algorithm 5.5 SQP with quasi-Newton approximation

Inputs:
    x_0: Starting point
    τ_{opt}: Optimality tolerance
    τ_{feas}: Feasibility tolerance

Outputs:
    x^*: Optimal point
    $f(x^*)$: Corresponding function value

$\lambda_0 = 0, \sigma_0 = 0$ Initial Lagrange multipliers
$\alpha_{init} = 1$ For line search
Evaluate functions (f, g, h) and derivatives $(\nabla f, J_g, J_h)$
$\nabla_x \mathcal{L} = \nabla f + J_h^T \lambda + J_g^T \sigma$
$k = 0$
while $\|\nabla_x \mathcal{L}\|_\infty > \tau_{opt}$ or $\|h\|_\infty > \tau_{feas}$ **do**
    if $k = 0$ or reset = **true then**
        $\tilde{H}_{\mathcal{L}_0} = I$ Initialize to identity matrix or scaled version (Eq. 4.95)
    else
        Update $\tilde{H}_{\mathcal{L}_{k+1}}$ Compute damped BFGS (Eqs. 5.91 to 5.94)
    end if
    Solve QP subproblem (Eq. 5.76) for p_x, p_λ

$$\begin{aligned} \text{minimize} \quad & \tfrac{1}{2} p_x^T \tilde{H}_\mathcal{L} p_x + \nabla_x \mathcal{L}^T p_x \\ \text{by varying} \quad & p_x \\ \text{subject to} \quad & J_h p_x + h = 0 \\ & J_g p_x + g \le 0 \end{aligned}$$

    $\lambda_{k+1} = \lambda_k + p_\lambda$
    $\alpha = \texttt{linesearch}(p_x, \alpha_{init})$ Use merit function or filter (Section 5.5.3)
    $x_{k+1} = x_k + \alpha p_k$ Update step
    $\mathcal{W}_{k+1} = \mathcal{W}_k$ Active set becomes initial working set for next QP
    Evaluate functions (f, g, h) and derivatives $(\nabla f, J_g, J_h)$
    $\nabla_x \mathcal{L} = \nabla f + J_h^T \lambda + J_g^T \sigma$
    $k = k + 1$
end while

Example 5.12 SQP applied to equality constrained problem

We now solve Ex. 5.2 using the SQP method (Alg. 5.5). We start at $x_0 = [2, 1]$ with an initial Lagrange multiplier $\lambda = 0$ and an initial estimate of the Lagrangian Hessian as $\tilde{H}_{\mathcal{L}} = I$ for simplicity. The line search uses an augmented Lagrangian merit function with a fixed penalty parameter ($\mu = 1$) and a quadratic bracketed search as described in Section 4.3.2. The choice between a merit function and line search has only a small effect in this simple problem. The gradient of the equality constraint is

$$J_h = \begin{bmatrix} \frac{1}{2}x_1 & 2x_2 \end{bmatrix} = \begin{bmatrix} 1 & 2 \end{bmatrix},$$

and differentiating the Lagrangian with respect to x yields

$$\nabla_x \mathcal{L} = \begin{bmatrix} 1 + \frac{1}{2}\lambda x_1 \\ 2 + 2\lambda x_2 \end{bmatrix} = \begin{bmatrix} 1 \\ 2 \end{bmatrix}.$$

The KKT system to be solved (Eq. 5.62) in the first iteration is

$$\begin{bmatrix} 1 & 0 & 1 \\ 0 & 1 & 2 \\ 1 & 2 & 0 \end{bmatrix} \begin{bmatrix} s_{x_1} \\ s_{x_2} \\ s_\lambda \end{bmatrix} = \begin{bmatrix} -1 \\ -2 \\ -1 \end{bmatrix}.$$

The solution of this system is $s = [-0.2, -0.4, -0.8]$. Using $p = [-0.2, -0.4]$, the full step $\alpha = 1$ satisfies the strong Wolfe conditions, so for the new iteration we have $x_1 = [1.8, 0.6]$, $\lambda_1 = -0.8$.

To update the approximate Hessian $\tilde{H}_\mathcal{L}$ using the damped BFGS update (Eq. 5.93), we need to compare the values of $s_0^T y_0 = -0.272$ and $s_0^T W_0 s_0 = 0.2$. Because $s_k^T y_k < 0.2 s_k^T \tilde{H}_{\mathcal{L}_k} s_k$, we need to compute the scalar $\theta = 0.339$ using Eq. 5.94. This results in a partial BFGS update to maintain positive definiteness. After a few iterations, $\theta = 1$ for the remainder of the optimization, corresponding to a full BFGS update. The initial estimate for the Lagrangian Hessian is poor (just a scaled identity matrix), so some damping is necessary. However, the estimate is greatly improved after a few iterations. Using the quasi-Newton update in Eq. 5.91, we get the approximate Hessian for the next iteration as

$$\tilde{H}_{\mathcal{L}_1} = \begin{bmatrix} 1.076 & -0.275 \\ -0.275 & 0.256 \end{bmatrix}.$$

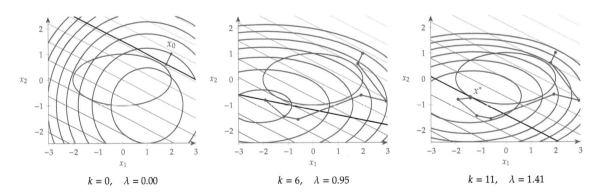

Fig. 5.42 SQP algorithm iterations.

We repeat this process for subsequent iterations, as shown in Figure 5.42. The gray contours show the QP subproblem (Eq. 5.72) solved at each iteration: the quadratic objective appears as elliptical contours and the linearized

5.5 Sequential Quadratic Programming

constraint as a straight line. The starting point is infeasible, and the iterations remain infeasible until the last few iterations.

This behavior is common for SQP because although it satisfies the linear approximation of the constraints at each step, it does not necessarily satisfy the constraints of the actual problem, which is nonlinear. As the constraint approximation becomes more accurate near the solution, the nonlinear constraint is then satisfied. Figure 5.43 shows the convergence of the Lagrangian gradient norm, with the characteristic quadratic convergence at the end.

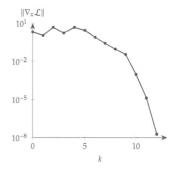

Fig. 5.43 Convergence history of the norm of the Lagrangian gradient.

Example 5.13 SQP applied to inequality constrained problem

We now solve the inequality constrained version of the previous example (Ex. 5.4) with the same initial conditions and general approach. The only difference is that rather than solving the linear system of equations Eq. 5.62, we have to solve an active-set QP problem at each iteration, as outlined in Alg. 5.4. The iteration history and convergence of the norm of the Lagrangian gradient are plotted in Figs. 5.44 and 5.45, respectively.

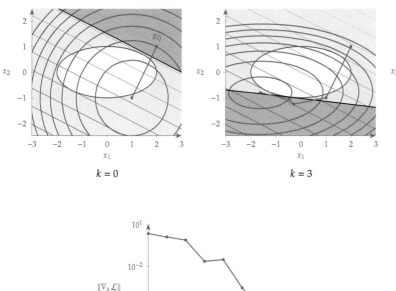

Fig. 5.44 Iteration history of SQP applied to an inequality constrained problem, with the Lagrangian and the linearized constraint overlaid (with a darker infeasible region).

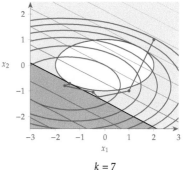

Fig. 5.45 Convergence history of the norm of the Lagrangian gradient.

> **Tip 5.6** How to handle maximum and minimum constraints
>
> Constraints that take the maximum or minimum of a set of quantities are often desired. For example, the stress in a structure may be evaluated at many points, and we want to make sure the maximum stress does not exceed a specified yield stress, such that
>
> $$\max(\sigma) \leq \sigma_{\text{yield}}.$$
>
> However, the maximum function is not continuously differentiable (because the maximum can switch elements between iterations), which may cause difficulties when using gradient-based optimization. The constraint aggregation methods from Section 5.7 can enforce such conditions with a smooth function. Nevertheless, it is challenging for an optimizer to find a point that satisfies the KKT conditions because the information is reduced to one constraint.
>
> Instead of taking the maximum, you should consider constraining the stress at all n_σ points as follows
>
> $$\sigma_j \leq \sigma_{\text{yield}}, \quad j = 1, \ldots, n_\sigma.$$
>
> Now all constraints are continuously differentiable. The optimizer has n_σ constraints instead of 1, but that generally provides more information and makes it easier for the optimizer to satisfy the KKT conditions with more than one Lagrange multiplier. Even though we have added more constraints, an active set method makes this efficient because it considers only the critical constraints.

5.6 Interior-Point Methods

Interior-point methods use concepts from both SQP and interior penalty methods.* These methods form an objective similar to the interior penalty but with the key difference that instead of penalizing the constraints directly, they add slack variables to the set of optimization variables and penalize the slack variables. The resulting formulation is as follows:

$$\begin{aligned}
\underset{x,s}{\text{minimize}} \quad & f(x) - \mu_b \sum_{j=1}^{n_g} \ln s_j \\
\text{subject to} \quad & h(x) = 0 \\
& g(x) + s = 0.
\end{aligned} \quad (5.95)$$

This formulation turns the inequality constraints into equality constraints and thus avoids the combinatorial problem.

Similar to SQP, we apply Newton's method to solve for the KKT conditions. However, instead of solving the KKT conditions of the

---

*The name *interior point* stems from early methods based on interior penalty methods that assumed that the initial point was feasible. However, modern interior-point methods can start with infeasible points.

5.6 Interior-Point Methods

original problem (Eq. 5.59), we solve the KKT conditions of the interior-point formulation (Eq. 5.95).

These slack variables in Eq. 5.95 do not need to be squared, as was done in deriving the KKT conditions, because the logarithm is only defined for positive s values and acts as a barrier preventing negative values of s (although we need to prevent the line search from producing negative s values, as discussed later). Because s is always positive, that means that $g(x^*) < 0$ at the solution, which satisfies the inequality constraints.

Like penalty method formulations, the interior-point formulation (Eq. 5.95) is only equivalent to the original constrained problem in the limit, as $\mu_b \to 0$. Thus, as in the penalty methods, we need to solve a sequence of solutions to this problem where μ_b approaches zero.

First, we form the Lagrangian for this problem as

$$\mathcal{L}(x, \lambda, \sigma, s) = f(x) - \mu_b e^\mathsf{T} \ln s + h(x)^\mathsf{T} \lambda + (g(x) + s)^\mathsf{T} \sigma, \quad (5.96)$$

where $\ln s$ is an n_g-vector whose components are the logarithms of each component of s, and $e = [1, \ldots, 1]$ is an n_g-vector of 1s introduced to express the sum in vector form. By taking derivatives with respect to x, λ, σ, and s, we derive the KKT conditions for this problem as

$$\begin{aligned} \nabla f(x) + J_h(x)^\mathsf{T} \lambda + J_g(x)^\mathsf{T} \sigma &= 0 \\ h &= 0 \\ g + s &= 0 \\ -\mu_b S^{-1} e + \sigma &= 0, \end{aligned} \quad (5.97)$$

where S is a diagonal matrix whose diagonal entries are given by the slack variable vector, and therefore $S^{-1}_{kk} = 1/s_k$. The result is a set of $n_x + n_h + 2n_g$ equations and the same number of variables.

To get a system of equations that is more favorable for Newton's method, we multiply the last equation by S to obtain

$$\begin{aligned} \nabla f(x) + J_h(x)^\mathsf{T} \lambda + J_g(x)^\mathsf{T} \sigma &= 0 \\ h &= 0 \\ g + s &= 0 \\ -\mu_b e + S\sigma &= 0. \end{aligned} \quad (5.98)$$

We now have a set of residual equations to which we can apply Newton's method, just like we did for SQP. Taking the Jacobian of the residuals in Eq. 5.98, we obtain the linear system

$$\begin{bmatrix} H_{\mathcal{L}}(x) & J_h(x)^\mathsf{T} & J_g(x)^\mathsf{T} & 0 \\ J_h(x) & 0 & 0 & 0 \\ J_g(x) & 0 & 0 & I \\ 0 & 0 & S & \Sigma \end{bmatrix} \begin{bmatrix} s_x \\ s_\lambda \\ s_\sigma \\ s_s \end{bmatrix} = - \begin{bmatrix} \nabla_x \mathcal{L}(x, \lambda, \sigma) \\ h(x) \\ g(x) + s \\ S\sigma - \mu_b e \end{bmatrix}, \quad (5.99)$$

where Σ is a diagonal matrix whose entries are given by σ, and I is the identity matrix. For numerical efficiency, we make the matrix symmetric by multiplying the last equation by S^{-1} to get the symmetric linear system, as follows:

$$\begin{bmatrix} H_{\mathcal{L}}(x) & J_h(x)^\mathsf{T} & J_g(x)^\mathsf{T} & 0 \\ J_h(x) & 0 & 0 & 0 \\ J_g(x) & 0 & 0 & I \\ 0 & 0 & I & S^{-1}\Sigma \end{bmatrix} \begin{bmatrix} s_x \\ s_\lambda \\ s_\sigma \\ s_s \end{bmatrix} = - \begin{bmatrix} \nabla_x \mathcal{L}(x, \lambda, \sigma) \\ h(x) \\ g(x) + s \\ \sigma - \mu_b S^{-1} e \end{bmatrix}. \quad (5.100)$$

The advantage of this equivalent system is that we can use a linear solver specialized for symmetric matrices, which is more efficient than a solver for general linear systems. If we had applied Newton's method to the original KKT system (Eq. 5.97) and then made it symmetric, we would have obtained a term with S^{-2}, which would make the system more challenging than with the S^{-1} term in Eq. 5.100. Figure 5.46 shows the structure and block sizes of the matrix.

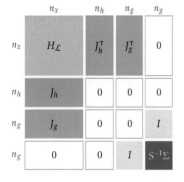

Fig. 5.46 Structure and shape of the interior-point system matrix from Eq. 5.100.

5.6.1 Modifications to the Basic Algorithm

We can reuse many of the concepts covered under SQP, including quasi-Newton estimates of the Lagrangian Hessian and line searches with merit functions or filters. The merit function would usually be modified to a form more consistent with the formulation used in Eq. 5.95. For example, we could write a merit function as follows:

$$\hat{f}(x) = f(x) - \mu_b \sum_{i=1}^{n_g} \ln s_i + \frac{1}{2}\mu_p \left(\|h(x)\|^2 + \|g(x) + s\|^2 \right), \quad (5.101)$$

where μ_b is the barrier parameter from Eq. 5.95, and μ_p is the penalty parameter. Additionally, we must enforce an $\alpha_{\max}$ in the line search so that the implicit constraint on $s > 0$ remains enforced. The maximum allowed step size can be computed prior to the line search because we know the value of s and p_s and require that

$$s + \alpha p_s \geq 0. \quad (5.102)$$

5.6 Interior-Point Methods

In practice, we enforce a fractional tolerance so that we do not get too close to zero. For example, we could enforce the following:

$$s + \alpha_{\max} p_s = \tau s, \quad (5.103)$$

where τ is a small value (e.g., $\tau = 0.005$). The maximum step size is the smallest positive value that satisfies this equation for all entries in s. A possible algorithm for determining the maximum step size for feasibility is shown in Alg. 5.6.

Algorithm 5.6 Maximum step size for feasibility

Inputs:
  s: Current slack values
  p_s: Proposed step
  τ: Fractional tolerance (e.g., 0.005)

Outputs:
  $\alpha_{\max}$: Maximum feasible step length

$\alpha_{\max} = 1$
for $i = 1$ to n_g **do**
  $\alpha = (\tau - 1)\dfrac{s_i}{p_{s_i}}$
  if $\alpha > 0$ **then**
    $\alpha_{\max} = \min(\alpha_{\max}, \alpha)$
  end if
end for

The line search typically uses a simple backtracking approach because we must enforce a maximum step length. After the line search, we can update x and s as follows:

$$x_{k+1} = x_k + \alpha_k p_x, \quad \text{where } \alpha_k \in (0, \alpha_{\max}] \quad (5.104)$$

$$s_{k+1} = s_k + \alpha_k p_s. \quad (5.105)$$

The Lagrange multipliers σ must also remain positive, so the procedure in Alg. 5.6 is repeated for σ to find the maximum step length for the Lagrange multipliers α_σ. Enforcing a maximum step size for Lagrange multiplier updates was not necessary for the SQP method because the QP subproblem handled the enforcement of nonnegative Lagrange multipliers. We then update both sets of Lagrange multipliers using this step size:

$$\lambda_{k+1} = \lambda_k + \alpha_\sigma p_\lambda \quad (5.106)$$

$$\sigma_{k+1} = \sigma_k + \alpha_\sigma p_\sigma. \quad (5.107)$$

Finally, we need to update the barrier parameter μ_b. The simplest approach is to decrease it by a multiplicative factor:

$$\mu_{b_{k+1}} = \rho \mu_{b_k}, \qquad (5.108)$$

where ρ is typically around 0.2. Better methods are adaptive based on how well the optimizer is progressing. There are other implementation details for improving robustness that can be found in the literature.[104,105]

The steps for a basic interior-point method are detailed in Alg. 5.7.[†] This version focuses on a line search approach, but there are variations of interior-point methods that use the trust-region approach.

104. Wächter and Biegler, *On the implementation of an interior-point filter line-search algorithm for large-scale nonlinear programming*, 2005.

105. Byrd et al., *An interior point algorithm for large-scale nonlinear programming*, 1999.

[†]IPOPT is an open-source nonlinear interior-point method.[106] The commercial packages Knitro[102] and fmincon mentioned earlier also include interior-point methods.

106. Wächter and Biegler, *On the implementation of a primal-dual interior point filter line search algorithm for large-scale nonlinear programming*, 2006.

Algorithm 5.7 Interior-point method with a quasi-Newton approximation

Inputs:
 x_0: Starting point
 τ_{opt}: Optimality tolerance
 τ_{feas}: Feasibility tolerance

Outputs:
 x^*: Optimal point
 $f(x^*)$: Optimal function value

$\lambda_0 = 0;\ \sigma_0 = 0$ — Initial Lagrange multipliers
$s_0 = 1$ — Initial slack variables
$\tilde{H}_{\mathcal{L}_0} = I$ — Initialize Hessian of Lagrangian approximation to identity matrix
$k = 0$
while $\|\nabla_x \mathcal{L}\|_\infty > \tau_{\text{opt}}$ or $\|h\|_\infty > \tau_{\text{feas}}$ **do**
 Evaluate $J_h, J_g, \nabla_x \mathcal{L}$
 Solve the KKT system (Eq. 5.100) for p

$$\begin{bmatrix} \tilde{H}_{\mathcal{L}_k} & J_h^T & J_g^T & 0 \\ J_h(x) & 0 & 0 & 0 \\ J_g(x) & 0 & 0 & I \\ 0 & 0 & I & S^{-1}\Sigma \end{bmatrix} \begin{bmatrix} p_x \\ p_\lambda \\ p_\sigma \\ p_s \end{bmatrix} = - \begin{bmatrix} \nabla_x \mathcal{L}(x,\lambda,\sigma) \\ h(x) \\ g(x) + s \\ \sigma - \mu S^{-1} e \end{bmatrix}$$

 $\alpha_{\max} = \mathtt{alphamax}(s, p_s)$ — Use Alg. 5.6
 $\alpha_k = \mathtt{backtrack}(p_x, p_s, \alpha_{\max})$ Line search (Alg. 4.2) with merit function (Eq. 5.101)
 $x_{k+1} = x_k + \alpha_k p_x$ — Update design variables
 $s_{k+1} = s_k + \alpha_k p_s$ — Update slack variables
 $\alpha_\sigma = \mathtt{alphamax}(\sigma, p_\sigma)$
 $\lambda_{k+1} = \lambda_k + \alpha_\sigma s_\lambda$ — Update equality Lagrange multipliers
 $\sigma_{k+1} = \sigma_k + \alpha_\sigma s_\sigma$ — Update inequality Lagrange multipliers
 Update $\tilde{H}_{\mathcal{L}_{k+1}}$ — Compute quasi-Newton approximation using Eq. 5.91
 $\mu_b = \rho \mu_b$ — Reduce barrier parameter
 $k = k + 1$
end while

5.6.2 SQP Comparisons and Examples

Both interior-point methods and SQP are considered state-of-the-art approaches for solving nonlinear constrained optimization problems. Each of these two methods has its strengths and weaknesses. The KKT system structure is identical at each iteration for interior-point methods, so we can exploit this structure for improved computational efficiency. SQP is not as amenable to this because changes in the working set cause the system's structure to change between iterations. The downside of the interior-point structure is that turning all constraints into equalities means that all constraints must be included at every iteration, even if they are inactive. In contrast, active-set SQP only needs to consider a subset of the constraints, reducing the subproblem size.

Active-set SQP methods are generally more effective for medium-scale problems, whereas interior-point methods are more effective for large-scale problems. Interior-point methods are usually more sensitive to the initial starting point and the scaling of the problem. Therefore, SQP methods are usually more suitable for solving sequences of warm-started problems.[79,107] These are just general guidelines; both approaches should be considered and tested for a given problem of interest.

79. Nocedal and Wright, *Numerical Optimization*, 2006.

107. Gill et al., *On the performance of SQP methods for nonlinear optimization*, 2015.

Example 5.14 Numerical solution of graphical solution example

Recall the constrained problem with a quadratic objective and quadratic constraints introduced in Ex. 5.1. Instead of finding an approximate solution graphically or trying to solve this analytically, we can now solve this numerically using SQP or the interior-point method. The resulting optimization paths are shown in Fig. 5.47. These results are only illustrative; paths and iterations can vary significantly with the starting point and algorithmic parameters.

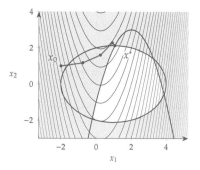

Sequential quadratic programming

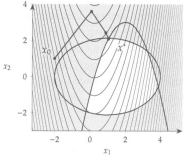

Interior-point method

Fig. 5.47 Numerical solution of problem solved graphically in Ex. 5.1.

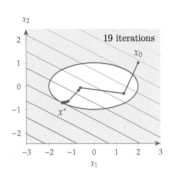

Fig. 5.48 Interior-point algorithm iterations.

Example 5.15 Interior-point method applied to inequality constrained problem

Here we solve Ex. 5.4 using the interior-point method (Alg. 5.7) starting from $x_0 = [2, 1]$. The initial Lagrange multiplier is $\sigma = 0$, and the initial slack variable is $s = 1$. Starting with a penalty parameter of $\mu = 20$ results in the iterations shown in Fig. 5.48.

For the first iteration, differentiating the Lagrangian with respect to x yields

$$\nabla_x \mathcal{L}(x_1, x_2) = \begin{bmatrix} 1 + \frac{1}{2}\sigma x_1 \\ 2 + 2\sigma x_2 \end{bmatrix} = \begin{bmatrix} 1 \\ 2 \end{bmatrix},$$

and the gradient of the constraint is

$$\nabla g(x_1, x_2) = \begin{bmatrix} \frac{1}{2} x_1 \\ 2 x_2 \end{bmatrix} = \begin{bmatrix} 1 \\ 2 \end{bmatrix}.$$

The interior-point system of equations (Eq. 5.100) at the starting point is

$$\begin{bmatrix} 1 & 0 & 1 & 0 \\ 0 & 1 & 2 & 0 \\ 1 & 2 & 0 & 1 \\ 0 & 0 & 1 & 0 \end{bmatrix} \begin{bmatrix} s_{x_1} \\ s_{x_2} \\ s_\sigma \\ s_s \end{bmatrix} = \begin{bmatrix} -1 \\ -2 \\ -2 \\ 20 \end{bmatrix}.$$

The solution is $s = [-21, -42, 20, 103]$. Performing a line search in the direction $p = [-21, -42]$ yields $x_1 = [1.34375, -0.3125]$. The Lagrange multiplier and slack variable are updated to $\sigma_1 = 20$ and $s_1 = 104$, respectively.

To update the approximate Hessian $\tilde{H}_{\mathcal{L}_k}$, we use the damped BFGS update (Eq. 5.93) to ensure that $\tilde{H}_{\mathcal{L}_k}$ is positive definite. By comparing $s_0^\mathsf{T} y_0 = 73.21$ and $s_0^\mathsf{T} \tilde{H}_{\mathcal{L}_0} s_0 = 2.15$, we can see that $s_k^\mathsf{T} y_k \geq 0.2 s_k^\mathsf{T} \tilde{H}_{\mathcal{L}_k} s_k$, and therefore, we do a full BFGS update with $\theta_0 = 1$ and $r_0 = y_0$. Using the quasi-Newton update (Eq. 5.91), we get the approximate Hessian:

$$\tilde{H}_{\mathcal{L}_1} = \begin{bmatrix} 1.388 & 4.306 \\ 4.306 & 37.847 \end{bmatrix}.$$

We reduce the barrier parameter μ by a factor of 2 at each iteration. This process is repeated for subsequent iterations.

The starting point is infeasible, but the algorithm finds a feasible point after the first iteration. From then on, it approaches the optimum from within the feasible region, as shown in Fig. 5.48.

Example 5.16 Constrained spring system

Consider the spring system from Ex. 4.17, which is an unconstrained optimization problem. We can constrain the spring system by attaching two cables as shown in Fig. 5.49, where $\ell_{c_1} = 9$ m, $\ell_{c_2} = 6$ m, $y_c = 2$ m, $x_{c_1} = 7$ m, and $x_{c_2} = 3$ m.

5.7 Constraint Aggregation

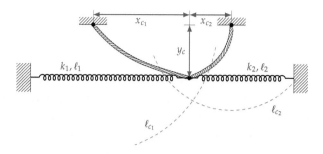

Fig. 5.49 Spring system constrained by two cables.

Because the cables do not resist compression forces, they correspond to inequality constraints, yielding the following problem:

$$\underset{x_1,x_2}{\text{minimize}} \quad \frac{1}{2}k_1\left(\sqrt{(\ell_1+x_1)^2+x_2^2}-\ell_1\right)^2 + \frac{1}{2}k_2\left(\sqrt{(\ell_2-x_1)^2+x_2^2}-\ell_2\right)^2 - mgx_2$$

$$\text{subject to} \quad \sqrt{(x_1+x_{c_1})^2+(x_2+y_c)^2} \leq \ell_{c_1}$$

$$\sqrt{(x_1-x_{c_2})^2+(x_2+y_c)^2} \leq \ell_{c_2}.$$

The optimization paths for SQP and the interior-point method are shown in Fig. 5.50.

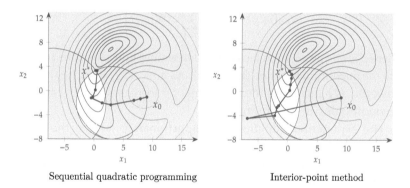

Fig. 5.50 Optimization of constrained spring system.

5.7 Constraint Aggregation

As will be discussed in Chapter 6, some derivative computation methods are efficient for problems with many inputs and few outputs, and others are advantageous for problems with few inputs and many outputs. Thus, if we have many design variables and many constraints, there is no efficient way to compute the required constraint Jacobian.

One workaround is to *aggregate* the constraints and solve the optimization problem with a new set of constraints. Each aggregation would have the form

$$\bar{g}(x) \equiv \bar{g}(g(x)) \leq 0, \quad (5.109)$$

where $\bar{g}$ is a scalar, and g is the vector of constraints we want to aggregate. One of the properties we want for the aggregation function is that if any of the original constraints are violated, then $\bar{g} > 0$.

One way to aggregate constraints would be to define the aggregated constraint function as the maximum of all constraints,

$$\bar{g}(x) = \max(g(x)). \quad (5.110)$$

If $\max(g(x)) \leq 0$, then we know that all of components of $g(x) \leq 0$. However, the maximum function is not differentiable, so it is not desirable for gradient-based optimization. In the rest of this section, we introduce several viable functions for constraint aggregation that are differentiable.

The Kreisselmeier–Steinhauser (KS) aggregation was one of the first aggregation functions proposed for optimization and is defined as follows:[108]

108. Kreisselmeier and Steinhauser, *Systematic control design by optimizing a vector performance index*, 1979.

$$\bar{g}_{KS}(\rho, g) = \frac{1}{\rho} \ln\left(\sum_{j=1}^{n_g} \exp(\rho g_j)\right), \quad (5.111)$$

where ρ is an aggregation factor that determines how close this function is to the maximum function (Eq. 5.110). As $\rho \to \infty$, $\bar{g}_{KS}(\rho, g) \to \max(g)$. However, as ρ increases, the curvature of $\bar{g}$ increases, which can cause ill-conditioning in the optimization.

The exponential function disproportionately weighs the higher positive values in the constraint vector, but it does so in a smooth way. Because the exponential function can easily result in overflow, it is preferable to use the alternate (but equivalent) form of the KS function,

$$\bar{g}_{KS}(\rho, g) = \max_j g_j + \frac{1}{\rho} \ln\left(\sum_{j=1}^{n_g} \exp\left(\rho\left(g_j - \max_j g_j\right)\right)\right). \quad (5.112)$$

The value of ρ should be tuned for each problem, but $\rho = 100$ works well for many problems.

Example 5.17 Constrained spring system with aggregated constraints

Consider the constrained spring system from Ex. 5.16. Aggregating the two constraints using the KS function, we can formulate a single constraint as

$$\bar{g}_{KS}(x_1, x_2) = \frac{1}{\rho} \ln\left(\exp\left(\rho g_2(x_1, x_2)\right) + \exp\left(\rho g_2(x_1, x_2)\right)\right),$$

5.7 Constraint Aggregation

where

$$g_1(x_1, x_2) = \sqrt{(x_1 + x_{c_1})^2 + (x_2 + y_c)^2} - \ell_{c_1}$$

$$g_2(x_1, x_2) = \sqrt{(x_1 - x_{c_2})^2 + (x_2 + y_c)^2} - \ell_{c_2}.$$

Figure 5.51 shows the contour of $\tilde{g}_{KS} = 0$ for increasing values of the aggregation parameter ρ.

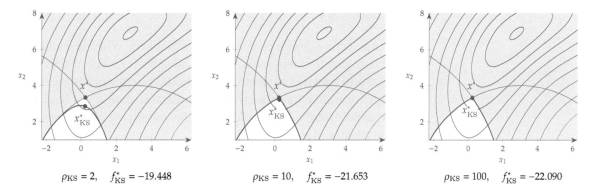

$\rho_{KS} = 2$, $f^*_{KS} = -19.448$ $\qquad$ $\rho_{KS} = 10$, $f^*_{KS} = -21.653$ $\qquad$ $\rho_{KS} = 100$, $f^*_{KS} = -22.090$

For the lowest value of ρ, the feasible region is reduced, resulting in a conservative optimum. For the highest value of ρ, the optimum obtained with constraint aggregation is graphically indistinguishable, and the objective function value approaches the true optimal value of -22.1358.

Fig. 5.51 KS function aggregation of two constraints. The optimum of the problem with aggregated constraints, x^*_{KS}, approaches the true optimum as the aggregation parameter ρ_{KS} increases.

The p-norm aggregation function is another option for aggregation and is defined as follows:[109]

$$\tilde{g}_{PN}(\rho) = \max_j |g_j| \left(\sum_{j=1}^{n_g} \left| \frac{g_j}{\max_j g_j} \right|^p \right)^{\frac{1}{p}}. \qquad (5.113)$$

109. Duysinx and Bendsøe, *Topology optimization of continuum structures with local stress constraints*, 1998.

The absolute value in this equation can be an issue if g can take both positive and negative values because the function is not differentiable in regions where g transitions from positive to negative.

A class of aggregation functions known as *induced functions* was designed to provide more accurate estimates of $\max(g)$ for a given value of ρ than the KS and induced norm functions.[110] There are two main types of induced functions: one uses exponentials, and the other uses powers. The induced exponential function is given by

110. Kennedy and Hicken, *Improved constraint-aggregation methods*, 2015.

$$g_{IE}(\rho) = \frac{\sum_{j=1}^{n_g} g_j \exp(\rho g_j)}{\sum_{j=1}^{n_g} \exp(\rho g_j)}. \qquad (5.114)$$

The induced power function is given by

$$g_{\text{IP}}(\rho) = \frac{\sum_{j=1}^{n_g} g_i^{\rho+1}}{\sum_{j=1}^{n_g} g_i^{\rho}}. \tag{5.115}$$

The induced power function is only applicable if $g_j \geq 0$ for $j = 1, \ldots, n_g$.

5.8 Summary

Most engineering design problems are constrained. When formulating a problem, practitioners should be critical of their choice of objective function and constraints. Metrics that should be constraints are often wrongly formulated as objectives. A constraint should not limit the design unnecessarily and should reflect the underlying physical reason for that constraint as much as possible.

The first-order optimality conditions for constrained problems—the KKT conditions—require the gradient of the objective to be a linear combination of the gradients of the constraints. This ensures that there is no feasible descent direction. Each constraint is associated with a Lagrange multiplier that quantifies how significant that constraint is at the optimum. For inequality constraints, a Lagrange multiplier that is zero means that the corresponding constraint is inactive. For inequality constraints, slack variables quantify how close a constraint is to becoming active; a slack variable that is zero means that the corresponding constraint is active. Lagrange multipliers and slack variables are unknowns that need to be solved together with the design variables. The complementary slackness condition introduces a combinatorial problem that is challenging to solve.

Penalty methods solve constrained problems by adding a metric to the objective function quantifying how much the constraints are violated. These methods are helpful as a conceptual model and are used in gradient-free optimization algorithms (Chapter 7). However, penalty methods only find approximate solutions and are subject to numerical issues when used with gradient-based optimization.

Methods based on the KKT conditions are preferable. The most widely used among such methods are SQP and interior-point methods. These methods apply Newton's method to the KKT conditions. One primary difference between these two methods is in the treatment of inequality constraints. SQP methods distinguish between active and inactive constraints, treating potentially active constraints as equality constraints and ignoring the potentially inactive ones. Interior-point methods add slack variables to force all constraints to behave like equality constraints.

Problems

5.1 Answer *true* or *false* and correct the false statements.

   a. Penalty methods are among the most effective methods for constrained optimization.

   b. For an equality constraint in n-dimensional space, all feasible directions about a point are perpendicular to the constraint gradient at that point and define a hyperplane with dimension $n - 1$.

   c. The feasible directions about a point on an inequality constraint define an open half-space whose dividing hyperplane is perpendicular to the gradient of the constraint at that point.

   d. A point is optimal if there is only one feasible direction that is also a descent direction.

   e. For an inequality constrained problem, if we replace the inequalities that are active at the optimum with equality constraints and ignore the inactive constraints, we get the same optimum.

   f. For a point to be optimal, the Lagrange multipliers for both the equality constraint and the active inequality constraints must be positive.

   g. The complementary slackness conditions are easy to solve for because either the Lagrange multiplier is zero or the slack variable is zero.

   h. At the optimum of a constrained problem, the Hessian of the Lagrangian function must be positive semidefinite.

   i. The Lagrange multipliers represent the change in the objective function we would get for a perturbation in the constraint value.

   j. SQP seeks to find the solution of the KKT system.

   k. Interior-point methods must start with a point in the interior of the feasible region.

   l. Constraint aggregation combines multiple constraints into a single constraint that is equivalent.

5.2 Let us modify Ex. 5.2 so that the equality constraint is the negative of the original one—that is,

$$h(x_1, x_2) = -\frac{1}{4}x_1^2 - x_2^2 + 1 = 0.$$

Classify the critical points and compare them with the original solution. What does that tell you about the significance of the Lagrange multiplier sign?

5.3 Similar to the previous exercise, consider Ex. 5.4 and modify it so that the inequality constraint is the negative of the original one—that is,

$$h(x_1, x_2) = -\frac{1}{4}x_1^2 - x_2^2 + 1 \leq 0.$$

Classify the critical points and compare them with the original solution.

5.4 Consider the following optimization problem:

$$\begin{aligned}
\text{minimize} \quad & x_1^2 + 3x_2^2 + 4 \\
\text{by varying} \quad & x_1, x_2 \\
\text{subject to} \quad & x_2 \geq 1 \\
& x_1^2 + 4x_2^2 \leq 4.
\end{aligned} \tag{5.116}$$

Find the optimum analytically.

5.5 Find the rectangle of maximum area that can be inscribed in an ellipse. Give your answer in terms of the ratio of the two areas. Check that your answer is intuitively correct for the special case of a rectangle inscribed in a circle.

5.6 In Section 2.1, we mentioned that Euclid showed that among rectangles of a given perimeter, the square has the largest area. Formulate the problem and solve it analytically. What are the units in this problem, and what is the physical interpretation of the Lagrange multiplier? *Exploration*: Show that if you minimize the perimeter with an area constrained to the optimal value you found previously, you get the same solution.

5.7 *Column in compression.* Consider a thin-walled tubular column subjected to a compression force, as shown in Fig. 5.52. We want to minimize the mass of the column while ensuring that the structure does not yield or buckle under a compression force of magnitude F. The design variables are the radius of the tube (R) and the wall thickness (t). This design optimization problem can

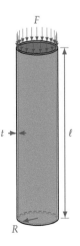

Fig. 5.52 Slender tubular column in compression.

be stated as follows:

$$\text{minimize} \quad 2\rho\ell\pi Rt \qquad \text{mass}$$
$$\text{by varying} \quad R, t \qquad \text{radius, wall thickness}$$
$$\text{subject to} \quad \frac{F}{2\pi Rt} - \sigma_{\text{yield}} \leq 0 \qquad \text{yield stress}$$
$$F - \frac{\pi^3 E R^3 t}{4\ell^2} \leq 0 \qquad \text{buckling load}$$

In the formula for the mass in this objective, ρ is the material density, and we assume that $t \ll R$. The first constraint is the compressive stress, which is simply the force divided by the cross-sectional area. The second constraint uses Euler's critical buckling load formula, where E is the material Young's modulus, and the second moment of area is replaced with the one corresponding to a circular cross section ($I = \pi R^3 t$).

Find the optimum R and t as a function of the other parameters. Pick reasonable values for the parameters, and verify your solution graphically. Plot the gradients of the objective and constraints at the optimum, and verify the Lagrange multipliers graphically.

5.8 *Beam with H section.* Consider a cantilevered beam with an H-shaped cross section composed of a web and flanges subject to a transverse load, as shown in Fig. 5.53. The objective is to minimize the structural weight by varying the web thickness t_w and the flange thickness t_b, subject to stress constraints. The other cross-sectional parameters are fixed; the web height h is 250 mm, and the flange width b is 125 mm. The axial stress in the flange and the shear stress in the web should not exceed the corresponding yield values ($\sigma_{\text{yield}} = 200$ MPa, and $\tau_{\text{yield}} = 116$ MPa, respectively). The optimization problem can be stated as follows:

$$\text{minimize} \quad 2bt_b + ht_w \qquad \text{mass}$$
$$\text{by varying} \quad t_b, t_w \qquad \text{flange and web thicknesses}$$
$$\text{subject to} \quad \frac{P\ell h}{2I} - \sigma_{\text{yield}} \leq 0 \qquad \text{axial stress}$$
$$\frac{1.5P}{ht_w} - \tau_{\text{yield}} \leq 0 \qquad \text{shear stress}$$

The second moment of area for the H section is

$$I = \frac{h^3}{12} t_w + \frac{b}{6} t_b^3 + \frac{h^2 b}{2} t_b \,.$$

Find the optimal values of t_b and t_w by solving the KKT conditions analytically. Plot the objective contours and constraints to verify your result graphically.

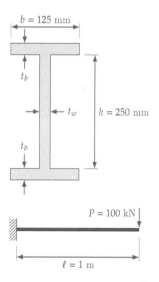

Fig. 5.53 Cantilever beam with H section.

5.9 *Penalty method implementation.* Program one or more penalty methods from Section 5.4.

    a. Solve the constrained problem from Ex. 5.6 as a first test of your implementation. Use an existing software package for the optimization subproblem or the unconstrained optimizer you implemented in Prob. 4.9. How far can you push the penalty parameter until the optimizer fails? How close can you get to the exact optimum? Try different starting points and verify that the algorithm always converges to the same optimum.

    b. Solve Prob. 5.3.

    c. Solve Prob. 5.11.

    d. *Exploration*: Solve any other problem from this section or a problem of your choosing.

5.10 *Constrained optimizer implementation.* Program an SQP or interior-point algorithm. You may repurpose the BFGS algorithm that you implemented in Prob. 4.9. For SQP, start by implementing only equality constraints, reformulating test problems with inequality constraints as problems with only equality constraints.

    a. Reproduce the results from Ex. 5.12 (SQP) or Ex. 5.15 (interior point).

    b. Solve Prob. 5.3.

    c. Solve Prob. 5.11.

    d. Compare the computational cost, precision, and robustness of your optimizer with those of an existing software package.

Fig. 5.54 Ellipsoid fuel tank.

5.11 *Aircraft fuel tank.* A jet aircraft needs to carry a streamlined external fuel tank with a required volume. The tank shape is approximated as an ellipsoid (Fig. 5.54). We want to minimize the drag of the fuel tank by varying its length and diameter—that is:

$$\begin{aligned}\text{minimize} \quad & D(\ell, d) \\ \text{by varying} \quad & \ell, d \\ \text{subject to} \quad & V_{\text{req}} - V(\ell, d) \leq 0\,.\end{aligned}$$

The drag is given by

$$D = \frac{1}{2}\rho v^2 C_D S,$$

where the air density is $\rho = 0.55$ kg/m^3, and the aircraft speed is $v = 300$ m/s. The drag coefficient of an ellipsoid can be estimated as*

$$C_D = C_f \left[1 + 1.5 \left(\frac{d}{\ell}\right)^{3/2} + 7 \left(\frac{d}{\ell}\right)^3\right].$$

We assume a friction coefficient of $C_f = 0.0035$. The drag is proportional to the surface area of the tank, which, for an ellipsoid, is

$$S = \frac{\pi}{2} d^2 \left(1 + \frac{\ell}{de} \arcsin e\right),$$

where $e = \sqrt{1 - d^2/\ell^2}$. The volume of the fuel tank is

$$V = \frac{\pi}{6} d^2 \ell,$$

and the required volume is $V_{req} = 2.5$ m^3.

Find the optimum tank length and diameter numerically using your own optimizer or a software package. Verify your solution graphically by plotting the objective function contours and the constraint.

5.12 Solve a variation of Ex. 5.16 where we replace the system of cables with a cable and a rod that resists both tension and compression. The cable is positioned above the spring, as shown in Fig. 5.55, where $x_c = 2$ m, and $y_c = 3$ m, with a maximum length of $\ell_c = 7.0$ m. The rod is positioned at $x_r = 2$ m and $y_r = 4$ m, with a length of $\ell_r = 4.5$ m. How does this change the problem

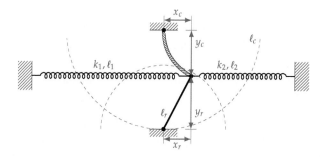

Fig. 5.55 Spring system constrained by two cables.

formulation? Does the optimum change?

5.13 *Three-bar truss.* Consider the truss shown in Fig. 5.56. The truss is subjected to a load P, and we want to minimize the mass of the structure subject to stress and buckling constraints.† The axial

†This is a well-known optimization problem formulated by Schmit[32] when he first proposed integrating numerical optimization with finite-element structural analysis.

32. Schmit, *Structural design by systematic synthesis*, 1960.

stresses in each bar are

$$\sigma_1 = \frac{1}{\sqrt{2}} \left(\frac{P \cos\theta}{A_o} + \frac{P \sin\theta}{A_o + \sqrt{2} A_m} \right)$$

$$\sigma_2 = \frac{\sqrt{2} P \sin\theta}{A_o + \sqrt{2} A_m}$$

$$\sigma_3 = \frac{1}{\sqrt{2}} \left(\frac{P \sin\theta}{A_o + \sqrt{2} A_m} - \frac{P \cos\theta}{A_o} \right),$$

where A_o is the cross-sectional area of the outer bars 1 and 3, and A_m is the cross-sectional area of the middle bar 2. The full optimization problem for the three-bar truss is as follows:

$$\begin{array}{lll}
\text{minimize} & \rho\left(\ell(2\sqrt{2} A_o + A_m)\right) & \text{mass} \\
\text{by varying} & A_o, A_m & \text{cross-sectional areas} \\
\text{subject to} & A_{\min} - A_o \leq 0 & \text{area lower bound} \\
& A_{\min} - A_m \leq 0 & \\
& \sigma_{\text{yield}} - \sigma_1 \leq 0 & \text{stress constraints} \\
& \sigma_{\text{yield}} - \sigma_2 \leq 0 & \\
& \sigma_{\text{yield}} - \sigma_3 \leq 0 & \\
& -\sigma_1 - \dfrac{\pi^2 E \beta A_o}{2\ell^2} \leq 0 & \text{buckling constraints} \\
& -\sigma_2 - \dfrac{\pi^2 E \beta A_m}{2\ell^2} \leq 0 & \\
& -\sigma_3 - \dfrac{\pi^2 E \beta A_o}{2\ell^2} \leq 0 &
\end{array}$$

In the buckling constraints, β relates the second moment of area to the area ($I = \beta A^2$) and is dependent on the cross-sectional shape of the bars. Assuming a square cross section, $\beta = 1/12$. The bars are made out of an aluminum alloy with the following properties: $\rho = 2710$ kg/m^3, $E = 69$ GPa, $\sigma_{\text{yield}} = 110$ MPa.

Find the optimal bar cross-sectional areas using your own optimizer or a software package. Which constraints are active? Verify your result graphically. *Exploration*: Try different combinations of unit magnitudes (e.g., Pa versus MPa for the stresses) for the functions of interest and the design variables to observe the effect of scaling.

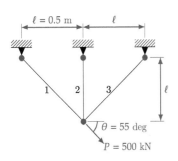

Fig. 5.56 Three-bar truss elements.

5.14 Solve the same three-bar truss optimization problem in Prob. 5.13 by aggregating all the constraints into a single constraint. Try

5.8 Summary

different aggregation parameters and see how close you can get to the solution you obtained for Prob. 5.13.

5.15 *Ten-bar truss.* Consider the 10-bar truss structure described in Appendix D.2.2. The full design optimization problem is as follows:

$$\begin{aligned}
&\text{minimize} && \rho \sum_{i=1}^{10} A_i \ell_i && \text{mass} \\
&\text{by varying} && A_i, \quad i = 1, \ldots, 10 && \text{cross-sectional areas} \\
&\text{subject to} && A_i \geq A_{\min} && \text{minimum area} \\
& && |\sigma_i| \leq \sigma_{y_i} \quad i = 1, \ldots, 10 && \text{stress constraints}
\end{aligned}$$

Find the optimal mass and corresponding cross-sectional areas using your own optimizer or a software package. Show a convergence plot. Report the number of function evaluations and the number of major iterations. *Exploration:* Restart from different starting points. Do you get more than one local minimum? What can you conclude about the multimodality of the design space?

5.16 Solve the same 10-bar truss optimization problem of Prob. 5.15 by aggregating all the constraints into a single constraint. Try different aggregation parameters and see how close you can get to the solution you obtained for Prob. 5.15.

5.17 Consider the aircraft wing design problem described in Appendix D.1.6. Now we will add a constraint on the bending stress at the root of the wing, as described in Ex. 1.3.

We derive the bending stress using the one-dimensional beam bending theory. Assuming that the lift distribution is uniform, the load per unit length is L/b. We can consider the wing as a cantilever of length $b/2$. The bending moment at the wing root is

$$M = \frac{(L/b)(b/2)^2}{2} = \frac{Lb}{8}.$$

Now we assume that the wing structure has the H-shaped cross section from Prob. 5.8 with a constant thickness of $t_w = t_b = 4\,\text{mm}$. We relate the cross-section height h_{sec} and width b_{sec} to the chord as $h_{\text{sec}} = 0.1c$ and $b_{\text{sec}} = 0.4c$. With these assumptions, we can compute the second moment of area I in terms of c.

The maximum bending stress is then

$$\sigma_{\max} = \frac{M h_{\text{sec}}}{2I}.$$

Considering the safety factor of 1.5 and the ultimate load factor of 2.5, the stress constraint is

$$2.5\sigma_{max} - \frac{\sigma_{yield}}{1.5} \leq 0,$$

where $\sigma_{yield} = 200$ MPa.

Solve this problem and compare the solution with the unconstrained optimum. Plot the objective contours and constraint to verify your result graphically.

Computing Derivatives

6

The gradient-based optimization methods introduced in Chapters 4 and 5 require the derivatives of the objective and constraints with respect to the design variables, as illustrated in Fig. 6.1. Derivatives also play a central role in other numerical algorithms. For example, the Newton-based methods introduced in Section 3.8 require the derivatives of the residuals.

The accuracy and computational cost of the derivatives are critical for the success of these methods. Gradient-based methods are only efficient when the derivative computation is also efficient. The computation of derivatives can be the bottleneck in the overall optimization procedure, especially when the model solver needs to be called repeatedly. This chapter introduces the various methods for computing derivatives and discusses the relative advantages of each method.

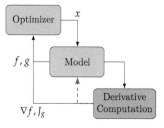

Fig. 6.1 Efficient derivative computation is crucial for the overall efficiency of gradient-based optimization.

By the end of this chapter you should be able to:

1. List the methods for computing derivatives.
2. Explain the pros and cons of these methods.
3. Implement the methods for some computational models.
4. Understand how the methods are connected through the unified derivatives equation.

6.1 Derivatives, Gradients, and Jacobians

The derivatives we focus on are *first-order* derivatives of one or more functions of interest (f) with respect to a vector of variables (x). In the engineering optimization literature, the term *sensitivity analysis* is often used to refer to the computation of derivatives, and derivatives are sometimes referred to as *sensitivity derivatives* or *design sensitivities*. Although these terms are not incorrect, we prefer to use the more specific and concise term *derivative*.

For the sake of generality, we do not specify which functions we want to differentiate in this chapter (which could be an objective, constraints, residuals, or any other function). Instead, we refer to the functions being differentiated as the *functions of interest* and represent them as a vector-valued function, $f = [f_1, f_2, \ldots, f_{n_f}]$. Neither do we specify the variables with respect to which we differentiate (which could be design variables, state variables, or any other independent variable).

The derivatives of all the functions of interest with respect to all the variables form a Jacobian matrix,

$$J_f = \frac{\partial f}{\partial x} = \begin{bmatrix} \nabla f_1^\mathsf{T} \\ \vdots \\ \nabla f_{n_f}^\mathsf{T} \end{bmatrix} = \underbrace{\begin{bmatrix} \frac{\partial f_1}{\partial x_1} & \cdots & \frac{\partial f_1}{\partial x_{n_x}} \\ \vdots & \ddots & \vdots \\ \frac{\partial f_{n_f}}{\partial x_1} & \cdots & \frac{\partial f_{n_f}}{\partial x_{n_x}} \end{bmatrix}}_{(n_f \times n_x)}, \quad (6.1)$$

which is an $(n_f \times n_x)$ rectangular matrix where each row corresponds to the gradient of each function with respect to all the variables. Row i of the Jacobian is the gradient of function f_i. Each column in the Jacobian is called the *tangent* with respect to a given variable x_j. The Jacobian can be related to the ∇ operator as follows:

$$J_f = f \nabla^\mathsf{T} = \begin{bmatrix} f_1 \\ \vdots \\ f_{n_f} \end{bmatrix} \begin{bmatrix} \frac{\partial}{\partial x_1} & \cdots & \frac{\partial}{\partial x_{n_x}} \end{bmatrix} = \begin{bmatrix} \frac{\partial f_1}{\partial x_1} & \cdots & \frac{\partial f_1}{\partial x_{n_x}} \\ \vdots & \ddots & \vdots \\ \frac{\partial f_{n_f}}{\partial x_1} & \cdots & \frac{\partial f_{n_f}}{\partial x_{n_x}} \end{bmatrix}. \quad (6.2)$$

Example 6.1 Jacobian of a vector-valued function

Consider the following function with two variables and two functions of interest:

$$f(x) = \begin{bmatrix} f_1(x_1, x_2) \\ f_2(x_1, x_2) \end{bmatrix} = \begin{bmatrix} x_1 x_2 + \sin x_1 \\ x_1 x_2 + x_2^2 \end{bmatrix}.$$

We can differentiate this symbolically to obtain exact reference values:

$$\frac{\partial f}{\partial x} = \begin{bmatrix} x_2 + \cos x_1 & x_1 \\ x_2 & x_1 + 2x_2 \end{bmatrix}.$$

We evaluate this at $x = (\pi/4, 2)$, which yields

$$\frac{\partial f}{\partial x} = \begin{bmatrix} 2.707 & 0.785 \\ 2.000 & 4.785 \end{bmatrix}.$$

6.2 Overview of Methods for Computing Derivatives

We can classify the methods for computing derivatives according to the representation used for the numerical model. There are three possible representations, as shown in Fig. 6.2. In one extreme, we know nothing about the model and consider it a black box where we can only control the inputs and observe the outputs (Fig. 6.2, left). In this chapter, we often refer to x as the *input variables* and f as the *output variables*. When this is the case, we can only compute derivatives using finite differences (Section 6.4).

In the other extreme, we have access to the source code used to compute the functions of interest and perform the differentiation line by line (Fig. 6.2, right). This is the essence of the algorithmic differentiation approach (Section 6.6). The complex-step method (Section 6.5) is related to algorithmic differentiation, as explained in Section 6.6.5.

In the intermediate case, we consider the model residuals and states (Fig. 6.2, middle), which are the quantities required to derive and implement implicit analytic methods (Section 6.7). When the model can be represented with multiple components, we can use a coupled derivative approach (Section 13.3) where any of these derivative computation methods can be used for each component.

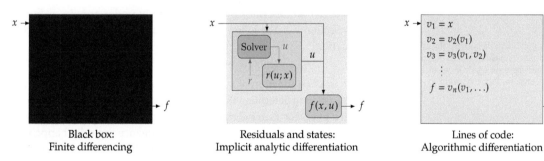

Black box:
Finite differencing

Residuals and states:
Implicit analytic differentiation

Lines of code:
Algorithmic differentiation

Fig. 6.2 Derivative computation methods can consider three different levels of information: function values (left), model states (middle), and lines of code (right).

Tip 6.1 Identify and mitigate the sources of numerical noise

As mentioned in Tip 3.2, it is vital to determine the level of numerical noise in your model. This is especially important when computing derivatives of the model because taking the derivative can amplify the noise. There are several common sources of model numerical noise, some of which we can mitigate.

Iterative solvers can introduce numerical noise when the convergence tolerance is too high or when they have an inherent limit in their precision (see Section 3.5.3). When we do not have enough precision, we can reduce the convergence tolerance or increase the iteration limit.

Another possible source of error is file input and output. Many legacy codes are driven by reading and writing input and output files. However, the numbers in the files usually have fewer digits than the code's working precision. The ideal solution is to modify the code to be called directly and pass the data through memory. Another solution is to increase the precision in the files.

6.3 Symbolic Differentiation

Symbolic differentiation is well known and widely used in calculus, but it is of limited use in the numerical optimization of most engineering models. Except for the most straightforward cases (e.g., Ex. 6.1), many engineering models involve a large number of operations, utilize loops and various conditional logic, are implicitly defined, or involve iterative solvers (see Chapter 3). Although the mathematical expressions within these iterative procedures is explicit, it is challenging, or even impossible, to use symbolic differentiation to obtain closed-form mathematical expressions for the derivatives of the procedure. Even when it is possible, these expressions are almost always computationally inefficient.

Example 6.2 Symbolic differentiation leading to expression swell

Kepler's equation describes the orbit of a body under gravity, as briefly discussed in Section 2.2. The following implicit equation can be obtained from Kepler's equation:*

*Here, f is the difference between the eccentric and mean anomalies, x is the mean anomaly, and the eccentricity is set to 1. For more details, see Probs. 3.6 and 6.6.

$$f = \sin(x + f).$$

Thus, f is an implicit function of x. As a simple numerical procedure, we use fixed-point iteration to determine the value of f for a given input x. That means we start with a guess for f on the right-hand side of that expression to estimate a new value for f, and repeat. In this case, convergence typically happens in about 10 iterations. Arbitrarily, we choose x as the initial guess for f, resulting in the following computational procedure:

> **Input:** x
> $f = x$
> **for** $i = 1$ **to** 10 **do**
> $f = \sin(x + f)$
> **end for**
> **return** f

Now that we have a computational procedure, we would like to compute the derivative df/dx. We can use a symbolic math toolbox to find the following closed-form expression for this derivative:

```
dfdx =
cos(x + sin(x + sin(x + sin(x + sin(x + sin(x + sin(x + sin(x +
   sin(2*x))))))))))*(cos(x + sin(x + sin(x + sin(x + sin(x +
   sin(x + sin(x + sin(2*x))))))))))*(cos(x + sin(x + sin(x +
   sin(x + sin(x + sin(x + sin(2*x))))))))*(cos(x + sin(x + sin
   (x + sin(x + sin(x + sin(2*x))))))) *(cos(x + sin(x + sin(x +
   sin(x + sin(2*x))))))*(cos(x + sin(x + sin(x + sin(2*x)))))
   *(cos(x + sin(x + sin(2*x))))*(cos(x + sin(x + sin(2*x)))*(
   cos(x + sin(2*x))*(2*cos(2*x) + 1) + 1) + 1) + 1) + 1) + 1) +
   1) + 1)
```

This expression is long and is full of redundant calculations. This problem becomes exponentially worse as the number of iterations in the loop is increased, so this approach is intractable for computational models of even moderate complexity—this is known as *expression swell*. Therefore, we dedicate the rest of this chapter to methods for computing derivatives numerically.

Symbolic differentiation is still valuable for obtaining derivatives of simple explicit components within a larger model. Furthermore, algorithm differentiation (discussed in a later section) relies on symbolic differentiation to differentiate each line of code in the model.

6.4 Finite Differences

Because of their simplicity, finite-difference methods are a popular approach to computing derivatives. They are versatile, requiring nothing more than function values. Finite differences are the only viable option when we are dealing with black-box functions because they do not require any knowledge about how the function is evaluated. Most gradient-based optimization algorithms perform finite differences by default when the user does not provide the required gradients. However, finite differences are neither accurate nor efficient.

6.4.1 Finite-Difference Formulas

Finite-difference approximations are derived by combining Taylor series expansions. It is possible to obtain finite-difference formulas that estimate an arbitrary order derivative with any order of truncation error by using the right combinations of these expansions. The simplest finite-difference formula can be derived directly from a Taylor series expansion in the jth direction,

$$f(x + h\hat{e}_j) = f(x) + h\frac{\partial f}{\partial x_j} + \frac{h^2}{2!}\frac{\partial^2 f}{\partial x_j^2} + \frac{h^3}{3!}\frac{\partial^3 f}{\partial x_j^3} + \ldots, \quad (6.3)$$

where $\hat{e}_j$ is the unit vector in the jth direction. Solving this for the first derivative, we obtain the finite-difference formula,

$$\frac{\partial f}{\partial x_j} = \frac{f(x + h\hat{e}_j) - f(x)}{h} + O(h), \tag{6.4}$$

where h is a small scalar called the *finite-difference step size*. This approximation is called the *forward difference* and is directly related to the definition of a derivative because

$$\frac{\partial f}{\partial x_j} = \lim_{h \to 0} \frac{f(x + h\hat{e}_j) - f(x)}{h} \approx \frac{f(x + h\hat{e}_j) - f(x)}{h}. \tag{6.5}$$

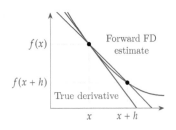

Fig. 6.3 Exact derivative compared with a forward finite-difference approximation (Eq. 6.4).

The truncation error is $O(h)$, and therefore this is a first-order approximation. The difference between this approximation and the exact derivative is illustrated in Fig. 6.3.

The backward-difference approximation can be obtained by replacing h with $-h$ to yield

$$\frac{\partial f}{\partial x_j} = \frac{f(x) - f(x - h\hat{e}_j)}{h} + O(h), \tag{6.6}$$

which is also a first-order approximation.

Assuming each function evaluation yields the full vector f, the previous formulas compute the jth column of the Jacobian in Eq. 6.1. To compute the full Jacobian, we need to loop through each direction $\hat{e}_j$, add a step, recompute f, and compute a finite difference. Hence, the cost of computing the complete Jacobian is proportional to the number of input variables of interest, n_x.

For a second-order estimate of the first derivative, we can use the expansion of $f(x - h\hat{e}_j)$ to obtain

$$f(x - h\hat{e}_j) = f(x) - h\frac{\partial f}{\partial x_j} + \frac{h^2}{2!}\frac{\partial^2 f}{\partial x_j^2} - \frac{h^3}{3!}\frac{\partial^3 f}{\partial x_j^3} + \dots \tag{6.7}$$

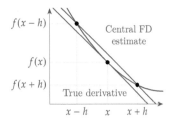

Fig. 6.4 Exact derivative compared with a central finite-difference approximation (Eq. 6.8).

Then, if we subtract this from the expansion in Eq. 6.3 and solve the resulting equation for the derivative of f, we get the *central-difference* formula,

$$\frac{\partial f}{\partial x_j} = \frac{f(x + h\hat{e}_j) - f(x - h\hat{e}_j)}{2h} + O(h^2). \tag{6.8}$$

The stencil of points for this formula is shown in Fig. 6.4, where we can see that this estimate is closer to the actual derivative than the forward difference.

Even more accurate estimates can be derived by combining different Taylor series expansions to obtain higher-order truncation error

6.4 Finite Differences

terms. This technique is widely used in finite-difference methods for solving differential equations, where higher-order estimates are desirable. However, finite-precision arithmetic eventually limits the achievable accuracy for our purposes (as discussed in the next section). With double-precision arithmetic, there are not enough significant digits to realize a significant advantage beyond central difference.

We can also estimate second derivatives (or higher) by combining Taylor series expansions. For example, adding the expansions for $f(x+h)$ and $f(x-h)$ cancels out the first derivative and third derivative terms, yielding the second-order approximation to the second-order derivative,

$$\frac{\partial^2 f}{\partial x_j^2} = \frac{f(x + 2h\hat{e}_j) - 2f(x) + f(x - 2h\hat{e}_j)}{4h^2} + O\left(h^2\right). \qquad (6.9)$$

The finite-difference method can also be used to compute directional derivatives, which are the scalar projection of the gradient into a given direction. To do this, instead of stepping in orthogonal directions to get the gradient, we need to step in the direction of interest, p, as shown in Fig. 6.5. Using the forward difference, for example,

$$\nabla_p f = \frac{f(x + hp) - f(x)}{h} + O(h). \qquad (6.10)$$

One application of directional derivatives is to compute the slope in line searches (Section 4.3).

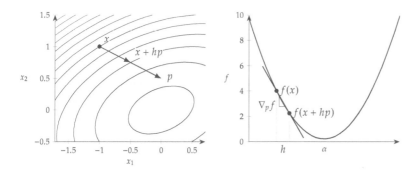

Fig. 6.5 Computing a directional derivative using a forward finite difference.

6.4.2 The Step-Size Dilemma

When estimating derivatives using finite-difference formulas, we are faced with the *step-size dilemma*. Because each estimate has a truncation error of $O(h)$ (or $O(h^2)$ when second order), we would like to choose as small of a step size as possible to reduce this error. However, as the step size reduces, *subtractive cancellation* (a roundoff error introduced in Section 3.5.1) becomes dominant. Given the opposing trends of these errors, there is an optimal step size for which the sum of the two errors is at a minimum.

Theoretically, the optimal step size for the forward finite difference is approximately $\sqrt{\varepsilon_f}$, where ε_f is the precision of f. The error bound is also about $\sqrt{\varepsilon_f}$. For the central difference, the optimal step size scales approximately with $\varepsilon_f^{1/3}$, with an error bound of $\varepsilon_f^{2/3}$. These step and error bound estimates are just approximate and assume well-scaled problems.

Example 6.3 Accuracy of finite differences

To demonstrate the step-size dilemma, consider the following function:

$$f(x) = \frac{e^x}{\sqrt{\sin^3 x + \cos^3 x}}.$$

The exact derivative at $x = 1.5$ is computed to 16 digits based on symbolic differentiation as a reference value.

In Fig. 6.6, we show the derivatives given by the forward difference, where we can see that as we decrease the step size, the derivative approaches the exact value, but then it worsens and becomes zero for a small enough step size.

We plot the relative error of the forward- and central-difference formulas for a *decreasing* step size in Fig. 6.7. As the step decreases, the forward-difference estimate initially converges at a linear rate because its truncation error is $O(h)$, whereas the central difference converges quadratically. However, as the step reduces below a particular value (about 10^{-8} for the forward difference and 10^{-5} for the central difference), subtractive cancellation errors become increasingly significant. These values match the theoretical predictions for the optimal step and error bounds when we set $\varepsilon_f = 10^{-16}$. When h is so small that no difference exists in the output (for steps smaller than 10^{-16}), the finite-difference estimates yield zero (and $\varepsilon = 1$), which corresponds to 100 percent error.

Table 6.1 lists the data for the forward difference, where we can see the number of digits in the difference Δf decreasing with decreasing step size until the difference is zero (for $h = 10^{-17}$).

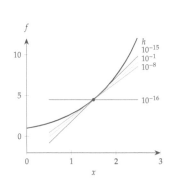

Fig. 6.6 The forward-difference derivative initially improves as the step decreases but eventually gives a zero derivative for a small enough step size.

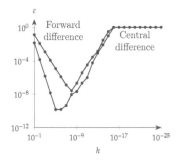

Fig. 6.7 As the step size h decreases, the total error in the finite-difference estimates initially decreases because of a reduced truncation error. However, subtractive cancellation takes over when the step is small enough and eventually yields an entirely wrong derivative.

Tip 6.2 When using finite differencing, always perform a step-size study

In practice, most gradient-based optimizers use finite differences by default to compute the gradients. Given the potential for inaccuracies, finite differences are often the culprit in cases where gradient-based optimizers fail to converge. Although some of these optimizers try to estimate a good step size, there is no substitute for a step-size study by the user. The step-size study must be

h	$f(x+h)$	Δf	df/dx
10^{-1}	4.9562638252880662	0.4584837713419043	4.58483771
10^{-2}	4.5387928890592475	0.0410128351130856	4.10128351
10^{-4}	4.4981854440562818	0.0004053901101200	4.05390110
10^{-6}	4.4977841073787870	0.0000040534326251	4.05343263
10^{-8}	4.4977800944804409	0.0000000405342790	4.05342799
10^{-10}	4.4977800543515052	0.0000000004053433	4.05344203
10^{-12}	4.4977800539502155	0.0000000000040536	4.05453449
10^{-14}	4.4977800539462027	0.0000000000000409	4.17443857
10^{-16}	4.4977800539461619	0.0000000000000000	0.00000000
10^{-18}	4.4977800539461619	0.0000000000000000	0.00000000
Exact	4.4977800539461619		4.05342789

Table 6.1 Subtractive cancellation leads to a loss of precision and, ultimately, inaccurate finite-difference estimates.

performed for all variables and does not necessarily apply to the whole design space. Therefore, repeating this study for other values of x might be required.

Because we do not usually know the exact derivative, we cannot plot the error as we did in Fig. 6.7. However, we can always tabulate the derivative estimates as we did in Table 6.1. In the last column, we can see from the pattern of digits that match the previous step size that $h = 10^{-8}$ is the best step size in this case.

Finite-difference approximations are sometimes used with larger steps than would be desirable from an accuracy standpoint to help smooth out numerical noise or discontinuities in the model. This approach sometimes works, but it is better to address these problems within the model whenever possible. Figure 6.8 shows an example of this effect. For this noisy function, the larger step ignores the noise and gives the correct trend, whereas the smaller step results in an estimate with the wrong sign.

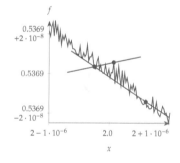

Fig. 6.8 Finite-differencing noisy functions can either smooth the derivative estimates or result in estimates with the wrong trends.

6.4.3 Practical Implementation

Algorithm 6.1 details a procedure for computing a Jacobian using forward finite differences. It is usually helpful to scale the step size based on the value of x_j, unless x_j is too small. Therefore, we combine the relative and absolute quantities to obtain the following step size:

$$\Delta x_j = h \left(1 + |x_j|\right) . \qquad (6.11)$$

This is similar to the expression for the convergence criterion in Eq. 4.24. Although the absolute step size usually differs for each x_j, the relative step size h is often the same and is user-specified.

> **Algorithm 6.1** Forward finite-difference gradient computation of a vector-valued function $f(x)$
>
> **Inputs:**
> x: Point about which to compute the gradient
> f: Vector of functions of interest
> **Outputs:**
> J: Jacobian of f with respect to x
>
> ---
>
> $f_0 = f(x)$ — Evaluate reference values
> $h = 10^{-6}$ — Relative step size (value should be tuned)
> **for** $j = 1$ **to** n_x **do**
> $\quad \Delta x = h(1 + |x_j|)$ — Step size should be scaled but not smaller than h
> $\quad x_j = x_j + \Delta x$ — Modify in place for efficiency, but copying vector is also an option
> $\quad f_+ = f(x)$ — Evaluate function at perturbed point
> $\quad J_{*,j} = \dfrac{f_+ - f_0}{\Delta x}$ — Finite difference yields one column of Jacobian at a time
> $\quad x_j = x_j - \Delta x$ — Do not forget to reset!
> **end for**

6.5 Complex Step

The complex-step derivative approximation, strangely enough, computes derivatives of real functions using complex variables. Unlike finite differences, the complex-step method requires access to the source code and cannot be applied to black-box components. The complex-step method is accurate but no more efficient than finite differences because the computational cost still scales linearly with the number of variables.

6.5.1 Theory

The complex-step method can also be derived using a Taylor series expansion. Rather than using a real step h, as we did to derive the finite-difference formulas, we use a pure imaginary step, ih.* If f is a real function in real variables and is also analytic (differentiable in the complex domain), we can expand it in a Taylor series about a real point x as follows:

$$f(x + ih\hat{e}_j) = f(x) + ih\frac{\partial f}{\partial x_j} - \frac{h^2}{2}\frac{\partial^2 f}{\partial x_j^2} - i\frac{h^3}{6}\frac{\partial^3 f}{\partial x_j^3} + \ldots . \quad (6.12)$$

Taking the imaginary parts of both sides of this equation, we have

$$\operatorname{Im}\left(f(x + ih\hat{e}_j)\right) = h\frac{\partial f}{\partial x_j} - \frac{h^3}{6}\frac{\partial^3 f}{\partial x_j^3} + \ldots . \quad (6.13)$$

---

*This method originated with the work of Lyness and Moler,[112] who developed formulas that use complex arithmetic for computing the derivatives of real functions of arbitrary order with arbitrary order truncation error, much like the Taylor series combination approach in finite differences. Later, Squire and Trapp[49] observed that the simplest of these formulas was convenient for computing first derivatives.

49. Squire and Trapp, *Using complex variables to estimate derivatives of real functions*, 1998.

112. Lyness and Moler, *Numerical differentiation of analytic functions*, 1967.

6.5 Complex Step

Dividing this by h and solving for $\partial f/\partial x_j$ yields the *complex-step derivative approximation*,[†]

$$\frac{\partial f}{\partial x_j} = \frac{\text{Im}\left(f(x + ih\hat{e}_j)\right)}{h} + O(h^2), \qquad (6.14)$$

[†] This approximation can also be derived from one of the Cauchy–Riemann equations, which are fundamental in complex analysis and express complex differentiability.[50]

50. Martins et al., *The complex-step derivative approximation*, 2003.

which is a second-order approximation. To use this approximation, we must provide a complex number with a perturbation in the imaginary part, compute the original function using complex arithmetic, and take the imaginary part of the output to obtain the derivative.

In practical terms, this means that we must convert the function evaluation to take complex numbers as inputs and compute complex outputs. Because we have assumed that $f(x)$ is a real function of a real variable in the derivation of Eq. 6.14, the procedure described here does not work for models that already involve complex arithmetic. In Section 6.5.2, we explain how to convert programs to handle the required complex arithmetic for the complex-step method to work in general. The complex-step method has been extended to compute exact second derivatives as well.[113,114]

Unlike finite differences, this formula has no subtraction operation and thus no subtractive cancellation error. The only source of numerical error is the truncation error. However, the truncation error can be eliminated if h is decreased to a small enough value (say, 10^{-200}). Then, the precision of the complex-step derivative approximation (Eq. 6.14) matches the precision of f. This is a tremendous advantage over the finite-difference approximations (Eqs. 6.4 and 6.8).

113. Lantoine et al., *Using multicomplex variables for automatic computation of high-order derivatives*, 2012.

114. Fike and Alonso, *Automatic differentiation through the use of hyper-dual numbers for second derivatives*, 2012.

Like the finite-difference approach, each evaluation yields a column of the Jacobian ($\partial f/\partial x_j$), and the cost of computing all the derivatives is proportional to the number of design variables. The cost of the complex-step method is comparable to that of a central difference because we compute a real and an imaginary part for every number in our code.

If we take the real part of the Taylor series expansion (Eq. 6.12), we obtain the value of the function on the real axis,

$$f(x) = \text{Re}\left(f(x + ih\hat{e}_j)\right) + O(h^2). \qquad (6.15)$$

Similar to the derivative approximation, we can make the truncation error disappear by using a small enough h. This means that no separate evaluation of $f(x)$ is required to get the original real value of the function; we can simply take the real part of the complex evaluation.

What is a "small enough h"? When working with finite-precision arithmetic, the error can be eliminated entirely by choosing an h so small that all h^2 terms become zero because of underflow (i.e., h^2 is smaller than the smallest representable number, which is approximately 10^{-324} when using double precision; see Section 3.5.1). Eliminating these squared terms does not affect the accuracy of the derivative carried in the imaginary part because the squared terms only appear in the error terms of the complex-step approximation.

At the same time, h must be large enough that the imaginary part ($h \cdot \partial f / \partial x$) does not underflow. Suppose that μ is the smallest representable number. Then, the two requirements result in the following bounds:

$$\mu \left| \frac{\partial f}{\partial x} \right|^{-1} < h < \sqrt{\mu}. \tag{6.16}$$

A step size of 10^{-200} works well for double-precision functions.

Example 6.4 Complex-step accuracy compared with finite differences

To show how the complex-step method works, consider the function in Ex. 6.3. In addition to the finite-difference relative errors from Fig. 6.7, we plot the complex-step error in Fig. 6.9.

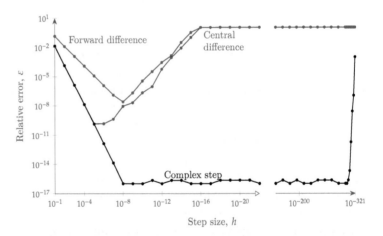

Fig. 6.9 Unlike finite differences, the complex-step method is not subject to subtractive cancellation. Therefore, the error is the same as that of the function evaluation (machine zero in this case).

The complex-step estimate converges quadratically with decreasing step size, as predicted by the truncation error term. The relative error reduces to machine precision at around $h = 10^{-8}$ and stays at that level. The error eventually increases when h is so small that the imaginary parts get affected by underflow (around $h = 10^{-308}$ in this case).

The real parts and the derivatives of the complex evaluations are listed in Table 6.2. For a small enough step, the real part is identical to the original real function evaluation, and the complex-step method yields derivatives that match to machine precision.

Comparing the best accuracy of each of these approaches, we can see that by using finite differences, we only achieve a fraction of the accuracy that is obtained by using the complex-step approximation.

6.5 Complex Step

h	Re (f)	Im $(f)/h$
10^{-1}	4.4508662116993065	4.0003330384671729
10^{-2}	4.4973069409015318	4.0528918144659292
10^{-4}	4.4977800066307951	4.0534278402854467
10^{-6}	4.4977800539414297	4.0534278938932582
10^{-8}	4.4977800539461619	4.0534278938986201
10^{-10}	4.4977800539461619	4.0534278938986201
10^{-12}	4.4977800539461619	4.0534278938986201
10^{-14}	4.4977800539461619	4.0534278938986210
10^{-16}	4.4977800539461619	4.0534278938986201
10^{-18}	4.4977800539461619	4.0534278938986210
10^{-200}	4.4977800539461619	4.0534278938986201
Exact	4.4977800539461619	4.0534278938986201

Table 6.2 For a small enough step, the real part of the complex evaluation is identical to the real evaluation, and the derivative matches to machine precision.

6.5.2 Complex-Step Implementation

We can use the complex-step method even when the evaluation of f involves the solution of numerical models through computer programs.[50] The outer loop for computing the derivatives of multiple functions with respect to all variables (Alg. 6.2) is similar to the one for finite differences. A reference function evaluation is not required, but now the function must handle complex numbers correctly.

50. Martins et al., *The complex-step derivative approximation*, 2003.

Algorithm 6.2 Computing the gradients of a vector-valued function $f(x)$ using the complex-step method

Inputs:
  x: Point about which to compute the gradient
  f: Function of interest

Outputs:
  J: Jacobian of f about point x

$h = 10^{-200}$ Typical "small enough" step size
for $j = 1$ **to** n_x **do**
    $x_j = x_j + ih$ Add complex step to variable j
    $f_+ = f(x)$ Evaluate function with complex perturbation
    $J_{*,j} = \dfrac{\text{Im}(f_+)}{h}$ Extract derivatives from imaginary part
    $x_j = x_j - ih$ Reset perturbed variable
end for

The complex-step method can be applied to any model, but modifications might be required. We need the source code for the model to make sure that the program can handle complex numbers and the associated arithmetic, that it handles logical operators consistently, and that certain functions yield the correct derivatives.

First, the program may need to be modified to use complex numbers. In programming languages like Fortran or C, this involves changing real-valued type declarations (e.g., double) to complex type declarations (e.g., double complex). In some languages, such as MATLAB, Python, and Julia, this is unnecessary because functions are overloaded to automatically accept either type.

Second, some changes may be required to preserve the correct logical flow through the program. Relational logic operators (e.g., "greater than", "less than", "if", and "else") are usually not defined for complex numbers. These operators are often used in programs, together with conditional statements, to redirect the execution thread. The original algorithm and its "complexified" version should follow the same execution thread. Therefore, defining these operators to compare only the real parts of the arguments is the correct approach. Functions that choose one argument, such as the maximum or the minimum values, are based on relational operators. Following the previous argument, we should determine the maximum and minimum values based on the real parts alone.

Third, some functions need to be redefined for complex arguments. The most common function that needs to be redefined is the absolute value function. For a complex number, $z = x + iy$, the absolute value is defined as

$$|z| = \sqrt{x^2 + y^2}, \tag{6.17}$$

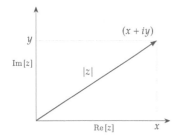

Fig. 6.10 The usual definition of a complex absolute value returns a real number (the length of the vector), which is not compatible with the complex-step method.

as shown in Fig. 6.10. This definition is not complex analytic, which is required in the derivation of the complex-step derivative approximation.

As shown in Fig. 6.11, the correct derivatives for the real absolute value function are $+1$ and -1, depending on whether x is greater than or less than zero. The following complex definition of the absolute value yields the correct derivatives:

$$\operatorname{abs}(x + iy) = \begin{cases} -x - iy, & \text{if } x < 0 \\ +x + iy, & \text{if } x \geq 0. \end{cases} \tag{6.18}$$

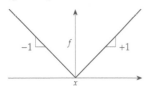

Fig. 6.11 The absolute value function needs to be redefined such that the imaginary part yields the correct derivatives.

Setting the imaginary part to $y = h$ and dividing by h corresponds to the slope of the absolute value function. There is an exception at $x = 0$, where the function is not analytic, but a derivative does not exist in any case. We use the "greater or equal" in the logic so that the approximation yields the correct right-sided derivative at that point.

> **Tip 6.3** Test complexified code by running it with $h = 0$
>
> Once you have made your code complex, the first test you should perform is to run your code with no imaginary perturbation and verify that no variable ends up with a nonzero imaginary part. If any number in the code acquires a nonzero imaginary part, something is wrong, and you must trace the source of the error. This is a necessary but not sufficient test.

Depending on the programming language, we may need to redefine some trigonometric functions. This is because some default implementations, although correct, do not maintain accurate derivatives for small complex-step sizes. We must replace these with mathematically equivalent implementations that avoid numerical issues.

Fortunately, we can automate most of these changes by using scripts to process the codes, and in most programming languages, we can easily redefine functions using operator overloading.‡

‡ For more details on the problematic functions and how to implement the complex-step method in various programming languages, see Martins et al.[50] A summary, implementation guide, and scripts are available at: http://bit.ly/complexstep

50. Martins et al., *The complex-step derivative approximation*, 2003.

> **Tip 6.4** Check the convergence of the imaginary part
>
> When the solver that computes f is iterative, it might be necessary to change the convergence criterion so that it checks for the convergence of the imaginary part, in addition to the existing check on the real part. The imaginary part, which contains the derivative information, often lags relative to the real part in terms of convergence, as shown in Fig. 6.12. Therefore, if the solver only checks for the real part, it might yield a derivative with a precision lower than the function value. In this example, f is the drag coefficient given by a computational fluid dynamics solver and ε is the relative error for each part.

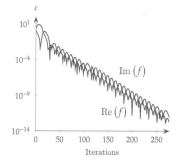

Fig. 6.12 The imaginary parts of the variables often lag relative to the real parts in iterative solvers.

6.6 Algorithmic Differentiation

Algorithmic differentiation (AD)—also known as *computational differentiation* or *automatic differentiation*—is a well-known approach based on the systematic application of the chain rule to computer programs.[115,116] The derivatives computed with AD can match the precision of the function evaluation. The cost of computing derivatives with AD can be proportional to either the number of variables or the number of functions, depending on the type of AD, making it flexible.

Another attractive feature of AD is that its implementation is largely automatic, thanks to various AD tools. To explain AD, we start by

115. Griewank, *Evaluating Derivatives*, 2000.

116. Naumann, *The Art of Differentiating Computer Programs—An Introduction to Algorithmic Differentiation*, 2011.

6.6.1 Variables and Functions as Lines of Code

The basic concept of AD is as follows. Even long, complicated codes consist of a sequence of basic operations (e.g., addition, multiplication, cosine, exponentiation). Each operation can be differentiated symbolically with respect to the variables in the expression. AD performs this symbolic differentiation and adds the code that computes the derivatives for each variable in the code. The derivatives of each variable accumulate in what amounts to a numerical version of the chain rule.

The fundamental building blocks of a code are unary and binary operations. These operations can be combined to obtain more elaborate explicit functions, which are typically expressed in one line of computer code. We represent the variables in the computer code as the sequence $v = [v_1, \ldots, v_i, \ldots, v_n]$, where n is the total number of variables assigned in the code. One or more of these variables at the start of this sequence are given and correspond to x, and one or more of the variables toward the end of the sequence are the outputs of interest, f, as illustrated in Fig. 6.13. In general, a variable assignment corresponding to a line of code can involve any other variable, including itself, through an explicit function,

$$v_i = v_i(v_1, v_2, \ldots, v_i, \ldots, v_n). \tag{6.19}$$

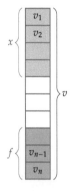

Fig. 6.13 AD considers all the variables in the code, where the inputs x are among the first variables, and the outputs f are among the last.

Except for the most straightforward codes, many of the variables in the code are overwritten as a result of iterative loops.

To understand AD, it is helpful to imagine a version of the code where all the loops are *unrolled*. Instead of overwriting variables, we create new versions of those variables, as illustrated in Fig. 6.14. Then, we can represent the computations in the code in a sequence with no loops such that each variable in this larger set only depends on *previous* variables, and then

$$v_i = v_i(v_1, v_2, \ldots, v_{i-1}). \tag{6.20}$$

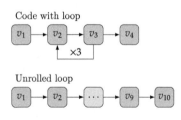

Fig. 6.14 Unrolling of loops is a useful mental model to understand the derivative propagation in the AD of general code.

Given such a sequence of operations and the derivatives for each operation, we can apply the chain rule to obtain the derivatives for the entire sequence. Unrolling the loops is just a mental model for understanding how the chain rule operates, and it is not explicitly done in practice.

The chain rule can be applied in two ways. In the *forward mode*, we choose one input variable and work forward toward the outputs until

6.6.2 Forward-Mode AD

The chain rule for the forward mode can be written as

$$\frac{dv_i}{dv_j} = \sum_{k=j}^{i-1} \frac{\partial v_i}{\partial v_k} \frac{dv_k}{dv_j}, \quad (6.21)$$

where each partial derivative is obtained by symbolically differentiating the explicit expression for v_i. The total derivatives are the derivatives with respect to the chosen input v_j, which can be computed using this chain rule.

Using the forward mode, we evaluate a sequence of these expressions by fixing j in Eq. 6.21 (effectively choosing one input v_j) and incrementing i to get the derivative of each variable v_i. We only need to sum up to $i - 1$ because of the form of Eq. 6.20, where each v_i only depends on variables that precede it.

For a more convenient notation, we define a new variable that represents the total derivative of variable i with respect to a fixed input j as $\dot{v}_i \equiv dv_i/dv_j$ and rewrite the chain rule as

$$\dot{v}_i = \sum_{k=j}^{i-1} \frac{\partial v_i}{\partial v_k} \dot{v}_k. \quad (6.22)$$

The chosen input j corresponds to the *seed*, which we set to $\dot{v}_j = 1$ (using the definition for $\dot{v}_j$, we see that means setting $dv_j/dv_j = 1$). This chain rule then propagates the total derivatives forward, as shown in Fig. 6.15, affecting all the variables that depend on the seeded variable.

Once we are done applying the chain rule (Eq. 6.22) for the chosen input variable v_j, we end up with the total derivatives dv_i/dv_j for all $i > j$. The sum in the chain rule (Eq. 6.22) only needs to consider the nonzero partial derivative terms. If a variable k does not explicitly appear in the expression for v_i, then $\partial v_i/\partial v_k = 0$, and there is no need to consider the corresponding term in the sum. In practice, this means that only a small number of terms is considered for each sum.

Suppose we have four variables $v_1, v_2, v_3,$ and v_4, and $x \equiv v_1, f \equiv v_4$, and we want df/dx. We assume that each variable depends explicitly on all the previous ones. Using the chain rule (Eq. 6.22), we set $j = 1$ (because we want the derivative with respect to $x \equiv v_1$) and increment

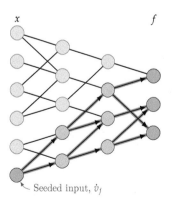

Fig. 6.15 The forward mode propagates derivatives to all the variables that depend on the seeded input variable.

in i to get the sequence of derivatives:

$$\begin{aligned}
\dot{v}_1 &= 1 \\
\dot{v}_2 &= \frac{\partial v_2}{\partial v_1}\dot{v}_1 \\
\dot{v}_3 &= \frac{\partial v_3}{\partial v_1}\dot{v}_1 + \frac{\partial v_3}{\partial v_2}\dot{v}_2 \\
\dot{v}_4 &= \frac{\partial v_4}{\partial v_1}\dot{v}_1 + \frac{\partial v_4}{\partial v_2}\dot{v}_2 + \frac{\partial v_4}{\partial v_3}\dot{v}_3 \equiv \frac{\mathrm{d}f}{\mathrm{d}x}.
\end{aligned} \qquad (6.23)$$

In each step, we just need to compute the partial derivatives of the current operation v_i and then multiply using the total derivatives $\dot{v}$ that have already been computed. We move forward by evaluating the partial derivatives of v in the same sequence to evaluate the original function. This is convenient because all of the unknowns are *partial* derivatives, meaning that we only need to compute derivatives based on the operation at hand (or line of code).

In this abstract example with four variables that depend on each other sequentially, the Jacobian of the variables with respect to themselves is as follows:

$$J_v = \begin{bmatrix} 1 & 0 & 0 & 0 \\ \frac{\mathrm{d}v_2}{\mathrm{d}v_1} & 1 & 0 & 0 \\ \frac{\mathrm{d}v_3}{\mathrm{d}v_1} & \frac{\mathrm{d}v_3}{\mathrm{d}v_2} & 1 & 0 \\ \frac{\mathrm{d}v_4}{\mathrm{d}v_1} & \frac{\mathrm{d}v_4}{\mathrm{d}v_2} & \frac{\mathrm{d}v_4}{\mathrm{d}v_3} & 1 \end{bmatrix}. \qquad (6.24)$$

By setting the seed $\dot{v}_1 = 1$ and using the forward chain rule (Eq. 6.22), we have computed the first column of J_v from top to bottom. This column corresponds to the tangent with respect to v_1. Using forward-mode AD, obtaining derivatives for other outputs is free (e.g., $\mathrm{d}v_3/\mathrm{d}v_1 \equiv \dot{v}_3$ in Eq. 6.23).

However, if we want the derivatives with respect to additional inputs, we would need to set a different seed and evaluate an entire set of similar calculations. For example, if we wanted $\mathrm{d}v_4/\mathrm{d}v_2$, we would set the seed as $\dot{v}_2 = 1$ and evaluate the equations for $\dot{v}_3$ and $\dot{v}_4$, where we would now have $\mathrm{d}v_4/\mathrm{d}v_2 = \dot{v}_4$. This would correspond to computing the second column in J_v (Eq. 6.24).

Thus, the cost of the forward mode scales linearly with the number of inputs we are interested in and is independent of the number of outputs.

6.6 Algorithmic Differentiation

Example 6.5 Forward-mode AD

Consider the function with two inputs and two outputs from Ex. 6.1. We could evaluate the explicit expressions in this function using only two lines of code. However, to make the AD process more apparent, we write the code such that each line has a single unary or binary operation, which is how a computer ends up evaluating the expression:

$$
\begin{aligned}
v_1 &= v_1(v_1) = x_1 \\
v_2 &= v_2(v_2) = x_2 \\
v_3 &= v_3(v_1, v_2) = v_1 v_2 \\
v_4 &= v_4(v_1) = \sin v_1 \\
v_5 &= v_5(v_3, v_4) = v_3 + v_4 &= f_1 \\
v_6 &= v_6(v_2) = v_2^2 \\
v_7 &= v_7(v_3, v_6) = v_3 + v_6 &= f_2.
\end{aligned}
$$

Using the forward mode, set the seed $\dot{v}_1 = 1$, and $\dot{v}_2 = 0$ to obtain the derivatives with respect to x_1. When using the chain rule (Eq. 6.22), only one or two partial derivatives are nonzero in each sum because the operations are either unary or binary in this case. For example, the addition operation that computes v_5 does not depend explicitly on v_2, so $\partial v_5/\partial v_2 = 0$. To further elaborate, when evaluating the operation $v_5 = v_3 + v_4$, we do not need to know how v_3 was computed; we just need to know the value of the two numbers we are adding. Similarly, when evaluating the derivative $\partial v_5/\partial v_2$, we do not need to know how or whether v_3 and v_4 depended on v_2; we just need to know how this one operation depends on v_2. So even though symbolic derivatives are involved in individual operations, the overall process is distinct from symbolic differentiation. We do not combine all the operations and end up with a symbolic derivative. We develop a computational procedure to compute the derivative that ends up with a number for a given input—similar to the computational procedure that computes the functional outputs and does not produce a symbolic functional output.

Say we want to compute df_2/dx_1, which in our example corresponds to dv_7/dv_1. The evaluation point is the same as in Ex. 6.1: $x = (\pi/4, 2)$. Using the chain rule (Eq. 6.22) and considering only the nonzero partial derivative terms, we get the following sequence:

$$
\begin{aligned}
\dot{v}_1 &= 1 \\
\dot{v}_2 &= 0 \\
\dot{v}_3 &= \frac{\partial v_3}{\partial v_1}\dot{v}_1 + \frac{\partial v_3}{\partial v_2}\dot{v}_2 = v_2 \cdot \dot{v}_1 + v_1 \cdot 0 = 2 \\
\dot{v}_4 &= \frac{\partial v_4}{\partial v_1}\dot{v}_1 = \cos v_1 \cdot \dot{v}_1 = 0.707\ldots \\
\dot{v}_5 &= \frac{\partial v_5}{\partial v_3}\dot{v}_3 + \frac{\partial v_5}{\partial v_4}\dot{v}_4 = 1 \cdot \dot{v}_3 + 1 \cdot \dot{v}_4 = 2.707\ldots \equiv \frac{\partial f_1}{\partial x_1}
\end{aligned}
$$

$$\dot{v}_6 = \frac{\partial v_6}{\partial v_2}\dot{v}_2 = 2v_2 \cdot \dot{v}_2 = 0$$
$$\dot{v}_7 = \frac{\partial v_7}{\partial v_3}\dot{v}_3 + \frac{\partial v_7}{\partial v_6}\dot{v}_6 = 1 \cdot \dot{v}_3 + 1 \cdot \dot{v}_6 = 2 \equiv \frac{\partial f_2}{\partial x_1}. \quad (6.25)$$

This sequence is illustrated in matrix form in Fig. 6.16. The procedure is equivalent to performing forward substitution in this linear system.

We now have a *procedure* (not a symbolic expression) for computing df_2/dx_1 for any (x_1, x_2). The dependencies of these operations are shown in Fig. 6.17 as a *computational graph*.

Although we set out to compute df_2/dx_1, we also obtained df_1/dx_1 as a by-product. We can obtain the derivatives for all outputs with respect to one input for the same cost as computing the outputs. If we wanted the derivative with respect to the other input, df_1/dx_2, a new sequence of calculations would be necessary.

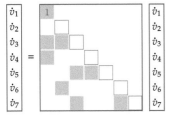

Fig. 6.16 Dependency used in the forward chain rule propagation in Eq. 6.25. The forward mode is equivalent to solving a lower triangular system by forward substitution, where the system is sparse.

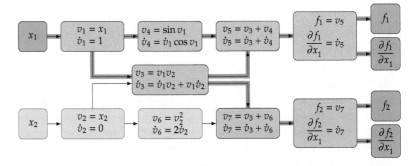

Fig. 6.17 Computational graph for the numerical example evaluations, showing the forward propagation of the derivative with respect to x_1.

So far, we have assumed that we are computing derivatives with respect to each component of x. However, just like for finite differences, we can also compute directional derivatives using forward-mode AD. We do so by setting the appropriate seed in the $\dot{v}$'s that correspond to the inputs in a vectorized manner. Suppose we have $x \equiv [v_1, \ldots, v_{n_x}]$. To compute the derivative with respect to x_j, we would set the seed as the unit vector $\dot{v} = \hat{e}_j$ and follow a similar process for the other elements. To compute a directional derivative in direction p, we would set the seed as $\dot{v} = p/\|p\|$.

Tip 6.5 Use a directional derivative for quick verification

We can use a directional derivative in arbitrary directions to verify the gradient computation. The directional derivative is the scalar projection of the gradient in the chosen direction, that is, $\nabla f^\top p$. We can use the directional derivative to verify the gradient computed by some other method, which is especially useful when the evaluation of f is expensive and we have many gradient elements. We can verify a gradient by projecting it into some direction

6.6 Algorithmic Differentiation

(say, $p = [1, \ldots, 1]$) and then comparing it to the directional derivative in that direction. If the result matches the reference, then all the gradient elements are most likely correct (it is good practice to try a couple more directions just to be sure). However, if the result does not match, this directional derivative does not reveal which gradient elements are incorrect.

6.6.3 Reverse-Mode AD

The *reverse mode* is also based on the chain rule but uses the alternative form:

$$\frac{dv_i}{dv_j} = \sum_{k=j+1}^{i} \frac{\partial v_k}{\partial v_j} \frac{dv_i}{dv_k}, \qquad (6.26)$$

where the summation happens in reverse (starts at i and decrements to $j + 1$). This is less intuitive than the forward chain rule, but it is equally valid. Here, we fix the index i corresponding to the output of interest and decrement j until we get the desired derivative.

Similar to the forward-mode total derivative notation (Eq. 6.22), we define a more convenient notation for the variables that carry the total derivatives with a fixed i as $\bar{v}_j \equiv dv_i/dv_j$, which are sometimes called *adjoint* variables. Then we can rewrite the chain rule as

$$\bar{v}_j = \sum_{k=j+1}^{i} \frac{\partial v_k}{\partial v_j} \bar{v}_k. \qquad (6.27)$$

This chain rule propagates the total derivatives backward after setting the reverse seed $\bar{v}_i = 1$, as shown in Fig. 6.18. This affects all the variables on which the seeded variable depends.

The reverse-mode variables $\bar{v}$ represent the derivatives of one output, i, with respect to all the input variables (instead of the derivatives of all the outputs with respect to one input, j, in the forward mode). Once we are done applying the reverse chain rule (Eq. 6.27) for the chosen output variable v_i, we end up with the total derivatives dv_i/dv_j for all $j < i$.

Applying the reverse mode to the same four-variable example as before, we get the following sequence of derivative computations (we set $i = 4$ and decrement j):

$$\bar{v}_4 = 1$$

$$\bar{v}_3 = \frac{\partial v_4}{\partial v_3} \bar{v}_4$$

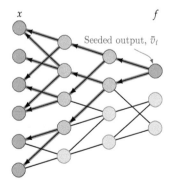

Fig. 6.18 The reverse mode propagates derivatives to all the variables on which the seeded output variable depends.

$$\bar{v}_2 = \frac{\partial v_3}{\partial v_2}\bar{v}_3 + \frac{\partial v_4}{\partial v_2}\bar{v}_4$$
$$\bar{v}_1 = \frac{\partial v_2}{\partial v_1}\bar{v}_2 + \frac{\partial v_3}{\partial v_1}\bar{v}_3 + \frac{\partial v_4}{\partial v_1}\bar{v}_4 \equiv \frac{\mathrm{d}f}{\mathrm{d}x}.$$
(6.28)

The partial derivatives of v must be computed for v_4 first, then v_3, and so on. Therefore, we have to traverse the code in reverse. In practice, not every variable depends on every other variable, so a computational graph is created during code evaluation. Then, when computing the adjoint variables, we traverse the computational graph in reverse. As before, the derivatives we need to compute in each line are only partial derivatives.

Recall the Jacobian of the variables,

$$J_v = \begin{bmatrix} 1 & 0 & 0 & 0 \\ \frac{\mathrm{d}v_2}{\mathrm{d}v_1} & 1 & 0 & 0 \\ \frac{\mathrm{d}v_3}{\mathrm{d}v_1} & \frac{\mathrm{d}v_3}{\mathrm{d}v_2} & 1 & 0 \\ \frac{\mathrm{d}v_4}{\mathrm{d}v_1} & \frac{\mathrm{d}v_4}{\mathrm{d}v_2} & \frac{\mathrm{d}v_4}{\mathrm{d}v_3} & 1 \end{bmatrix}.$$
(6.29)

By setting $\bar{v}_4 = 1$ and using the reverse chain rule (Eq. 6.27), we have computed the last row of J_v from right to left. This row corresponds to the gradient of $f \equiv v_4$. Using the reverse mode of AD, obtaining derivatives with respect to additional inputs is free (e.g., $dv_4/dv_2 \equiv \bar{v}_2$ in Eq. 6.28).

However, if we wanted the derivatives of additional outputs, we would need to evaluate a different sequence of derivatives. For example, if we wanted dv_3/dv_1, we would set $\bar{v}_3 = 1$ and evaluate the expressions for $\bar{v}_2$ and $\bar{v}_1$ in Eq. 6.28, where $dv_3/dv_1 \equiv \bar{v}_1$. Thus, the cost of the reverse mode scales linearly with the number of outputs and is independent of the number of inputs.

One complication with the reverse mode is that the resulting sequence of derivatives requires the values of the variables, starting with the last ones and progressing in reverse. For example, the partial derivative in the second operation of Eq. 6.28 might involve v_3. Therefore, the code needs to run in a forward pass first, and all the variables must be stored for use in the reverse pass, which increases memory usage.

Example 6.6 Reverse-mode AD

Suppose we want to compute $\partial f_2/\partial x_1$ for the function from Ex. 6.5. First, we need to run the original code (a forward pass) and store the values of all the variables because they are necessary in the reverse chain rule (Eq. 6.26) to compute the numerical values of the partial derivatives. Furthermore, the

6.6 Algorithmic Differentiation

reverse chain rule requires the information on all the dependencies to determine which partial derivatives are nonzero. The forward pass and dependencies are represented by the computational graph shown in Fig. 6.19.

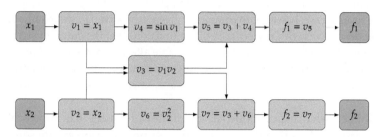

Fig. 6.19 Computational graph for the function.

Using the chain rule (Eq. 6.26) and setting the seed for the desired variable $\bar{v}_7 = 1$, we get

$$\begin{aligned}
\bar{v}_7 &= 1 \\
\bar{v}_6 &= \frac{\partial v_7}{\partial v_6}\bar{v}_7 & &= \bar{v}_7 = 1 \\
\bar{v}_5 &= & &= = 0 \\
\bar{v}_4 &= \frac{\partial v_5}{\partial v_4}\bar{v}_5 = \bar{v}_5 = 0 \\
\bar{v}_3 &= \frac{\partial v_7}{\partial v_3}\bar{v}_7 + \frac{\partial v_5}{\partial v_3}\bar{v}_5 = \bar{v}_7 + \bar{v}_5 = 1 \\
\bar{v}_2 &= \frac{\partial v_6}{\partial v_2}\bar{v}_6 + \frac{\partial v_3}{\partial v_2}\bar{v}_3 = 2v_2\bar{v}_6 + v_1\bar{v}_3 = 4.785 = \frac{\partial f_2}{\partial x_2} \\
\bar{v}_1 &= \frac{\partial v_4}{\partial v_1}\bar{v}_4 + \frac{\partial v_3}{\partial v_1}\bar{v}_3 = (\cos v_1)\bar{v}_4 + v_2\bar{v}_3 = 2 = \frac{\partial f_2}{\partial x_1}.
\end{aligned} \qquad (6.30)$$

After running the forward evaluation and storing the elements of v, we can run the reverse pass shown in Fig. 6.20. This reverse pass is illustrated in matrix form in Fig. 6.21. The procedure is equivalent to performing back substitution in this linear system.

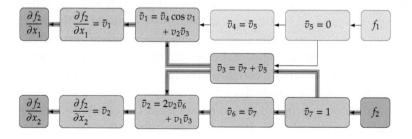

Fig. 6.20 Computational graph for the reverse mode, showing the backward propagation of the derivative of f_2.

Although we set out to evaluate df_2/dx_1, we also computed df_2/dx_2 as a by-product. For each output, the derivatives of all inputs come at the cost of

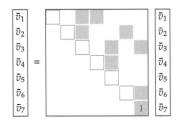

Fig. 6.21 Dependency used in the reverse chain rule propagation in Eq. 6.30. The reverse mode is equivalent to solving an upper triangular system by backward substitution, where the system is sparse.

evaluating only one more line of code. Conversely, if we want the derivatives of f_1, a whole new set of computations is needed.

In forward mode, the computation of a given derivative, $\dot{v}_i$, requires the partial derivatives of the line of code that computes v_i with respect to its inputs. In the reverse case, however, to compute a given derivative, $\bar{v}_j$, we require the partial derivatives with respect to v_j of the functions that the current variable v_j affects. Knowledge of the function a variable affects is not encoded in that variable computation, and that is why the computational graph is required.

Unlike with forward-mode AD and finite differences, it is impossible to compute a directional derivative by setting the appropriate seeds. Although the seeds in the forward mode are associated with the inputs, the seeds for the reverse mode are associated with the outputs. Suppose we have multiple functions of interest, $f \equiv [v_{n-n_f}, \ldots, v_n]$. To find the derivatives of f_1 in a vectorized operation, we would set $\bar{v} = [1, 0, \ldots, 0]$. A seed with multiple nonzero elements computes the derivatives of a *weighted* function with respect to all the variables, where the weight for each function is determined by the corresponding $\bar{v}$ value.

6.6.4 Forward Mode or Reverse Mode?

Our goal is to compute J_f, the $(n_f \times n_x)$ matrix of derivatives of all the functions of interest f with respect to all the input variables x. However, AD computes many other derivatives corresponding to intermediate variables. The complete Jacobian for all the intermediate variables, $v_i = v_i(v_1, v_2, \ldots, v_i, \ldots, v_n)$, assuming that the loops have been unrolled, has the structure shown in Figs. 6.22 and 6.23.

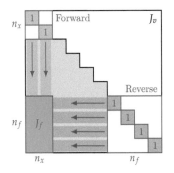

Fig. 6.22 When $n_x < n_f$, the forward mode is advantageous.

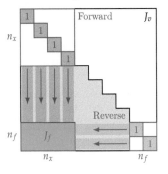

Fig. 6.23 When $n_x > n_f$, the reverse mode is advantageous.

The input variables x are among the first entries in v, whereas the functions of interest f are among the last entries of v. For simplicity, let us assume that the entries corresponding to x and f are contiguous, as previously shown in Fig. 6.13. Then, the derivatives we want (J_f) are a block located on the lower left in the much larger matrix (J_v), as shown in Figs. 6.22 and 6.23. Although we are only interested in this block, AD requires the computation of additional intermediate derivatives.

The main difference between the forward and the reverse approaches is that the forward mode computes the Jacobian column by column, whereas the reverse mode does it row by row. Thus, the cost of the forward mode is proportional to n_x, whereas the cost of the reverse mode is proportional to n_f. If we have more outputs (e.g., objective and constraints) than inputs (design variables), the forward mode is more efficient, as illustrated in Fig. 6.22. On the other hand, if we have many more inputs than outputs, then the reverse mode is more efficient, as illustrated in Fig. 6.23. If the number of inputs is similar to the number of outputs, neither mode has a significant advantage.

In both modes, each forward or reverse pass costs less than 2–3 times the cost of running the original code in practice. However, because the reverse mode requires storing a large amount of data, memory costs also need to be considered. In principle, the required memory is proportional to the number of variables, but there are techniques that can reduce the memory usage significantly.*

*One of the main techniques for reducing the memory usage of reverse AD is *checkpointing*; see Chapter 12 in Griewank.[115]

115. Griewank, *Evaluating Derivatives*, 2000.

6.6.5 AD Implementation

There are two main ways to implement AD: by *source code transformation* or by *operator overloading*. The function we used to demonstrate the issues with symbolic differentiation (Ex. 6.2) can be differentiated much more easily with AD. In the examples that follow, we use this function to demonstrate how the forward and reverse mode work using both source code transformation and operator overloading.

Source Code Transformation

AD tools that use source code transformation process the whole source code automatically with a parser and add lines of code that compute the derivatives. The added code is highlighted in Exs. 6.7 and 6.8.

Example 6.7 Source code transformation for forward mode

Running an AD source transformation tool on the code from Ex. 6.2 produces the code that follows.

Input: $x, \dot{x}$ — Set seed $\dot{x} = 1$ to get df/dx
$f = x$
$\dot{f} = \dot{x}$ — Automatically added by AD tool
for $i = 1$ **to** 10 **do**
    $f = \sin(x + f)$
    $\dot{f} = (\dot{x} + \dot{f}) \cdot \cos(x + f)$ — Automatically added by AD tool
end for
return $f, \dot{f}$ — df/dx is given by $\dot{f}$

The AD tool added a new line after each variable assignment that computes the corresponding derivative. We can then set the seed, $\dot{x} = 1$ and run the code. As the loops proceed, $\dot{f}$ accumulates the derivative as f is successively updated.

Example 6.8 Source code transformation for reverse mode

The reverse-mode AD version of the code from Ex. 6.2 follows.

Input: $x, \bar{f}$ Set $\bar{f} = 1$ to get df/dx
$f = x$
for $i = 1$ **to** 10 **do**
    push(f) Save current value of f on top of stack
    $f = \sin(x + f)$
end for
for $i = 10$ **to** 1 **do**   Reverse loop added by AD tool
    $f = \text{pop}()$ Get value of f from top of stack
    $\bar{f} = \cos(x + f) \cdot \bar{f}$
end for
$\bar{x} = \bar{f}$
return $f, \bar{x}$ df/dx is given by $\bar{x}$

The first loop is identical to the original code except for one line. Because the derivatives that accumulate in the reverse loop depend on the intermediate values of the variables, we need to store all the variables in the forward loop. We store and retrieve the variables using a stack, hence the call to "push".[†]

The second loop, which runs in reverse, is where the derivatives are computed. We set the reverse seed, $\bar{f} = 1$, and then the adjoint variables accumulate the derivatives back to the start.

[†] A stack, also known as *last in, first out* (LIFO), is a data structure that stores a one-dimensional array. We can only add an element to the top of the stack (push) and take the element from the top of the stack (pop).

Operator Overloading

The operator overloading approach creates a new augmented data type that stores both the variable value and its derivative. Every floating-point number v is replaced by a new type with two parts $(v, \dot{v})$, commonly referred to as a *dual number*. All standard operations (e.g., addition, multiplication, sine) are *overloaded* such that they compute v according to the original function value and $\dot{v}$ according to the derivative of that function. For example, the multiplication operation, $x_1 \cdot x_2$, would be defined for the dual-number data type as

$$(x_1, \dot{x}_1) \cdot (x_2, \dot{x}_2) = (x_1 x_2, \ x_1 \dot{x}_2 + \dot{x}_1 x_2), \tag{6.31}$$

where we compute the original function value in the first term, and the second term carries the derivative of the multiplication.

Although we wrote the two parts explicitly in Eq. 6.31, the source code would only show a normal multiplication, such as $v_3 = v_1 \cdot v_2$. However, each of these variables would be of the new type and carry the corresponding $\dot{v}$ quantities. By overloading all the required operations,

6.6 Algorithmic Differentiation

the computations happen "behind the scenes", and the source code does not have to be changed, except to declare all the variables to be of the new type and to set the seed. Example 6.9 lists the original code from Ex. 6.2 with notes on the actual computations that are performed as a result of overloading.

Example 6.9 Operator overloading for forward mode

Using the derived data types and operator overloading approach in forward mode does not change the code listed in Ex. 6.2. The AD tool provides overloaded versions of the functions in use, which in this case are assignment, addition, and sine. These functions are overloaded as follows:

$$v_2 = v_1 \quad \Rightarrow \quad (v_2, \dot{v}_2) = (v_1, \dot{v}_1)$$
$$v_1 + v_2 \quad \Rightarrow \quad (v_1, \dot{v}_1) + (v_2, \dot{v}_2) \equiv (v_1 + v_2, \dot{v}_1 + \dot{v}_2)$$
$$\sin(v) \quad \Rightarrow \quad \sin(v, \dot{v}) \equiv (\sin(v), \cos(v)\dot{v}) .$$

In this case, the source code is unchanged, but additional computations occur through the overloaded functions. We reproduce the code that follows with notes on the hidden operations that take place.

Input: x x is of a new data type with two components $(x, \dot{x})$
$f = x$ $(f, \dot{f}) = (x, \dot{x})$ through the overloading of the "=" operation
for $i = 1$ **to** 10 **do**
    $f = \sin(x + f)$ Code is unchanged, but overloading computes the derivative[‡]
end for
return f The new data type includes $\dot{f}$, which is df/dx

[‡] The overloading of "+" computes $(v, \dot{v}) = \left(x + f, \dot{x} + \dot{f}\right)$ and then the overloading of "sin" computes $\left(f, \dot{f}\right) = (\sin(v), \cos(v)\dot{v})$.

We set the seed, $\dot{x} = 1$, and for each function assignment, we add the corresponding derivative line. As the loops are repeated, $\dot{f}$ accumulates the derivative as f is successively updated.

The implementation of the reverse mode using operating overloading is less straightforward and is not detailed here. It requires a new data type that stores the information from the computational graph and the variable values when running the forward pass. This information can be stored using the *taping* technique. After the forward evaluation of using the new type, the "tape" holds the sequence of operations, which is then evaluated in reverse to propagate the reverse-mode seed.[§]

[§] See Sec. 5.4 in Griewank[115] for more details on reverse mode using operating overloading.

115. Griewank, *Evaluating Derivatives*, 2000.

Connection of AD with the Complex-Step Method

The complex-step method from Section 6.5 can be interpreted as forward-mode AD using operator overloading, where the data type is the complex number $(x, y) \equiv x + iy$, and the imaginary part y carries the derivative. To see this connection more clearly, let us write the complex multiplication operation as

$$f = (x_1 + iy_1)(x_2 + iy_2) = (x_1 x_2 - y_1 y_2) + i(y_1 x_2 + x_1 y_2). \quad (6.32)$$

This equation is similar to the overloaded multiplication (Eq. 6.31). The only difference is that the real part includes the term $-y_1 y_2$, which corresponds to the second-order error term in Eq. 6.15. In this case, the complex part gives the exact derivative, but a second-order error might appear for other operations. As argued before, these errors vanish in finite-precision arithmetic if the complex step is small enough.

Tip 6.6 AD tools

There are AD tools available for most programming languages, including Fortran,[117,118] C/C++,[119] Python,[120,121], Julia,[122] and MATLAB.[123] These tools have been extensively developed and provide the user with great functionality, including the calculation of higher-order derivatives, multivariable derivatives, and reverse-mode options. Although some AD tools can be applied recursively to yield higher-order derivatives, this approach is not typically efficient and is sometimes unstable.[124]

---

117. Utke et al., *OpenAD/F: A modular open-source tool for automatic differentiation of Fortran codes*, 2008.

118. Hascoet and Pascual, *The Tapenade automatic differentiation tool: Principles, model, and specification*, 2013.

119. Griewank et al., *Algorithm 755: ADOL-C: A package for the automatic differentiation of algorithms written in C/C++*, 1996.

120. Wiltschko et al., *Tangent: automatic differentiation using source code transformation in Python*, 2017.

121. Bradbury et al., *JAX: Composable Transformations of Python+NumPy Programs*, 2018.

122. Revels et al., *Forward-mode automatic differentiation in Julia*, 2016.

123. Neidinger, *Introduction to automatic differentiation and MATLAB object-oriented programming*, 2010.

124. Betancourt, *A geometric theory of higher-order automatic differentiation*, 2018.

Source Code Transformation versus Operator Overloading

The source code transformation and the operator overloading approaches each have their relative advantages and disadvantages. The overloading approach is much more elegant because the original code stays practically the same and can be maintained directly. On the other hand, the source transformation approach enlarges the original code and results in less readable code, making it hard to work with. Still, it is easier to see what operations take place when debugging. Instead of maintaining source code transformed by AD, it is advisable to work with the original source and devise a workflow where the parser is rerun before compiling a new version.

One advantage of the source code transformation approach is that it tends to yield faster code and allows more straightforward compile-time optimizations. The overloading approach requires a language that supports user-defined data types and operator overloading, whereas source transformation does not. Developing a source transformation AD tool is usually more challenging than developing the overloading approach because it requires an elaborate parser that understands the source syntax.

6.6.6 AD Shortcuts for Matrix Operations

The efficiency of AD can be dramatically increased with manually implemented shortcuts. When the code involves matrix operations, manual implementation of a higher-level differentiation of those operations is more efficient than the line-by-line AD implementation. Giles[125] documents the forward and reverse differentiation of many matrix elementary operations.

125. Giles, *An extended collection of matrix derivative results for forward and reverse mode algorithmic differentiation*, 2008.

For example, suppose that we have a matrix multiplication $C = AB$. Then, the forward mode yields

$$\dot{C} = \dot{A}B + A\dot{B}. \qquad (6.33)$$

The idea is to use $\dot{A}$ and $\dot{B}$ from the AD code preceding the operation and then manually implement this formula (bypassing any AD of the code that performs that operation) to obtain $\dot{C}$, as shown in Fig. 6.24. Then we can use $\dot{C}$ to seed the remainder of the AD code.

The reverse mode of the multiplication yields

$$\bar{A} = \bar{C}B^\mathsf{T}, \quad \bar{B} = A^\mathsf{T}\bar{C}. \qquad (6.34)$$

Similarly, we take $\bar{C}$ from the reverse AD code and implement the formula manually to compute $\bar{A}$ and $\bar{B}$, which we can use in the remaining AD code in the reverse procedure.

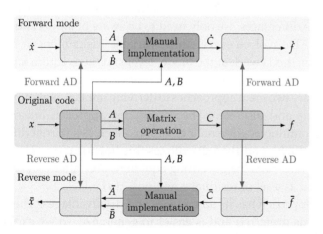

Fig. 6.24 Matrix operations, including the solution of linear systems, can be differentiated manually to bypass more costly AD code.

One particularly useful (and astounding!) result is the differentiation of the matrix inverse product. If we have a linear solver such that $C = A^{-1}B$, we can bypass the solver in the AD process by using the following results:

$$\dot{C} = A^{-1}\left(\dot{B} - \dot{A}C\right) \qquad (6.35)$$

for the forward mode and

$$\bar{B} = A^{-\mathsf{T}}\bar{C}, \quad \bar{A} = -\bar{B}C^{\mathsf{T}} \tag{6.36}$$

for the reverse mode.

In addition to deriving the formulas just shown, Giles[125] derives formulas for the matrix derivatives of the inverse, determinant, norms, quadratic, polynomial, exponential, eigenvalues and eigenvectors, and singular value decomposition. Taking shortcuts as described here applies more broadly to any case where a part of the process can be differentiated manually to produce a more efficient derivative computation.

125. Giles, *An extended collection of matrix derivative results for forward and reverse mode algorithmic differentiation*, 2008.

6.7 Implicit Analytic Methods—Direct and Adjoint

Direct and adjoint methods—which we refer to jointly as *implicit analytic methods*—linearize the model governing equations to obtain a system of linear equations whose solution yields the desired derivatives. Like the complex-step method and AD, implicit analytic methods compute derivatives with a precision matching that of the function evaluation. The direct method is analogous to forward-mode AD, whereas the adjoint method is analogous to reverse-mode AD.

Analytic methods can be thought of as lying in between the finite-difference method and AD in terms of the number of variables involved. With finite differences, we only need to be aware of inputs and outputs, whereas AD involves every single variable assignment in the code. Analytic methods work at the model level and thus require knowledge of the governing equations and the corresponding state variables.

There are two main approaches to deriving implicit analytic methods: continuous and discrete. The continuous approach linearizes the original continuous governing equations, such as a partial differential equation (PDE), and then discretizes this linearization. The discrete approach linearizes the governing equations only after they have been discretized as a set of residual equations, $r(u) = 0$.

Each approach has its advantages and disadvantages. The discrete approach is preferred and is easier to generalize, so we explain this approach exclusively. One of the primary reasons the discrete approach is preferred is that the resulting derivatives are consistent with the function values because they use the same discretization. The continuous approach is only consistent in the limit of a fine discretization. The resulting inconsistencies can mislead the optimization.*

*Peter and Dwight[126] compare the continuous and discrete adjoint approaches in more detail.

126. Peter and Dwight, *Numerical sensitivity analysis for aerodynamic optimization: A survey of approaches*, 2010.

6.7.1 Residuals and Functions

As mentioned in Chapter 3, a discretized numerical model can be written as a system of residuals,

$$r(u; x) = 0, \quad (6.37)$$

where the semicolon denotes that the design variables x are fixed when these equations are solved for the state variables u. Through these equations, u is an *implicit function* of x. This relationship is represented by the box containing the solver and residual equations in Fig. 6.25.

The functions of interest, $f(x, u)$, are typically explicit functions of the state variables and the design variables. However, because u is an implicit function of x, f is ultimately an implicit function of x as well. To compute f for a given x, we must first find u such that $r(u; x) = 0$. This is usually the most computationally costly step and requires a solver (see Section 3.6). The residual equations could be nonlinear and involve many state variables. In PDE-based models it is common to have millions of states. Once we have solved for the state variables u, we can compute the functions of interest f. The computation of f for a given u and x is usually much cheaper because it does not require a solver. For example, in PDE-based models, computing such functions typically involves an integration of the states over a surface, or some other transformation of the states.

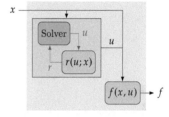

Fig. 6.25 Relationship between functions and design variables for a system involving a solver. The implicit equations $r(u; x) = 0$ define the states u for a given x, so the functions of interest f depend explicitly and implicitly on the design variables x.

To compute df/dx using finite differences, we would have to use the solver to find u for each perturbation of x. That means that we would have to run the solver n_x times, which would not scale well when the solution is costly. AD also requires the propagation of derivatives through the solution process. As we will see, implicit analytic methods avoid involving the potentially expensive nonlinear solution in the derivative computation.

Example 6.10 Residuals and functions in structural analysis

Recall Ex. 3.2, where we introduced the structural model of a truss structure. The residuals in this case are the linear equations,

$$r(u) \equiv K(x)u - q = 0, \quad (6.38)$$

where the state variables are the displacements, u. Solving for the displacement requires only a linear solver in this case, but it is still the most costly part of the analysis. Suppose that the design variables are the cross-sectional areas of the truss members. Then, the stiffness matrix is a function of x, but the external forces are not.

Suppose that the functions of interest are the stresses in each of the truss members. This is an explicit function of the displacements, which is given by the matrix multiplication

$$f(x, u) \equiv \sigma(u) = Su,$$

where S is a matrix that depends on x. This is a much cheaper computation than solving the linear system (Eq. 6.38).

6.7.2 Direct and Adjoint Derivative Equations

The derivatives we ultimately want to compute are the ones in the Jacobian df/dx. Given the explicit and implicit dependence of f on x, we can use the chain rule to write the total derivative Jacobian of f as

$$\frac{df}{dx} = \frac{\partial f}{\partial x} + \frac{\partial f}{\partial u}\frac{du}{dx}, \tag{6.39}$$

where the result is an $(n_f \times n_x)$ matrix.[†]

[†]This chain rule can be derived by writing the total differential of f as

$$df = \frac{\partial f}{\partial x} dx + \frac{\partial f}{\partial u} du$$

and then "dividing" it by dx. See Appendix A.2 for more background on differentials.

In this context, the total derivatives, df/dx, take into account the change in u that is required to keep the residuals of the governing equations (Eq. 6.37) equal to zero. The partial derivatives in Eq. 6.39 represent the variation of $f(x, u)$ with respect to changes in x or u without regard to satisfying the governing equations.

To better understand the difference between total and partial derivatives in this context, imagine computing these derivatives using finite differences with small perturbations. For the total derivatives, we would perturb x, re-solve the governing equations to obtain u, and then compute f, which would account for both dependency paths in Fig. 6.25. To compute the partial derivatives $\partial f/\partial x$ and $\partial f/\partial u$, however, we would perturb x or u and recompute f without re-solving the governing equations. In general, these partial derivative terms are cheap to compute numerically or can be obtained symbolically.

To find the total derivative du/dx, we need to consider the governing equations. Assuming that we are at a point where $r(x, u) = 0$, any perturbation in x must be accompanied by a perturbation in u such that the governing equations remain satisfied. Therefore, the differential of the residuals can be written as

$$dr = \frac{\partial r}{\partial x} dx + \frac{\partial r}{\partial u} du = 0. \tag{6.40}$$

This constraint is illustrated in Fig. 6.26 in two dimensions, but keep in mind that x, u, and r are vectors in the general case. The governing equations (Eq. 6.37) map an n_x-vector x to an n_u-vector u. This mapping defines a hypersurface (also known as a *manifold*) in the x–u space.

6.7 Implicit Analytic Methods—Direct and Adjoint

The total derivative df/dx that we ultimately want to compute represents the effect that a perturbation on x has on f subject to the constraint of remaining on this hypersurface, which can be achieved with the appropriate variation in u.

To obtain a more useful equation, we rearrange Eq. 6.40 to get the linear system

$$\frac{\partial r}{\partial u}\frac{du}{dx} = -\frac{\partial r}{\partial x}, \qquad (6.41)$$

where $\partial r/\partial x$ and du/dx are both $(n_u \times n_x)$ matrices, and $\partial r/\partial u$ is a square matrix of size $(n_u \times n_u)$. This linear system is useful because if we provide the partial derivatives in this equation (which are cheap to compute), we can solve for the total derivatives du/dx (whose computation would otherwise require re-solving $r(u) = 0$). Because du/dx is a matrix with n_x columns, this linear system needs to be solved for each x_i with the corresponding column of the right-hand-side matrix $\partial r/\partial x_i$.

Now let us assume that we can invert the matrix in the linear system (Eq. 6.41) and substitute the solution for du/dx into the total derivative equation (Eq. 6.39). Then we get

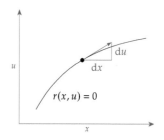

Fig. 6.26 The governing equations determine the values of u for a given x. Given a point that satisfies the equations, the appropriate differential in u must accompany a differential of x about that point for the equations to remain satisfied.

$$\frac{df}{dx} = \frac{\partial f}{\partial x} - \frac{\partial f}{\partial u}\frac{\partial r}{\partial u}^{-1}\frac{\partial r}{\partial x}, \qquad (6.42)$$

where all the derivative terms on the right-hand side are partial derivatives. The partial derivatives in this equation can be computed using any of the methods that we have described earlier: symbolic differentiation, finite differences, complex step, or AD. Equation 6.42 shows two ways to compute the total derivatives, which we call the *direct method* and the *adjoint method*.

The direct method (already outlined earlier) consists of solving the linear system (Eq. 6.41) and substituting du/dx into Eq. 6.39. Defining $\phi \equiv -du/dx$, we can rewrite Eq. 6.41 as

$$\frac{\partial r}{\partial u}\phi = \frac{\partial r}{\partial x}. \qquad (6.43)$$

After solving for ϕ (one column at the time), we can use it in the total derivative equation (Eq. 6.39) to obtain,

$$\frac{df}{dx} = \frac{\partial f}{\partial x} - \frac{\partial f}{\partial u}\phi. \qquad (6.44)$$

This is sometimes called the *forward mode* because it is analogous to forward-mode AD.

Solving the linear system (Eq. 6.43) is typically the most computationally expensive operation in this procedure. The cost of this approach scales with the number of inputs n_x but is essentially independent of the number of outputs n_f. This is the same scaling behavior as finite differences and forward-mode AD. However, the constant of proportionality is typically much smaller in the direct method because we only need to solve the nonlinear equations $r(u; x) = 0$ once to obtain the states.

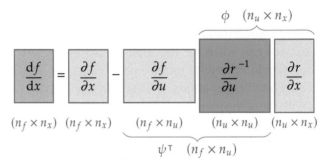

Fig. 6.27 The total derivatives (Eq. 6.42) can be computed either by solving for ϕ (direct method) or by solving for ψ (adjoint method).

The adjoint method changes the linear system that is solved to compute the total derivatives. Looking at Fig. 6.27, we see that instead of solving the linear system with $\partial r/\partial x$ on the right-hand side, we can solve it with $\partial f/\partial u$ on the right-hand side. This corresponds to replacing the two Jacobians in the middle with a new matrix of unknowns,

$$\psi^T \equiv \frac{\partial f}{\partial u}\frac{\partial r}{\partial u}^{-1}, \qquad (6.45)$$

where the columns of ψ are called the *adjoint vectors*. Multiplying both sides of Eq. 6.45 by $\partial r/\partial u$ on the right and taking the transpose of the whole equation, we obtain the *adjoint equation*,

$$\frac{\partial r}{\partial u}^T \psi = \frac{\partial f}{\partial u}^T. \qquad (6.46)$$

This linear system has no dependence on x. Each adjoint vector is associated with a function of interest f_j and is found by solving the adjoint equation (Eq. 6.46) with the corresponding row $\partial f_j/\partial u$. The solution (ψ) is then used to compute the total derivative

$$\frac{df}{dx} = \frac{\partial f}{\partial x} - \psi^T \frac{\partial r}{\partial x}. \qquad (6.47)$$

This is sometimes called the *reverse mode* because it is analogous to reverse-mode AD.

As we will see in Section 6.9, the adjoint vectors are equivalent to the total derivatives df/dr, which quantify the change in the function of interest given a perturbation in the residual that gets zeroed out by an appropriate change in u.‡

6.7.3 Direct or Adjoint?

Similar to the direct method, the solution of the adjoint linear system (Eq. 6.46) tends to be the most expensive operation. Although the linear system is of the same size as that of the direct method, the cost of the adjoint method scales with the number of outputs n_f and is essentially independent of the number of inputs n_x. The comparison between the computational cost of the direct and adjoint methods is summarized in Table 6.3 and illustrated in Fig. 6.28.

Similar to the trade-offs between forward- and reverse-mode AD, if the number of outputs is greater than the number of inputs, the direct (forward) method is more efficient (Fig. 6.28, top). On the other hand, if the number of inputs is greater than the number of outputs, it is more efficient to use the adjoint (reverse) method (Fig. 6.28, bottom). When the number of inputs and outputs is large and similar, neither method has an advantage, and the cost of computing the full total derivative Jacobian might be prohibitive. In this case, aggregating the outputs and using the adjoint method might be effective, as explained in Tip 6.7.

In practice, the adjoint method is implemented much more often than the direct method. Although both methods require a similar implementation effort, the direct method competes with methods that are much more easily implemented, such as finite differencing, complex step, and forward-mode AD. On the other hand, the adjoint method only competes with reverse-mode AD, which is plagued by the memory issue.

‡The adjoint vector can also be interpreted as a Lagrange multiplier vector associated with equality constraints $r = 0$. Defining the Lagrangian $\mathcal{L}(x, u) = f + \psi^\mathsf{T} r$ and differentiating it with respect to x, we get

$$\frac{\partial \mathcal{L}}{\partial x} = \frac{\partial f}{\partial x} + \psi^\mathsf{T} \frac{\partial r}{\partial x}.$$

Thus, the total derivatives df/dx are the derivatives of this Lagrangian.

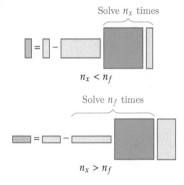

Fig. 6.28 Two possibilities for the size of df/dx in Fig. 6.27. When $n_x < n_f$, it is advantageous to solve the linear system with the vector to the right of the square matrix because it has fewer columns. When $n_x > n_f$, it is advantageous to solve the transposed linear system with the vector to the left because it has fewer rows.

Step	Direct	Adjoint
Partial derivative computation	Same	Same
Linear solution	n_x times	n_f times
Matrix multiplications	Same	Same

Table 6.3 Cost comparison of computing derivatives with direct and adjoint methods.

Another reason why the adjoint method is more widely used is that many optimization problems have a few functions of interest (one objective and a few constraints) and many design variables. The adjoint method has made it possible to solve optimization problems involving computationally intensive PDE models.§

Although implementing implicit analytic methods is labor intensive,

§One widespread application of the adjoint method has been in aerodynamic and hydrodynamic shape optimization.[127]

127. Martins, *Perspectives on aerodynamic design optimization*, 2020.

it is worthwhile if the differentiated code is used frequently and in applications that demand repeated evaluations. For such applications, analytic differentiation with partial derivatives computed using AD is the recommended approach for differentiating code because it combines the best features of these methods.

Example 6.11 Differentiating an implicit function

Consider the following simplified equation for the natural frequency of a beam:
$$f = \lambda m^2, \tag{6.48}$$
where λ is a function of m through the following relationship:
$$\frac{\lambda}{m} + \cos \lambda = 0.$$

Figure 6.29 shows the equivalent of Fig. 6.25 in this case. Our goal is to compute the derivative df/dm. Because λ is an implicit function of m, we cannot find an explicit expression for λ as a function of m, substitute that expression into Eq. 6.48, and then differentiate normally. Fortunately, the implicit analytic methods allow us to compute this derivative.

Referring back to our nomenclature,
$$f(x, u) \equiv f(m, \lambda) = \lambda m^2,$$
$$r(u; x) \equiv r(\lambda; m) = \frac{\lambda}{m} + \cos \lambda = 0,$$

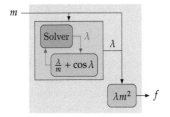

Fig. 6.29 Model for Ex. 6.11.

where m is the design variable and λ is the state variable. The partial derivatives that we need for the total derivative computation (Eq. 6.42) are as follows:

$$\frac{\partial f}{\partial x} = \frac{\partial f}{\partial m} = 2\lambda m, \qquad \frac{\partial f}{\partial u} = \frac{\partial f}{\partial \lambda} = m^2$$
$$\frac{\partial r}{\partial x} = \frac{\partial r}{\partial m} = -\frac{\lambda}{m^2}, \qquad \frac{\partial r}{\partial u} = \frac{\partial r}{\partial \lambda} = \frac{1}{m} - \sin \lambda.$$

Because this is a problem of only one function of interest and one design variable, there is no distinction between the direct and adjoint methods (forward and reverse), and the linear system solution is simply a division. Substituting these partial derivatives into the total derivative equation (Eq. 6.42) yields

$$\frac{df}{dm} = 2\lambda m + \frac{\lambda}{\frac{1}{m} - \sin \lambda}.$$

Thus, we obtained the desired derivative despite the implicitly defined function. Here, it was possible to get an explicit expression for the total derivative, but generally, it is only possible to get a numeric value.

Example 6.12 Direct and adjoint methods applied to structural analysis

Consider the structural analysis we reintroduced in Ex. 6.10. Let us compute the derivatives of the stresses with respect to the cross-sectional truss member areas and denote the number of degrees of freedom as n_u and the number of truss members as n_t. Figure 6.30 shows the equivalent of Fig. 6.25 for this case.

We require four Jacobians of partial derivatives: $\partial r/\partial x$, $\partial r/\partial u$, $\partial \sigma/\partial x$, and $\partial \sigma/\partial u$. When differentiating the governing equations with respect to an area x_i, neither the displacements nor the external forces depend directly on the areas,¶ so we obtain

$$\frac{\partial r}{\partial x_i} = \frac{\partial}{\partial x_i}(Ku - q) = \frac{\partial}{\partial x_i}(Ku) = \frac{\partial K}{\partial x_i}u.$$

This is a vector of size n_u corresponding to one column of $\partial r/\partial x$. We can compute this term by symbolically differentiating the equations that assemble the stiffness matrix. Alternatively, we could use AD on the function that computes the stiffness matrix or use finite differencing. Using AD, we can employ the techniques described in Section 6.7.4 for an efficient implementation.

It is more efficient to compute the derivative of the product Ku directly instead of differentiating K and then multiplying by u. This avoids storing and subtracting the entire perturbed matrix. We can apply a forward finite difference to the product as follows:

$$\frac{\partial r}{\partial x_i} \approx \frac{K(x + h\hat{e}_i)u - K(x)u}{h}.$$

Because the external forces do not depend on the displacements in this case,∥ the partial derivatives of the governing equations with respect to the displacements are given by

$$\frac{\partial r}{\partial u} = K.$$

We already have the stiffness matrix, so this term does not require any further computations.

The partial derivative of the stresses with respect to the areas is zero ($\partial \sigma/\partial x = 0$) because there is no direct dependence.** Thus, the partial derivative of the stress with respect to displacements is

$$\frac{\partial \sigma}{\partial u} = S,$$

which is an ($n_t \times n_u$) matrix that we already have from the stress computation.

Now we can use either the direct or adjoint method by replacing the partial derivatives in the respective equations. The direct linear system (Eq. 6.43) yields

$$K\phi_i = \frac{\partial}{\partial x_i}(Ku),$$

where i corresponds to each truss member area. Once we have ϕ_i, we can use it to compute the total derivatives of all the stresses with respect to member area i with Eq. 6.44, as follows:

$$\frac{d\sigma}{dx_i} = -S\phi_i.$$

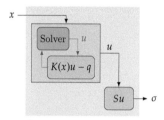

Fig. 6.30 Model for Ex. 6.12

¶ The displacements do change with the areas but only through the solution of the governing equations, which are not considered when taking partial derivatives.

∥ This is not true for large displacements, but we assume small displacements.

** Although ultimately, the areas do change the stresses, they do so only through changes in the displacements.

The adjoint linear system (Eq. 6.46) yields[††]

$$K\psi_j = S_{j,*}^T,$$

where j corresponds to each truss member, and $S_{j,*}$ is the jth row of S. Once we have ψ_j, we can use it to compute the total derivative of the stress in member j with respect to all truss member areas with Eq. 6.47, as follows:

$$\frac{d\sigma_j}{dx} = -\psi_j^T \frac{\partial}{\partial x}(Ku).$$

In this case, there is no advantage in using one method over the other because the number of areas is the same as the number of stresses. However, if we aggregated the stresses as suggested in Tip 6.7, the adjoint would be advantageous.

[††] Usually, the stiffness matrix is symmetric, and $K^T = K$. This means that the solver for displacements can be repurposed for adjoint computation by setting the right-hand side shown here instead of the loads. For that reason, this right-hand side is sometimes called a *pseudo-load*.

Tip 6.7 Aggregate outputs to reduce the cost of adjoint or reverse methods

For problems with many outputs and many inputs, there is no efficient way of computing the Jacobian. This is common in some structural optimization problems, where the number of stress constraints is similar to the number of design variables because they are both associated with each structural element (see Ex. 6.12).

We can address this issue by aggregating the functions of interest as described in Section 5.7 and then implementing the adjoint method to compute the gradient. In Ex. 6.12, we would aggregate the stresses in one or more groups to reduce the number of required adjoint solutions.

We can use these techniques to aggregate any outputs, but in principle, these outputs should have some relation to each other. For example, they could be the stresses in a structure (see Ex. 6.12).[‡‡]

[‡‡] Lambe et al.[128] provide recommendations on constraint aggregation for structural optimization.

128. Lambe et al., *An evaluation of constraint aggregation strategies for wing box mass minimization*, 2017.

6.7.4 Adjoint Method with AD Partial Derivatives

Implementing the implicit analytic methods for models involving long, complicated code requires significant development effort. In this section, we focus on implementing the adjoint method because it is more widely used, as explained in Section 6.7.3. We assume that $n_f < n_x$, so that the adjoint method is advantageous.

To ease the implementation of adjoint methods, we recommend a *hybrid adjoint* approach where the reverse mode of AD computes the partial derivatives in the adjoint equations (Eq. 6.46) and total derivative equation (Eq. 6.47).[§§]

[§§] Kenway et al.[129] provide more details on this approach and its applications.

129. Kenway et al., *Effective Adjoint Approaches for Computational Fluid Dynamics*, 2019.

The partials terms $\partial f/\partial x$ form an $(n_f \times n_x)$ matrix and $\partial f/\partial u$ is an $(n_f \times n_u)$ matrix. These partial derivatives can be computed by identifying the section of the code that computes f for a given x and u and running the AD tool for that section. This produces code that takes $\bar{f}$ as an input and outputs $\bar{x}$ and $\bar{u}$, as shown in Fig. 6.31. Recall that we must first run the entire original code that computes u and f. Then we can run the AD code with the desired seed. Suppose we want the derivative of the jth component of f. We would set $\bar{f}_j = 1$ and the other elements to zero. After running the AD code, we obtain $\bar{x}$ and $\bar{u}$, which correspond to the rows of the respective matrix of partial terms, that is,

$$\bar{x} = \frac{\partial f_j}{\partial x}, \quad \bar{u} = \frac{\partial f_j}{\partial u}. \tag{6.49}$$

Thus, with each run of the AD code, we obtain the derivatives of one function with respect to all design variables and all state variables. One run is required for each element of f. The reverse mode is advantageous if $n_f < n_{x,}$.

The Jacobian $\partial r/\partial u$ can also be computed using AD. Because $\partial r/\partial u$ is typically sparse, the techniques covered in Section 6.8 significantly increase the efficiency of computing this matrix. This is a square matrix, so neither AD mode has an advantage over the other if we explicitly compute and store the whole matrix.

However, reverse-mode AD is advantageous when using an iterative method to solve the adjoint linear system (Eq. 6.46). When using an iterative method, we do not form $\partial r/\partial u$. Instead, we require products of the transpose of this matrix with some vector v,[¶¶]

$$\frac{\partial r}{\partial u}^T v. \tag{6.50}$$

The elements of v act as weights on the residuals and can be interpreted as a projection onto the direction of v. Suppose we have the reverse AD code for the residual computation, as shown in Fig. 6.32. This code requires a reverse seed $\bar{r}$, which determines the weights we want on each residual. Typically, a seed would have only one nonzero entry to find partial derivatives (e.g., setting $\bar{r} = [1, 0, \ldots, 0]$ would yield the first row of the Jacobian, $\bar{u} \equiv \partial r_1/\partial u$). However, to get the product in Eq. 6.50, we require the seed to be weighted as $\bar{r} = v$. Then, we can compute the product by running the reverse AD code once to obtain $\bar{u} \equiv [\partial r/\partial u]^T v$.

The final term needed to compute total derivatives with the adjoint method is the last term in Eq. 6.47, which can be written as

$$\psi^T \frac{\partial r}{\partial x} = \left(\frac{\partial r}{\partial x}^T \psi\right)^T. \tag{6.51}$$

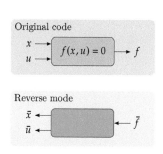

Fig. 6.31 Applying reverse AD to the code that computes f produces code that computes the partial derivatives of f with respect to x and u.

[¶¶]See Appendix B.4 for more details on iterative solvers.

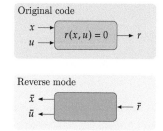

Fig. 6.32 Applying reverse AD to the code that computes r produces code that computes the partial derivatives of r with respect to x and u.

This is yet another transpose vector product that can be obtained using the same reverse AD code for the residuals, except that now the residual seed is $\bar{r} = \psi$, and the product we want is given by $\bar{x}$.

In sum, it is advantageous to use reverse-mode AD to compute the partial derivative terms for the adjoint equations, especially if the adjoint equations are solved using an iterative approach that requires only matrix-vector products. Similar techniques and arguments apply for the direct method, except that in that case, forward-mode AD is advantageous for computing the partial derivatives.

Tip 6.8 Verifying the implementation of derivative computations

Always compare your derivative computation against a different implementation. You can compare analytic derivatives with finite-difference derivatives, but that is only a partial verification because finite differences are not accurate enough. Comparing against the complex-step method or AD is preferable. Still, finite differences are recommended as an additional check. If you can only use finite differences, compare two different finite difference approximations.

You should use unit tests to verify each partial derivative term as you are developing the code (see Tip 3.4) instead of just hoping it all works together at the end (it usually does not!). One necessary but not sufficient test for the verification of analytic methods is the *dot-product test*. For analytic methods, the dot-product test can be derived from Eq. 6.42. For a chosen variable x_i and function f_j, we have the following equality:

$$\psi_j^\mathsf{T} \frac{\partial r}{\partial x_i} = \frac{\partial f_j}{\partial u} \phi_i. \tag{6.52}$$

Each side of this equation yields a scalar that should match to working precision. The dot-product test verifies that your partial derivatives and the solutions for the direct and adjoint linear systems are consistent. For AD, the dot-product test for a code with inputs x and outputs f is as follows:

$$\dot{x}^\mathsf{T} \bar{x} = \dot{x}^\mathsf{T} \left(\frac{\partial f}{\partial x}^\mathsf{T} \bar{f} \right) = \left(\dot{x}^\mathsf{T} \frac{\partial f}{\partial x}^\mathsf{T} \right) \bar{f} = \dot{f}^\mathsf{T} \bar{f}. \tag{6.53}$$

6.8 Sparse Jacobians and Graph Coloring

In this chapter, we have discussed various ways to compute a Jacobian of a model. If the Jacobian has many zero elements, it is said to be *sparse*. In many cases, we can take advantage of that sparsity to significantly reduce the computational time required to construct the Jacobian.

When applying a forward approach (forward-mode AD, finite differencing, or complex step), the cost of computing the Jacobian scales

6.8 Sparse Jacobians and Graph Coloring

with n_x. Each forward pass re-evaluates the model to compute one column of the Jacobian. For example, when using finite differencing, n_x evaluations would be required. To compute the jth column of the Jacobian, the input vector would be

$$[x_1, x_2, \ldots, x_j + h, \ldots, x_{n_x}]. \tag{6.54}$$

We can significantly reduce the cost of computing the Jacobian depending on its sparsity pattern. As a simple example, consider a square diagonal Jacobian:

$$\frac{df}{dx} \equiv \begin{bmatrix} J_{11} & 0 & 0 & 0 & 0 \\ 0 & J_{22} & 0 & 0 & 0 \\ 0 & 0 & J_{33} & 0 & 0 \\ 0 & 0 & 0 & J_{44} & 0 \\ 0 & 0 & 0 & 0 & J_{55} \end{bmatrix}. \tag{6.55}$$

For this scenario, the Jacobian can be constructed with one evaluation rather than n_x evaluations. This is because a given output f_i depends on only one input x_i. We could think of the outputs as n_x independent functions. Thus, for finite differencing, rather than requiring n_x input vectors with n_x function evaluations, we can use one input vector, as follows:

$$[x_1 + h, x_2 + h, \ldots, x_5 + h], \tag{6.56}$$

allowing us to compute all the nonzero entries in one pass.*

Although the diagonal case is easy to understand, it is a special situation. To generalize this concept, let us consider the following (5×6) matrix as an example:

$$\begin{bmatrix} J_{11} & 0 & 0 & J_{14} & 0 & J_{16} \\ 0 & 0 & J_{23} & J_{24} & 0 & 0 \\ J_{31} & J_{32} & 0 & 0 & 0 & 0 \\ 0 & 0 & 0 & 0 & J_{45} & 0 \\ 0 & 0 & J_{53} & 0 & J_{55} & J_{56} \end{bmatrix}. \tag{6.57}$$

A subset of columns that does not have more than one nonzero in any given row are said to be *structurally orthogonal*. In this example, the following sets of columns are structurally orthogonal: (1, 3), (1, 5), (2, 3), (2, 4, 5), (2, 6), and (4, 5). Structurally orthogonal columns can be combined, forming a smaller Jacobian that reduces the number of forward passes required. This reduced Jacobian is referred to as *compressed*. There is more than one way to compress this Jacobian, but in this case, the minimum number of compressed columns—referred to as *colors*—is three. In the following compressed Jacobian, we combine

*Curtis et al.[130] were the first to show that the number of function evaluations could be reduced for sparse Jacobians.

130. Curtis et al., *On the estimation of sparse Jacobian matrices*, 1974.

columns 1 and 3 (blue); columns 2, 4, and 5 (red); and leave column 6 on its own (black):

$$\begin{bmatrix} J_{11} & 0 & 0 & J_{14} & 0 & J_{16} \\ 0 & 0 & J_{23} & J_{24} & 0 & 0 \\ J_{31} & J_{32} & 0 & 0 & 0 & 0 \\ 0 & 0 & 0 & 0 & J_{45} & 0 \\ 0 & 0 & J_{53} & 0 & J_{55} & J_{56} \end{bmatrix} \Rightarrow \begin{bmatrix} J_{11} & J_{14} & J_{16} \\ J_{23} & J_{24} & 0 \\ J_{31} & J_{32} & 0 \\ 0 & J_{45} & 0 \\ J_{53} & J_{55} & J_{56} \end{bmatrix}. \quad (6.58)$$

For finite differencing, complex step, and forward-mode AD, only compression among columns is possible. Reverse mode AD allows compression among the rows. The concept is the same, but instead, we look for structurally orthogonal rows. One such compression is as follows:

$$\begin{bmatrix} J_{11} & 0 & 0 & J_{14} & 0 & J_{16} \\ 0 & 0 & J_{23} & J_{24} & 0 & 0 \\ J_{31} & J_{32} & 0 & 0 & 0 & 0 \\ 0 & 0 & 0 & 0 & J_{45} & 0 \\ 0 & 0 & J_{53} & 0 & J_{55} & J_{56} \end{bmatrix} \Rightarrow \begin{bmatrix} J_{11} & 0 & 0 & J_{14} & J_{45} & J_{16} \\ 0 & 0 & J_{23} & J_{24} & 0 & 0 \\ J_{31} & J_{32} & J_{53} & 0 & J_{55} & J_{56} \end{bmatrix}.$$

(6.59)

AD can also be used even more flexibly when both modes are used: forward passes to evaluate groups of structurally orthogonal columns and reverse passes to evaluate groups of structurally orthogonal rows. Rather than taking incremental steps in each direction as is done in finite differencing, we set the AD seed vector with 1s in the directions we wish to evaluate, similar to how the seed is set for directional derivatives, as discussed in Section 6.6.

For these small Jacobians, it is straightforward to determine how to compress the matrix in the best possible way. For a large matrix, this is not so easy. One approach is to use *graph coloring*. This approach starts by building a graph where the vertices represent the row and column indices, and the edges represent nonzero entries in the Jacobian. Then, algorithms are applied to this graph that estimate the fewest number of "colors" (orthogonal columns) using heuristics. Graph coloring is a large field of research, where derivative computation is one of many applications.[†]

[†]Gebremedhin et al.[131] provide a review of graph coloring in the context of computing derivatives. Gray et al.[132] show how to use graph coloring to compute total coupled derivatives.

131. Gebremedhin et al., *What color is your Jacobian? Graph coloring for computing derivatives*, 2005.

132. Gray et al., *OpenMDAO: An open-source framework for multidisciplinary design, analysis, and optimization*, 2019.

Example 6.13 Speed up from sparse derivatives

In static aerodynamic analyses, the forces and moments produced at two different flow conditions are independent. If there are many different flow conditions of interest, the resulting Jacobian is sparse. Examples include evaluating the power produced by a wind turbine at different wind speeds or assessing an aircraft's performance throughout a flight envelope. Many other engineering analyses have a similar structure.

Consider a typical wind turbine blade optimization. The Jacobian of the functions of interest is fully dense with respect to geometry changes. However, the part of the Jacobian that contains the derivatives with respect to the various flow conditions is diagonal, as illustrated on left side of Fig. 6.33. Blank blocks represent derivatives that are zero. We can compress the diagonal part of the Jacobian as shown on the right side of Fig. 6.33.

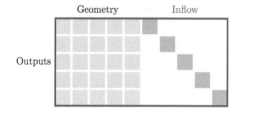
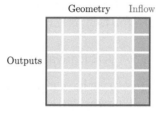

Fig. 6.33 Jacobian structure for wind turbine problem. The original Jacobian (left) can be replaced with a compressed one (right).

To illustrate the potential benefits of using a sparse representation, we time the Jacobian computation for various sizes of inflow conditions using forward AD with and without graph coloring (Fig. 6.34). For more than 100 inflow conditions, the difference in time required exceeds one order of magnitude (note the log-log scale). Because Jacobians are needed at every iteration in the optimization, this is a tremendous speedup, enabled by exploiting the sparsity pattern.[133]

133. Ning, *Using blade element momentum methods with gradient-based design optimization*, 2021.

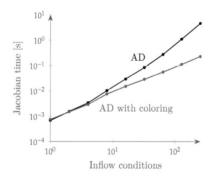

Fig. 6.34 Jacobian computational time with and without coloring.

6.9 Unified Derivatives Equation

Now that we have introduced all the methods for computing derivatives, we will see how they are connected. For example, we have mentioned that the direct and adjoint methods are analogous to the forward and reverse mode of AD, respectively, but we did not show

134. Martins and Hwang, *Review and unification of methods for computing derivatives of multidisciplinary computational models*, 2013.

this mathematically. The unified derivatives equation (UDE) expresses both methods.[134] Also, the implicit analytic methods from Section 6.7 assumed one set of implicit equations ($r = 0$) and one set of explicit functions (f). The UDE formulates the derivative computation for systems with mixed sets of implicit and explicit equations.

We first derive the UDE from basic principles and give an intuitive explanation of the derivative terms. Then, we show how we can use the UDE to handle implicit and explicit equations. We also show how the UDE can retrieve the direct and adjoint equations. Finally, we show how the UDE is connected to AD.

6.9.1 UDE Derivation

Suppose we have a set of n residual equations with the same number of unknowns,

$$r_i(u_1, u_2, \ldots, u_n) = 0, \quad i = 1, \ldots, n, \qquad (6.60)$$

and that there is at least one solution u^* such that $r(u^*) = 0$. Such a solution can be visualized for $n = 2$, as shown in Fig. 6.35.

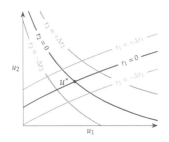

Fig. 6.35 Solution of a system of two equations expressed by residuals.

These residuals are general: each one can depend on any subset of the variables u and can be truly implicit functions or explicit functions converted to the implicit form (see Section 3.3 and Ex. 3.3). The total differentials for these residuals are

$$\mathrm{d}r_i = \frac{\partial r_i}{\partial u_1} \mathrm{d}u_1 + \ldots + \frac{\partial r_i}{\partial u_n} \mathrm{d}u_n, \quad i = 1, \ldots, n. \qquad (6.61)$$

These represent first-order changes in r due to perturbations in u. The differentials of u can be visualized as perturbations in the space of the variables. The differentials of r can be visualized as linear changes to the surface defined by $r = 0$, as illustrated in Fig. 6.36.

We can write the differentials (Eq. 6.61) in matrix form as

$$\begin{bmatrix} \dfrac{\partial r_1}{\partial u_1} & \cdots & \dfrac{\partial r_1}{\partial u_n} \\ \vdots & \ddots & \vdots \\ \dfrac{\partial r_n}{\partial u_1} & \cdots & \dfrac{\partial r_n}{\partial u_n} \end{bmatrix} \begin{bmatrix} \mathrm{d}u_1 \\ \vdots \\ \mathrm{d}u_n \end{bmatrix} = \begin{bmatrix} \mathrm{d}r_1 \\ \vdots \\ \mathrm{d}r_n \end{bmatrix}. \qquad (6.62)$$

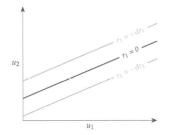

Fig. 6.36 The differential $\mathrm{d}r$ can be visualized as a linearized (first-order) change of the contour value.

The partial derivatives in the matrix are derivatives of the expressions for r with respect to u that can be obtained symbolically, and they are in general functions of u. The vector of differentials $\mathrm{d}u$ represents perturbations in u that can be solved for a given vector of changes $\mathrm{d}r$.

6.9 Unified Derivatives Equation

Now suppose that we are at a solution u^*, such that $r(u^*) = 0$. All the partial derivatives $(\partial r/\partial u)$ can be evaluated at u^*. When all entries in dr are zero, then the solution of this linear system yields $du = 0$. This is because if there is no disruption in the residuals that are already zero, the variables do not need to change either.

How is this linear system useful? With these differentials, we can choose different combinations of dr to obtain any total derivatives that we want. For example, we can get the total derivatives of u with respect to a single residual r_i by keeping dr_i while setting all the other differentials to zero ($dr_{j \neq i} = 0$). The visual interpretation of this total derivative is shown in Fig. 6.37 for $n = 2$ and $i = 1$. Setting $dr = [0, \ldots, 0, dr_i, 0, \ldots, 0]$ in Eq. 6.62 and moving dr_i to the denominator, we obtain the following linear system:*

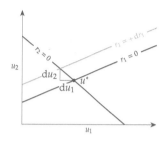

Fig. 6.37 The total derivatives du_1/dr_1 and du_2/dr_1 represent the first-order changes needed to satisfy a perturbation $r_1 = dr_1$ while keeping $r_2 = 0$.

*As explained in Appendix A.2, we take the liberty of treating differentials algebraically and skip a more rigorous and lengthy proof.

$$\begin{bmatrix} \frac{\partial r_1}{\partial u_1} & \cdots & \frac{\partial r_1}{\partial u_i} & \cdots & \frac{\partial r_1}{\partial u_n} \\ \vdots & \ddots & \vdots & & \vdots \\ \frac{\partial r_i}{\partial u_1} & \cdots & \frac{\partial r_i}{\partial u_i} & \cdots & \frac{\partial r_i}{\partial u_n} \\ \vdots & & \vdots & \ddots & \vdots \\ \frac{\partial r_n}{\partial u_1} & \cdots & \frac{\partial r_n}{\partial u_i} & \cdots & \frac{\partial r_n}{\partial u_n} \end{bmatrix} \begin{bmatrix} \frac{du_1}{dr_i} \\ \vdots \\ \frac{du_i}{dr_i} \\ \vdots \\ \frac{du_n}{dr_i} \end{bmatrix} = \begin{bmatrix} 0 \\ \vdots \\ 1 \\ \vdots \\ 0 \end{bmatrix}. \quad (6.63)$$

Doing the same for all $i = 1, \ldots, n$, we get the following n linear systems:

$$\begin{bmatrix} \frac{\partial r_1}{\partial u_1} & \cdots & \frac{\partial r_1}{\partial u_n} \\ \vdots & \ddots & \vdots \\ \frac{\partial r_n}{\partial u_1} & \cdots & \frac{\partial r_n}{\partial u_n} \end{bmatrix} \begin{bmatrix} \frac{du_1}{dr_1} & \cdots & \frac{du_1}{dr_n} \\ \vdots & \ddots & \vdots \\ \frac{du_n}{dr_1} & \cdots & \frac{du_n}{dr_n} \end{bmatrix} = \begin{bmatrix} 1 & \cdots & 0 \\ \vdots & \ddots & \vdots \\ 0 & \cdots & 1 \end{bmatrix}. \quad (6.64)$$

Solving these linear systems yields the total derivatives of all the elements of u with respect to all the elements of r. We can write this more compactly in matrix form as

$$\frac{\partial r}{\partial u} \frac{du}{dr} = I. \quad (6.65)$$

This is the forward form of the UDE.

The total derivatives du/dr might not seem like the derivatives in which we are interested. Based on the implicit analytic methods derived in Section 6.7.2, these look like derivatives of states with respect to residuals, not the derivatives that we ultimately want to compute (df/dx). However, we will soon see that with the appropriate choice of r and u, we can obtain a linear system that solves for the total derivatives we want.

With Eq. 6.65, we can solve one column at a time. Similar to AD, we can also solve for the rows instead by transposing the systems as [†]

†Normally, for two matrices A and B, $(AB)^\mathsf{T} = B^\mathsf{T} A^\mathsf{T}$, but in this case, $AB = I \Rightarrow B = A^{-1} \Rightarrow B^\mathsf{T} = A^{-\mathsf{T}} \Rightarrow A^\mathsf{T} B^\mathsf{T} = I$.

$$\frac{\partial r}{\partial u}^\mathsf{T} \frac{du}{dr}^\mathsf{T} = I, \qquad (6.66)$$

which is the reverse form of the UDE. Now, each column j yields du_j/dr—the total derivative of one variable with respect to all the residuals. This total derivative is interpreted visually in Fig. 6.38.

The usefulness of the total derivative Jacobian du/dr might still not be apparent. In the next section, we explain how to set up the UDE to include df/dx in the UDE unknowns (du/dr).

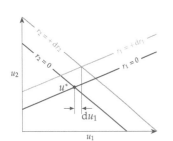

Fig. 6.38 The total derivatives du_1/dr_1 and du_1/dr_2 represent the first-order change in u_1 resulting from perturbations dr_1 and dr_2.

Example 6.14 Computing and interpreting du/dr

Suppose we want to find the rectangle that is inscribed in the ellipse given by

$$r_1(u_1, u_2) = \frac{u_1^2}{4} + u_2^2 - 1 = 0.$$

A change in this residual represents a change in the size of the ellipse without changing its proportions. Of all the possible rectangles that can be inscribed in the ellipse, we want the rectangle with an area that is half of that of this ellipse, such that

$$r_2(u_1, u_2) = 4u_1 u_2 - \pi = 0.$$

A change in this residual represents a change in the area of the rectangle. There are two solutions, as shown in the left pane of Fig. 6.39. These solutions can be found using Newton's method, which converges to one solution or the other, depending on the starting guess. We will pick the one on the right, which is $[u_1, u_2] = [1.79944, 0.43647]$. The solution represents the coordinates of the rectangle corner that touches the ellipse.

Taking the partial derivatives, we can write the forward UDE (Eq. 6.65) for this problem as follows:

$$\begin{bmatrix} u_1/2 & 2u_2 \\ 4u_2 & 4u_1 \end{bmatrix} \begin{bmatrix} \dfrac{du_1}{dr_1} & \dfrac{du_1}{dr_2} \\ \dfrac{du_2}{dr_1} & \dfrac{du_2}{dr_2} \end{bmatrix} = \begin{bmatrix} 1 & 0 \\ 0 & 1 \end{bmatrix}. \qquad (6.67)$$

6.9 Unified Derivatives Equation

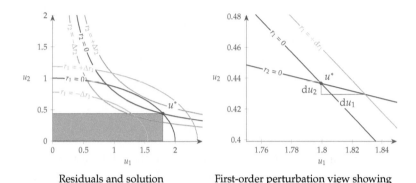

Fig. 6.39 Rectangle inscribed in ellipse problem.

Solving this linear system for each of the two right-hand sides, we get

$$\begin{bmatrix} \dfrac{du_1}{dr_1} & \dfrac{du_1}{dr_2} \\ \dfrac{du_2}{dr_1} & \dfrac{du_2}{dr_2} \end{bmatrix} = \begin{bmatrix} 1.45353 & -0.17628 \\ -0.35257 & 0.18169 \end{bmatrix}. \tag{6.68}$$

These derivatives reflect the change in the coordinates of the point where the rectangle touches the ellipse as a result of a perturbation in the size of the ellipse, dr_1, and the area of the rectangle dr_2. The right side of Fig. 6.39 shows the visual interpretation of du/dr_1 as an example.

6.9.2 UDE for Mixed Implicit and Explicit Components

In the previous section, the UDE was derived based on residual equations. The equations were written in implicit form, but there was no assumption on whether the equations were implicit or explicit. Because we can write an explicit equation in implicit form (see Section 3.3 and Ex. 3.3), the UDE allows a mix of implicit and explicit set of equations, which we now call *components*.

To derive the implicit analytic equations in Section 6.7, we considered two components: an implicit component that determines u by solving $r(u; x) = 0$ and an explicit component that computes the functions of interest, $f(x, u)$.

We can recover the implicit analytic differentiation equations (direct and adjoint) from the UDE by defining a set of variables that concatenates the state variables with inputs and outputs as follows:

$$\hat{u} \equiv \begin{bmatrix} x \\ u \\ f \end{bmatrix}. \tag{6.69}$$

This is a vector with $(n_x + n_u + n_f)$ variables. For the residuals, we need a vector with the same size. We can obtain this by realizing that the residuals associated with the inputs and outputs are just explicit functions that can be written in implicit form. Then, we have

$$\hat{r} \equiv \begin{bmatrix} x - \check{x} \\ r - \check{r}(x,u) \\ f - \check{f}(x,u) \end{bmatrix} = 0. \quad (6.70)$$

Here, we distinguish x (the actual variable in the UDE system) from $\check{x}$ (a given input) and f (the variable) from $\check{f}$ (an explicit function of x and u). Similarly, r is the vector of variables associated with the residual and $\check{r}$ is the residual function itself. Taking the differential of the residuals, and considering only one of them to be nonzero at a time, we obtain,

$$d\hat{r} \equiv \begin{bmatrix} dx \\ dr \\ df \end{bmatrix}. \quad (6.71)$$

Using these variable and residual definitions in Eqs. 6.65 and 6.66 yields the full UDE shown in Fig. 6.40, where the block we ultimately want to compute is df/dx.

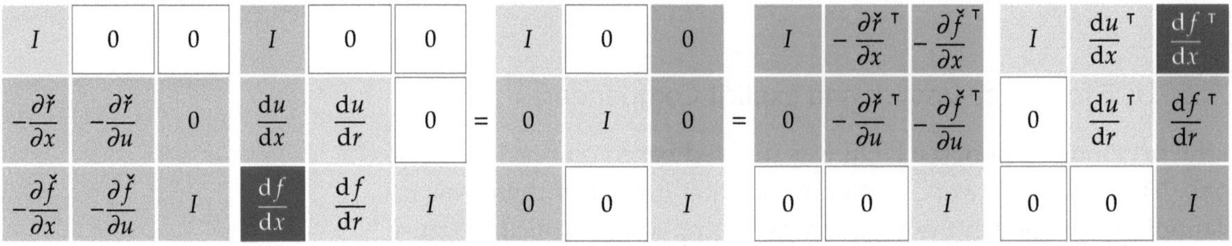

$$\begin{bmatrix} I & 0 & 0 \\ -\dfrac{\partial \check{r}}{\partial x} & -\dfrac{\partial \check{r}}{\partial u} & 0 \\ -\dfrac{\partial \check{f}}{\partial x} & -\dfrac{\partial \check{f}}{\partial u} & I \end{bmatrix} \begin{bmatrix} I & 0 & 0 \\ \dfrac{du}{dx} & \dfrac{du}{dr} & 0 \\ \dfrac{df}{dx} & \dfrac{df}{dr} & I \end{bmatrix} = \begin{bmatrix} I & 0 & 0 \\ 0 & I & 0 \\ 0 & 0 & I \end{bmatrix} = \begin{bmatrix} I & -\dfrac{\partial \check{r}}{\partial x}^T & -\dfrac{\partial \check{f}}{\partial x}^T \\ 0 & -\dfrac{\partial \check{r}}{\partial u}^T & -\dfrac{\partial \check{f}}{\partial u}^T \\ 0 & 0 & I \end{bmatrix} \begin{bmatrix} I & \dfrac{du}{dx}^T & \dfrac{df}{dx}^T \\ 0 & \dfrac{du}{dr}^T & \dfrac{df}{dr}^T \\ 0 & 0 & I \end{bmatrix}$$

Fig. 6.40 The direct and adjoint methods can be recovered from the UDE.

To compute df/dx using the forward UDE (left-hand side of the equation in Fig. 6.40, we can ignore all but three blocks in the total derivatives matrix: I, du/dx, and df/dx. By multiplying these blocks and using the definition $\phi \equiv -du/dx$, we recover the direct linear system (Eq. 6.43) and the total derivative equation (Eq. 6.44).

To compute df/dx using the reverse UDE (right-hand side of the equation in Fig. 6.40), we can ignore all but three blocks in the total derivatives matrix: I, df/dr, and df/dx. By multiplying these blocks and defining $\psi \equiv -df/dr$, we recover the adjoint linear system (Eq. 6.46) and the corresponding total derivative equation (Eq. 6.47). The definition of ψ here is significant because, as mentioned in Section 6.7.2, the adjoint vector is the total derivative of the objective function with respect to the governing equation residuals.

6.9 Unified Derivatives Equation

By defining one implicit component (associated with u) and two explicit components (associated with x and f), we have retrieved the direct and adjoint methods from the UDE. In general, we can define an arbitrary number of components, so the UDE provides a mathematical framework for computing the derivatives of coupled systems. Furthermore, each component can be implicit or explicit, so the UDE can handle an arbitrary mix of components. All we need to do is to include the desired states in the UDE augmented variables vector (Eq. 6.69) and the corresponding residuals in the UDE residuals vector (Eq. 6.70). We address coupled systems in Section 13.3.3 and use the UDE in Section 13.2.6, where we extend it to coupled systems with a hierarchy of components.

Example 6.15 Computing arbitrary derivatives with the UDE

Say we want to compute the total derivatives of the perimeter of the rectangle from Ex. 6.14 with respect to the axes of the ellipse. The equation for the ellipse can be rewritten as

$$r_3(u_1, u_2) = \frac{u_1^2}{x_1^2} + \frac{u_2^2}{x_2^2} - 1 = 0,$$

where x_1 and x_2 are the semimajor and semiminor axes of the ellipse, respectively. The baseline values are $[x_1, x_2] = [2, 1]$. The residual for the rectangle area is

$$r_4(u_1, u_2) = 4u_1 u_2 - \frac{\pi}{2} x_1 x_2 = 0.$$

To add the independent variables x_1 and x_2, we write them as residuals in implicit form as

$$r_1(x_1) = x_1 - 2 = 0, \quad r_2(x_2) = x_2 - 1 = 0.$$

The perimeter can be written in implicit form as

$$r_5(u_1, u_2) = f - 4(u_1 + u_2) = 0.$$

Now we have a system of five equations and five variables, with the dependencies shown in Fig. 6.41. The first two variables in x are given, and we can compute u and f using a solver as before.

Taking all the partial derivatives, we get the following forward system:

$$\begin{bmatrix} 1 & 0 & 0 & 0 & 0 \\ 0 & 1 & 0 & 0 & 0 \\ -\frac{2u_1^2}{x_1^3} & -\frac{2u_2^2}{x_2^3} & \frac{2u_1}{x_1^2} & \frac{2u_2}{x_2^2} & 0 \\ -\frac{\pi}{2}x_2 & -\frac{\pi}{2}x_1 & 4u_2 & 4u_1 & 0 \\ 0 & 0 & -4 & -4 & 1 \end{bmatrix} \begin{bmatrix} 1 & 0 & 0 & 0 & 0 \\ 0 & 1 & 0 & 0 & 0 \\ \frac{du_1}{dx_1} & \frac{du_1}{dx_2} & \frac{du_1}{dr_3} & \frac{du_1}{dr_4} & 0 \\ \frac{du_2}{dx_1} & \frac{du_2}{dx_2} & \frac{du_2}{dr_3} & \frac{du_2}{dr_4} & 0 \\ \frac{df}{dx_1} & \frac{df}{dx_2} & \frac{df}{dr_3} & \frac{df}{dr_4} & 1 \end{bmatrix} = I.$$

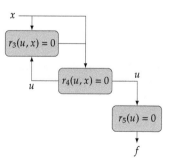

Fig. 6.41 Dependencies of the residuals.

We only want the two df/dx terms in this equation. We can either solve this linear system twice to compute the first two columns, or we can compute both terms with a single solution of the reverse (transposed) system. Transposing the system, substituting the numerical values for x and u, and removing the total derivative terms that we do not need, we get the following system:

$$\begin{bmatrix} 1 & 0 & -0.80950 & -1.57080 & 0 \\ 0 & 1 & -0.38101 & -3.14159 & 0 \\ 0 & 0 & 0.89972 & 1.74588 & -4 \\ 0 & 0 & 0.87294 & 7.19776 & -4 \\ 0 & 0 & 0 & 0 & 1 \end{bmatrix} \begin{bmatrix} \dfrac{df}{dx_1} \\ \dfrac{df}{dx_2} \\ \dfrac{df}{dr_3} \\ \dfrac{df}{dr_4} \\ 1 \end{bmatrix} = \begin{bmatrix} 0 \\ 0 \\ 0 \\ 0 \\ 1 \end{bmatrix}.$$

Solving this linear system, we obtain

$$\begin{bmatrix} \dfrac{df}{dx_1} \\ \dfrac{df}{dx_2} \\ \dfrac{df}{dr_3} \\ \dfrac{df}{dr_4} \end{bmatrix} = \begin{bmatrix} 3.59888 \\ 1.74588 \\ 4.40385 \\ 0.02163 \end{bmatrix}.$$

The total derivatives of interest are shown in Fig. 6.42.

We could have obtained the same solution using the adjoint equations from Section 6.7.2. The only difference is the nomenclature because the adjoint vector in this case is $\psi = -[df/dr_3, df/dr_4]$. We can interpret these terms as the change of f with respect to changes in the ellipse size and rectangle area, respectively.

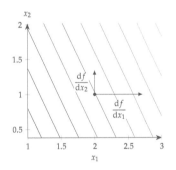

Fig. 6.42 Contours of f as a function of x and the total derivatives at $x = [2, 1]$.

6.9.3 Recovering AD

Now we will see how we can recover AD from the UDE. First, we define the UDE variables associated with each operation or line of code (assuming all loops have been unrolled), such that $u \equiv v$ and

$$v_i = \check{v}_i(v_1, \ldots, v_{i-1}), \quad i = 1, \ldots, n. \tag{6.72}$$

Recall from Section 6.6.1 that each variable is an *explicit* function of the previous ones.

6.9 Unified Derivatives Equation

To define the appropriate residuals, we use the same technique from before to convert an explicit function into implicit form by moving all the terms in the left-hand side to obtain

$$r_i = v_i - \breve{v}_i(v_1, \ldots, v_{i-1}). \tag{6.73}$$

The distinction between v and $\breve{v}$ is that v represents variables that are considered independent in the UDE, whereas $\breve{v}$ represents the explicit expressions. Of course, the values for these become equal once the system is solved. Similar to the differentials in Eq. 6.71, $dr \equiv dv$

Taking the partial derivatives of the residuals (Eq. 6.73) with respect to v (Eq. 6.72), and replacing the total derivatives in the forward form of the UDE (Eq. 6.65) with the new symbols yields

$$\begin{bmatrix} 1 & 0 & \cdots & 0 \\ -\dfrac{\partial \breve{v}_2}{\partial v_1} & 1 & \ddots & \vdots \\ \vdots & \ddots & \ddots & 0 \\ -\dfrac{\partial \breve{v}_n}{\partial v_1} & \cdots & -\dfrac{\partial \breve{v}_n}{\partial v_{n-1}} & 1 \end{bmatrix} \begin{bmatrix} \dfrac{dv_1}{dv_1} & 0 & \cdots & 0 \\ \dfrac{dv_2}{dv_1} & \dfrac{dv_2}{dv_2} & \ddots & \vdots \\ \vdots & \ddots & \ddots & 0 \\ \dfrac{dv_n}{dv_1} & \cdots & \dfrac{dv_n}{dv_{n-1}} & \dfrac{dv_n}{dv_n} \end{bmatrix} = I. \tag{6.74}$$

This equation is the matrix form of the AD forward chain rule (Eq. 6.21), where each column of the total derivative matrix corresponds to the tangent vector ($\dot{v}$) for the chosen input variable. As observed in Fig. 6.16, the partial derivatives form a lower triangular matrix. The Jacobian we ultimately want to compute (df/dx) is composed of a subset of derivatives in the bottom-left corner near the dv_n/dv_1 term. To compute these derivatives, we need to perform forward substitution and compute one column of the total derivative matrix at a time, where each column is associated with the inputs of interest.

Similarly, the reverse form of the UDE (Eq. 6.66) yields the transpose of Eq. 6.74,

$$\begin{bmatrix} 1 & -\dfrac{\partial \breve{v}_2}{\partial v_1} & \cdots & -\dfrac{\partial \breve{v}_n}{\partial v_1} \\ 0 & 1 & \ddots & \vdots \\ \vdots & \ddots & \ddots & -\dfrac{\partial \breve{v}_n}{\partial v_{n-1}} \\ 0 & \cdots & 0 & 1 \end{bmatrix} \begin{bmatrix} \dfrac{dv_1}{dv_1} & \dfrac{dv_2}{dv_1} & \cdots & \dfrac{dv_n}{dv_1} \\ 0 & \dfrac{dv_2}{dv_2} & \ddots & \vdots \\ \vdots & \ddots & \dfrac{dv_{n-1}}{dv_{n-1}} & \dfrac{dv_n}{dv_{n-1}} \\ 0 & \cdots & 0 & \dfrac{dv_n}{dv_n} \end{bmatrix} = I. \tag{6.75}$$

This is equivalent to the AD reverse chain rule (Eq. 6.26), where each column of the total derivative matrix corresponds to the gradient vector

($\bar{v}$) for the chosen output variable. The partial derivatives now form an upper triangular matrix, as previously shown in Fig. 6.21. The derivatives of interest are now near the top-right corner of the total derivative matrix near the dv_n/dv_1 term. To compute these derivatives, we need to perform back substitutions, computing one column of the matrix at a time.

> **Tip 6.9** Scaling affects the derivatives
>
> When scaling a problem (Tips 4.4 and 5.3), you should be aware that the scale changes also affect the derivatives. You can apply the derivative methods of this chapter to the scaled function directly. However, scaling is often done outside the model because the scaling is specific to the optimization problem. In this case, you may want to use the original functions and derivatives and make the necessary modifications in an outer function that provides the objectives and constraints.
>
> Using the nomenclature introduced in Tip 4.4, we represent the scaled design variables given to the optimizer as $\bar{x}$. Then, the unscaled variables are $x = s_x \odot \bar{x}$. Thus, the required scaled derivatives are
>
> $$\frac{d\bar{f}}{d\bar{x}} = \frac{df}{dx} \odot \frac{s_x}{s_f}. \tag{6.76}$$

> **Tip 6.10** Provide your own derivatives and use finite differences only as a last resort
>
> Because of the step-size dilemma, finite differences are often the cause of failed optimizations. To put it more dramatically, *finite differences are the root of all evil*. Most gradient-based optimization software uses finite differences internally as a default if you do not provide your own gradients. Although some software packages try to find reasonable finite-difference steps, it is easy to get inaccurate derivatives, which then causes optimization difficulties or total failure. This is the top reason why beginners give up on gradient-based optimization!
>
> Instead, you should provide gradients computed using one of the other methods described in this chapter. In contrast with finite differences, the derivatives computed by the other methods are usually as accurate as the function computation. You should also avoid using finite-difference derivatives as a reference for a definitive verification of the other methods.
>
> If you have to use finite differences as a last resort, make sure to do a step-size study (see Tip 6.2). You should then provide your own finite-difference derivatives to the optimization or make sure that the optimizer finite-difference estimates are acceptable.

6.10 Summary

Derivatives are useful in many applications beyond optimization. This chapter introduced the methods available to compute the first derivatives of the outputs of a model with respect to its inputs. In optimization, the outputs are usually the objective function and the constraint functions, and the inputs are the design variables. The typical characteristics of the available methods are compared in Table 6.4.

	Accuracy	Scalability	Ease of implementation	Implicit functions
Symbolic	●		Hard	
Finite difference			Easy	●
Complex step	●		Intermediate	●
AD	●	●	Intermediate	●
Implicit analytic	●	●	Hard	●

Table 6.4 Characteristics of the various derivative computation methods. Some of these characteristics are problem or implementation dependent, so these are not universal.

Symbolic differentiation is accurate but only scalable for simple, explicit functions of low dimensionality. Therefore, it is necessary to compute derivatives numerically. Although it is generally intractable or inefficient for many engineering models, symbolic differentiation is used by AD at each line of code and in implicit analytic methods to derive expressions for computing the required partial derivatives.

Finite-difference approximations are popular because they are easy to implement and can be applied to any model, including black-box models. The downsides are that these approximations are not accurate, and the cost scales linearly with the number of variables. Many of the issues practitioners experience with gradient-based optimization can be traced to errors in the gradients when algorithms automatically compute these gradients using finite differences.

The complex-step method is accurate and relatively easy to implement. It usually requires some changes to the analysis source code, but this process can be scripted. The main advantage of the complex-step method is that it produces analytically accurate derivatives. However, like the finite-difference method, the cost scales linearly with the number of inputs, and each simulation requires more effort because of the complex arithmetic.

AD produces analytically accurate derivatives, and many implementations can be fully automated. The implementation requires access to the source code but is still relatively straightforward to apply. The computational cost of forward-mode AD scales with the number of inputs, and the reverse mode scales with the number of outputs. The

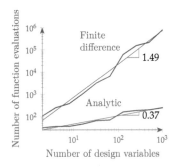

Fig. 6.43 Efficient gradient computation with an analytic method improves the scalability of gradient-based algorithms compared to finite differencing. In this case, we show the results for the n-dimensional Rosenbrock, where the cost of computing the derivatives analytically is independent of n.

scaling factor for the forward mode is generally lower than that for finite differences. The cost of reverse-mode AD is independent of the number of design variables.

Implicit analytic methods (direct and adjoint) are accurate and scalable but require significant implementation effort. These methods are exact (depending on how the partial derivatives are obtained), and like AD, they provide both forward and reverse modes with the same scaling advantages. Gradient-based optimization using the adjoint method is a powerful combination that scales well with the number of variables, as shown in Fig. 6.43. The disadvantage is that because implicit methods are intrusive, they require considerable development effort.

A hybrid approach where the partial derivatives for the implicit analytic equations are computed with AD is generally recommended. This hybrid approach is computationally more efficient than AD while reducing the implementation effort of implicit analytic methods and ensuring accuracy.

The UDE encapsulates all the derivative computation methods in a single linear system. Using the UDE, we can formulate the derivative computation for an arbitrary system of mixed explicit and implicit components. This will be used in Section 13.2.6 to develop a mathematical framework for solving coupled systems and computing the corresponding derivatives.

Problems

6.1 Answer *true* or *false* and justify your answer.

  a. A first-order derivative is only one of many types of sensitivity analysis.

  b. Each column of the Jacobian matrix represents the gradient of one of the functions of interest with respect to all the variables.

  c. You can only compute derivatives of models for which you have the source code or, at the very least, understand how the model computes the functions of interest.

  d. Symbolic differentiation is intractable for all but the simplest models because of expression swell.

  e. Finite-difference approximations can compute first derivatives with a precision matching that of the function being differentiated.

  f. The complex-step method can only be used to compute derivatives of real functions.

  g. AD via source code transformation uses a code parser to differentiate each line of code symbolically.

  h. The forward mode of AD computes the derivatives of all outputs with respect to one input, whereas the reverse mode computes the derivative of one output with respect to all inputs.

  i. The adjoint method requires the same partial derivatives as the direct method.

  j. Of the two implicit analytic methods, the direct method is more widely used than the adjoint method because most problems have more design variables than functions of interest.

  k. Graph coloring makes Jacobians sparse by selectively replacing small-valued entries with zeros to trade accuracy for speed.

  l. The unified derivatives equation can represent implicit analytic approaches but not AD.

6.2 Reproduce the comparison between the complex-step and finite-difference methods from Ex. 6.4. Do you get any complex-step derivatives with zero error compared with the analytic reference?

What does that mean, and how should you show those points on the plot?

6.3 Compute the derivative using symbolic differentiation and using algorithmic differentiation (either forward or reverse mode) for the iterative code in Ex. 6.2. Use a package to facilitate the AD portion. Most scientific computing languages have AD packages (see Tip 6.6).

6.4 Write a forward-mode-AD script that computes the derivative of the function in Ex. 6.3 using operator overloading. You need to define your own type and provide it with overloaded functions for exp, sin, cos, sqrt, addition, division, and exponentiation.

6.5 Suppose you have two airplanes that are flying in a horizontal plane defined by x and y coordinates. Both airplanes start at $y = 0$, but airplane 1 starts at $x = 0$, whereas airplane 2 has a head start of $x = \Delta x$. The airplanes fly at a constant velocity. Airplane 1 has a velocity of v_1 in the direction of the positive x-axis, and airplane 2 has a velocity of v_2 at an angle γ with the x-axis. The functions of interest are the distance (d) and the angle (θ) between the two airplanes as a function of time. The independent variables are $\Delta x, \gamma, v_1, v_2$, and t. Write the code that computes the functions of interest (outputs) for a given set of independent variables (inputs). Use AD to differentiate the code. Choose a set of inputs, compute the derivatives of all the outputs with respect to the inputs, and verify them against the complex-step method.

6.6 Kepler's equation, which we mentioned in Section 2.2, defines the relationship between a planet's polar coordinates and the time elapsed from a given initial point through the implicit equation

$$E - e \sin(E) = M,$$

where M is the mean anomaly (a parameterization of time), E is the eccentric anomaly (a parameterization of the polar angle), and e is the eccentricity of the elliptical orbit. Suppose that the function of interest is the difference between the eccentric and mean anomalies,

$$f(E, M) = E - M.$$

Derive an analytic expression for df/de and df/dM. Verify your result against the complex-step method or AD (you will need a solver for Kepler's equation, which was the subject of Prob. 3.6).

6.7 Compute the derivatives for the 10-bar truss problem described in Appendix D.2.2 using the direct and adjoint implicit differentiation methods. Compute the derivatives of the objective (mass) with respect to the design variables (10 cross-sectional areas), and the derivatives of the constraints (stresses in all 10 bars) with respect to the design variables (a 10×10 Jacobian matrix). Compute the derivatives using the following:

  a. A finite-difference formula of your choice
  b. The complex-step derivative method
  c. AD
  d. The implicit analytic method (direct and adjoint)

  Report the errors relative to the implicit analytic methods. Discuss your findings and the relative merits of each approach.

6.8 You can now solve the 10-bar truss problem (previously solved in Prob. 5.15) using the derivatives computed in Prob. 6.7. Solve this optimization problem using both finite-difference derivatives and derivatives computed using an implicit analytic method. Report the following:

  a. Convergence plot with two curves for the different derivative computation approaches on the same plot
  b. Number of function calls required to converge for each method (This metric is more meaningful if you use more than one starting point and average the results.)

  Discuss your findings.

6.9 Aggregate the constraints for the 10-bar truss problem and extend the code from Prob. 6.7 to compute the required constraint derivatives using the implicit analytic method that is most advantageous in this case. Verify your derivatives against the complex-step method. Solve the optimization problem and compare your results to the ones you obtained in Prob. 6.8. How close can you get to the reference solution?

7 Gradient-Free Optimization

Gradient-free algorithms fill an essential role in optimization. The gradient-based algorithms introduced in Chapter 4 are efficient in finding local minima for high-dimensional nonlinear problems defined by continuous smooth functions. However, the assumptions made for these algorithms are not always valid, which can render these algorithms ineffective. Also, gradients might not be available when a function is given as a black box.

In this chapter, we introduce only a few popular representative gradient-free algorithms. Most are designed to handle unconstrained functions only, but they can be adapted to solve constrained problems by using the penalty or filtering methods introduced in Chapter 5. We start by discussing the problem characteristics relevant to the choice between gradient-free and gradient-based algorithms and then give an overview of the types of gradient-free algorithms.

> By the end of this chapter you should be able to:
>
> 1. Identify problems that are well suited for gradient-free optimization.
> 2. Describe the characteristics and approach of more than one gradient-free optimization method.
> 3. Use gradient-free optimization algorithms to solve real engineering problems.

7.1 When to Use Gradient-Free Algorithms

Gradient-free algorithms can be useful when gradients are not available, such as when dealing with black-box functions. Although gradients can always be approximated with finite differences, these approximations suffer from potentially significant inaccuracies (see Section 6.4.2). Gradient-based algorithms require a more experienced user because they take more effort to set up and run. Overall, gradient-free algorithms are easier to get up and running but are much less efficient, particularly as the dimension of the problem increases.

One significant advantage of gradient-free algorithms is that they do not assume function continuity. For gradient-based algorithms, function smoothness is essential when deriving the optimality conditions, both for unconstrained functions and constrained functions. More specifically, the Karush–Kuhn–Tucker (KKT) conditions (Eq. 5.11) require that the function be continuous in value (C^0), gradient (C^1), and Hessian (C^2) in at least a small neighborhood of the optimum. If, for example, the gradient is discontinuous at the optimum, it is undefined, and the KKT conditions are not valid. Away from optimum points, this requirement is not as stringent. Although gradient-based algorithms work on the same continuity assumptions, they can usually tolerate the occasional discontinuity, as long as it is away from an optimum point. However, for functions with excessive numerical noise and discontinuities, gradient-free algorithms might be the only option.

Many considerations are involved when choosing between a gradient-based and a gradient-free algorithm. Some of these considerations are common sources of misconception. One problem characteristic often cited as a reason for choosing gradient-free methods is multimodality. Design space multimodality can be a result of an objective function with multiple local minima. In the case of a constrained problem, the multimodality can arise from the constraints that define disconnected or nonconvex feasible regions.

As we will see shortly, some gradient-free methods feature a global search that increases the likelihood of finding the global minimum. This feature makes gradient-free methods a common choice for multimodal problems. However, not all gradient-free methods are global search methods; some perform only a local search. Additionally, even though gradient-based methods are by themselves local search methods, they are often combined with global search strategies, as discussed in Tip 4.8. It is not necessarily true that a global search, gradient-free method is more likely to find a global optimum than a multistart gradient-based method. As always, problem-specific testing is needed.

Furthermore, it is assumed far too often that any complex problem is multimodal, but that is frequently not the case. Although it might be impossible to prove that a function is unimodal, it is easy to prove that a function is multimodal simply by finding another local minimum. Therefore, we should not make any assumptions about the multimodality of a function until we show definite multiple local minima. Additionally, we must ensure that perceived local minima are not artificial minima arising from numerical noise.

Another reason often cited for using a gradient-free method is multiple objectives. Some gradient-free algorithms, such as the genetic algorithm discussed in this chapter, can be naturally applied to multiple objectives. However, it is a misconception that gradient-free methods are always preferable just because there are multiple objectives. This topic is introduced in Chapter 9.

Another common reason for using gradient-free methods is when there are discrete design variables. Because the notion of a derivative with respect to a discrete variable is invalid, gradient-based algorithms cannot be used directly (although there are ways around this limitation, as discussed in Chapter 8). Some gradient-free algorithms can handle discrete variables directly.

The preceding discussion highlights that although multimodality, multiple objectives, or discrete variables are commonly mentioned as reasons for choosing a gradient-free algorithm, these are not necessarily automatic decisions, and careful consideration is needed. Assuming a choice exists (i.e., the function is not too noisy), one of the most relevant factors when choosing between a gradient-free and a gradient-based approach is the dimension of the problem.

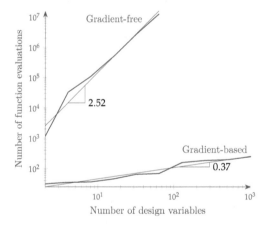

Fig. 7.1 Cost of optimization for increasing number of design variables in the n-dimensional Rosenbrock function. A gradient-free algorithm is compared with a gradient-based algorithm, with gradients computed analytically. The gradient-based algorithm has much better scalability.

Figure 7.1 shows how many function evaluations are required to minimize the n-dimensional Rosenbrock function for varying numbers of design variables. Two classes of algorithms are shown in the plot: gradient-free and gradient-based algorithms. The gradient-based algorithm uses analytic gradients in this case. Although the exact numbers are problem dependent, similar scaling has been observed in large-scale computational fluid dynamics–based optimization.[135] The general takeaway is that for small-size problems (usually ≤ 30 variables[136]), gradient-free methods can be useful in finding a solution.

135. Yu et al., *On the influence of optimization algorithm and starting design on wing aerodynamic shape optimization*, 2018.

136. Rios and Sahinidis, *Derivative-free optimization: a review of algorithms and comparison of software implementations*, 2013.

Furthermore, because gradient-free methods usually take much less developer time to use, a gradient-free solution may even be preferable for these smaller problems. However, if the problem is large in dimension, then a gradient-based method may be the only viable method despite the need for more developer time.

> **Tip 7.1** Choose your bounds carefully for global algorithms
>
> Unlike gradient-based methods, which usually do not require design variable bounds, global algorithms require these bounds to be set. Because the global search tends to explore the whole design space within the specified bounds, the algorithm's effectiveness diminishes considerably if the variable bounds are unnecessarily wide.

7.2 Classification of Gradient-Free Algorithms

There is a much wider variety of gradient-free algorithms compared with their gradient-based counterparts. Although gradient-based algorithms tend to perform local searches, have a mathematical rationale, and be deterministic, gradient-free algorithms exhibit different combinations of these characteristics. We list some of the most widely known gradient-free algorithms in Table 7.1 and classify them according to the characteristics introduced in Fig. 1.22.*

*Rios and Sahinidis[136] review and benchmark a large selection of gradient-free optimization algorithms.

136. Rios and Sahinidis, *Derivative-free optimization: a review of algorithms and comparison of software implementations*, 2013.

Table 7.1 Classification of gradient-free optimization methods using the characteristics of Fig. 1.22.

	Search		Algorithm		Function evaluation		Stochasticity	
	Local	Global	Mathematical	Heuristic	Direct	Surrogate	Deterministic	Stochastic
Nelder–Mead	•		•		•		•	
GPS		•	•		•		•	
MADS		•	•		•			•
Trust region	•		•			•	•	
Implicit filtering	•		•			•	•	
DIRECT		•	•		•		•	
MCS		•	•		•		•	
EGO		•	•			•	•	
Hit and run		•		•	•			•
Evolutionary		•		•	•			•

7.2 Classification of Gradient-Free Algorithms

Local search, gradient-free algorithms that use direct function evaluations include the Nelder–Mead algorithm, generalized pattern search (GPS), and mesh-adaptive direct search (MADS). Although classified as local search in the table, the latter two methods are frequently used with globalization approaches. The Nelder–Mead algorithm (which we detail in Section 7.3) is heuristic, whereas the other two are not.

GPS and MADS (discussed in Section 7.4) are examples of *derivative-free optimization* (DFO) algorithms, which, despite the name, do not include all gradient-free algorithms. DFO algorithms are understood to be largely heuristic-free and focus on local search.[†] GPS is a family of methods that iteratively seek an improvement using a set of points around the current point. In its simplest versions, GPS uses a pattern of points based on the coordinate directions, but more sophisticated versions use a more general set of vectors. MADS improves GPS algorithms by allowing a much larger set of such vectors and improving convergence.

Model-based, local search algorithms include trust-region algorithms and implicit filtering. The model is an analytic approximation of the original function (also called a *surrogate model*), and it should be smooth, easy to evaluate, and accurate in the neighborhood of the current point. The trust-region approach detailed in Section 4.5 can be considered gradient-free if the surrogate model is constructed using just evaluations of the original function without evaluating its gradients. This does not prevent the trust-region algorithm from using gradients of the surrogate model, which can be computed analytically. Implicit filtering methods extend the trust-region method by adding a surrogate model of the function gradient to guide the search. This effectively becomes a gradient-based method applied to the surrogate model instead of evaluating the function directly, as done for the methods in Chapter 4.

Global search algorithms can be broadly classified as deterministic or stochastic, depending on whether they include random parameter generation within the optimization algorithm.

Deterministic, global search algorithms can be either direct or model based. Direct algorithms include Lipschitzian-based partitioning techniques—such as the "divide a hyperrectangle" (DIRECT) algorithm detailed in Section 7.5 and branch-and-bound search (discussed in Chapter 8)—and multilevel coordinate search (MCS). The DIRECT algorithm selectively divides the space of the design variables into smaller and smaller n-dimensional boxes—hyperrectangles. It uses mathematical arguments to decide which boxes should be subdivided. Branch-and-bound search also partitions the design space,

---

[†] The textbooks by Conn et al.[137] and Audet and Hare[138] provide a more extensive treatment of gradient-free optimization algorithms that are based on mathematical criteria. Kokkolaras[139] presents a succinct discussion on when to use DFO.

137. Conn et al., *Introduction to Derivative-Free Optimization*, 2009.

138. Audet and Hare, *Derivative-Free and Blackbox Optimization*, 2017.

139. Kokkolaras, *When, why, and how can derivative-free optimization be useful to computational engineering design?* 2020.

but it estimates lower and upper bounds for the optimum by using the function variation between partitions. MCS is another algorithm that partitions the design space into boxes, where a limit is imposed on how small the boxes can get based on the number of times it has been divided.

Global-search algorithms based on surrogate models are similar to their local search counterparts. However, they use surrogate models to reproduce the features of a multimodal function instead of convex surrogate models. One of the most widely used of these algorithms is efficient global optimization (EGO), which employs kriging surrogate models and uses the idea of expected improvement to maximize the likelihood of finding the optimum more efficiently (surrogate modeling techniques, including kriging are introduced in Chapter 10, which also described EGO). Other algorithms use radial basis functions (RBFs) as the surrogate model and also maximize the probability of improvement at new iterates.

Stochastic algorithms rely on one or more nondeterministic procedures; they include hit-and-run algorithms and the broad class of evolutionary algorithms. When performing benchmarks of a stochastic algorithm, you should run a large enough number of optimizations to obtain statistically significant results.

Hit-and-run algorithms generate random steps about the current iterate in search of better points. A new point is accepted when it is better than the current one, and this process repeats until the point cannot be improved.

What constitutes an evolutionary algorithm is not well defined.[‡] Evolutionary algorithms are inspired by processes that occur in nature or society. There is a plethora of evolutionary algorithms in the literature, thanks to the fertile imagination of the research community and a never-ending supply of phenomena for inspiration.[§] These algorithms are more of an analogy of the phenomenon than an actual model. They are, at best, simplified models and, at worst, barely connected to the phenomenon. Nature-inspired algorithms tend to invent a specific terminology for the mathematical terms in the optimization problem. For example, a design point might be called a "member of the population", or the objective function might be the "fitness".

The vast majority of evolutionary algorithms are population based, which means they involve a set of points at each iteration instead of a single one (we discuss a genetic algorithm in Section 7.6 and a particle swarm method in Section 7.7). Because the population is spread out in the design space, evolutionary algorithms perform a global search. The stochastic elements in these algorithms contribute to global exploration

[‡]Simon[140] provides a more comprehensive review of evolutionary algorithms.

140. Simon, *Evolutionary Optimization Algorithms*, 2013.

[§] These algorithms include the following: ant colony optimization, artificial bee colony algorithm, artificial fish swarm, artificial flora optimization algorithm, bacterial foraging optimization, bat algorithm, big bang–big crunch algorithm, biogeography-based optimization, bird mating optimizer, cat swarm optimization, cockroach swarm optimization, cuckoo search, design by shopping paradigm, dolphin echolocation algorithm, elephant herding optimization, firefly algorithm, flower pollination algorithm, fruit fly optimization algorithm, galactic swarm optimization, gray wolf optimizer, grenade explosion method, harmony search algorithm, hummingbird optimization algorithm, hybrid glowworm swarm optimization algorithm, imperialist competitive algorithm, intelligent water drops, invasive weed optimization, mine bomb algorithm, monarch butterfly optimization, moth-flame optimization algorithm, penguin search optimization algorithm, quantum-behaved particle swarm optimization, salp swarm algorithm, teaching–learning-based optimization, whale optimization algorithm, and water cycle algorithm.

and reduce the susceptibility to getting stuck in local minima. These features increase the likelihood of getting close to the global minimum but by no means guarantee it. The algorithm may only get close because heuristic algorithms have a poor convergence rate, especially in higher dimensions, and because they lack a first-order mathematical optimality criterion.

This chapter covers five gradient-free algorithms: the Nelder–Mead algorithm, GPS, the DIRECT method, genetic algorithms, and particle swarm optimization. A few other algorithms that can be used for continuous gradient-free problems (e.g., simulated annealing and branch and bound) are covered in Chapter 8 because they are more frequently used to solve discrete problems. In Chapter 10, on surrogate modeling, we discuss kriging and efficient global optimization.

7.3 Nelder–Mead Algorithm

The simplex method of Nelder and Mead[28] is a deterministic, direct-search method that is among the most cited gradient-free methods. It is also known as the *nonlinear simplex*—not to be confused with the simplex algorithm used for linear programming, with which it has nothing in common. To avoid ambiguity, we will refer to it as the Nelder–Mead algorithm.

28. Nelder and Mead, *A simplex method for function minimization*, 1965.

The Nelder–Mead algorithm is based on a simplex, which is a geometric figure defined by a set of $n + 1$ points in the design space of n variables, $X = \{x^{(0)}, x^{(1)}, \ldots, x^{(n)}\}$. Each point $x^{(i)}$ represents a design (i.e., a full set of design variables). In two dimensions, the simplex is a triangle, and in three dimensions, it becomes a tetrahedron (see Fig. 7.2).

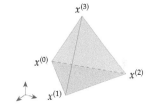

Fig. 7.2 A simplex for $n = 3$ has four vertices.

Each optimization iteration corresponds to a different simplex. The algorithm modifies the simplex at each iteration using five simple operations. The sequence of operations to be performed is chosen based on the relative values of the objective function at each of the points.

The first step of the simplex algorithm is to generate $n + 1$ points based on an initial guess for the design variables. This could be done by simply adding steps to each component of the initial point to generate n new points. However, this will generate a simplex with different edge lengths, and equal-length edges are preferable. Suppose we want the length of all sides to be l and that the first guess is $x^{(0)}$. The remaining points of the simplex, $\{x^{(1)}, \ldots, x^{(n)}\}$, can be computed by

$$x^{(i)} = x^{(0)} + s^{(i)}, \qquad (7.1)$$

where $s^{(i)}$ is a vector whose components j are defined by

$$s_j^{(i)} = \begin{cases} \frac{1}{n\sqrt{2}}\left(\sqrt{n+1}-1\right) + \frac{1}{\sqrt{2}}, & \text{if } j = i \\ \frac{1}{n\sqrt{2}}\left(\sqrt{n+1}-1\right), & \text{if } j \neq i. \end{cases} \quad (7.2)$$

Figure 7.3 shows a starting simplex for a two-dimensional problem.

At any given iteration, the objective f is evaluated for every point, and the points are ordered based on the respective values of f, from the lowest to the highest. Thus, in the ordered list of simplex points $X = \{x^{(0)}, x^{(1)}, \ldots, x^{(n-1)}, x^{(n)}\}$, the best point is $x^{(0)}$, and the worst one is $x^{(n)}$.

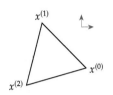

Fig. 7.3 Starting simplex for $n = 2$.

The Nelder–Mead algorithm performs five main operations on the simplex to create a new one: *reflection, expansion, outside contraction, inside contraction*, and *shrinking*, as shown in Fig. 7.4. Except for shrinking, each of these operations generates a new point,

$$x = x_c + \alpha \left(x_c - x^{(n)}\right), \quad (7.3)$$

where α is a scalar, and x_c is the centroid of all the points except for the worst one, that is,

$$x_c = \frac{1}{n} \sum_{i=0}^{n-1} x^{(i)}. \quad (7.4)$$

This generates a new point along the line that connects the worst point, $x^{(n)}$, and the centroid of the remaining points, x_c. This direction can be seen as a possible descent direction.

Each iteration aims to replace the worst point with a better one to form a new simplex. Each iteration always starts with reflection, which generates a new point using Eq. 7.3 with $\alpha = 1$, as shown in Fig. 7.4. If the reflected point is better than the best point, then the "search direction" was a good one, and we go further by performing an expansion using Eq. 7.3 with $\alpha = 2$. If the reflected point is between the second-worst and the worst point, then the direction was not great, but it improved somewhat. In this case, we perform an outside contraction ($\alpha = 1/2$). If the reflected point is worse than our worst point, we try an inside contraction instead ($\alpha = -1/2$). Shrinking is a last-resort operation that we can perform when nothing along the line connecting $x^{(n)}$ and x_c produces a better point. This operation consists of reducing the size of the simplex by moving all the points closer to the best point,

$$x^{(i)} = x^{(0)} + \gamma \left(x^{(i)} - x^{(0)}\right) \quad \text{for} \quad i = 1, \ldots, n, \quad (7.5)$$

where $\gamma = 0.5$.

7.3 Nelder–Mead Algorithm

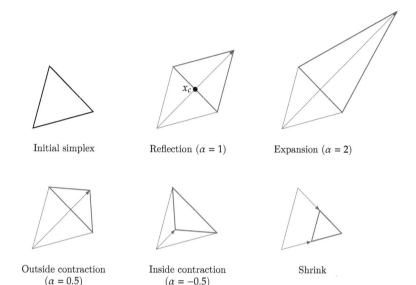

Fig. 7.4 Nelder–Mead algorithm operations for $n = 2$.

Algorithm 7.1 details how a new simplex is obtained for each iteration. In each iteration, the focus is on replacing the worst point with a better one instead of improving the best. The corresponding flowchart is shown in Fig. 7.5.

The cost for each iteration is one function evaluation if the reflection is accepted, two function evaluations if an expansion or contraction is performed, and $n + 2$ evaluations if the iteration results in shrinking. Although we could parallelize the n evaluations when shrinking, it would not be worthwhile because the other operations are sequential.

There several ways to quantify the convergence of the simplex method. One straightforward way is to use the size of simplex, that is,

$$\Delta_x = \sum_{i=0}^{n-1} \left\| x^{(i)} - x^{(n)} \right\|, \qquad (7.6)$$

and specify that it must be less than a certain tolerance. Another measure of convergence we can use is the standard deviation of the function value,

$$\Delta_f = \sqrt{\frac{\sum_{i=0}^{n} \left(f^{(i)} - \bar{f} \right)^2}{n + 1}}, \qquad (7.7)$$

where $\bar{f}$ is the mean of the $n + 1$ function values. Another possible convergence criterion is the difference between the best and worst value in the simplex. Nelder–Mead is known for occasionally converging to non-stationary points, so you should check the result if possible.

Algorithm 7.1 Nelder–Mead algorithm

Inputs:
$x^{(0)}$: Starting point
τ_x: Simplex size tolerances
τ_f: Function value standard deviation tolerances

Outputs:
x^*: Optimal point

---

for $j = 1$ to n **do**   *Create a simplex with edge length l*
  $x^{(j)} = x^{(0)} + s^{(j)}$ *$s^{(j)}$ given by Eq. 7.2*
end for

while $\Delta_x > \tau_x$ or $\Delta_f > \tau_f$ **do**   *Simplex size (Eq. 7.6) and standard deviation (Eq. 7.7)*
  Sort $\{x^{(0)}, \ldots, x^{(n-1)}, x^{(n)}\}$ *Order from the lowest (best) to the highest $f(x^{(j)})$*
  $x_c = \frac{1}{n} \sum_{i=0}^{n-1} x^{(i)}$ *The centroid excluding the worst point $x^{(n)}$ (Eq. 7.4)*
  $x_r = x_c + \left(x_c - x^{(n)}\right)$ *Reflection, Eq. 7.3 with $\alpha = 1$*

  if $f(x_r) < f(x^{(0)})$ **then**   *Is reflected point is better than the best?*
   $x_e = x_c + 2\left(x_c - x^{(n)}\right)$ *Expansion, Eq. 7.3 with $\alpha = 2$*
   if $f(x_e) < f(x^{(0)})$ **then**   *Is expanded point better than the best?*
    $x^{(n)} = x_e$ *Accept expansion and replace worst point*
   else
    $x^{(n)} = x_r$ *Accept reflection*
   end if
  else if $f(x_r) \leq f(x^{(n-1)})$ **then**   *Is reflected better than second worst?*
   $x^{(n)} = x_r$ *Accept reflected point*
  else
   if $f(x_r) > f(x^{(n)})$ **then**   *Is reflected point worse than the worst?*
    $x_{ic} = x_c - 0.5\left(x_c - x^{(n)}\right)$ *Inside contraction, Eq. 7.3 with $\alpha = -0.5$*
    if $f(x_{ic}) < f(x^{(n)})$ **then**   *Inside contraction better than worst?*
     $x^{(n)} = x_{ic}$ *Accept inside contraction*
    else
     for $j = 1$ to n **do**
      $x^{(j)} = x^{(0)} + 0.5\left(x^{(j)} - x^{(0)}\right)$ *Shrink, Eq. 7.5 with $\gamma = 0.5$*
     end for
    end if
   else
    $x_{oc} = x_c + 0.5\left(x_c - x^{(n)}\right)$ *Outside contraction, Eq. 7.3 with $\alpha = 0.5$*
    if $f(x_{oc}) < f(x_r)$ **then**   *Is contraction better than reflection?*
     $x^{(n)} = x_{oc}$ *Accept outside contraction*
    else
     for $j = 1$ to n **do**
      $x^{(j)} = x^{(0)} + 0.5\left(x^{(j)} - x^{(0)}\right)$ *Shrink, Eq. 7.5 with $\gamma = 0.5$*

7.3 NELDER–MEAD ALGORITHM

```
            end for
         end if
      end if
   end if
end while
```

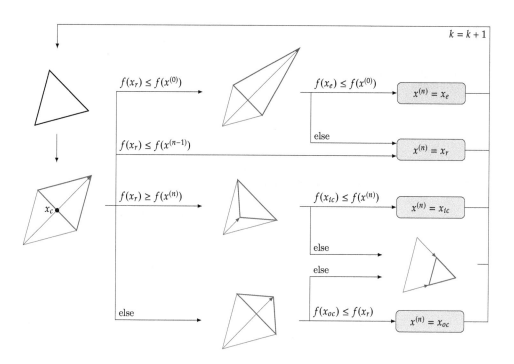

Fig. 7.5 Flowchart of Nelder–Mead (Alg. 7.1).

Like most direct-search methods, Nelder–Mead cannot directly handle constraints. One approach to handling constraints would be to use a penalty method (discussed in Section 5.4) to form an unconstrained problem. In this case, the penalty does not need not be differentiable, so a linear penalty method would suffice.

Example 7.1 Nelder–Mead algorithm applied to the bean function

Figure 7.6 shows the sequence of simplices that results when minimizing the bean function using a Nelder–Mead simplex. The initial simplex on the upper left is equilateral. The first iteration is an expansion, followed by an inside contraction, another reflection, and an inside contraction before the shrinking. The simplices then shrink dramatically in size, slowly converging to the minimum.

Using a convergence tolerance of 10^{-6} in the difference between f_{best} and f_{worst}, the problem took 68 function evaluations.

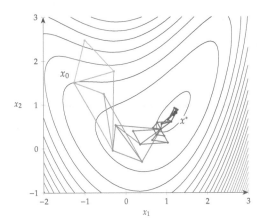

Fig. 7.6 Sequence of simplices that minimize the bean function.

7.4 Generalized Pattern Search

GPS builds upon the ideas of a coordinate search algorithm. In coordinate search, we evaluate points along a mesh aligned with the coordinate directions, move toward better points, and shrink the mesh when we find no improvement nearby. Consider a two-dimensional coordinate search for an unconstrained problem. At a given point x_k, we evaluate points that are a distance Δ_k away in all coordinate directions, as shown in Fig. 7.7. If the objective function improves at any of these points (four points in this case), we recenter with x_{k+1} at the most improved point, keep the mesh size the same at $\Delta_{k+1} = \Delta_k$, and start with the next iteration. Alternatively, if none of the points offers an improvement, we keep the same center ($x_{k+1} = x_k$) and shrink the mesh to $\Delta_{k+1} < \Delta_k$. This process repeats until it meets some convergence criteria.

We now explore various ways in which GPS improves coordinate search. Coordinate search moves along coordinate directions, but this is not necessarily desirable. Instead, the GPS search directions only need to form a *positive spanning set*. Given a set of directions $D = \{d_1, d_2, \ldots, d_{n_d}\}$, the set D is a positive spanning set if the vectors are linearly independent and a nonnegative linear combination of these vectors spans the n-dimensional space.* Coordinate vectors fulfill this requirement, but there is an infinite number of options. The vectors d are referred to as *positive spanning directions*. We only consider linear combinations with positive multipliers, so in two dimensions, the unit coordinate vectors $\hat{e}_1$ and $\hat{e}_2$ are not sufficient to span two-dimensional space; however, $\hat{e}_1, \hat{e}_2, -\hat{e}_1$, and $-\hat{e}_2$ are sufficient.

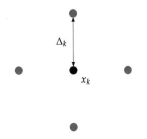

Fig. 7.7 Local mesh for a two-dimensional coordinate search at iteration k.

*Section 5.2 discusses the concept of span and polyhedral cones; Fig. 5.6 is particularly relevant.

7.4 Generalized Pattern Search

For a given dimension n, the largest number of vectors that could be used while remaining linearly independent (known as the *maximal set*) is $2n$. Conversely, the minimum number of possible vectors needed to span the space (known as the *minimal set*) is $n + 1$. These sizes are necessary but not sufficient conditions.

Some algorithms randomly generate a positive spanning set, whereas other algorithms require the user to specify a set based on knowledge of the problem. The positive spanning set need not be fixed throughout the optimization. A common default for a maximal set is the set of coordinate directions $\pm \hat{e}_i$. In three dimensions, this would be:

$$D = \{d_1, \ldots, d_6\}, \quad \text{where} \quad \begin{cases} d_1 &= [1,0,0] \\ d_2 &= [0,1,0] \\ d_3 &= [0,0,1] \\ d_4 &= [-1,0,0] \\ d_5 &= [0,-1,0] \\ d_6 &= [0,0,-1] \end{cases}. \tag{7.8}$$

A potential default minimal set is the positive coordinate directions $+\hat{e}_i$ and a vector filled with -1 (or more generally, the negative sum of the other vectors). As an example in three dimensions:

$$D = \{d_1, \ldots, d_4\}, \quad \text{where} \quad \begin{cases} d_1 &= [1,0,0] \\ d_2 &= [0,1,0] \\ d_3 &= [0,0,1] \\ d_4 &= [-1,-1,-1] \end{cases}. \tag{7.9}$$

Figure 7.8 shows an example maximal set (four vectors) and minimal set (three vectors) for a two-dimensional problem.

These direction vectors are then used to create a *mesh*. Given a current center point x_k, which is the best point found so far, and a mesh size Δ_k, the mesh is created as follows:

$$x_k + \Delta_k d \quad \text{for all } d \in D. \tag{7.10}$$

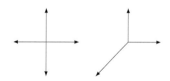

Fig. 7.8 A maximal set of positive spanning vectors in two dimensions (left) and a minimal set (right).

For example, in two dimensions, if the current point is $x_k = [1,1]$, the mesh size is $\Delta_k = 0.5$, and we use the coordinate directions for d, then the mesh points would be $\{[1,1.5], [1,0.5], [0.5,1], [1.5,1]\}$.

The evaluation of points in the mesh is called *polling* or a *poll*. In the coordinate search example, we evaluated every point in the mesh, which is usually inefficient. More typically, we use *opportunistic polling*, which terminates polling at the first point that offers an improvement.

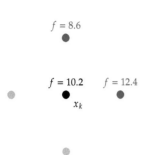

Fig. 7.9 A two-dimensional example of opportunistic polling with $d_1 = [1, 0], d_2 = [0, 1], d_3 = [-1, 0], d_4 = [0, -1]$. An improvement in f was found at d_2, so we do not evaluate d_3 and d_4 (shown with a faded color).

Figure 7.9 shows a two-dimensional example where the order of evaluation is $d_1 = [1, 0], d_2 = [0, 1], d_3 = [-1, 0], d_4 = [0, -1]$. Because we found an improvement at d_2, we do not continue evaluating d_3 and d_4. Opportunistic polling may not yield the best point in the mesh, but the improvement in efficiency is usually worth the trade-off. Some algorithms add a user option for utilizing a full poll, in which case all points in the mesh are evaluated. If more than one point offers a reduction, the best one is accepted. Another approach that is sometimes used is called *dynamic polling*. In this approach, a successful poll reorders the direction vectors so that the direction that was successful last time is checked first in the next poll.

GPS consists of two phases: a search phase and a poll phase. The search phase is global, whereas the poll phase is local. The search phase is intended to be flexible and is not specified by the GPS algorithm. Common options for the search phase include the following:

- No search phase.
- A mesh search, similar to polling but with large spacing across the domain.
- An alternative solver, such as Nelder–Mead or a genetic algorithm.
- A surrogate model, which could then use any number of solvers that include gradient-based methods. This approach is often used when the function is expensive, and a lower-fidelity surrogate can guide the optimizer to promising regions of the larger design space.
- Random evaluation using a space-filling method (see Section 10.2).

The type of search can change throughout the optimization. Like the polling phase, the goal of the search phase is to find a better point (i.e., $f(x_{k+1}) < f(x_k)$) but within a broader domain. We begin with a search at every iteration. If the search fails to produce a better point, we continue with a poll. If a better point is identified in either phase, the iteration ends, and we begin a new search. Optionally, a successful poll could be followed by another poll. Thus, at each iteration, we might perform a search and a poll, just a search, or just a poll.

We describe one option for a search procedure based on the same mesh ideas as the polling step. The concept is to extend the mesh throughout the entire domain, as shown in Fig. 7.10. In this example, the mesh size Δ_k is shared between the search and poll phases. However, it is usually more effective if these sizes are independent. Mathematically, we can define the global mesh as the set

$$G = \{x_k + \Delta_k Dz \text{ for all } z_i \in \mathbb{Z}^+\}, \tag{7.11}$$

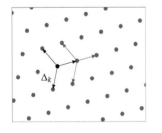

Fig. 7.10 Meshing strategy extended across the domain. The same directions (and potentially spacing) are repeated at each mesh point, as indicated by the lighter arrows throughout the entire domain.

7.4 Generalized Pattern Search

where D is a matrix whose columns contain the basis vectors d. The vector z consists of nonnegative integers, and we consider all possible combinations of integers that fall within the bounds of the domain.

We choose a fixed number of search evaluation points and randomly select points from the global mesh for the search strategy. If improvement is found among that set, then we recenter x_{k+1} at this improved point, grow the mesh ($\Delta_{k+1} > \Delta_k$), and end the iteration (and then restart the search). A simple search phase along these lines is described in Alg. 7.2 and the main GPS algorithm is shown in Alg. 7.3.

Algorithm 7.2 An example search phase for GPS

Inputs:
  x_k: Center point
  Δ_k: Mesh size
  $\underline{x}, \overline{x}$: Lower and upper bounds
  D: Column vectors representing positive spanning set
  n_s: Number of search points
  f_k: The function previously evaluated at $f(x_k)$

Outputs:
  success: True if successful in finding improved point
  x_{k+1}: New center point
  f_{k+1}: Corresponding function value

---

success = false
$x_{k+1} = x_k$
$f_{k+1} = f_k$
Construct global mesh G, using directions D, mesh size Δ_k, and bounds $\underline{x}, \overline{x}$
for $i = 1$ **to** n_s **do**
    Randomly select $s \in G$
    Evaluate $f_s = f(s)$
    if $f_s < f_k$ **then**
        $x_{k+1} = s$
        $f_{k+1} = f_s$
        success = true
        break
    end if
end for

The convergence of the GPS algorithm is often determined by a user-specified maximum number of iterations. However, other criteria are also used, such as a threshold on mesh size or a threshold on the improvement in the function value over previous iterations.

Algorithm 7.3 Generalized Pattern Search

Inputs:
 x_0: Starting point
 $\underline{x}, \overline{x}$: Lower and upper bounds
 Δ_0: Starting mesh size
 n_s: Number of search points
 $k_{\max}$: Maximum number of iterations

Outputs:
 x^*: Best point
 f^*: Corresponding function value

$D = [I, -I]$ where I is $(n \times n)$ A coordinate aligned maximal positive spanning set (for example)
$k = 0$
$x_k = x_0$
Evaluate $f_k = f(x_k)$
while $k < k_{\max}$ **do**   Or other termination criteria
    search_success, $x_{k+1}, f_{k+1} = \text{search}(x_k, \Delta_k, f_k)$ Any search strategy
    if search_success **then**
        $\Delta_{k+1} = \min(2\Delta_k, \Delta_{\max})$ Or some other growth rate
        $k = k + 1$
        continue   Move on to next iteration
    else   Poll
        poll_success = false
        for $j = 1$ to n_d **do**
            $s = x_k + \Delta_k d_j$ Where d_j is a column of D
            Evaluate $f_s = f(s)$
            if $f_s < f_k$ **then**
                $x_{k+1} = s$
                $f_{k+1} = f_s$
                $\Delta_{k+1} = \Delta_k$
                $k = k + 1$
                poll_success = true
                break
            end if
        end for
    end if
    if not poll_success **then**
        $x_{k+1} = x_k$
        $f_{k+1} = f_k$
        $\Delta_{k+1} = 0.5\Delta_k$ Shrink
    end if
    $k = k + 1$
end while

7.4 Generalized Pattern Search

GPS can handle linear and nonlinear constraints. For linear constraints, one effective strategy is to change the positive spanning directions so that they align with any linear constraints that are nearby (Fig. 7.11). For nonlinear constraints, penalty approaches (Section 5.4) are applicable, although the filter method (Section 5.5.3) is another effective approach.

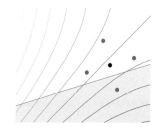

Fig. 7.11 Mesh direction changed during optimization to align with linear constraints when close to the constraint.

Example 7.2 Minimization of a multimodal function with GPS

In this example, we optimize the Jones function (Appendix D.1.4). We start at $x = [0, 0]$ with an initial mesh size of $\Delta = 0.1$. We evaluate two search points at each iteration and run for 12 iterations. The iterations are plotted in Fig. 7.12.

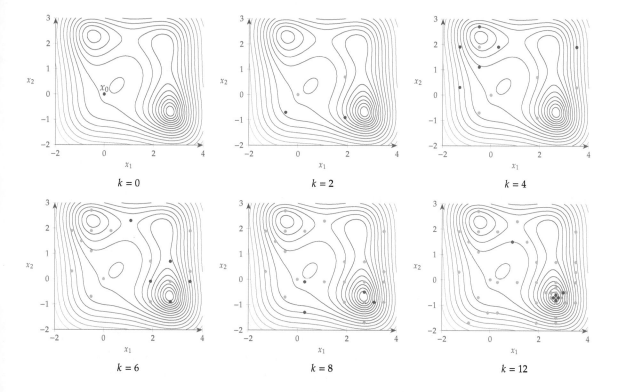

Fig. 7.12 Convergence history of a GPS algorithm on the multimodal Jones function. Faded points indicate past iterations.

MADS is a well-known extension of GPS. The main difference between these two methods is in the number of possibilities for polling directions.[141] In GPS, the polling directions are relatively restrictive (e.g., left side of Fig. 7.13 for a minimal basis in two dimensions). MADS adds a new sizing parameter called the *poll size parameter* (Δ_k^p) that can be varied independently from the existing mesh size parameter (Δ_k^m). These sizes are constrained by $\Delta_k^p \geq \Delta_k^m$ so the mesh sizing can become

141. Audet and J. E. Dennis, *Mesh adaptive direct search algorithms for constrained optimization*, 2006.

smaller while allowing the poll size (which dictates the maximum magnitude of the step) to remain large. This provides a much denser set of options in poll directions (e.g., the grid points on the right panel of Fig. 7.13). MADS randomly chooses the polling directions from this much larger set of possibilities while maintaining a positive spanning set.†

† The NOMAD software is an open-source implementation of MADS.¹⁴²

142. Le Digabel, *Algorithm 909: NOMAD: Nonlinear optimization with the MADS algorithm*, 2011.

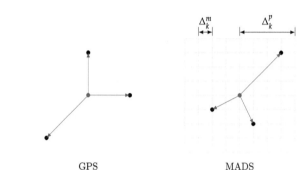

Fig. 7.13 Typical GPS spanning directions (left). In contrast, MADS randomly selects from many potential spanning directions by utilizing a finer mesh (right).

7.5 DIRECT Algorithm

*Jones et al.⁵² developed this method, aiming for a global search that did not rely on tunable parameters (e.g., population size in genetic algorithms).⁵³

52. Jones et al., *Lipschitzian optimization without the Lipschitz constant*, 1993.

53. Jones and Martins, *The DIRECT algorithm—25 years later*, 2021.

143. Jones, *Direct Global Optimization Algorithm*, 2009.

The DIRECT algorithm is different from the other gradient-free optimization algorithms in this chapter in that it is based on mathematical arguments.* This is a deterministic method guaranteed to converge to the global optimum under conditions that are not too restrictive (although it might require a prohibitive number of function evaluations). DIRECT has been extended to handle constraints without relying on penalty or filtering methods, but here we only explain the algorithm for unconstrained problems.¹⁴³

One way to ensure that we find the global optimum within a finite design space is by dividing this space into a regular rectangular grid and evaluating every point in this grid. This is called an *exhaustive search*, and the precision of the minimum depends on how fine the grid is. The cost of this brute-force strategy is high and goes up exponentially with the number of design variables.

The DIRECT method relies on a grid, but it uses an adaptive meshing scheme that dramatically reduces the cost. It starts with a single n-dimensional hypercube that spans the whole design space—like many other gradient-free methods, DIRECT requires upper and lower bounds on all the design variables. Each iteration divides this hypercube into smaller ones and evaluates the objective function at the center of each of these. At each iteration, the algorithm only divides rectangles determined to be *potentially optimal*. The fundamental strategy in the

7.5 DIRECT Algorithm

DIRECT method is how it determines this subset of potentially optimal rectangles, which is based on the mathematical concept of *Lipschitz continuity*.

We start by explaining Lipschitz continuity and then describe an algorithm for finding the global minimum of a one-dimensional function using this concept—Shubert's algorithm. Although Shubert's algorithm is not practical in general, it will help us understand the mathematical rationale for the DIRECT algorithm. Then we explain the DIRECT algorithm for one-dimensional functions before generalizing it for n dimensions.

Lipschitz Constant

Consider the single-variable function $f(x)$ shown in Fig. 7.14. For a trial point x^*, we can draw a cone with slope L by drawing the lines

$$f_+(x) = f(x^*) + L(x - x^*), \quad (7.12)$$
$$f_-(x) = f(x^*) - L(x - x^*), \quad (7.13)$$

to the left and right, respectively. We show this cone in Fig. 7.14 (left), as well as cones corresponding to other values of k.

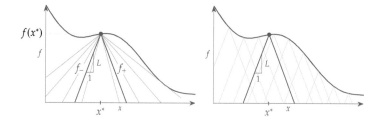

Fig. 7.14 From a given trial point x^*, we can draw a cone with slope L (left). For a function to be Lipschitz continuous, we need all cones with slope L to lie under the function for all points in the domain (right).

A function f is said to be *Lipschitz continuous* if there is a cone slope L such that the cones for all possible trial points in the domain remain under the function. This means that there is a positive constant k such that

$$|f(x) - f(x^*)| \le L|x - x^*|, \quad \text{for all} \quad x, x^* \in D, \quad (7.14)$$

where D is the function domain. Graphically, this condition means that we can draw a cone with slope L from any trial point evaluation $f(x^*)$ such that the function is always bounded by the cone, as shown in Fig. 7.14 (right). Any k that satisfies Eq. 7.14 is a *Lipschitz constant* for the corresponding function.

Shubert's Algorithm

If a Lipschitz constant for a single-variable function is known, Shubert's algorithm can find the global minimum of that function. Because the

Lipschitz constant is not available in the general case, the DIRECT algorithm is designed to not require this constant. However, we explain Shubert's algorithm first because it provides some of the basic concepts used in the DIRECT algorithm.

Shubert's algorithm starts with a domain within which we want to find the global minimum—$[a, b]$ in Fig. 7.15. Using the property of the Lipschitz constant L defined in Eq. 7.14, we know that the function is always above a cone of slope L evaluated at any point in the domain.

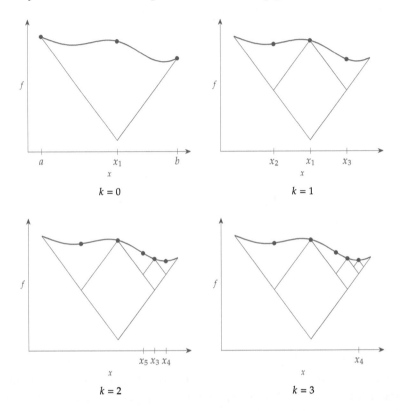

Fig. 7.15 Shubert's algorithm requires an initial domain and a valid Lipschitz constant and then increases the lower bound of the global minimum with each successive iteration.

Shubert's algorithm starts by sampling the endpoints of the interval within which we want to find the global minimum ($[a, b]$ in Fig. 7.15). We start by establishing a first lower bound on the global minimum by finding the cone's intersection (x_1 in Fig. 7.15, $k = 0$) for the extremes of the domain. We evaluate the function at x_1 and can now draw a cone about this point to find two more intersections (x_2 and x_3). Because these two points always intersect at the same objective lower bound value, they both need to be evaluated. Each subsequent iteration of Shubert's algorithm adds two new points to either side of the current point. These two points are evaluated, and the lower bounding function

7.5 DIRECT Algorithm

is updated with the resulting new cones. We then iterate by finding the two points that minimize the new lower bounding function, evaluating the function at these points, updating the lower bounding function, and so on.

The lowest bound on the function increases at each iteration and ultimately converges to the *global* minimum. At the same time, the segments in x decrease in size. The lower bound can switch from distinct regions as the lower bound in one region increases beyond the lower bound in another region.

The two significant shortcomings of Shubert's algorithm are that (1) a Lipschitz constant is usually not available for a general function, and (2) it is not easily extended to n dimensions. The DIRECT algorithm addresses these two shortcomings.

One-Dimensional DIRECT

Before explaining the n-dimensional DIRECT algorithm, we introduce the one-dimensional version based on principles similar to those of the Shubert algorithm.

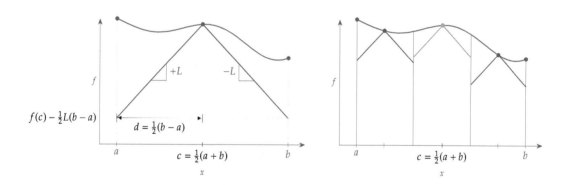

Fig. 7.16 The DIRECT algorithm evaluates the middle point (left), and each successive iteration trisects the segments that have the greatest potential (right).

Like Shubert's method, DIRECT starts with the domain $[a, b]$. However, instead of sampling the endpoints a and b, it samples the midpoint. Consider the closed domain $[a, b]$ shown in Fig. 7.16 (left). For each segment, we evaluate the objective function at the segment's midpoint. In the first segment, which spans the whole domain, the midpoint is $c_0 = (a + b)/2$. Assuming some value of L, which is not known and that we will not need, the lower bound on the minimum would be $f(c) - L(b - a)/2$.

We want to increase this lower bound on the function minimum by dividing this segment further. To do this in a regular way that reuses previously evaluated points and can be repeated indefinitely, we divide it into three segments, as shown in Fig. 7.16 (right). Now we have increased the lower bound on the minimum. Unlike the Shubert algorithm, the lower bound is a discontinuous function across the segments, as shown in Fig. 7.16 (right).

Instead of continuing to divide every segment into three other segments, we only divide segments selected according to a *potentially optimal* criterion. To better understand this criterion, consider a set of segments $[a_i, b_i]$ at a given DIRECT iteration, where segment i has a half-length $d_i = (b_i - a_i)/2$ and a function value $f(c_i)$ evaluated at the segment center $c_i = (a_i + b_i)/2$. If we plot $f(c_i)$ versus d_i for a set of segments, we get the pattern shown in Fig. 7.17.

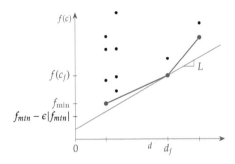

Fig. 7.17 Potentially optimal segments in the DIRECT algorithm are identified by the lower convex hull of this plot.

The overall rationale for the potentially optimal criterion is that two metrics quantify this potential: the size of the segment and the function value at the center of the segment. The larger the segment is, the greater the potential for that segment to contain the global minimum. The lower the function value, the greater that potential is as well. For a set of segments of the same size, we know that the one with the lowest function value has the best potential and should be selected. If two segments have the same function value and different sizes, we should select the one with the largest size. For a general set of segments with various sizes and value combinations, there might be multiple segments that can be considered potentially optimal.

We identify potentially optimal segments as follows. If we draw a line with a slope corresponding to a Lipschitz constant L from any point in Fig. 7.17, the intersection of this line with the vertical axis is a bound on the objective function for the corresponding segment. Therefore, the lowest bound for a given L can be found by drawing a line through the point that achieves the lowest intersection.

However, we do not know L, and we do not want to assume a value because we do not want to bias the search. If L were high, it would favor dividing the larger segments. Low values of L would result in dividing the smaller segments. The DIRECT method hinges on considering all

7.5 DIRECT Algorithm

possible values of L, effectively eliminating the need for this constant.

To eliminate the dependence on L, we select *all the points for which there is a line with slope L that does not go above any other point*. This corresponds to selecting the points that form a lower convex hull, as shown by the piecewise linear function in Fig. 7.17. This establishes a lower bound on the function for each segment size.

Mathematically, a segment j in the set of current segments S is said to be potentially optimal if there is a $L \geq 0$ such that

$$f(c_j) - Ld_j \leq f(c_i) - Ld_i \quad \text{for all} \quad i \in S \quad (7.15)$$
$$f(c_j) - Ld_j \leq f_{min} - \varepsilon |f_{min}|, \quad (7.16)$$

where f_{min} is the best current objective function value, and ε is a small positive parameter. The first condition corresponds to finding the points in the lower convex hull mentioned previously.

The second condition in Eq. 7.16 ensures that the potential minimum is better than the lowest function value found so far by at least a small amount. This prevents the algorithm from becoming too local, wasting function evaluations in search of smaller function improvements. The parameter ε balances the search between local and global. A typical value is $\varepsilon = 10^{-4}$, and its range is usually such that $10^{-7} \leq \varepsilon \leq 10^{-2}$.

There are efficient algorithms for finding the convex hull of an arbitrary set of points in two dimensions, such as the Jarvis march.[144] These algorithms are more than we need because we only require the lower part of the convex hull, so the algorithms can be simplified for our purposes.

As in the Shubert algorithm, the division might switch from one part of the domain to another, depending on the new function values. Compared with the Shubert algorithm, the DIRECT algorithm produces a discontinuous lower bound on the function values, as shown in Fig. 7.18.

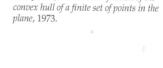

144. Jarvis, *On the identification of the convex hull of a finite set of points in the plane*, 1973.

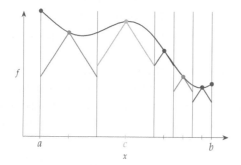

Fig. 7.18 The lower bound for the DIRECT method is discontinuous at the segment boundaries.

DIRECT in n Dimensions

The n-dimensional DIRECT algorithm is similar to the one-dimensional version but becomes more complex.† The main difference is that we deal with *hyperrectangles* instead of segments. A hyperrectangle can be defined by its center-point position c in n-dimensional space and a half-length in each direction i, δe_i, as shown in Fig. 7.19. The DIRECT algorithm assumes that the initial dimensions are normalized so that we start with a hypercube.

†In this chapter, we present an improved version of DIRECT.[143]

143. Jones, *Direct Global Optimization Algorithm*, 2009.

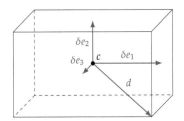

Fig. 7.19 Hyperrectangle in three dimensions, where d is the maximum distance between the center and the vertices, and δe_i is the half-length in each direction i.

To identify the *potentially optimal rectangles* at a given iteration, we use exactly the same conditions in Eqs. 7.15 and 7.16, but c_i is now the center of the hyperrectangle, and d_i is the maximum distance from the center to a vertex. The explanation illustrated in Fig. 7.17 still applies in the n-dimensional case and still involves simply finding the lower convex hull of a set of points with different combinations of f and d.

The main complication introduced in the n-dimensional case is the division (trisection) of a selected hyperrectangle. The question is which directions should be divided first. The logic to handle this in the DIRECT algorithm is to prioritize reducing the dimensions with the maximum length, ensuring that hyperrectangles do not deviate too much from the proportions of a hypercube. First, we select the set of the longest dimensions for division (there are often multiple dimensions with the same length). Among this set of the longest dimensions, we select the direction that has been divided the least over the whole history of the search. If there are still multiple dimensions in the selection, we simply select the one with the lowest index. Algorithm 7.4 details the full algorithm.‡

‡Alg. 7.4 follows the revised version of DIRECT,[143] which differs from the original version.[145] The original version trisected all the long sides of the selected rectangles instead of just one side.

143. Jones, *Direct Global Optimization Algorithm*, 2009.

145. Jones et al., *Efficient global optimization of expensive black-box functions*, 1998.

Figure 7.20 shows the first three iterations for a two-dimensional example and the corresponding visualization of the conditions expressed in Eqs. 7.15 and 7.16. We start with a square that contains the whole domain and evaluate the center point. The value of this point is plotted on the f–d plot on the far right.

The first iteration trisects the starting square along the first dimension and evaluates the two new points. The values for these three points

7.5 DIRECT Algorithm

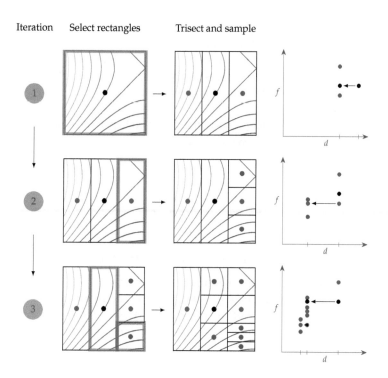

Fig. 7.20 DIRECT iterations for two-dimensional case (left) and corresponding identification of potentially optimal rectangles (right).

are plotted in the second column from the right in the f–d plot, where the center point is reused, as indicated by the arrow and the matching color. At this iteration, we have two points that define the convex hull. In the second iteration, we have three rectangles of the same size, so we divide the one with the lowest value and evaluate the centers of the two new rectangles (which are squares in this case). We now have another column of points in the f–d plot corresponding to a smaller d and an additional point that defines the lower convex hull. Because the convex hull now has two points, we trisect two different rectangles in the third iteration.

Algorithm 7.4 DIRECT in n-dimensions

Inputs:
  $\underline{x}, \overline{x}$: Lower and upper bounds
Outputs:
  x^*: Optimal point

$k = 0$ Initialize iteration counter
Normalize bounded space to hypercube and evaluate its center, c_0
$f_{\min} = f(c_0)$ Stores the minimum function value so far
Initialize $t(i) = 0$ for $i = 1, \ldots, n$ Counts the times dimension i has been trisected
while not converged **do**

Find set S of potentially optimal hyperrectangles
for each hyperrectangle in S **do**
    Find the set I of dimensions with maximum side length
    Select i in I with the lowest $t(i)$, breaking ties in favor of lower i
    Divide the rectangle into thirds along dimension i
    $t(i) = t(i) + 1$
    Evaluate the center points of the outer two hyperrectangles
    Update $f_{\min}$ based on these evaluations
end for
$k = k + 1$ Increment iteration counter
end while

Example 7.3 Minimization of multimodal function with DIRECT

Consider the multimodal Jones function (Appendix D.1.4). Applying the DIRECT method to this function, we get the f-d plot shown in Fig. 7.21, where the final points and convex hull are highlighted. The sequence of rectangles is shown in Fig. 7.22. The algorithm converges to the global minimum after dividing the rectangles around the other local minima a few times.

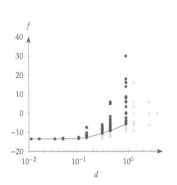

Fig. 7.21 Potentially optimal rectangles for the DIRECT iterations shown in Fig. 7.22.

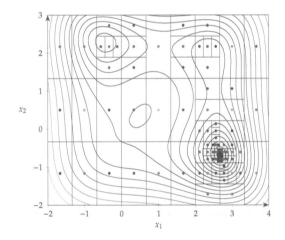

Fig. 7.22 The DIRECT method quickly determines the region with the global minimum of the Jones function after briefly exploring the regions with other minima.

7.6 Genetic Algorithms

Genetic algorithms (GAs) are the most well-known and widely used type of evolutionary algorithm. They were also among the earliest to have been developed.* Like many evolutionary algorithms, GAs are *population based*. The optimization starts with a set of design points (the

*The first GA software was written in 1954, followed by other seminal work.[146] Initially, these GAs were not developed to perform optimization but rather to model the evolutionary process. GAs were eventually applied to optimization.[147]

146. Barricelli, *Esempi numerici di processi di evoluzione*, 1954.

147. Jong, *An analysis of the behavior of a class of genetic adaptive systems*, 1975.

7.6 Genetic Algorithms

population) rather than a single starting point, and each optimization iteration updates this set in some way. Each GA iteration is called a *generation*, each of which has a population with n_p points. Each point is represented by a *chromosome*, which contains the values for all the design variables, as shown in Fig. 7.23. Each design variable is represented by a *gene*. As we will see later, there are different ways for genes to represent the design variables.

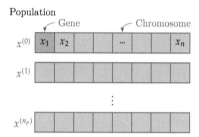

Fig. 7.23 Each GA iteration involves a population of design points, where each design is represented by a chromosome, and each design variable is represented by a gene.

GAs evolve the population using an algorithm inspired by biological reproduction and evolution using three main steps: (1) selection, (2) crossover, and (3) mutation. Selection is based on natural selection, where members of the population that acquire favorable adaptations are more likely to survive longer and contribute more to the population gene pool. Crossover is inspired by chromosomal crossover, which is the exchange of genetic material between chromosomes during sexual reproduction. Mutation mimics genetic mutation, which is a permanent change in the gene sequence that occurs naturally.

Algorithm 7.5 and Fig. 7.24 show how these three steps perform optimization. Although most GAs follow this general procedure, there is a great degree of flexibility in how the steps are performed, leading to many variations in GAs. For example, there is no single method specified for the generation of the initial population, and the size of that population varies. Similarly, there are many possible methods for selecting the parents, generating the offspring, and selecting the survivors. Here, the new population (P_{k+1}) is formed exclusively by the offspring generated from the crossover. However, some GAs add an extra selection process that selects a surviving population of size n_p among the population of parents and offspring.

In addition to the flexibility in the various operations, GAs use different methods for representing the design variables. The design variable representation can be used to classify genetic algorithms into two broad categories: *binary-encoded* and *real-encoded* genetic algorithms. Binary-encoded algorithms use bits to represent the design variables, whereas the real-encoded algorithms keep the same real value representation

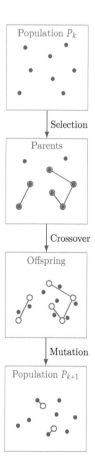

Fig. 7.24 GA iteration steps.

used in most other algorithms. The details of the operations in Alg. 7.5 depend on whether we are using one or the other representation, but the principles remain the same. In the rest of this section, we describe a particular way of performing these operations for each of the possible design variable representations.

Algorithm 7.5 Genetic algorithm

Inputs:
  $\underline{x}, \overline{x}$: Lower and upper bounds
Outputs:
  x^*: Best point
  f^*: Corresponding function value

$k = 0$
$P_k = \left\{ x^{(1)}, x^{(2)}, \ldots, x^{(n_p)} \right\}$ Generate initial population
while $k < k_{\max}$ **do**
    Compute $f(x)$ for all $x \in P_k$ Evaluate objective function
    Select $n_p/2$ parent pairs from P_k for crossover   Selection
    Generate a new population of n_p offspring (P_{k+1})   Crossover
    Randomly mutate some points in the population   Mutation
    $k = k + 1$
end while

7.6.1 Binary-Encoded Genetic Algorithms

The original genetic algorithms were based on binary encoding because they more naturally mimic chromosome encoding. Binary-coded GAs are applicable to discrete or mixed-integer problems.[†] When using binary encoding, we represent each variable as a binary number with m bits. Each bit in the binary representation has a *location*, i, and a *value*, b_i (which is either 0 or 1). If we want to represent a real-valued variable, we first need to consider a finite interval $x \in [\underline{x}, \overline{x}]$, which we can then divide into $2^m - 1$ intervals. The size of the interval is given by

$$\Delta x = \frac{\overline{x} - \underline{x}}{2^m - 1}. \qquad (7.17)$$

To have a more precise representation, we must use more bits.

When using binary-encoded GAs, we do not need to encode the design variables because they are generated and manipulated directly in the binary representation. Still, we do need to decode them before providing them to the evaluation function. To decode a binary

[†] One popular binary-encoded genetic algorithm implementation is the elitist nondominated sorting genetic algorithm (NSGA-II; discussed in Section 9.3.4 in connection with multiobjective optimization).[148]

148. Deb et al., *A fast and elitist multiobjective genetic algorithm: NSGA-II*, 2002.

7.6 Genetic Algorithms

representation, we use

$$x = \underline{x} + \sum_{i=0}^{m-1} b_i 2^i \Delta x. \tag{7.18}$$

Example 7.4 Binary representation of a real number

Suppose we have a continuous design variable x that we want to represent in the interval $[-20, 80]$ using 12 bits. Then, we have $2^{12} - 1 = 4{,}095$ intervals, and using Eq. 7.17, we get $\Delta x \approx 0.0244$. This interval is the error in this finite-precision representation. For the following sample binary representation:

i	1	2	3	4	5	6	7	8	9	10	11	12
b_i	0	0	0	1	0	1	1	0	0	0	0	1

We can use Eq. 7.18 to compute the equivalent real number, which turns out to be $x \approx 32.55$.

Initial Population

The first step in a genetic algorithm is to generate an initial set (population) of points. As a rule of thumb, the population size should be approximately one order of magnitude larger than the number of design variables, and this size should be tuned.

One popular way to choose the initial population is to do it at random. Using binary encoding, we can assign each bit in the representation of the design variables a 50 percent chance of being either 1 or 0. This can be done by generating a random number $0 \leq r \leq 1$ and setting the bit to 0 if $r \leq 0.5$ and 1 if $r > 0.5$. For a population of size n_p, with n design variables, where each variable is encoded using m bits, the total number of bits that needs to be generated is $n_p \times n \times m$.

To achieve better spread in a larger dimensional space, the sampling methods described in Section 10.2 are generally more effective than random populations.

Although we then need to evaluate the function across many points (a population), these evaluations can be performed in parallel.

Selection

In this step, we choose points from the population for reproduction in a subsequent step (called a *mating pool*). On average, it is desirable to choose a mating pool that improves in fitness (thus mimicking the concept of natural selection), but it is also essential to maintain diversity. In total, we need to generate $n_p/2$ pairs.

The simplest selection method is to randomly select two points from the population until the requisite number of pairs is complete. This approach is not particularly effective because there is no mechanism to move the population toward points with better objective functions.

Tournament selection is a better method that randomly pairs up n_p points and selects the best point from each pair to join the mating pool. The same pairing and selection process is repeated to create $n_p/2$ more points to complete a mating pool of n_p points.

Example 7.5 Tournament selection process

Figure 7.25 illustrates the process with a small population. Each member of the population ends up in the mating pool zero, one, or two times, with better points more likely to appear in the pool. The best point in the population will always end up in the pool twice, whereas the worst point in the population is always eliminated.

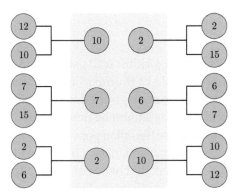

Fig. 7.25 Tournament selection example. The best point in each randomly selected pair is moved into the mating pool.

Another standard method is *roulette wheel selection*. This concept is patterned after a roulette wheel used in a casino. Better points are allocated a larger sector on the roulette wheel to have a higher probability of being selected.

First, the objective function for all the points in the population must be converted to a fitness value because the roulette wheel needs all positive values and is based on maximizing rather than minimizing. To achieve that, we first perform the following conversion to fitness:

$$F = \frac{-f_i + \Delta F}{\max(1, \Delta F - f_{\text{low}})}, \tag{7.19}$$

7.6 Genetic Algorithms

where $\Delta F = 1.1 f_{\text{high}} - 0.1 f_{\text{low}}$ is based on the highest and lowest function values in the population, and the denominator is introduced to scale the fitness.

Then, to find the sizes of the sectors in the roulette wheel selection, we take the normalized cumulative sum of the scaled fitness values to compute an interval for each member in the population j as

$$S_j = \frac{\sum_{i=1}^{j} F_i}{\sum_{i=1}^{n_p} F_i}. \tag{7.20}$$

We can now create a mating pool of n_p points by turning the roulette wheel n_p times. We do this by generating a random number $0 \leq r \leq 1$ at each turn. The jth member is copied to the mating pool if

$$S_{j-1} < r \leq S_j. \tag{7.21}$$

This ensures that the probability of a member being selected for reproduction is proportional to its scaled fitness value.

Example 7.6 Roulette wheel selection process

Assume that $F = [5, 10, 20, 45]$. Then $S = [0.25, 0.3125, 0.875, 1]$, which divides the "wheel" into four segments, shown graphically in Fig. 7.26. We would then draw four random numbers (say, 0.6, 0.2, 0.9, 0.7), which would correspond to the following $n_p/2$ pairs: $(x_3 \text{ and } x_1)$, $(x_4 \text{ and } x_3)$.

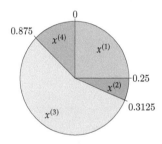

Fig. 7.26 Roulette wheel selection example. Fitter members receive a proportionally larger segment on the wheel.

Crossover

In the reproduction operation, two points (offspring) are generated from a pair of points (parents). Various strategies are possible in genetic algorithms. *Single-point crossover* usually involves generating a random integer $1 \leq k \leq m - 1$ that defines the *crossover point*. This is illustrated in Fig. 7.27. For one of the offspring, the first k bits are taken from parent 1 and the remaining bits from parent 2. For the second offspring, the first k bits are taken from parent 2 and the remaining ones from parent 1. Various extensions exist, such as two-point crossover or n-point crossover.

Mutation

Mutation is a random operation performed to change the genetic information and is needed because even though selection and reproduction

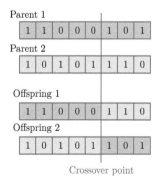

Fig. 7.27 The crossover point determines which parts of the chromosome from each parent get inherited by each offspring.

effectively recombine existing information, occasionally some useful genetic information might be lost. The mutation operation protects against such irrecoverable loss and introduces additional diversity into the population.

When using bit representation, every bit is assigned a small permutation probability, say $p = 0.005 \sim 0.05$. This is done by generating a random number $0 \leq r \leq 1$ for each bit, which is changed if $r < p$. An example is illustrated in Fig. 7.28.

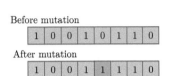

Fig. 7.28 Mutation randomly switches some of the bits with low probability.

7.6.2 Real-Encoded Genetic Algorithms

As the name implies, real-encoded GAs represent the design variables in their original representation as real numbers. This has several advantages over the binary-encoded approach. First, real encoding represents numbers up to machine precision rather than being limited by the number of bits chosen for the binary encoding. Second, it avoids the "Hamming cliff" issue of binary encoding, which is caused by the fact that many bits must change to move between adjacent real numbers (e.g., 0111 to 1000). Third, some real-encoded GAs can generate points outside the design variable bounds used to create the initial population; in many problems, the design variables are not bounded. Finally, it avoids the burden of binary coding and decoding. The main disadvantage is that integer or discrete variables cannot be handled. For continuous problems, a real-encoded GA is generally more efficient than a binary-encoded GA.[140] We now describe the required changes to the GA operations in the real-encoded approach.

140. Simon, *Evolutionary Optimization Algorithms*, 2013.

Initial Population

The most common approach is to pick the n_p points using random sampling within the provided design bounds. Each member is often chosen at random within some initial bounds. For each design variable x_i, with bounds such that $\underline{x}_i \leq x_i \leq \overline{x}_i$, we could use,

$$x_i = \underline{x}_i + r(\overline{x}_i - \underline{x}_i) \tag{7.22}$$

where r is a random number such that $0 \leq r \leq 1$. Again, the sampling methods described in Section 10.2 are more effective for higher-dimensional spaces.

Selection

The selection operation does not depend on the design variable encoding. Therefore, we can use one of the selection approaches described for the binary-encoded GA: tournament or roulette wheel selection.

Crossover

When using real encoding, the term *crossover* does not accurately describe the process of creating the two offspring from a pair of points. Instead, the approaches are more accurately described as a *blending*, although the name *crossover* is still often used.

There are various options for the reproduction of two points encoded using real numbers. A standard method is *linear crossover*, which generates two or more points in the line defined by the two parent points. One option for linear crossover is to generate the following two points:

$$x_{c_1} = 0.5 x_{p_1} + 0.5 x_{p_2}, \qquad (7.23)$$
$$x_{c_2} = 2 x_{p_2} - x_{p_1},$$

where parent 2 is more fit than parent 1 ($f(x_{p_2}) < f(x_{p_1})$). An example of this linear crossover approach is shown in Fig. 7.29, where we can see that child 1 is the average of the two parent points, whereas child 2 is obtained by extrapolating in the direction of the "fitter" parent.

Another option is a simple crossover like the binary case where a random integer is generated to split the vectors—for example, with a split after the first index:

$$x_{p_1} = [x_1, x_2, x_3, x_4]$$
$$x_{p_2} = [x_5, x_6, x_7, x_8]$$
$$\Downarrow \qquad (7.24)$$
$$x_{c_1} = [x_1, x_6, x_7, x_8]$$
$$x_{c_2} = [x_5, x_2, x_3, x_4].$$

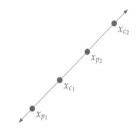

Fig. 7.29 Linear crossover produces two new points along the line defined by the two parent points.

This simple crossover does not generate as much diversity as the binary case and relies more heavily on effective mutation. Many other strategies have been devised for real-encoded GAs.[149]

149. Deb, *Multi-Objective Optimization Using Evolutionary Algorithms*, 2001.

Mutation

As with a binary-encoded GA, mutation should only occur with a small probability (e.g., $p = 0.005 \sim 0.05$). However, rather than changing each bit with probability p, we now change each design variable with probability p.

Many mutation methods rely on random variations around an existing member, such as a uniform random operator:

$$x_{\text{new}\,i} = x_i + (r_i - 0.5)\Delta_i, \quad \text{for} \quad i = 1, \ldots n, \qquad (7.25)$$

where r_i is a random number between 0 and 1, and Δ_i is a preselected maximum perturbation in the ith direction. Many nonuniform methods exist as well. For example, we can use a normal probability distribution

$$x_{\text{new}\,i} = x_i + \mathcal{N}(0, \sigma_i), \quad \text{for} \quad i = 1, \ldots n, \quad (7.26)$$

where σ_i is a preselected standard deviation, and random samples are drawn from the normal distribution. During the mutation operations, bound checking is necessary to ensure the mutations stay within the lower and upper limits.

Example 7.7 Genetic algorithm applied to the bean function

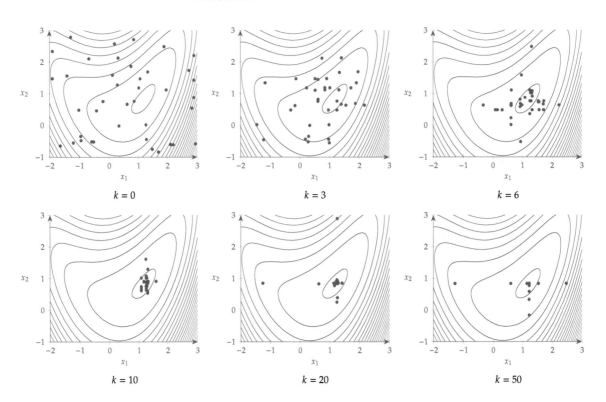

Fig. 7.30 Evolution of the population using a bit-encoded GA to minimize the bean function, where k is the generation number.

Figure 7.30 shows the evolution of the population when minimizing the bean function using a bit-encoded GA. The initial population size was 40, and the simulation was run for 50 generations. Figure 7.31 shows the evolution when using a real-encoded GA but otherwise uses the same parameters as the bit-encoded optimization. The real-encoded GA converges faster in this case.

7.6.3 Constraint Handling

Various approaches exist for handling constraints. Like the Nelder–Mead method, we can use a penalty method (e.g., augmented La-

7.6 Genetic Algorithms

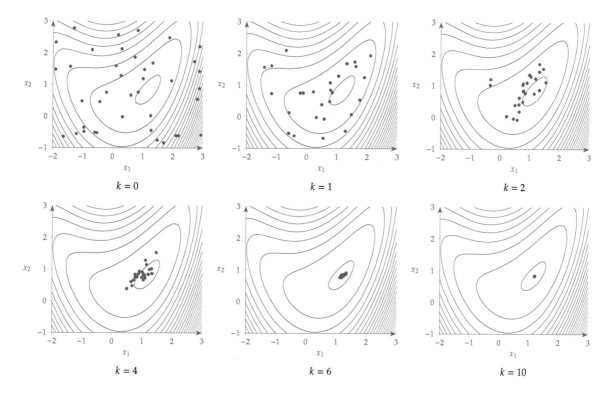

Fig. 7.31 Evolution of the population using a real-encoded GA to minimize the bean function, where k is the generation number.

grangian, linear penalty). However, there are additional options for GAs. In the tournament selection, we can use other selection criteria that do not depend on penalty parameters. One such approach for choosing the best selection among two competitors is as follows:

1. Prefer a feasible solution.
2. Among two feasible solutions, choose the one with a better objective.
3. Among two infeasible solutions, choose the one with a smaller constraint violation.

This concept is a lot like the filter methods discussed in Section 5.5.3.

7.6.4 Convergence

Rigorous mathematical convergence criteria, like those used in gradient-based optimization, do not apply to GAs. The most common way to terminate a GA is to specify a maximum number of iterations, which corresponds to a computational budget. Another similar approach is to let the algorithm run indefinitely until the user manually terminates the algorithm, usually by monitoring the trends in population fitness.

A more automated approach is to track a running average of the population's fitness. However, it can be challenging to decide what tolerance to apply to this criterion because we generally are not interested in the average performance. A more direct metric of interest is the fitness of the best member in the population. However, this can be a problematic criterion because the best member can disappear as a result of crossover or mutation. To avoid this and to improve convergence, many GAs employ *elitism*. This means that the fittest population member is retained to guarantee that the population does not regress. Even without this behavior, the best member often changes slowly, so the user should not terminate the algorithm unless the best member has not improved for several generations.

7.7 Particle Swarm Optimization

Like a GA, particle swarm optimization (PSO) is a stochastic population-based optimization algorithm based on the concept of "swarm intelligence". Swarm intelligence is the property of a system whereby the collective behaviors of unsophisticated agents interacting locally with their environment cause coherent global patterns. In other words: dumb agents, properly connected into a swarm, can yield smart results.*

*PSO was first proposed by Eberhart and Kennedy.[150] Eberhart was an electrical engineer, and Kennedy was a social psychologist.

150. Eberhart and Kennedy, *New optimizer using particle swarm theory*, 1995.

The "swarm" in PSO is a set of design points (*agents* or *particles*) that move in n-dimensional space, looking for the best solution. Although these are just design points, the history for each point is relevant to the PSO algorithm, so we adopt the term *particle*. Each particle moves according to a velocity. This velocity changes according to the past objective function values of that particle and the current objective values of the rest of the particles. Each particle remembers the location where it found its best result so far, and it exchanges information with the swarm about the location where the swarm has found the best result so far.

The position of particle i for iteration $k + 1$ is updated according to

$$x_{k+1}^{(i)} = x_k^{(i)} + v_{k+1}^{(i)} \Delta t, \qquad (7.27)$$

where Δt is a constant artificial time step. The velocity for each particle is updated as follows:

$$v_{k+1}^{(i)} = \alpha v_k^{(i)} + \beta \frac{x_{\text{best}}^{(i)} - x_k^{(i)}}{\Delta t} + \gamma \frac{x_{\text{best}} - x_k^{(i)}}{\Delta t}. \qquad (7.28)$$

The first component in this update is the "inertia", which determines how similar the new velocity is to the velocity in the previous iteration

7.7 Particle Swarm Optimization

through the parameter α. Typical values for the inertia parameter α are in the interval $[0.8, 1.2]$. A lower value of α reduces the particle's inertia and tends toward faster convergence to a minimum. A higher value of α increases the particle's inertia and tends toward increased exploration to potentially help discover multiple minima. Some methods are adaptive, choosing the value of α based on the optimizer's progress.[151]

151. Zhan et al., *Adaptive particle swarm optimization*, 2009.

The second term represents "memory" and is a vector pointing toward the best position particle i has seen in all its iterations so far, $x_{\text{best}}^{(i)}$. The weight in this term consists of a random number β in the interval $[0, \beta_{\max}]$ that introduces a stochastic component to the algorithm. Thus, β controls how much influence the best point found by the particle so far has on the next direction.

The third term represents "social" influence. It behaves similarly to the memory component, except that x_{best} is the best point the entire swarm has found so far, and γ is a random number between $[0, \gamma_{\max}]$ that controls how much of an influence this best point has in the next direction. The relative values of β and γ thus control the tendency toward local versus global search, respectively. Both $\beta_{\max}$ and $\gamma_{\max}$ are in the interval $[0, 2]$ and are typically closer to 2. Sometimes, rather than using the best point in the entire swarm, the best point is chosen within a neighborhood.

Because the time step is artificial, we can eliminate it by multiplying Eq. 7.28 by Δt to yield a step:

$$\Delta x_{k+1}^{(i)} = \alpha \Delta x_k^{(i)} + \beta \left(x_{\text{best}}^{(i)} - x_k^{(i)} \right) + \gamma \left(x_{\text{best}} - x_k^{(i)} \right). \tag{7.29}$$

We then use this step to update the particle position for the next iteration:

$$x_{k+1}^{(i)} = x_k^{(i)} + \Delta x_{k+1}^{(i)}. \tag{7.30}$$

The three components of the update in Eq. 7.29 are shown in Fig. 7.32 for a two-dimensional case.

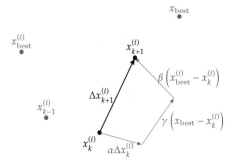

Fig. 7.32 Components of the PSO update.

The first step in the PSO algorithm is to initialize the set of particles (Alg. 7.6). As with a GA, the initial set of points can be determined randomly or can use a more sophisticated sampling strategy (see Section 10.2). The velocities are also randomly initialized, generally using some fraction of the domain size $(\bar{x} - \underline{x})$.

Algorithm 7.6 Particle swarm optimization algorithm

Inputs:
$\bar{x}$: Variable upper bounds
$\underline{x}$: Variable lower bounds
α: Inertia parameter
β_{max}: Self influence parameter
γ_{max}: Social influence parameter
Δx_{max}: Maximum velocity

Outputs:
x^*: Best point
f^*: Corresponding function value

$k = 0$
for $i = 1$ **to** n **do**   Loop to initialize all particles
    Generate position $x_0^{(i)}$ within specified bounds.
    Initialize "velocity" $\Delta x_0^{(i)}$
end for
while not converged **do**   Main iteration loop
    for $i = 1$ **to** n **do**
        if $f\left(x^{(i)}\right) < f\left(x_{best}^{(i)}\right)$ **then**   Best individual points
            $x_{best}^{(i)} = x^{(i)}$
        end if
        if $f(x^{(i)}) < f(x_{best})$ **then**   Best swarm point
            $x_{best} = x^{(i)}$
        end if
    end for
    for $i = 1$ **to** n **do**
        $\Delta x_{k+1}^{(i)} = \alpha \Delta x_k^{(i)} + \beta \left(x_{best}^{(i)} - x_k^{(i)}\right) + \gamma \left(x_{best} - x_k^{(i)}\right)$
        $\Delta x_{k+1}^{(i)} = \max\left(\min\left(\Delta x_{k+1}^{(i)}, \Delta x_{max}\right), -\Delta x_{max}\right)$ Limit velocity
        $x_{k+1}^{(i)} = x_k^{(i)} + \Delta x_{k+1}^{(i)}$ Update the particle position
        $x_{k+1}^{(i)} = \max\left(\min\left(x_{k+1}^{(i)}, \bar{x}\right), \underline{x}\right)$ Enforce bounds
    end for
    $k = k + 1$
end while

7.7 Particle Swarm Optimization

The main loop in the algorithm computes the steps to be added to each particle and updates their positions. Particles must be prevented from going beyond the bounds. If a particle reaches a boundary and has a velocity pointing out of bounds, it is helpful to reset to velocity to zero or reorient it toward the interior for the next iteration. It is also helpful to impose a maximum velocity. If the velocity is too large, the updated positions are unrelated to their previous positions, and the search is more random. The maximum velocity might also decrease across iterations to shift from exploration toward exploitation.

> **Example 7.8** PSO algorithm applied to the bean function
>
> Figure 7.33 shows the particle movements that result when minimizing the bean function using a particle swarm method. The initial population size was 40, and the optimization required 600 function evaluations. Convergence was assumed if the best value found by the population did not improve by more than 10^{-4} for three consecutive iterations.

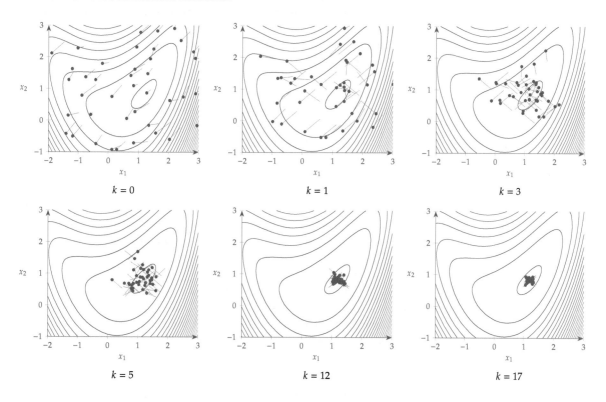

Fig. 7.33 Sequence of PSO iterations that minimize the bean function.

Several convergence criteria are possible, some of which are similar to the Nelder–Mead algorithm and GAs. Examples of convergence criteria include the distance (sum or norm) between each particle and the best particle, the best particle's objective function value changes for the last few generations, and the difference between the best and worst member. For PSO, another alternative is to check whether the velocities for all particles (as measured by a metric such as norm or mean) are below some tolerance. Some of these criteria that assume all the particles congregate (distance, velocities) do not work well for multimodal problems. In those cases, tracking only the best particle's objective function value may be more appropriate.

Tip 7.2 Compare optimization algorithms fairly

It is challenging to compare different algorithms fairly, especially when they use different convergence criteria. You can either compare the computational cost of achieving an objective with a specified accuracy or compare the objective achieved for a specified computational cost. To compare algorithms that use different convergence criteria, you can run them for as long as you can afford using the lowest convergence tolerance possible and tabulate the number of function evaluations and the respective objective function values. To compare the computational cost for a specified tolerance, you can determine the number of function evaluations that each algorithm requires to achieve a given number of digit agreement in the objective function. Alternatively, you can compare the objective achieved for the different algorithms for a given number of function evaluations. Comparison becomes more challenging for constrained problems because a better objective that is less feasible is not necessarily better. In that case, you need to make sure that all results are feasible to the same tolerance. When comparing algorithms that include stochastic procedures (e.g., GAs, PSO), you should run each optimization multiple times to get statistically significant data and compare the mean and variance of the performance metrics. Even for deterministic algorithms, results can vary significantly with starting points (or other parameters), so running multiple optimizations is preferable.

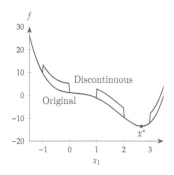

Fig. 7.34 Slice of the Jones function with the added checkerboard pattern.

Example 7.9 Comparison of algorithms for a multimodal discontinuous function

We now return to the Jones function (Appendix D.1.4), but we make it discontinuous by adding the following function:

$$\Delta f = 4 \lceil \sin(\pi x_1) \sin(\pi x_2) \rceil . \tag{7.31}$$

By taking the ceiling of the product of the two sine waves, this function creates a checkerboard pattern with 0s and 4s. In this latter case, each gradient evaluation is counted as an evaluation in addition to each function evaluation. Adding this function to the Jones function produces the discontinuous pattern shown in Fig. 7.34. This is a one-dimensional slice of constant x_2 through the optimum of

7.8 Summary

the Jones function; the full two-dimensional contour plot is shown in Fig. 7.35. The global optimum remains the same as the original function.

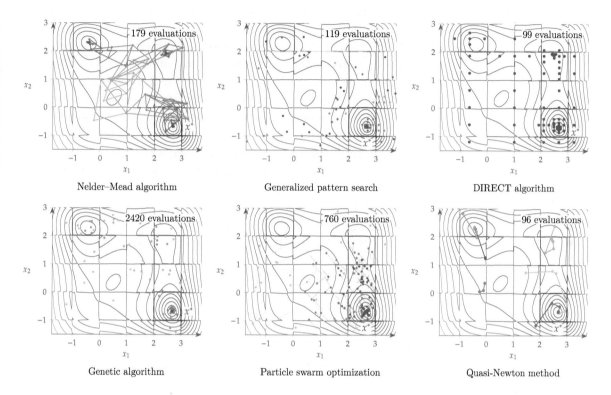

Fig. 7.35 Convergence path for gradient-free algorithms compared with gradient-based algorithms with multistart.

The resulting optimization paths demonstrate that some gradient-free algorithms effectively handle the discontinuities and find the global minimum. Nelder–Mead converges quickly, but not necessarily to the global minimum. GPS and DIRECT quickly converge to the global minimum. GAs and PSO also find the global minimum, but they require many more evaluations. The gradient-based algorithm (quasi-Newton) with multistart also converges the global minimum in two of the six random starts.

7.8 Summary

Gradient-free optimization algorithms are needed when the objective and constraint functions are not smooth enough or when it is not possible to compute derivatives with enough precision. One major advantage of gradient-free methods is that they tend to be robust to numerical noise and discontinuities, making them easier to use than gradient-based methods.

However, the overall cost of gradient-free optimization is sensitive to the cost of the function evaluations because they require many iterations for convergence, and the number of iterations scales poorly with the number of design variables.

There is a wide variety of gradient-free methods. They can perform a local or global search, use mathematical or heuristic criteria, and be deterministic or stochastic. A global search does not guarantee convergence to the global optimum but increases the likelihood of such convergence. We should be wary when heuristics establish convergence because the result might not correspond to the actual mathematical optimum. Heuristics in the optimization algorithm also limit the rate of convergence compared with algorithms based on mathematical principles.

In this chapter, we covered only a small selection of popular gradient-free algorithms. The Nelder–Mead algorithm is a local search algorithm based on heuristics and is easy to implement. GPS and DIRECT are based on mathematical criteria.

Evolutionary algorithms are global search methods based on the evolution of a population of designs. They stem from appealing heuristics inspired by natural or societal phenomena and have some stochastic element in their algorithms. The GAs and PSO algorithms covered in this chapter are only two examples from the plethora of evolutionary algorithms that have been invented.

Many of the methods presented in this chapter do not directly address constrained problems; in those cases, penalty or filtering methods are typically used to enforce constraints.

Problems

7.1 Answer *true* or *false* and justify your answer.

   a. Gradient-free optimization algorithms are not as efficient as gradient-based algorithms, but they converge to the global optimum.

   b. None of the gradient-free algorithms checks the KKT conditions for optimality.

   c. The Nelder–Meade algorithm is a deterministic local search algorithm using heuristic criteria and direct function evaluations.

   d. The simplex is a geometric figure defined by a set of n points, where n is the dimensionality of the design variable space.

   e. The DIRECT algorithm is a deterministic global search algorithm using mathematical criteria and direct function evaluations.

   f. The DIRECT method favors small rectangles with better function values over large rectangles with worse function values.

   g. Evolutionary algorithms are stochastic global search algorithms based on heuristics and direct function evaluations.

   h. GAs start with a population of designs that gradually decreases to a single individual design at the optimum.

   i. Each design in the initial population of a GA should be carefully selected to ensure a successful optimization.

   j. Stochastic procedures are necessary in the GAs to maintain population diversity and therefore reduce the likelihood of getting stuck in local minima.

   k. PSO follows a model developed by biologists in the research of how bee swarms search for pollen and nectar.

   l. All evolutionary algorithms are based on either evolutionary genetics or animal behavior.

7.2 Program the Nelder–Mead algorithm and perform the following studies:

   a. Reproduce the bean function results shown in Ex. 7.1.

   b. Add random noise to the function with a magnitude of 10^{-4} using a normal distribution and see if that makes a difference in the convergence of the Nelder–Mead algorithm. Compare the results to those of a gradient-based algorithm.

c. Consider the following function:

$$f(x_1, x_2, x_3) = |x_1| + 2|x_2| + x_3^2. \qquad (7.32)$$

Minimize this function with the Nelder–Mead algorithm and a gradient-based algorithm. Discuss your results.

d. *Exploration*: Study the logic of the Nelder–Mead algorithm and devise possible improvements. For example, is it a good idea to be greedier and do multiple expansions?

7.3 Program the DIRECT algorithm and perform the following studies:

a. Reproduce the Jones function results shown in Ex. 7.3.

b. Use a gradient-based algorithm with a multistart strategy to minimize the same function. On average, how many different starting points do you need to find the global minimum?

c. Minimize the Hartmann function (defined in Appendix D.1.5) using both methods. Compare and discuss your results.

d. *Exploration*: Develop a hybrid approach that starts with DIRECT and then switches to the gradient-based algorithm. Are you able to reduce the computational cost of DIRECT significantly while converging to the global minimum?

7.4 Program a GA and perform the following studies:

a. Reproduce the bean function results shown in Ex. 7.7.

b. Use your GA to minimize the Harmann function. Estimate the rate of convergence and compare the performance of the GA with a gradient-based algorithm.

c. Study the effect of adding checkerboard steps (Eq. 7.31) with a suitable magnitude to this function. How does this affect the performance of the GA and the gradient-based algorithm compared with the smooth case? Study the effect of reducing the magnitude of the steps.

d. *Exploration*: Experiment with different population sizes, types of crossover, and mutation probability. Can you improve on your original algorithm? Is that improvement still observed for other problems?

7.5 Program the PSO algorithm and perform the following studies:

a. Reproduce the bean function results shown in Ex. 7.8.

7.8 Summary

b. Use your PSO to minimize the n-dimensional Rosenbrock function (defined in Appendix D.1.2) with $n = 4$. Estimate the convergence rate and discuss the performance of PSO compared with a gradient-based algorithm.

c. Study the effect of adding noise to the objective function for both algorithms (see Prob. 7.2). Experiment with different levels of noise.

d. *Exploration*: Experiment with different population sizes and with the values of the coefficients in Eq. 7.29. Are you able to improve the performance of your implementation for multiple problems?

7.6 Study the effect of increased problem dimensionality using the n-dimensional Rosenbrock function defined in Appendix D.1.2. Solve the problem using three approaches:

a. Gradient-free algorithm

b. Gradient-based algorithm with gradients computed using finite differences

c. Gradient-based algorithm with exact gradients

You can either use an off-the-shelf optimizer or your own implementation. In each case, repeat the minimization for $n = 2, 4, 8, 16, \ldots$ up to at least 128, and see how far you can get with each approach. Plot the number of function calls required as a function of the problem dimension (n) for all three methods on one figure. Discuss any differences in optimal solutions found by the various algorithms and dimensions. Compare and discuss your results.

7.7 Consider the aircraft wing design problem described in Appendix D.1.6. We add a wrinkle to the drag computation to make the objective discontinuous. Previously, the approximation for the skin friction coefficient assumed that the boundary layer on the wing was fully turbulent. In this assignment, we assume that the boundary layer is fully laminar when the wing chord Reynolds number is less or equal to $Re = 6 \times 10^5$. The laminar skin friction coefficient is given by

$$C_f = \frac{1.328}{\sqrt{Re}}.$$

For $Re > 6 \times 10^5$, the boundary layer is assumed to be fully turbulent, and the previous skin friction coefficient approximation (Eq. D.14) holds.

Minimize the power with respect to span and chord by doing the following:

a. Implement one gradient-free algorithm of your choice, or alternatively, make up your own algorithm (and give it a good name!)

b. Use the quasi-Newton method you programmed in Prob. 4.9.

c. Use an existing optimizer

Discuss the relative performance of these methods as applied to this problem.

Discrete Optimization   8

Most algorithms in this book assume that the design variables are continuous. However, sometimes design variables must be discrete. Common examples of discrete optimization include scheduling, network problems, and resource allocation. This chapter introduces some techniques for solving discrete optimization problems.

> By the end of this chapter you should be able to:
>
> 1. Identify problems where you can avoid using discrete variables.
> 2. Convert problems with integer variables to ones with binary variables.
> 3. Understand the basics of various discrete optimization algorithms (branch and bound, greedy, dynamic programming, simulated annealing, binary genetic).
> 4. Identify which algorithms are likely to be most suitable for a given problem.

8.1 Binary, Integer, and Discrete Variables

Discrete optimization variables can be of three types: binary (sometimes called *zero-one*), integer, and discrete. A light switch, for example, can only be on or off and is a *binary* decision variable that is either 0 or 1. The number of wheels on a vehicle is an *integer* design variable because it is not useful to build a vehicle with half a wheel. The material in a structure that is restricted to titanium, steel, or aluminum is an example of a *discrete* variable. These cases can all be represented as integers (including the discrete categories, which can be mapped to integers). An optimization problem with integer design variables is referred to as *integer programming*, *discrete optimization*, or *combinatorial optimization*.* Problems with both continuous and discrete variables are referred to as *mixed-integer programming*.

*Sometimes subtle differences in meaning are intended, but typically, and in this chapter, these terms can be used interchangeably.

Unfortunately, discrete optimization is nondeterministic polynomial-time complete, or NP-complete, which means that we can easily verify a solution, but there is no known approach to find a solution efficiently. Furthermore, the time required to solve the problem becomes much worse as the problem size grows.

Example 8.1 The drawback of an exhaustive search

The scaling issue in discrete optimization is illustrated by a well-known discrete optimization problem: the traveling salesperson problem. Consider a set of cities represented graphically on the left of Fig. 8.1. The problem is to find the shortest possible route that visits each city exactly once and returns to the starting city. The path on the right of Fig. 8.1 shows one such solution (not necessarily the optimum). If there were only a handful of cities, you could imagine doing an exhaustive search. You would enumerate all possible paths, evaluate them, and return the one with the shortest distance. Unfortunately, this is not a scalable algorithm. The number of possible paths is $(n - 1)!$, where n is the number of cities. If, for example, we used all 50 U.S. state capitals as the set of cities, then there would be $49! = 6.08 \times 10^{62}$ possible paths! This is such a large number that we cannot evaluate all paths using an exhaustive search.

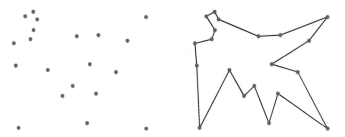

Fig. 8.1 Example of the traveling salesperson problem.

It is possible to construct algorithms that find the global optimum of discrete problems, such as exhaustive searches. Exhaustive search ideas can also be used for continuous problems (see Section 7.5, for example, but the cost is much higher). Although an exhaustive search may eventually arrive at the correct answer, executing that algorithm to completion is often not practical, as Ex. 8.1 highlights. Discrete optimization algorithms aim to search the large combinatorial space more efficiently, often using heuristics and approximate solutions.

8.2 Avoiding Discrete Variables

Even though a discrete optimization problem limits the options and thus conceptually sounds easier to solve, discrete optimization problems are usually much more challenging to solve than continuous problems. Thus, it is often desirable to find ways to avoid using discrete design variables. There are a few approaches to accomplish this.

> **Tip 8.1** Avoid discrete variables when possible
>
> Unless your optimization problem fits specific forms that are well suited to discrete optimization, your problem is likely expensive to solve, and it may be helpful to consider approaches to avoid discrete variables.

The first approach is an *exhaustive search*. We just discussed how exhaustive search scales poorly, but sometimes we have many continuous variables and only a few discrete variables with few options. In that case, enumerating all options is possible. For each combination of discrete variables, the optimization is repeated using all continuous variables. We then choose the best feasible solution among all the optimizations. This approach yields the global optimum, assuming that the continuous optimization finds the global optimum in the continuous variable space.

> **Example 8.2** Evaluate discrete variables exhaustively when the number of combinations is small
>
> Consider the optimization of a propeller. Although most of the design variables are continuous (e.g., propeller blade shape, twist, and chord distributions), the number of blades on a propeller is not. Fortunately, the number of blades falls within a reasonably small set (e.g., two to six). Assuming there are no other discrete variables, we could just perform five optimizations corresponding to each option and choose the best solution among the optima.

A second approach is *rounding*. We can optimize the discrete design variables for some problems as if they were continuous and round the optimal design variable values to integer values afterward. This can be a reasonable approach if the magnitude of the design variables is large or if there are many continuous variables and few discrete variables. After rounding, it is best to repeat the optimization once more, allowing only the continuous design variables to vary. This process might not lead to the true optimum, and the solution might not even be feasible. Furthermore, if the discrete variables are binary, rounding is generally too crude. However, rounding is an effective and practical approach for many problems.

Dynamic rounding is a variation of the rounding approach. Rather than rounding all continuous variables at once, dynamic rounding is an iterative process. It rounds only one or a subset of the discrete variables, fixes them, and re-optimizes the remaining variables using continuous optimization. The process is repeated until all discrete variables are fixed, followed by one last optimization with the continuous variables.

A third approach to avoiding discrete variables is to change the parameterization. For example, one approach in wind farm layout optimization is to parametrize the wind turbine locations as a discrete set of points on a grid. To turn this into a continuous problem, we could parametrize the position of each turbine using continuous coordinates. The trade-off of this continuous parameterization is that we can no longer change the number of turbines, which is still discrete. To re-parameterize, sometimes a continuous alternative is readily apparent, but more often, it requires a good deal of creativity.

Sometimes, an exhaustive search is not feasible, rounding is unacceptable, and a continuous representation is impossible. Fortunately, there are several techniques for solving discrete optimization problems.

8.3 Branch and Bound

A popular method for solving integer optimization problems is the *branch-and-bound* method. Although it is not always the most efficient method,* it is popular because it is robust and applies to a wide variety of discrete problems. One case where the branch-and-bound method is especially effective is solving convex integer programming problems where it is guaranteed to find the global optimum. The most common convex integer problem is a linear integer problem (where all the objectives and constraints are linear in the design variables). This method can be extended to nonconvex integer optimization problems, but it is generally far less effective for those problems and is not guaranteed to find the global optimum. In this section, we assume linear mixed-integer problems but include a short discussion on nonconvex problems.

*Better methods may exist that leverage the specific problem structure, some of which are discussed in this chapter.

A linear mixed-integer optimization problem can be expressed as follows:

$$
\begin{aligned}
\text{minimize} \quad & c^\mathsf{T} x \\
\text{subject to} \quad & \hat{A}x \leq \hat{b} \\
& Ax + b = 0 \\
& x_i \in \mathbb{Z}^+ \text{ for some or all } i,
\end{aligned}
\tag{8.1}
$$

where $\mathbb{Z}^+$ represents the set of all positive integers, including zero.

8.3.1 Binary Variables

Before tackling the integer variable case, we explore the binary variable case, where the discrete entries in x_i must be 0 or 1. Most integer problems can be converted to binary problems by adding additional variables and constraints. Although the new problem is larger, it is usually far easier to solve.

Example 8.3 Converting an integer problem to a binary one

Consider a problem where an engineering device may use one of n different materials: $y \in (1 \ldots n)$. Rather than having one design variable y, we can convert the problem to have n binary variables x_i, where $x_i = 0$ if material i is not selected and $x_i = 1$ if material i is selected. We would also need to add an additional linear constraint to make sure that one (and only one) material is selected:

$$\sum_{i=1}^{n} x_i = 1.$$

The key to a successful branch-and-bound problem is a good *relaxation*. Relaxation aims to construct an approximation of the original optimization problem that is easier to solve. Such approximations are often accomplished by removing constraints.

Many types of relaxation are possible for a given problem, but for *linear* mixed-integer programming problems, the most natural relaxation is to change the integer constraints to continuous bound constraints ($0 \leq x_i \leq 1$). In other words, we solve the corresponding continuous linear programming problem, also known as an *LP* (discussed in Section 11.2). If the solution to the original LP happens to return all binary values, that is the final solution, and we terminate the search. If the LP returns fractional values, then we need to branch.

Branching is done by adding constraints and solving the new optimization problems. For example, we could branch by adding constraints on x_1 to the relaxed LP, creating two new optimization problems: one with the constraint $x_1 = 0$ and another with the constraint $x_1 = 1$. This procedure is then repeated with additional branching as needed.

Figure 8.2 illustrates the branching concept for binary variables. If we explored all of those branches, this would amount to an exhaustive search. The main benefit of the branch-and-bound algorithm is that we can find ways to eliminate branches (referred to as *pruning*) to narrow down the search scope.

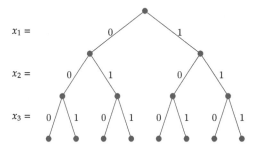

Fig. 8.2 Enumerating the options for a binary problem with branching.

There are two ways to prune. If any of the relaxed problems is infeasible, we know that everything from that node downward (i.e., that branch) is also infeasible. Adding more constraints cannot make an infeasible problem feasible again. Thus, that branch is pruned, and we go back up the tree. We can also eliminate branches by determining that a better solution cannot exist on that branch. The algorithm keeps track of the best solution to the problem found so far.

If one of the relaxed problems returns an objective that is worse than the best we have found, we can prune that branch. We know this because adding constraints always leads to a solution that is either the same or worse, never better (assuming that we find the global optimum, which is guaranteed for LP problems).

The solution from a relaxed problem provides a lower bound—the best that could be achieved if continuing on that branch. The logic for these various possibilities is summarized in Alg. 8.1.

The initial best known solution can be set as $f_{\text{best}} = \infty$ if nothing is known, but if a known feasible solution exists (or can be found quickly by some heuristic), providing any finite best point can speed up the optimization.

Many variations exist for the branch-and-bound algorithm. One variation arises from the choice of which variables to branch on at a given node.

One common strategy is to branch on the variable with the largest fractional component. For example, if $\hat{x} = [1.0, 0.4, 0.9, 0.0]$, we could branch on x_2 or x_3 because both are fractional. We hypothesize that we are more likely to force the algorithm to make faster progress by branching on variables that are closer to midway between integers. In this case, that value would be $x_2 = 0.4$. We would choose to branch on the value closest to 0.5, that is,

$$\min_i |x_i - 0.5|. \tag{8.2}$$

Another variation of branch and bound arises from how the tree search is performed. Two common strategies are *depth-first* and *breadth-*

8.3 Branch and Bound

first. A depth-first strategy continues as far down as possible (e.g., by always branching left) until it cannot go further, and then it follows right branches. A breadth-first strategy explores all nodes on a given level before increasing depth. Various other strategies exist. In general, we do not know beforehand which one is best for a given problem.

Depth-first is a common strategy because, in the absence of more information about a problem, it is most likely to be the fastest way to find a solution—reaching the bottom of the tree generally forces a solution. Finding a solution quickly is desirable because its solution can then be used as a lower bound on other branches.

The depth-first strategy requires less memory storage because breadth-first must maintain a longer history as the number of levels increases. In contrast, depth-first only requires node storage equal to the number of levels.

Algorithm 8.1 Branch-and-bound algorithm

Inputs:
  f_{best}: Best known solution, if any; otherwise $f_{best} = \infty$
Outputs:
  x^*: Optimal point
  $f(x^*)$: Optimal function value

Let S be the set of indices for binary constrained design variables
while branches remain **do**
    Solve relaxed problem for $\hat{x}, \hat{f}$
    if relaxed problem is infeasible **then**
        Prune this branch, back up tree
    else
        if $\hat{x}_i \in \{0, 1\}$ for all $i \in S$ **then**   *A solution is found*
            $f_{best} = \min(f_{best}, \hat{f})$, back up tree
        else
            if $\hat{f} > f_{best}$ **then**
                Prune this branch, back up tree
            else   *A better solution might exist*
                Branch further
            end if
        end if
    end if
end while

Example 8.4 A binary branch-and-bound optimization

Consider the following discrete problem with binary design variables:

$$\begin{aligned} \text{minimize} \quad & -2.5x_1 - 1.1x_2 - 0.9x_3 - 1.5x_4 \\ \text{subject to} \quad & 4.3x_1 + 3.8x_2 + 1.6x_3 + 2.1x_4 \le 9.2 \\ & 4x_1 + 2x_2 + 1.9x_3 + 3x_4 \le 9 \\ & x_i \in \{0, 1\} \text{ for all } i. \end{aligned}$$

To solve this problem, we begin at the first node by solving the linear relaxation. The binary constraint is removed and instead replaced with continuous bounds: $0 \le x_i \le 1$. The solution to this LP is as follows:

$$x^* = [1, 0.5274, 0.4975, 1]$$
$$f^* = -5.0279.$$

There are nonbinary values in the solution, so we need to branch. As mentioned previously, a typical choice is to branch on the variable with the most fractional component. In this case, that is x_3, so we create two additional problems, which add the constraints $x_3 = 0$ and $x_3 = 1$, respectively (Fig. 8.3).

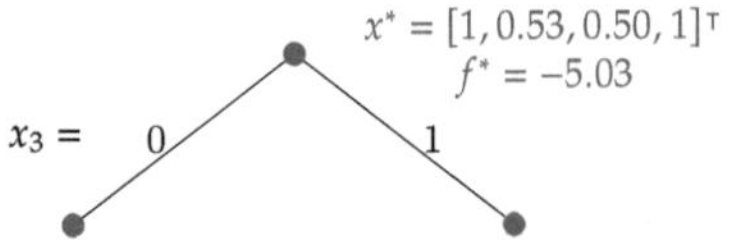

Fig. 8.3 Initial binary branch.

Although depth-first was recommended previously, in this example, we use breadth-first because it yields a more concise example. The depth-first tree is also shown at the end of the example. We solve both of the problems at this next level as shown in Fig. 8.4. Neither of these optimizations yields all binary values, so we have to branch both of them. In this case, the left node branches on x_2 (the only fractional component), and the right node also branches on x_2 (the most fractional component).

Fig. 8.4 Solutions along these two branches.

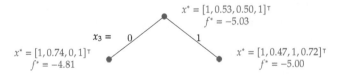

The first branch (see Fig. 8.5) yields a feasible binary solution! The corresponding function value $f = -4$ is saved as the best value so far. There is no need to continue on this branch because the solution cannot be improved on this particular branch.

We continue solving along the rest of this row (Fig. 8.6). The third node in this row yields another binary solution. In this case, the function value is $f = -4.9$, which is better, so this becomes the new best value so far. The second

8.3 Branch and Bound

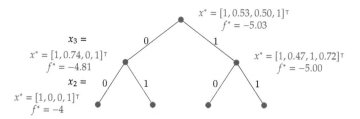

Fig. 8.5 The first feasible solution.

and fourth nodes do not yield a solution. Typically, we would have to branch these further, but they have a lower bound that is worse than the best solution so far. Thus, we can prune both of these branches.

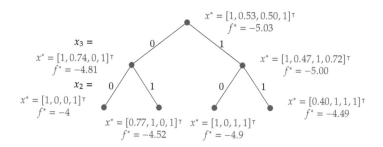

Fig. 8.6 The rest of the solutions on this row.

All branches have been pruned, so we have solved the original problem:

$$x^* = [1, 0, 1, 1]$$
$$f^* = -4.9.$$

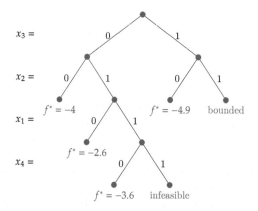

Fig. 8.7 Search path using a depth-first strategy.

Alternatively, we could have used a depth-first strategy. In this case, it is less efficient, but in general, the best strategy is not known beforehand. The depth-first tree for this same example is shown in Fig. 8.7. Feasible solutions to the problem are shown as f^*.

8.3.2 Integer Variables

If the problem cannot be cast in binary form, we can use the same procedure with integer variables. Instead of branching with two constraints ($x_i = 0$ and $x_i = 1$), we branch with two inequality constraints that encourage integer solutions. For example, if the variable we branched on was $x_i = 3.4$, we would branch with two new problems with the following constraints: $x_i \leq 3$ or $x_i \geq 4$.

Example 8.5 Branch and bound with integer variables

Consider the following problem:

$$\begin{aligned}
\text{minimize} \quad & -x_1 - 2x_2 - 3x_3 - 1.5x_4 \\
\text{subject to} \quad & x_1 + x_2 + 2x_3 + 2x_4 \leq 10 \\
& 7x_1 + 8x_2 + 5x_3 + x_4 = 31.5 \\
& x_i \in \mathbb{Z}^+ \text{ for } i = 1, 2, 3 \\
& x_4 \geq 0.
\end{aligned}$$

We begin by solving the LP relaxation, replacing the integer constraints with a lower bound constraint of zero ($x_i \geq 0$). The solution to that problem is

$$x^* = [0, 1.1818, 4.4091, 0], \quad f^* = -15.59.$$

We begin by branching on the most fractional value, which is x_3. We create two new branches:

- The original LP with the added constraint $x_3 \leq 4$
- The original LP with the added constraint $x_3 \geq 5$

Even though depth-first is usually more efficient, we use breadth-first because it is easier to display on a figure. The solution to that first problem is

$$x^* = [0, 1.4, 4, 0.3], \quad f^* = -15.25.$$

The second problem is infeasible, so we can prune that branch.

Recall that the last variable is allowed to be continuous, so we now branch on x_2 by creating two new problems with additional constraints: $x_2 \leq 1$ and $x_2 \geq 2$.

The problem continues using the same procedure shown in the breadth-first tree in Fig. 8.8. The figure gives some indication of why solving integer problems is more time-consuming than solving binary ones. Unlike the binary case, the same value is revisited with tighter constraints. For example, the constraint $x_3 \leq 4$ is enforced early on. Later, two additional problems are created with tighter bounds on the same variable: $x_3 \leq 2$ and $x_3 \geq 3$. In general, the same variable could be revisited many times as the constraints are slowly tightened, whereas in the binary case, each variable is only visited once because the values can only be 0 or 1.

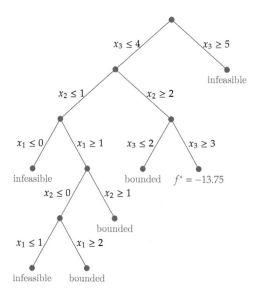

Fig. 8.8 A breadth search of the mixed-integer programming example.

Once all the branches are pruned, we obtain the solution:

$$x^* = [0, 2, 3, 0.5]$$
$$f^* = -13.75.$$

Nonconvex mixed-integer problems can also be used with the branch-and-bound method and generally use this latter strategy of forming two branches of continuous constraints. In this case, the relaxed problem is not a convex problem, so there is no guarantee that we have found a lower bound for that branch. Furthermore, the cost of each suboptimization problem is increased. Thus, for nonconvex discrete problems, this approach is usually only practical for a relatively small number of discrete design variables.

8.4 Greedy Algorithms

Greedy algorithms are among the simplest methods for discrete optimization problems. This method is more of a concept than a specific algorithm. The implementation varies with the application. The idea is to reduce the problem to a subset of smaller problems (often down to a single choice) and then make a locally optimal decision. That decision is locked in, and then the next small decision is made in the same manner. A greedy algorithm does not revisit past decisions and thus ignores much of the coupling between design variables.

> **Example 8.6** A weighted directed graph
>
> As an example, consider the *weighted directed graph* shown in Fig. 8.9. This graph might represent a transportation problem for shipping goods, information flow through a social network, or a supply chain problem. The objective is to traverse from node 1 to node 12 with the lowest possible total cost (the numbers above the path segments denote the cost of each path). A series of discrete choices must be made at each step, and those choices limit the available options in the next step.
>
>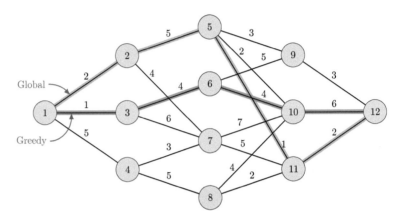
>
> A greedy algorithm simply makes the best choice assuming each decision is the only decision to be made. Starting at node 1, we first choose to move to node 3 because that is the lowest cost between the three options (node 2 costs 2, node 3 costs 1, node 4 costs 5). We then choose to move to node 6 because that is the smallest cost between the next two available options (node 6 costs 4, node 7 costs 6), and so on. The path selected by the greedy algorithm is highlighted in the figure and results in a total cost of 15. The global optimum is also highlighted in the figure and has a total cost of 10.

Fig. 8.9 The greedy algorithm in this weighted directed graph results in a total cost of 15, whereas the best possible cost is 10.

The greedy algorithm used in Ex. 8.6 is easy to apply and scalable but does not generally find the global optimum. To find that global optimum, we have to consider the impact of our choices on future decisions. A method to achieve this for certain problem structures is discussed in the next section.

Even for a fixed problem, there are many ways to construct a greedy algorithm. The advantage of the greedy approach is that the algorithms are easy to construct, and they bound the computational expense of the problem. One disadvantage of the greedy approach is that it usually does not find an optimal solution (and in some cases finds the worst solution![152]). Furthermore, the solution is not necessarily feasible.

152. Gutin et al., *Traveling salesman should not be greedy: domination analysis of greedy-type heuristics for the TSP*, 2002.

Despite the disadvantages, greedy algorithms can sometimes quickly find solutions reasonably close to an optimal solution.

Example 8.7 Greedy algorithms

A few other examples of greedy algorithms are listed below. For the traveling salesperson problem (Ex. 8.1), always select the nearest city as the next step. Consider the propeller problem (Ex. 8.2 but with additional discrete variables (number of blades, type of material, and number of shear webs). A greedy method could optimize the discrete variables one at a time, with the others fixed (i.e., optimize the number of blades first, fix that number, then optimize material, and so on). As a final example, consider the grocery store shopping problem discussed in a separate chapter (Ex. 11.1).* A few possible greedy algorithms for this problem include: always pick the cheapest food item next, or always pick the most nutritious food item next, or always pick the food item with the most nutrition per unit cost.

*This is a form of the knapsack problem, which is a classic problem in discrete optimization discussed in more detail in the following section.

8.5 Dynamic Programming

Dynamic programming is a valuable approach for discrete optimization problems with a particular structure. This structure can also be exploited in continuous optimization problems and problems beyond optimization. The required structure is that the problem can be posed as a *Markov chain* (for continuous problems, this is called a *Markov process*). A Markov chain or process satisfies the Markov property, where a future state can be predicted from the current state without needing to know the entire history. The concept can be generalized to a finite number of states (i.e., more than one but not the entire history) and is called a *variable-order Markov chain*.

If the Markov property holds, we can transform the problem into a recursive one. Using recursion, a smaller problem is solved first, and then larger problems are solved that use the solutions from the smaller problems.

This approach may sound like a greedy optimization, but it is not. We are not using a heuristic but fully solving the smaller problems. Because of the problem structure, we can reuse those solutions. We will illustrate this in examples. This approach has become particularly useful in optimal control and some areas of economics and computational biology. More general design problems, such as the propeller example (Ex. 8.2), do not fit this type of structure (i.e., choosing the number of blades cannot be broken up into a smaller problem separate from choosing the material).

A classic example of a Markov chain is the Fibonacci sequence, defined as follows:

$$f_0 = 0$$
$$f_1 = 1 \tag{8.3}$$
$$f_n = f_{n-1} + f_{n-2}.$$

We can compute the next number in the sequence using only the last two states.* We could implement the computation of this sequence using recursion, as shown algorithmically in Alg. 8.2 and graphically in Fig. 8.10 for f_5.

*We can also convert this to a standard first-order Markov chain by defining $g_n = f_{n-1}$ and considering our state to be (f_n, g_n). Then, each state only depends on the previous state.

Algorithm 8.2 Fibonacci with recursion

procedure fib(n)
    if $n \leq 1$ **then**
        return n
    else
        return fib($n-1$) + fib($n-2$)
    end if
end procedure

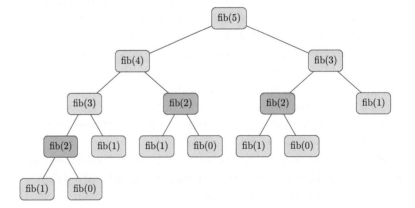

Fig. 8.10 Computing Fibonacci sequence using recursion. The function fib(2) is highlighted as an example to show the repetition that occurs in this recursive procedure.

Although this recursive procedure is simple, it is inefficient. For example, the calculation for fib(2) (highlighted in Fig. 8.10) is repeated multiple times. There are two main approaches to avoiding this inefficiency. The first is a top-down procedure called *memoization*, where we store previously computed values to avoid having to compute them again. For example, the first time we need fib(2), we call the fib function and store the result (the value 1). As we progress down the tree, if we need fib(2) again, we do not call the function but retrieve the stored value instead.

8.5 Dynamic Programming

A bottom-up procedure called *tabulation* is more common. This procedure is how we would typically show the Fibonacci sequence. We start from the bottom (f_0) and work our way forward, computing each new value using the previous states. Rather than using recursion, this involves a simple loop, as shown in Alg. 8.3. Whereas memoization fills entries on demand, tabulation systematically works its way up, filling in entries. In either case, we reduce the computational complexity of this algorithm from exponential complexity (approximately $O(2^n)$) to linear complexity ($O(n)$).

Algorithm 8.3 Fibonacci with tabulation

$f_0 = 0$
$f_1 = 1$
for $i = 2$ **to** n **do**
  $f_i = f_{i-1} + f_{i-2}$
end for

These procedures can be applied to optimization, but before introducing examples, we formalize the mathematics of the approach. One main difference in optimization is that we do not have a set formula like a Fibonacci sequence. Instead, we need to make a design decision at each state, which changes the next state. For example, in the problem shown in Fig. 8.9, we decide which path to take.

Mathematically, we express a given state as s_i and make a design decision x_i, which transitions us to the next state s_{i+1} (Fig. 8.11),

$$s_{i+1} = t_i(s_i, x_i), \quad (8.4)$$

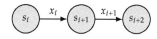

Fig. 8.11 Diagram of state transitions in a Markov chain.

where t is a transition function.† At each transition, we compute the cost function c.‡ For generality, we specify a cost function that may change at each iteration i:

$$c_i(s_i, x_i). \quad (8.5)$$

†For some problems, the transition function is stochastic.

‡It is common to use discount factors on future costs.

We want to make a set of decisions that minimize the sum of the current and future costs up to a certain time, which is called the *value function*,

$$v(s_i) = \underset{x_i,\dots,x_n}{\text{minimize}} \; (c_i(s_i, x_i) + c_{i+1}(s_{i+1}, x_{i+1}) + \dots + c_n(s_n, x_n)), \quad (8.6)$$

where n defines the *time horizon* up to which we consider the cost. For continuous problems, the time horizon may be infinite. The value function (Eq. 8.6) is the *minimum* cost, not just the cost for some arbitrary set of decisions.§

§We use v and c for the scalars in Eq. 8.6 instead of Greek letters because the connection to "value" and "cost" is clearer.

Bellman's principle of optimality states that because of the structure of the problem (where the next state only depends on the current state), we can determine the best solution at this iteration x_i^* if we already know all the optimal future decisions $x_{i+1}^*, \ldots, x_n^*$. Thus, we can recursively solve this problem from the back (bottom) by determining x_n^*, then x_{n-1}^*, and so on back to x_i^*. Mathematically, this recursive procedure is captured by the *Bellman equation*:

$$v(s_i) = \underset{x_i}{\text{minimize}}\,(c(s_i, x_i) + v(s_{i+1})) \,. \tag{8.7}$$

We can also express this equation in terms of our transition function to show the dependence on the current decision:

$$v(s_i) = \underset{x_i}{\text{minimize}}\,(c(s_i, x_i) + v(t_i(s_i, x_i))) \,. \tag{8.8}$$

Example 8.8 Dynamic programming applied to a graph problem

Let us solve the graph problem posed in Ex. 8.6 using dynamic programming. For convenience, we repeat a smaller version of the figure in Fig. 8.12. We use the tabulation (bottom-up) approach. To do this, we construct a table where we keep track of the cost to move from this node to the end (node 12) and which node we should move to next:

Node	1	2	3	4	5	6	7	8	9	10	11	12
Cost												
Next												

We start from the end. The last node is simple: there is no cost to move from node 12 to the end (we are already there), and there is no next node.

Node	1	2	3	4	5	6	7	8	9	10	11	12
Cost												0
Next												–

Now we move back one level to consider nodes 9, 10, and 11. These nodes all lead to node 12 and are thus straightforward. We need to be more careful with the formulas as we get to the more complicated cases next.

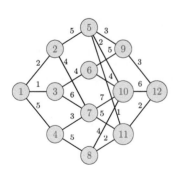

Fig. 8.12 Small version of Fig. 8.9 for convenience.

Node	1	2	3	4	5	6	7	8	9	10	11	12
Cost									3	6	2	0
Next									12	12	12	–

8.5 Dynamic Programming

Now we move back one level to nodes 5, 6, 7, and 8. Using the Bellman equation for node 5, the cost is

$$\text{cost}(5) = \min[3 + \text{cost}(9), 2 + \text{cost}(10), 1 + \text{cost}(11)]. \tag{8.9}$$

We have already computed the minimum value for cost(9), cost(10), and cost(11), so we just look up these values in the table. In this case, the minimum total value is 3 and is associated with moving to node 11. Similarly, the cost for node 6 is

$$\text{cost}(6) = \min[5 + \text{cost}(9), 4 + \text{cost}(10)]. \tag{8.10}$$

The result is 8, and it is realized by moving to node 9.

Node	1	2	3	4	5	6	7	8	9	10	11	12
Cost					3	8			3	6	2	0
Next					11	9			12	12	12	–

We repeat this process, moving back and reusing optimal solutions to find the global optimum. The completed table is as follows:

Node	1	2	3	4	5	6	7	8	9	10	11	12
Cost	10	8	12	9	3	8	7	4	3	6	2	0
Next	2	5	6	8	11	9	11	11	12	12	12	–

From this table, we see that the minimum cost is 10. This cost is achieved by moving first to node 2. Under node 2, we see that we next go to node 5, then 11, and finally 12. Thus, the tabulation gives us the global minimum for cost and the design decisions to achieve that.

To illustrate the concepts more generally, consider another classic problem in discrete optimization—the knapsack problem. In this problem, we have a fixed set of items we can select from. Each item has a weight w_i and a cost c_i. Because the knapsack problem is usually written as a maximization problem and cost implies minimization, we should use value instead. However, we proceed with cost to be consistent with our earlier notation. The knapsack has a fixed capacity K (a scalar) that cannot be exceeded.

The objective is to choose the items that yield the highest total cost subject to the capacity of our knapsack. The design variables x_i are either 1 or 0, indicating whether we take or do not take item i. This problem has many practical applications, such as shipping, data transfer, and investment portfolio selection.

The problem can be written as

$$\begin{aligned}\underset{x}{\text{maximize}} \quad & \sum_{i=1}^{n} c_i x_i \\ \text{subject to} \quad & \sum_{i=1}^{n} w_i x_i \leq K \\ & x_i \in \{0, 1\}.\end{aligned} \quad (8.11)$$

In its present form, the knapsack problem has a linear objective and linear constraints, so branch and bound would be a good approach. However, this problem can also be formulated as a Markov chain, so we can use dynamic programming. The dynamic programming version accommodates variations such as stochasticity and other constraints more easily.

To pose this problem as a Markov chain, we define the state as the remaining capacity of the knapsack k and the number of items we have already considered. In other words, we are interested in $v(k, i)$, where v is the value function (optimal value given the inputs), k is the remaining capacity in the knapsack, and i indicates that we have already considered items 1 through i (this does not mean we have added them all to our knapsack, only that we have considered them). We iterate through a series of decisions x_i deciding whether to take item i or not, which transitions us to a new state where i increases and k may decrease, depending on whether or not we took the item.

The real problem we are interested in is $v(K, n)$, which we solve using tabulation. Starting at the bottom, we know that $v(k, 0) = 0$ for any k. This means that no matter the capacity, the value is 0 if we have not considered any items yet. To work forward, consider a general case considering item i, with the assumption that we have already solved up to item $i - 1$ for any capacity. If item i cannot fit in our knapsack ($w_i > k$), then we cannot take the item. Alternatively, if the weight is less than the capacity, we need to decide whether to select item i or not. If we do not, then the value is unchanged, and $v(k, i) = v(k, i - 1)$. If we do select item i, then our value is c_i plus the best we could do with the previous items but with a capacity that was smaller by w_i: $v(k, i) = c_i + v(k - w_i, i - 1)$. Whichever of these decisions yields a better value is what we should choose.

To determine which items produce this cost, we need to add more logic. To keep track of the selected items, we define a selection matrix S of the same size as v (note that this matrix is indexed starting at zero in both dimensions). Every time we accept an item i, we register that in

8.5 Dynamic Programming

the matrix as $S_{k,i} = 1$. Algorithm 8.4 summarizes this process.

Algorithm 8.4 Knapsack with tabulation

Inputs:
c_i: Cost of item i
w_i: Weight of item i
K: Total available capacity

Outputs:
x^*: Optimal selections
$v(K, n)$: Corresponding cost, $v(k, i)$ is the optimal cost for capacity k considering items 1 through i; note that indexing starts at 0

---

for $k = 0$ to K do
    $v(k, 0) = 0$ *No items considered; value is zero for any capacity*
end for
for $i = 1$ to n do   *Iterate forward solving for one additional item at a time*
    for $k = 0$ to K do
        if $w_i > k$ then
            $v(k, i) = v(k, i - 1)$ *Weight exceeds capacity; value unchanged*
        else
            if $c_i + v(k - w_i, i - 1) > v(k, i - 1)$ then   *Take item*
                $v(k, i) = c_i + v(k - w_i, i - 1)$
                $S(k, i) = 1$
            else   *Reject item*
                $v(k, i) = v(k, i - 1)$
            end if
        end if
    end for
end for
$k = K$ *Initialize*
$x^* = \{\}$ *Initialize solution x^* as an empty set*
for $i = n$ to 1 by -1 do   *Loop to determine which items we selected*
    if $S_{k,i} = 1$ then
        add i to x^* *Item i was selected*
        $k = k - w_i$
    end if
end for

We fill all entries in the matrix $v[k, i]$ to extract the last value $v[K, n]$. For small numbers, filling this matrix (or table) is often illustrated manually, hence the name *tabulation*. As with the Fibonacci example, using dynamic programming instead of a fully recursive solution reduces the complexity from $O(2^n)$ to $O(Kn)$, which means it is pseudolinear. It is only pseudolinear because there is a dependence on the knapsack size. For small capacities, the problem scales well even with many items, but as the capacity grows, the problem scales much less efficiently. Note that the knapsack problem requires integer weights. Real numbers can be scaled up to integers (e.g., 1.2, 2.4 become 12, 24). Arbitrary precision floats are not feasible given the number of combinations to search across.

Example 8.9 Knapsack problem with dynamic programming

Consider five items with the following weights and costs:

$$w_i = [4, 5, 2, 6, 1]$$
$$c_i = [4, 3, 3, 7, 2].$$

The capacity of our knapsack is $K = 10$. Using Alg. 8.4, we find that the optimal cost is 12. The value matrix is as follows:

$$\begin{bmatrix} 0 & 0 & 0 & 0 & 0 & 0 \\ 0 & 0 & 0 & 0 & 0 & 2 \\ 0 & 0 & 0 & 3 & 3 & 3 \\ 0 & 0 & 0 & 3 & 3 & 5 \\ 0 & 4 & 4 & 4 & 4 & 5 \\ 0 & 4 & 4 & 4 & 4 & 6 \\ 0 & 4 & 4 & 7 & 7 & 7 \\ 0 & 4 & 4 & 7 & 7 & 9 \\ 0 & 4 & 4 & 7 & 10 & 10 \\ 0 & 4 & 7 & 7 & 10 & 12 \\ 0 & 4 & 7 & 7 & 11 & 12 \end{bmatrix}.$$

For this example, the selection matrix S is as follows:

$$S = \begin{bmatrix} 0 & 0 & 0 & 0 & 0 & 0 \\ 0 & 0 & 0 & 0 & 0 & 1 \\ 0 & 0 & 0 & 1 & 0 & 0 \\ 0 & 0 & 0 & 1 & 0 & 1 \\ 0 & 1 & 0 & 0 & 0 & 1 \\ 0 & 1 & 0 & 0 & 0 & 1 \\ 0 & 1 & 0 & 1 & 0 & 0 \\ 0 & 1 & 0 & 1 & 0 & 1 \\ 0 & 1 & 0 & 1 & 1 & 0 \\ 0 & 1 & 1 & 0 & 1 & 1 \\ 0 & 1 & 1 & 0 & 1 & 1 \end{bmatrix}.$$

Following this algorithm, we find that we selected items 3, 4, and 5 for a total cost of 12, as expected, and a total weight of 9.

Like greedy algorithms, dynamic programming is more of a technique than a specific algorithm. The implementation varies with the particular application.

8.6 Simulated Annealing

Simulated annealing* is a methodology designed for discrete optimization problems. However, it can also be effective for continuous multimodal problems, as we will discuss. The algorithm is inspired by the annealing process of metals. The atoms in a metal form a crystal lattice structure. If the metal is heated, the atoms move around freely. As the metal cools down, the atoms slow down, and if the cooling is slow enough, they reconfigure into a minimum-energy state. Alternatively, if the metal is quenched or cooled rapidly, the metal recrystallizes with a different higher-energy state (called an *amorphous metal*).

From statistical mechanics, the Boltzmann distribution (also called *Gibbs distribution*) describes the probability of a system occupying a given energy state:

$$P(e) \propto \exp\left(\frac{-e}{k_B T}\right), \tag{8.12}$$

where e is the energy level, T is the temperature, and k_B is Boltzmann's constant. This equation shows that as the temperature decreases, the probability of occupying a higher-energy state decreases, but it is not zero. Therefore, unlike in classical mechanics, an atom could jump to a higher-energy state with some small probability. This property imparts an exploratory nature to the optimization algorithm, which avoids premature convergence to a local minimum. The temperature level provides some control on the level of expected exploration.

An early approach to simulate this type of probabilistic thermodynamic model was the Metropolis algorithm.[155] In the Metropolis algorithm, the probability of transitioning from energy state e_1 to energy state e_2 is formulated as

$$P = \min\left(\exp\left(\frac{-(e_2 - e_1)}{k_B T}\right), 1\right), \tag{8.13}$$

where this probability is limited to be no greater than 1. This limit is needed because the exponent yields a value greater than 1 when $e_2 < e_1$, which would be nonsensical. Simulated annealing leverages this concept in creating an optimization algorithm.

In the optimization analogy, the objective function is the energy level. Temperature is a parameter controlled by the optimizer, which begins high and is slowly "cooled" to drive convergence. At a given iteration, the design variables are given by x, and the objective (or energy) is given by $f(x^{(k)})$. A new state x_{new} is selected at random in the neighborhood of x. If the energy level decreases, the new state is accepted. If the energy level increases, the new state might still be

*First developed by Kirkpatrick et al.[153] and Černý.[154]

153. Kirkpatrick et al., *Optimization by simulated annealing*, 1983.

154. Černý, *Thermodynamical approach to the traveling salesman problem: An efficient simulation algorithm*, 1985.

155. Metropolis et al., *Equation of state calculations by fast computing machines*, 1953.

accepted with probability

$$\exp\left(\frac{-\left(f(x_{\text{new}}) - f\left(x^{(k)}\right)\right)}{T}\right), \tag{8.14}$$

where Boltzmann's constant is removed because it is just an arbitrary scale factor in the optimization context. Otherwise, the state remains unchanged. Constraints can be handled in this algorithm without resorting to penalties by rejecting any infeasible step.

We must supply the optimizer with a function that provides a random *neighboring design* from the set of possible design configurations. A neighboring design is usually related to the current design instead of picking a pure random design from the entire set. In defining the neighborhood structure, we might wish to define transition probabilities so that all neighbors are not equally likely. This type of structure is common in Markov chain problems. Because the nature of different discrete problems varies widely, we cannot provide a generic neighbor-selecting algorithm, but an example is shown later for the specific case of a traveling salesperson problem.

Finally, we need to determine the *annealing schedule* (or *cooling schedule*), a process for decreasing the temperature throughout the optimization. A common approach is an exponential decrease:

$$T = T_0 \alpha^k, \tag{8.15}$$

where T_0 is the initial temperature, α is the cooling rate, and k is the iteration number. The cooling rate α is a number close to 1, such as 0.8–0.99. Another simple approach to iterate toward zero temperature is as follows:

$$T = T_0\left(1 - \frac{k}{k_{\max}}\right)^\beta, \tag{8.16}$$

where the exponent β is usually in the range of 1–4. A higher exponent corresponds to spending more time at low temperatures. In many approaches, the temperature is kept constant for a fixed number of iterations (or a fixed number of successful moves) before moving to the next decrease. Many methods are simple schedules with a predetermined rate, although more complex adaptive methods also exist.[†]
The annealing schedule can substantially impact the algorithm's performance, and some experimentation is required to select an appropriate schedule for the problem at hand. One essential requirement is that the temperature should start high enough to allow for exploration. This should be significantly higher than the maximum expected energy change (change in objective) but not so high that computational time is

[†] See Andresen and Gordon[156], for example.

156. Andresen and Gordon, *Constant thermodynamic speed for minimizing entropy production in thermodynamic processes and simulated annealing*, 1994.

8.6 Simulated Annealing

wasted with too much random searching. Also, cooling should occur slowly to improve the ability to recover from a local optimum, imitating the annealing process instead of the quenching process.

The algorithm is summarized in Alg. 8.5; for simplicity in the description, the annealing schedule uses an exponential decrease at every iteration.

Algorithm 8.5 Simulated Annealing

Inputs:
   x_0: Starting point
   T_0: Initial temperature
Outputs:
   x^*: Optimal point

---

for $k = 0$ to k_{max} do   Simple iteration; convergence metrics can be used instead
   $x_{new} = \text{neighbor}\left(x^{(k)}\right)$ Randomly generate from neighbors
   if $f(x_{new}) \leq f\left(x^{(k)}\right)$ then   Energy decreased; jump to new state
      $x^{(k+1)} = x_{new}$
   else
      $r \in \mathcal{U}[0, 1]$ Randomly draw from uniform distribution
      $P = \exp\left(\dfrac{-\left(f(x_{new}) - f\left(x^{(k)}\right)\right)}{T}\right)$
      if $P \geq r$ then   Probability high enough to jump
         $x^{(k+1)} = x_{new}$
      else
         $x^{(k+1)} = x^{(k)}$ Otherwise remain at current state
      end if
   end if
   $T = \alpha T$ Reduce temperature
end for

Example 8.10 Traveling salesperson with simulated annealing

This example sets up the traveling salesperson problem with 50 points randomly distributed (from uniform sampling) on a square grid with sides of length 1 (left of Fig. 8.13). The objective is the total Euclidean distance of a path that traverses all points and returns to the starting point. The design variables are a sequence of integers corresponding to the order in which the salesperson traverses the points.

We generate new neighboring designs using the technique from Lin,[157] where one of two options is randomly chosen at each iteration: (1) randomly choose two points and flip the direction of the path segments between those

[157] Lin, *Computer solutions of the traveling salesman problem*, 1965.

two points, or (2) randomly choose two points and move the path segments to follow another randomly chosen point. The distance traveled by the randomly generated initial set of points is 26.2.

We specify an iteration budget of 25,000 iterations, set the initial temperature to be 10, and decrease the temperature by a multiplicative factor of 0.95 at every 100 iterations. The right panel of Fig. 8.13 shows the final path, which has a length of 5.61. The final path might not be *the* global optimum (remember, these finite time methods are only approximations of the full combinatorial search), but the methodology is effective and fast for this problem in finding at least a near-optimal solution. Figure 8.14 shows the iteration history.

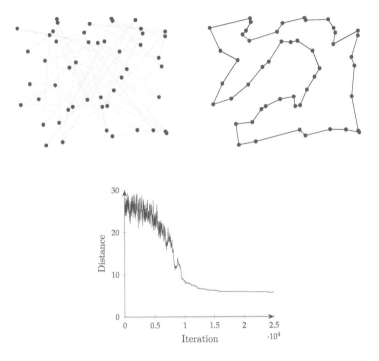

Fig. 8.13 Initial and final paths for traveling salesperson problem.

Fig. 8.14 Convergence history of the simulated annealing algorithm.

The simulated annealing algorithm can be applied to continuous multimodal problems as well. The motivation is similar because the initial high temperature permits the optimizer to escape local minima, whereas a purely descent-based approach would not. By slowly cooling, the initial exploration gives way to exploitation. The only real change in the procedure is in the `neighbor` function. A typical approach is to generate a random direction and choose a step size proportional to the temperature. Thus, smaller, more conservative steps are taken as the algorithm progresses. If bound constraints are present, they would be enforced at this step. Purely random step directions are not particularly efficient for many continuous problems, particularly when most directions are ill-conditioned (e.g., a narrow valley or near convergence). One variation adopts concepts from the Nelder–Mead algorithm (Section 7.3) to improve efficiency.[158] Overall, simulated annealing has made more impact on discrete problems compared with continuous ones.

158. Press et al., *Numerical Recipes in C: The Art of Scientific Computing*, 1992.

8.7 Binary Genetic Algorithms

The binary form of a genetic algorithm (GA) can be directly used with discrete variables. Because the binary form already requires a discrete representation for the population members, using discrete design variables is a natural fit. The details of this method were discussed in Section 7.6.1.

8.8 Summary

This chapter discussed various strategies for approaching discrete optimization problems. Some problems can be well approximated using rounding, can be reparameterized in a continuous way, or only have a few discrete combinations, allowing for explicit enumeration. For problems that can be posed as linear (or convex in general), branch and bound is effective. If the problem can be posed as a Markov chain, dynamic programming is a useful method.

If none of these categorizations are applicable, then a stochastic method, such as simulated annealing or GAs, may work well. These stochastic methods typically struggle as the dimensionality of the problem increases. However, simulated annealing can scale better for some problems if there are clever ways to quickly evaluate designs in the neighborhood, as is done with the traveling salesperson problem. An alternative to these various algorithms is to use a greedy strategy, which can scale well. Because this strategy is a heuristic, it usually results in a loss in solution quality.

Problems

8.1 Answer *true* or *false* and justify your answer.

  a. All discrete variables can be represented by integers.

  b. Discrete optimization algorithms sometimes use heuristics and find only approximate solutions.

  c. The rounding technique solves a discrete optimization problem with continuous variables and then rounds each resulting design variable, objective, and constraint to the nearest integer.

  d. Exhaustive search is the only way to be sure we have found the global minimum for a problem that involves discrete variables.

  e. The branch-and-bound method is guaranteed to find the global optimum for convex problems.

  f. When using the branch-and-bound method for binary variables, the same variable might have to be revisited.

  g. When using the branch-and-bound method, the breadth-first strategy requires less memory storage than the depth-first strategy.

  h. Greedy algorithms never reconsider a decision once it has been made.

  i. The Markov property applies when a future state can be predicted from the current state without needing to know any previous state.

  j. Both memoization and tabulation reduce the computational complexity of dynamic programming such that it no longer scales exponentially.

  k. Simulated annealing can be used to minimize smooth unimodal functions of continuous design variables.

  l. Simulated annealing, genetic algorithms, and dynamic programming include stochastic procedures.

8.2 *Branch and bound.* Solve the following problem using a *manual* branch-and-bound approach (i.e., show each LP subproblem), as is done in Ex. 8.4:

$$\text{maximize} \quad 0.5x_1 + 2x_2 + 3.5x_3 + 4.5x_4$$
$$\text{subject to} \quad 5.5x_1 + 0.5x_2 + 3.5x_3 + 2.3x_4 \leq 9.2$$
$$2x_1 + 4x_2 + 2x_4 \leq 8$$
$$1x_1 + 3x_2 + 3x_3 + 4x_4 \leq 4$$
$$x_i \in \{0, 1\} \text{ for all } i.$$

8.3 *Solve an integer linear programming problem.* A chemical company produces four types of products: A, B, C, and D. Each requires labor to produce and uses some combination of chlorine, sulfuric acid, and sodium hydroxide in the process. The production process can also produce these chemicals as a by-product, rather than just consuming them. The chemical mixture and labor required for the production of the three products are listed in the following table, along with the availability per day. The market values for one barrel of A, B, C, and D are $50, $30, $80, and $30, respectively. Determine the number of barrels of each to produce to maximize profit using three different approaches:

a. As a continuous linear programming problem with rounding
b. As an integer linear programming problem
c. Exhaustive search

	A	B	C	D	Limit
Chlorine	0.74	−0.05	1.0	−0.15	97
Sodium hydroxide	0.39	0.4	0.91	0.44	99
Sulfuric acid	0.86	0.89	0.09	0.83	52
Labor (person-hours)	5	7	7	6	1000

Discuss the results.

8.4 *Solve a dynamic programming problem.* Solve the knapsack problem with the following weights and costs:

$$w_i = [2, 5, 3, 4, 6, 1]$$
$$c_i = [5, 3, 1, 5, 7, 2]$$

and a capacity of $K = 12$. Maximize the cost subject to the capacity constraint. Use the following two approaches:

a. A greedy algorithm where you take the item with the best cost-to-weight ratio (that fits within the remaining capacity) at each iteration

b. Dynamic programming

8.5 *Simulated annealing.* Construct a traveling salesperson problem with 50 randomly generated points. Implement a simulated annealing algorithm to solve it.

8.6 *Binary genetic algorithm.* Solve the same problem as previously (traveling salesperson) with a binary genetic algorithm.

9 Multiobjective Optimization

Up to this point in the book, all of our optimization problem formulations have had a single objective function. In this chapter, we consider *multiobjective* optimization problems, that is, problems whose formulations have more than one objective function. Some common examples of multiobjective optimization include risk versus reward, profit versus environmental impact, acquisition cost versus operating cost, and drag versus noise.

> By the end of this chapter you should be able to:
>
> 1. Identify scenarios where multiobjective optimization is useful.
>
> 2. Understand the concept of dominance and identify a Pareto set.
>
> 3. Use various methods for performing multiobjective optimization and understand the pros and cons of the methods.

9.1 Multiple Objectives

Before discussing how to solve multiobjective problems, we must first explore what it means to have more than one objective. In some sense, there is no such thing as a multiobjective optimization problem. Although many metrics are important to the engineer, only one thing can be made best at a time. A common technique when presented with multiple objectives is to assign weights to the various objectives and combine them into a single objective.

More generally, multiobjective optimization helps explore the trade-offs between different metrics. Still, if we select one design from the presented options, we have indirectly chosen a single objective. However, the corresponding objective function may be difficult to formulate beforehand.

> **Tip 9.1 Are you sure you have multiple objectives?**
>
> A common pitfall for beginner optimization practitioners is to categorize a problem as multiobjective without critical evaluation. When considering whether you should use more than one objective, you should ask whether or not there is a more fundamental underlying objective or if some of the "objectives" are actually constraints. Solving a multiobjective problem is much more costly than solving a single objective one, so you should make sure you need multiple objectives.

> **Example 9.1 Selecting an objective**
>
> Determining the appropriate objective is often a real challenge. For example, in designing an aircraft, we may decide that minimizing drag and minimizing weight are important. However, these metrics compete with each other and cannot be minimized simultaneously. Instead, we may conclude that maximizing range (the distance the aircraft can fly) is the underlying metric that matters most for our application and appropriately balances the trade-offs between weight and drag. Or perhaps maximizing range is not the right metric. Range may be important, but only insofar as we reach some threshold. Increasing the range does not increase the value because the range is a constraint. The underlying objective in this scenario may be some other metric, such as operating costs.

Despite these considerations, there are still good reasons to pursue a multiobjective problem. A few of the most common reasons are as follows:

1. Multiobjective optimization can quantify the trade-off between different objectives. The benefits of this approach will become apparent when we discuss Pareto surfaces and can lead to important design insights.

2. Multiobjective optimization provides a "family" of designs rather than a single design. A family of options is desirable when decision-making needs to be deferred to a later stage as more information is gathered. For example, an executive team or higher-fidelity numerical simulations may be used to make later design decisions.

3. For some problems, the underlying objective is either unknown or too difficult to compute. For example, cost and environmental impact may be two important metrics for a new design. Although the latter could arguably be turned into a cost, doing so may

be too difficult to quantify and add an unacceptable amount of uncertainty (see Chapter 12).

Mathematically, the only change to our optimization problem formulation is that the objective statement,

$$\underset{x}{\text{minimize}} \quad f(x), \tag{9.1}$$

becomes

$$\underset{x}{\text{minimize}} \quad f(x) = \begin{bmatrix} f_1(x) \\ f_2(x) \\ \vdots \\ f_{n_f}(x) \end{bmatrix}, \quad \text{where} \quad n_f \geq 2. \tag{9.2}$$

The constraints are unchanged unless some of them have been reformulated as objectives. This multiobjective formulation might require trade-offs when trying to minimize all functions simultaneously because, beyond some point, further reduction in one objective can only be achieved by increasing one or more of the other objectives.

One exception occurs if the objectives are independent because they depend on different sets of design variables. Then, the objectives are said to be *separable*, and they can be minimized independently. If there are constraints, these need to be separable as well. However, separable objectives and constraints are rare because functions tend to be linked in engineering systems.

Given that multiobjective optimization requires trade-offs, we need a new definition of optimality. In the next section, we explain how there is an infinite number of optimal points, forming a surface in the space of objective functions. After defining optimality for multiple objectives, we present several possible methods for solving multiobjective optimization problems.

9.2 Pareto Optimality

With multiple objectives, we have to reconsider what it means for a point to be optimal. In multiobjective optimization, we use the concept of *Pareto optimality*.

Figure 9.1 shows three designs measured against two objectives that we want to minimize: f_1 and f_2. Let us first compare design A with design B. From the figure, we see that design A is better than design B in both objectives. In the language of multiobjective optimization, we say that design A *dominates* design B. One design is said to dominate

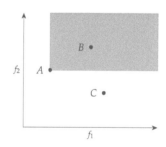

Fig. 9.1 Three designs, A, B, and C, are plotted against two objectives, f_1 and f_2. The region in the shaded rectangle highlights points that are dominated by design A.

another design if it is superior in all objectives (design A dominates any design in the shaded rectangle). Comparing design A with design C, we note that design A is better in one objective (f_1) but worse in the other objective (f_2). Neither design dominates the other.

A point is said to be *nondominated* if none of the other evaluated points dominate it (Fig. 9.2). If a point is nondominated by any point in the entire domain, then that point is called *Pareto optimal*. This does not imply that this point dominates all other points; it simply means no other point dominates it. The set of all Pareto optimal points is called the *Pareto set*. The Pareto set refers to the vector of points x^*, whereas the *Pareto front* refers to the vector of functions $f(x^*)$.

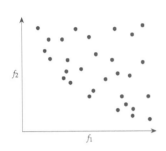

Fig. 9.2 A plot of all the evaluated points in the design space plotted against two objectives, f_1 and f_2. The set of red points is not dominated by any other and is thus the current approximation of the Pareto set.

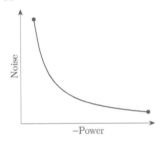

Fig. 9.3 A notional Pareto front representing power and noise trade-offs for a wind farm optimization problem.

Example 9.2 A Pareto front in wind farm optimization

The Pareto front is a valuable tool to produce design insights. Figure 9.3 shows a notional Pareto front for a wind farm optimization. The two objectives are maximizing power production (shown with a negative sign so that it is minimized) and minimizing noise.

The Pareto front is helpful to understand trade-off sensitivities. For example, the left endpoint represents the maximum power solution, and the right endpoint represents the minimum noise solution. The nature of the curve on the left side tells us how much power we have to sacrifice for a given reduction in noise. If the slope is steep, as is the case in the figure, we can see that a small sacrifice in maximum power production can be exchanged for significantly reduced noise. However, if more significant noise reductions are sought, then large power reductions are required. Conversely, if the left side of the figure had a flatter slope, we would know that small reductions in noise would require significant decreases in power. Understanding the magnitude of these trade-off sensitivities helps make high-level design decisions.

9.3 Solution Methods

Various solution methods exist for solving multiobjective problems. This chapter does not cover all methods but highlights some of the more commonly used approaches. These include the weighted-sum method, the epsilon-constraint method, the normal boundary intersection (NBI) method, and evolutionary algorithms.

9.3.1 Weighted Sum

The weighted-sum method is easy to use, but it is not particularly efficient. Other methods exist that are just as simple but have better performance. It is only introduced because it is well known and is frequently used. The idea is to combine all of the objectives into one objective using a weighted sum, which can be written as

$$\bar{f}(x) = \sum_{i=1}^{n_f} w_i f_i(x), \qquad (9.3)$$

where n_f is the number of objectives, and the weights are usually normalized such that

$$\sum_{i=1}^{n_f} w_i = 1. \qquad (9.4)$$

If we have two objectives, the objective reduces to

$$\bar{f}(x) = w f_1(x) + (1-w) f_2(x), \qquad (9.5)$$

where w is a weight in $[0, 1]$.

Consider a two-objective case. Points on the Pareto set are determined by choosing a weight w, completing the optimization for the composite objective, and then repeating the process for a new value of w. It is straightforward to see that at the extremes $w = 0$ and $w = 1$, the optimization returns the designs that optimize one objective while ignoring the other. The weighted-sum objective forms an equation for a line with the objectives as the ordinates. Conceptually, we can think of this method as choosing a slope for the line (by selecting w), then pushing that line down and to the left as far as possible until it is just tangent to the Pareto front (Fig. 9.4). With this form of the objective, the slope of the line would be

$$\frac{df_2}{df_1} = \frac{-w}{1-w}. \qquad (9.6)$$

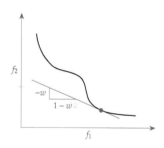

Fig. 9.4 The weighted-sum method defines a line for each value of w and finds the point tangent to the Pareto front.

This procedure identifies one point in the Pareto set, and the procedure must then be repeated with a new slope.

The main benefit of this method is that it is easy to use. However, the drawbacks are that (1) uniform spacing in w leads to nonuniform spacing along with the Pareto set, (2) it is not apparent which values of w should be used to sweep out the Pareto set evenly, and (3) this method can only return points on the convex portion of the Pareto front.

In Fig. 9.5, we highlight the convex portions of the Pareto front from Fig. 9.4. If we utilize the concept of pushing a line down and to the left, we see that these are the only portions of the Pareto front that can be found using a weighted-sum method.

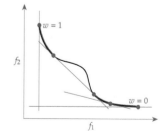

Fig. 9.5 The convex portions of this Pareto front are the portions highlighted.

9.3.2 Epsilon-Constraint Method

The epsilon-constraint method works by minimizing one objective while setting all other objectives as additional constraints:[159]

159. Haimes et al., *On a bicriterion formulation of the problems of integrated system identification and system optimization*, 1971.

$$\begin{aligned}
\underset{x}{\text{minimize}} \quad & f_i \\
\text{subject to} \quad & f_j \leq \varepsilon_j \quad \text{for all } j \neq i \\
& g(x) \leq 0 \\
& h(x) = 0.
\end{aligned} \tag{9.7}$$

Then, we must repeat this procedure for different values of ε_j.

This method is visualized in Fig. 9.6. In this example, we constrain f_1 to be less than or equal to a certain value and minimize f_2 to find the corresponding point on the Pareto front. We then repeat this procedure for different values of ε.

One advantage of this method is that the values of ε correspond directly to the magnitude of one of the objectives, so determining appropriate values for ε is more intuitive than selecting the weights in the previous method. However, we must be careful to choose values that result in a feasible problem. Another advantage is that this method reveals the nonconvex portions of the Pareto front. Both of these reasons strongly favor using the epsilon-constraint method over the weighted-sum method, especially because this method is not much harder to use. Its main limitation is that, like the weighted-sum method, a uniform spacing in ε does not generally yield a uniform spacing of the Pareto front (though it is usually much better spaced than weighted-sum), and therefore it might still be inefficient, particularly with more than two objectives.

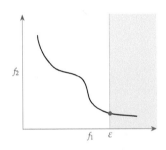

Fig. 9.6 The vertical line represents an upper bound constraint on f_1. The other objective, f_2, is minimized to reveal one point in the Pareto set. This procedure is then repeated for different constraints on f_1 to sweep out the Pareto set.

9.3.3 Normal Boundary Intersection

160. Das and Dennis, *Normal-boundary intersection: A new method for generating the Pareto surface in nonlinear multicriteria optimization problems*, 1998.

The NBI method is designed to address the issue of nonuniform spacing along the Pareto front.[160] We first find the extremes of the Pareto set; in other words, we minimize the objectives one at a time. These extreme points are referred to as *anchor points*. Next, we construct a plane that passes through the anchor points. We space points along this plane (usually evenly) and, starting from those points, solve optimization problems that search along directions normal to this plane.

This procedure is shown in Fig. 9.7 for a two-objective case. In this case, the plane that passes through the anchor points is a line. We now space points along this line by choosing a vector of weights b, as illustrated on the left-hand of Fig. 9.7. The weights are constrained such

9.3 Solution Methods

that $b_i \in [0, 1]$, and $\sum_i b_i = 1$. If we make $b_i = 1$ and all other entries zero, then this equation returns one of the anchor points, $f(x_i^*)$. For two objectives, we would set $b = [w, 1 - w]$ and vary w in equal steps between 0 and 1.

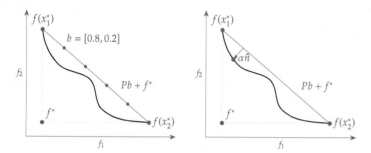

Fig. 9.7 A notional example of the NBI method. A plane is created that passes through the single-objective optima (the anchor points), and solutions are sought normal to that plane for a more evenly spaced Pareto front.

Starting with a specific value of b, we search along a direction perpendicular to the line defined by the anchor points, represented by $a\hat{n}$ in Fig. 9.7 (right). We seek to find the point along this direction that is the farthest away from the anchor points line (a maximization problem), with the constraint that the point is consistent with the objective functions. The resulting optimal point found along this direction is a point on the Pareto front. We then repeat this process for another set of weighting parameters in b.

We can see how this method is similar to the epsilon-constraint method, but instead of searching along lines parallel to one of the axes, we search along lines perpendicular to the plane defined by the anchor points. The idea is that even spacing along this plane is more likely to lead to even spacing along the Pareto front.

Mathematically, we start by determining the anchor points, which are just single-objective optimization problems. From the anchor points, we define what is called the *utopia* point. The utopia point is an ideal point that cannot be obtained, where every objective reaches its minimum simultaneously (shown in the lower-left corner of Fig. 9.7):

$$f^* = \begin{bmatrix} f_1(x_1^*) \\ f_2(x_2^*) \\ \vdots \\ f_n(x_n^*) \end{bmatrix}, \qquad (9.8)$$

where x_i^* denotes the design variables that minimize objective f_i. The utopia point defines the equation of a plane that passes through all anchor points,

$$Pb + f^*, \qquad (9.9)$$

where the ith column of P is $f(x_i^*) - f^*$. A single vector b, whose length is given by the number of objectives, defines a point on the plane.

We now define a vector ($\hat{n}$) that is normal to this plane, in the direction toward the origin. We search along this vector using a step length α, while maintaining consistency with our objective functions $f(x)$ yielding

$$f(x) = Pb + f^* + \alpha \hat{n}. \tag{9.10}$$

Computing the exact normal ($\hat{n}$) is involved, and the vector does not need to be exactly normal. As long as the vector points toward the Pareto front, then it will still yield well-spaced points. In practice, a quasi-normal vector is often used, such as,

$$\tilde{n} = -Pe, \tag{9.11}$$

where e is a vector of 1s.

We now solve the following optimization problem, for a given vector b, to yield a point on the Pareto front:

$$\begin{aligned}
\underset{x,\alpha}{\text{maximize}} \quad & \alpha \\
\text{subject to} \quad & Pb + f^* + \alpha \hat{n} = f(x) \\
& g(x) \leq 0 \\
& h(x) = 0.
\end{aligned} \tag{9.12}$$

This means that we find the point farthest away from the anchor-point plane, starting from a given value for b, while satisfying the original problem constraints. The process is then repeated for additional values of b to sweep out the Pareto front.

In contrast to the previously mentioned methods, this method yields a more uniformly spaced Pareto front, which is desirable for computational efficiency, albeit at the cost of a more complex methodology.

For most multiobjective design problems, additional complexity beyond the NBI method is unnecessary. However, even this method can still have deficiencies for problems with unusual Pareto fronts, and new methods continue to be developed. For example, the normal constraint method uses a very similar approach,[161] but with inequality constraints to address a deficiency in the NBI method that occurs when the normal line does not cross the Pareto front. This methodology has undergone various improvements, including better scaling through normalization.[162] A more recent improvement performs an even more efficient generation of the Pareto frontier by avoiding regions of the Pareto front where minimal trade-offs occur.[163]

161. Ismail-Yahaya and Messac, *Effective generation of the Pareto frontier using the normal constraint method*, 2002.

162. Messac and Mattson, *Normal constraint method with guarantee of even representation of complete Pareto frontier*, 2004.

163. Hancock and Mattson, *The smart normal constraint method for directly generating a smart Pareto set*, 2013.

9.3 Solution Methods

Example 9.3 A two-dimensional normal boundary interface problem

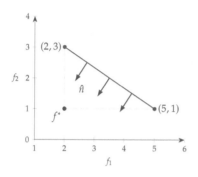

Fig. 9.8 Search directions are normal to the line connecting anchor points.

First, we optimize the objectives one at a time, which in our example results in the two anchor points shown in Fig. 9.8: $f(x_1^*) = (2, 3)$ and $f(x_2^*) = (5, 1)$. The utopia point is then

$$f^* = \begin{bmatrix} 2 \\ 1 \end{bmatrix}.$$

For the matrix P, recall that the ith column of P is $f(x_i^*) - f^*$:

$$P = \begin{bmatrix} 0 & 3 \\ 2 & 0 \end{bmatrix}.$$

Our quasi-normal vector is given by $-Pe$ (note that the true normal is $[-2, -3]$):

$$\tilde{n} = \begin{bmatrix} -3 \\ -2 \end{bmatrix}.$$

We now have all the parameters we need to solve Eq. 9.12.

9.3.4 Evolutionary Algorithms

Gradient-free methods can, and occasionally do, use all of the previously described methods. However, evolutionary algorithms also enable a fundamentally different approach. Genetic algorithms (GAs), a specific type of evolutionary algorithm, were introduced in Section 7.6.*

A GA is amenable to an extension that can handle multiple objectives because it keeps track of a large population of designs at each iteration. If we plot two objective functions for a given population of a GA iteration, we get something like that shown in Fig. 9.9. The points represent the current population, and the highlighted points in the lower left are the current nondominated set. As the optimization progresses, the nondominated set moves further down and to the left and eventually converges toward the actual Pareto front.

*The first application of an evolutionary algorithm for solving a multiobjective problem was by Schaffer.[164]

164. Schaffer, *Some experiments in machine learning using vector evaluated genetic algorithms.* 1984.

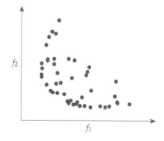

Fig. 9.9 Population for a multiobjective GA iteration plotted against two objectives. The nondominated set is highlighted at the bottom left and eventually converges toward the Pareto front.

In the multiobjective version of the GA, the reproduction and mutation phases are unchanged from the single-objective version. The primary difference is in determining the fitness and the selection procedure. Here, we provide an overview of one popular approach, the elitist nondominated sorting genetic algorithm (NSGA-II).[†]

[†]The NSGA-II algorithm was developed by Deb et al.[148] Some key developments include using the concept of domination in the selection process, preserving diversity among the nondominated set, and using elitism.[165]

148. Deb et al., *A fast and elitist multiobjective genetic algorithm: NSGA-II*, 2002.

165. Deb, *Introduction to evolutionary multiobjective optimization*, 2008.

166. Kung et al., *On finding the maxima of a set of vectors*, 1975.

A step in the algorithm is to find a nondominated set (i.e., the current approximation of the Pareto set), and several algorithms exist to accomplish this. In the following, we use the algorithm by Kung et al.,[166] which is one of the fastest. This procedure is described in Alg. 9.1, where "front" is a shorthand for the nondominated set (which is just the current approximation of the Pareto front). The algorithm recursively divides the population in half and finds the nondominated set for each half separately.

Algorithm 9.1 Find the nondominated set using Kung's algorithm

Inputs:
  p: A population sorted by the first objective
Outputs:
  f: The nondominated set for the population

```
procedure front(p)
    if length(p) = 1 then                    If there is only one point, it is the front
        return f
    end if
    Split population into two halves p_t and p_b
    ▷ Because input was sorted, p_t will be superior to p_b in the first objective
    t = front(p_t)                           Recursive call to find front for top half
    b = front(p_b)                           Recursive call to find front for bottom half
    Initialize f with the members from t                     merged population
    for i = 1 to length(b) do
        dominated = false                    Track whether anything in t dominates b_i
        for j = 1 to length(t) do
            if t_j dominates b_i then
                dominated = true
                break                        No need to continue search through t
            end if
        end for
        if not dominated then                b_i was not dominated by anything in t
            Add b_i to f
        end if
    end for
    return f
end procedure
```

9.3 Solution Methods

Before calling the algorithm, the population should be sorted by the first objective. First, we split the population into two halves, where the top half is superior to the bottom half in the first objective. Both populations are recursively fed back through the algorithm to find their nondominated sets. We then initialize a merged population with the members of the top half. All members in the bottom half are checked, and any that are nondominated by any member of the top half are added to the merged population. Finally, we return the merged population as the nondominated set.

With NSGA-II, in addition to determining the nondominated set, we want to rank all members by their *dominance depth*, which is also called *nondominated sorting*. In this approach, all nondominated points in the population (i.e., the current approximation of the Pareto set) are given a rank of 1. Those points are then removed from the set, and the next set of nondominated points is given a rank of 2, and so on. Figure 9.10 shows a sample population and illustrates the positions of the points with various rank values. There are alternative procedures that perform nondominated sorting directly, but we do not detail them here. This algorithm is summarized in Alg. 9.2.

The new population in the GA is filled by placing all rank 1 points in the new population, then all rank 2 points, and so on. At some point, an entire group of constant rank will not fit within the new population. Points with the same rank are all equivalent as far as Pareto optimality is concerned, so an additional sorting mechanism is needed to determine which members of this group to include.

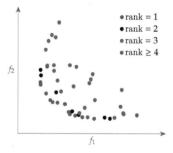

Fig. 9.10 Points in the population highlighted by rank.

Algorithm 9.2 Perform nondominated sorting

Inputs:
   p: A population

Outputs:
   rank: The rank for each member in the population

$r = 1$ Initialize current rank
$s = p$ Set subpopulation as entire population
while length(s) > 0 **do**
    $f = \mathrm{front}(\mathrm{sort}(s))$ Identify the current front
    Set rank for every member of f to r
    Remove all members of f from s
    $r = r + 1$ Increment rank
end while

We perform selection within a group that can only partially fit to preserve diversity. Points in this last group are ordered by their *crowding distance*, which is a measure of how spread apart the points are. The algorithm seeks to preserve points that are well spread. For each point, a hypercube in objective space is formed around it, which, in NSGA-II, is referred to as a *cuboid*. Figure 9.11 shows an example cuboid considering the rank 3 points from Fig. 9.10. The hypercube extends to the function values of its nearest neighbors in the function space. That does not mean that it necessarily touches its neighbors because the two closest neighbors can differ for each objective. The sum of the dimensions of this hypercube is the crowding distance.

When summing the dimensions, each dimension is normalized by the maximum range of that objective value. For example, considering only f_1 for the moment, if the objectives were in ascending order, then the contribution of point i to the crowding distance would be

$$d_{1,i} = \frac{f_{1_{i+1}} - f_{1_{i-1}}}{f_{1_{n_p}} - f_{1_1}}. \tag{9.13}$$

where n_p is the size of the population. Sometimes, instead of using the first and last points in the current objective set, user-supplied values are used for the min and max values of f that appear in that denominator. The anchor points (the single-objective optima) are assigned a crowding distance of infinity because we want to preference their inclusion. The algorithm for crowding distance is shown in Alg. 9.3.

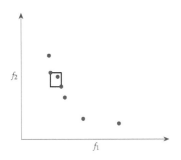

Fig. 9.11 A cuboid around one point, demonstrating the definition of crowding distance (except that the distances are normalized).

Algorithm 9.3 Crowding distance

Inputs:
 p: A population
Outputs:
 d: Crowding distances

Initialize d with zeros
for $i = 1$ **to** number of objectives **do**
  Set f as a vector containing the ith objective for each member in p
  $s = \text{sort}(f)$ and let I contain the corresponding indices ($s = f_I$)
  $d_{I_1} = \infty$ Anchor points receive an infinite crowding distance
  $d_{I_n} = \infty$
  for $j = 2$ **to** $n_p - 1$ **do**   Add distance for interior points
    $d_{I_j} = d_{I_j} + (s_{j+1} - s_{j-1})/(s_{n_p} - s_1)$
  end for
end for

9.3 Solution Methods

We can now put together the overall multiobjective GA, as shown in Alg. 9.4, where we use the components previously described (nondominated set, nondominated sorting, and crowding distance).

Algorithm 9.4 Elitist nondominated sorting genetic algorithm

Inputs:
 $\overline{x}$: Variable upper bounds
 $\underline{x}$: Variable lower bounds

Outputs:
 x^*: Best point

Generate initial population
while Stopping criterion is not satisfied **do**
 Using a parent population P, proceed as a standard GA for selection, crossover, and mutation, but use a crowded tournament selection to produced an offspring population O
 $C = P \cup O$ ◁ Combine populations
 Compute rank_i for $i = 1, 2, \ldots$ of C using Alg. 9.2
 ▷ Fill new parent population with as many whole ranks as possible
 $P = \emptyset$
 $r = 1$
 while true **do**
  set F as all C_i with $\text{rank}_i = r$
  if $\text{length}(P) + \text{length}(F) > n_p$ **then**
   break
  end if
  add F to P
  $r = r + 1$
 end while
 ▷ For last rank that does not fit, add by crowding distance
 if $\text{length}(P) < n_p$ **then** ◁ Population is not full
  $d = \text{crowding}(F)$ ◁ Alg. 9.3, using last F from terminated previous loop
  $m = n_p - \text{length}(P)$ ◁ Determine how many members to add
  Sort F by the crowding distance d in descending order
  Add the first m entries from F to P
 end if
end while

The crossover and mutation operations remain the same. Tournament selection (Fig. 7.25) is modified slightly to use this algorithm's ranking and crowding metrics. In the tournament, a member with a lower rank is superior. If two members have the same rank, then the one with the larger crowding distance is selected. This procedure is called *crowded tournament selection*.

After reproduction and mutation, instead of replacing the parent generation with the offspring generation, both the parent generation and the offspring generation are saved as candidates for the next generation. This strategy is called *elitism*, which means that the best member in the population is guaranteed to survive.

The population size is now twice its original size ($2n_p$), and the selection process must reduce the population back down to size n_p. This is done using the procedure explained previously. The new population is filled by including all rank 1 members, rank 2 members, and so on until an entire rank can no longer fit. Inclusion for members of that last rank is done in the order of the largest crowding distance until the population is filled. Many variations are possible, so although the algorithm is based on the concepts of NSGA-II, the details may differ somewhat.

The main advantage of this multiobjective approach is that if an evolutionary algorithm is appropriate for solving a given single-objective problem, then the extra information needed for a multiobjective problem is already there, and solving the multiobjective problem does not incur much additional computational cost. The pros and cons of this approach compared to the previous approaches are the same as those of gradient-based versus gradient-free methods, except that the multiobjective gradient-based approaches require solving multiple problems to generate the Pareto front. Still, solving multiple gradient-based problems may be more efficient than solving one gradient-free problem, especially for problems with a large number of design variables.

Example 9.4 Filling a new population in NSGA-II

After reproduction and mutation, we are left with a combined population of parents and offspring. In this small example, the combined population is of size 12, so we must reduce it back to 6. This example has two objectives, and the values for each member in the population are shown in the following table, where we assign a letter to each member. The population is plotted in Fig. 9.12.

	A	B	C	D	E	F	G	H	I	J	K	L
f_1	5	7	10	1	3	10	5	6	9	6	9	4
f_2	8	9	4	4	7	6	10	3	5	1	2	10

First, we compute the ranks using Alg. 9.2, resulting in the following output:

	A	B	C	D	E	F	G	H	I	J	K	L
	3	4	3	1	2	4	4	2	3	1	2	3

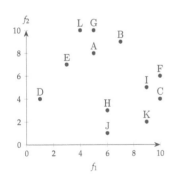

Fig. 9.12 Population for Ex. 9.4.

We see that the current nondominated set consists of points D and J and that there are four different ranks.

Next, we start filling the new population in the order of rank. Our maximum capacity is 6, so all rank 1 {D, J} and rank 2 {E, H, K} fit. We cannot add rank 3 {A, C, I, L} because the population size would be 9. So far, our new population consists of {D, J, E, H, K}. To choose which items from rank 3 continue forward, we compute the crowding distance for the members of rank 3:

A	C	I	L
1.67	∞	1.5	∞

We would then add, in order {C, L, A, I}, but we only have room for one, so we add C and complete this iteration with a new population of {D, J, E, H, K, C}.

9.4 Summary

Multiobjective optimization is particularly useful in quantifying trade-off sensitivities between critical metrics. It is also useful when we seek a family of potential solutions rather than a single solution. Some scenarios where a family of solutions might be preferred include when the models used in optimization are low fidelity and higher-fidelity design tools will be applied or when more investigation is needed. A multiobjective approach can produce candidate solutions for later refinement.

The presence of multiple objectives changes what it means for a design to be optimal. A design is Pareto optimal when it is nondominated by any other design. The weighted-sum method is perhaps the most well-known approach, but it is not recommended because other methods are just as easy and much more efficient. The epsilon-constraint method is still simple yet almost always preferable to the weighted-sum method. It typically provides a better spaced Pareto front and can resolve any nonconvex portions of the front. If we are willing to use a more complex approach, the normal boundary intersection method is even more efficient at capturing a well-spaced Pareto front.

Some gradient-free methods, such as a multiobjective GA, can also generate Pareto fronts. If a gradient-free approach is a good fit in the single objective version of the problem, adding multiple objectives can be done with little extra cost. Although gradient-free methods are sometimes associated with multiobjective problems, gradient-based algorithms may be more effective for many multiobjective problems.

Problems

9.1 Answer *true* or *false* and justify your answer.

a. The solution of multiobjective optimization problems is usually an infinite number of points.

b. It is advisable to include as many objectives as you can in your problem formulation to make sure you get the best possible design.

c. Multiobjective optimization can quantify trade-offs between objectives and constraints.

d. If the objectives are separable, that means that they can be minimized independently and that there is no Pareto front.

e. A point A dominates point B if it is better than B in at least one objective.

f. The Pareto set is the set of points that dominate all other points in the objective space.

g. When a point is Pareto optimal, you cannot make either of the objectives better.

h. The weighted-sum method obtains the Pareto front by solving optimization problems with different objective functions.

i. The epsilon-constraint method obtains the Pareto front by constraining one objective and minimizing all the others.

j. The utopia point is the point where every objective has a minimum value.

k. It is not possible to compute a Pareto front with a single-objective optimizer.

l. Because GAs optimize by evolving a population of diverse designs, they can be used for multiobjective optimization without modification.

9.2 Which of the following function value pairs would be Pareto optimal in a multiobjective minimization problem (there may be more than one)?

- (20, 4)
- (18, 5)
- (34, 2)
- (19, 6)

9.3 You seek to minimize the following two objectives:
$$f_1(x) = x_1^2 + x_2^2$$
$$f_2(x) = (x_1 - 1)^2 + 20(x_2 - 2)^2.$$

Identify the Pareto front using the weighted-sum method with 11 evenly spaced weights: $0, 0.1, 0.2, \ldots, 1$. If some parts of the front are underresolved, discuss how you might select weights for additional points.

9.4 Repeat Prob. 9.3 with the epsilon-constraint method. Constrain f_1 with 11 evenly spaced points between the anchor points. Contrast the Pareto front with that of the previous problem, and discuss whether improving the front with additional points will be easier with the previous method or with this method.

9.5 Repeat Prob. 9.3 with the normal boundary intersection method using the following 11 evenly spaced points:
$$b = [0, 1], [0.1, 0.9], [0.2, 0.8], \ldots, [1, 0].$$

9.6 Consider a two-objective population with the following combined parent/offspring population (objective values shown for all 16 members):

f_1	f_2
6.0	8.0
6.0	4.0
5.0	6.0
2.0	8.0
10.0	5.0
6.0	0.5
8.0	3.0
4.0	9.0
9.0	7.0
8.0	6.0
3.0	1.0
7.0	9.0
1.0	2.0
3.0	7.0
1.5	1.5
4.0	6.5

Develop code based on the NSGA-II procedure and determine the new population at the end of this iteration. Detail the results of each step during the process.

Surrogate-Based Optimization   10

A surrogate model, also known as a *response surface model* or *metamodel*, is an approximate model of a functional output that represents a "curve fit" to some underlying data. The goal of a surrogate model is to build a model that is much faster to compute than the original function, but that still retains sufficient accuracy away from known data points.

Surrogate-based optimization (SBO) performs optimization using the surrogate model, as shown in Fig. 10.1. When used in optimization, the surrogate might define the full optimization model (i.e., the inputs are design variables, and the outputs are objective and constraint functions), or the surrogate could be just a component of the overall model. SBO is more targeted than the broader field of surrogate modeling. Instead of aiming for a globally accurate surrogate, SBO just needs the surrogate model to be accurate enough to lead the optimizer to the true optimum.

In SBO, the surrogate model is usually improved during optimization as needed but can sometimes be constructed beforehand and remain fixed during optimization. Some optimization algorithms interrogate both the surrogate model and the original model, an approach that is sometimes called *surrogate-assisted optimization*.

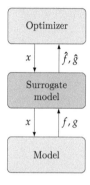

Fig. 10.1 Surrogate-based optimization replaces the original model with a surrogate model in the optimization process.

> By the end of this chapter you should be able to:
>
> 1. Identify and describe the steps in surrogate-based optimization.
> 2. Understand and use sampling methods.
> 3. Optimize parameters for a given surrogate model.
> 4. Perform cross-validation as part of model selection.
> 5. Describe different surrogate-based optimization approaches and the infill process.

10.1 When to Use a Surrogate Model

There are various scenarios for which surrogate models are helpful. One scenario is when the original model is computationally expensive. Surrogate models can be queried with minimal computational cost, but constructing them requires multiple evaluations of the original model. Suppose the number of evaluations needed to build a sufficiently accurate surrogate model is less than that needed to optimize the original model directly. In that case, SBO may be a worthwhile option. Constructing a surrogate model becomes even more compelling when it is reused in multiple optimizations.

Surrogate modeling can be effective in handling noisy models because they create a smooth representation of noisy data. This can be particularly advantageous when using gradient-based optimization.

One scenario that leads to both expensive evaluation and noisy output is experimental data. When the model data are experimental and the optimizer cannot query the experiment in an automated way, we can construct a surrogate model based on the experimental data. Then, the optimizer can query the surrogate model in the optimization.

Surrogate models are also helpful when we want to understand the design space, that is, how the objective and constraints (outputs) vary with respect to the design variables (inputs). By constructing a continuous model over discrete data, we obtain functional relationships that can be visualized more effectively.

When multiple sources of data are available, surrogate models can fuse the data to build a single model. The data could come from numerical models with different levels of fidelity or experimental data. For example, surrogate models can calibrate numerical model data using experimental data. This is helpful because experimental data is usually much more scarce than numerical data. The same reasoning applies to low- versus high-fidelity numerical data.

One potential issue with surrogate models is the *curse of dimensionality*, which refers to poor scalability with the number of inputs. The larger the number of inputs, the more model evaluations are needed to construct a surrogate model that is accurate enough. Therefore, the reasons for using surrogate models cited earlier might not be enough if the optimization problem has a large number of design variables.

The SBO process is shown in Fig. 10.2. First, we use sampling methods to choose the initial points to evaluate the function or conduct experiments. These points are sometimes referred to as *training data*. Next, we build a surrogate model from the sampled points. We can then perform optimization by querying the surrogate model. Based on the results of the optimization, we include additional points in the sample and reconstruct the surrogate (*infill*). We repeat this process until some convergence criterion or a maximum number of iterations is reached. In some procedures, infill is omitted; the surrogate is entirely constructed upfront and not subsequently updated.

The optimization step can be performed using any of the methods we covered previously. Because surrogate models are smooth and their gradients are easily computed, gradient-based optimization is preferred (see Chapter 4). However, some surrogate models can be highly multimodal, in which case a global search is preferred, either using gradient-based with multistart (see Tip 4.8) or a global gradient-free method (see Chapter 7).

This chapter discusses sampling, constructing a surrogate, and infill with some associated optimization strategies. We devote separate sections to two surrogate modeling methods that are more involved and widely used: kriging and deep neural nets. Many of the concepts discussed in this chapter have a wide range of applications beyond optimization.

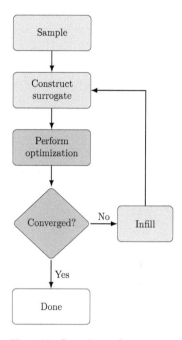

Fig. 10.2 Overview of surrogate-based optimization procedure.

Tip 10.1 Surrogate models can be useful *within* your model

In the context of SBO, we usually replace the function evaluation with a surrogate model, as shown in Fig. 10.1. However, it might not be advantageous to replace the whole model, but replace only part of that model instead. If a component of the model is evaluated frequently and does not have too many inputs, this approach might be worthwhile.

For example, when performing trajectory optimization of an aircraft, we need to evaluate the lift and drag of the aircraft at each point of the trajectory. This typically requires many points, and computing the lift and drag at each point might be prohibitive. Therefore, it might be worthwhile to use surrogate models that predict the lift and drag as functions of the angle of attack. If the optimization design variables include parameters that change the lift and drag characteristics, such as the wing shape, then the surrogate model needs to be rebuilt at every optimization iteration.

10.2 Sampling

Sampling methods, also known as *sampling plans*, select the evaluation points to construct the initial surrogate. These evaluation points must be chosen carefully. A straightforward approach is *full factorial* sampling, where we discretize each dimension and evaluate at all combinations of the resulting grid. This is not efficient because it scales exponentially with the number of input variables.

Example 10.1 Full factorial sampling is not scalable

Imagine a numerical model that computes the endurance of an aircraft. Suppose we only wanted to understand how endurance varied with one variable, such as wingspan. In that case, we could evaluate the model multiple times and fit a curve that could predict endurance at wingspans that we did not directly evaluate. If the model evaluation is computationally expensive, so that evaluating many points is prohibitive, we might use a relatively coarse sampling (say, 12 points). As long as the endurance changes are gradual across the domain, fitting a spline through these few points can generate a useful predictive model.

Now imagine that we have nine additional input variables that we care about: wing area, taper ratio, wing root twist, wingtip twist, wing dihedral, propeller spanwise position, battery size, tail area, and tail longitudinal position. If we discretized all 10 variables with the same coarse 12 intervals each, a full factorial sample would require 12^{10} model evaluations (approximately 62 billion) to assess all combinations. Thus, this type of sampling plan is not scalable.

Example 10.1 highlights one of the significant challenges of sampling methods: the curse of dimensionality. For SBO, even with better sampling plans, using a large number of variables is costly. We need to identify the most important or most influential variables. Knowledge of the particular domain is helpful, as is exploring the magnitude of the entries in a gradient vector (Chapter 6) across multiple points in the domain. We can use various strategies to help us decide which variables matter most, but for our purposes, we assume that the most influential variables have already been determined so that the dimensionality is reasonable. Having selected a set of variables, we are now interested in sampling methods that characterize the design space of interest more efficiently than full factorial sampling.

In addition to their use in SBO, the sampling strategies discussed in this section are useful in many other applications, including various applications discussed in this book: initializing a genetic algorithm (Section 7.6), particle swarm optimization (Section 7.7) or a multistart gradient-based algorithm (Tip 4.8), or choosing the points to run in a Monte Carlo simulation (Section 12.3.3). Because the function behavior at each sample is independent, we can efficiently parallelize the evaluation of the functions.

10.2.1 Latin Hypercube Sampling

Latin hypercube sampling (LHS) is a popular sampling method that is built on a random process but is more effective and efficient than pure random sampling. Random sampling scales better than full factorial searches, but it tends to exhibit clustering and requires many points to reach the desired distribution (i.e., the law of large numbers). For example, Fig. 10.3 compares 50 randomly generated points across uniform distributions in two dimensions versus Latin hypercube sampling. In random sampling, each sample is independent of past samples, but in LHS, we choose all samples beforehand to ensure a well-spread distribution.

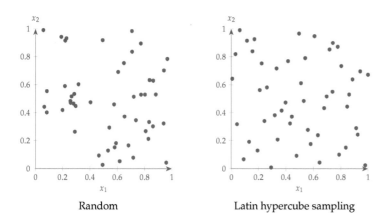

Fig. 10.3 Contrast between random and Latin hypercube sampling with 50 points using uniform distributions.

To describe the methodology, consider two random variables with bounds, whose design space we can represent as a square. Say we wanted only eight samples; we could divide the design space into eight intervals in each dimension, generating the grid of cells shown in Fig. 10.4.

A full factorial search would identify a point in each cell, but that does not scale well. To be as efficient as possible and still cover the variation, we would want each row and each column to have one sample in it. In other words, the projection of points onto each dimension should be uniform. For example, the left side of Fig. 10.5 shows the projection of a uniform LHS onto each dimension. We see that the points create a uniformly spread histogram.

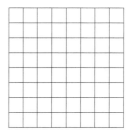

Fig. 10.4 A two-dimensional design space divided into eight intervals in each dimension.

The concept where one and only one point exists in any given row or column is called a *Latin square*, and the generalization to higher dimensions is called a *Latin hypercube*. There are many ways to achieve this, and some choices are better than others. Consider the sampling

Fig. 10.5 Example LHS with projections onto the axes.

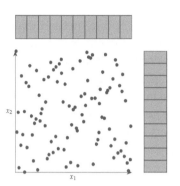
Uniform distribution in each direction

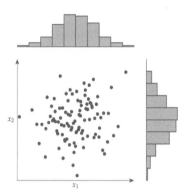
Normal distribution in each direction

plan shown on the left of Fig. 10.6. This plan meets our criteria but clearly does not fill the space and likely will not capture the relationships between design parameters well. Alternatively, the right side of Fig. 10.6 has a sample in each row and column while also spanning the space much more effectively.

Fig. 10.6 Contrasting sampling strategies that both fulfill the uniform projection requirement.

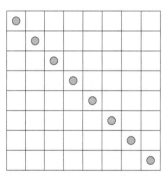
A sampling strategy whose projection uniformly spans each dimension but does not fill the space well

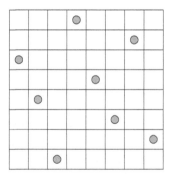
A sampling strategy whose projection uniformly spans each dimension and fills the space more effectively

LHS can be posed as an optimization problem where we seek to maximize the distance between the samples. The constraint is that the projection on each axis must follow a chosen probability distribution. The specified distribution is often uniform, as in the previous examples, but it could also be any distribution, such as a normal distribution, as shown on the right side of Fig. 10.5. This optimization problem does not have a unique solution, so random processes are used to determine the combination of points. Additionally, points are not usually placed in cell centers but at a random location within a given cell to allow for the possibility of reaching any point in the domain. The advantage of the

10.2 Sampling

LHS approach is that rather than relying on the law of large numbers to fill out our chosen probability distributions, we enforce it as a constraint. This method may still require many samples to characterize the design space accurately, but it usually requires far fewer than pure random sampling.

Instead of defining LHS as an optimization problem, a much simpler approach is typically used in which we ensure one sample per interval, but we rely on randomness to choose point combinations. Although this does not necessarily yield a maximum spread, it works well in practice and is simple to implement. Before discussing the algorithm, we discuss how to generate other distributions besides just uniform distributions.

We can convert from uniformly sampled points to an arbitrary distribution using a technique called *inversion sampling*. Assume that we want to generate samples x from an arbitrary probability density function (PDF) $p(x)$ or, equivalently, from the corresponding cumulative distribution function (CDF) $P(x)$.* The probability integral transform states that for any continuous CDF, $y = P(x)$, the variable y is uniformly distributed (a simple proof, but it is not shown here to avoid introducing additional notation). The procedure is to randomly sample from a uniform distribution (e.g., generate y), then compute the corresponding x such that $P(x) = y$, which we denote as $x = P^{-1}y$. This latter step is known as an *inverse CDF*, a *percent-point function*, or a *quantile function*. This process is depicted in Fig. 10.7 for a normal distribution. This same procedure allows us to use LHS with any distribution, simply by generating the samples on a uniform distribution.

*PDFs and CDFs are reviewed in Appendix A.9.

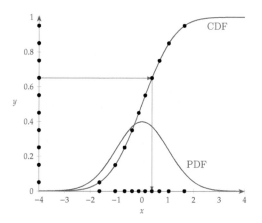

Fig. 10.7 An example of inversion sampling with a normal distribution. A few uniform samples are shown on the y-axis. The points are evaluated by the inverse CDF, represented by the arrows passing through the CDF for a normal distribution. If enough samples are drawn, the resulting distribution will be the PDF of a normal distribution.

A typical algorithm is described in Alg. 10.1. For each axis, we partition the CDF in n_s evenly spaced regions (evenly spaced along the CDF, which means that each region is equiprobable). We generate a random number within each evenly spaced interval, where 0 corresponds to the bottom of the interval and 1 to the top. We then evaluate the inverse CDF as described previously so that the points match our specified distribution (the CDF for a uniform distribution is just a line $P(x) = x$, so the output is not changed). Next, the column of points for that axis is randomly permuted. This process is repeated for each axis according to its specified probability distribution.

Algorithm 10.1 Latin hypercube sampling

Inputs:
 n_s: Number of samples
 n_d: Number of dimensions
 $P = \{P_1, \ldots, P_{n_d}\}$: (optionally) A set of cumulative distribution functions

Outputs:
 $X = \{x_1, \ldots, x_{n_s}\}$: Set of sample points

---

for $j = 1$ to n_d do
  for $i = 1$ to n_s do
   $V_{ij} = \dfrac{i}{n_s} - \dfrac{R_{ij}}{n_s}$ where $R_{ij} \in \mathbb{U}[0,1]$ Randomly choose a value in each equally spaced cell from uniform distribution
  end for
  $X_{*j} = P_j^{-1}(V_{*j})$ where P_j is a CDF   Evaluate inverse CDF
  Randomly permute the entries of this column X_{*j} Alternatively, permute the indices $1 \ldots n_s$ in the prior for loop
end for

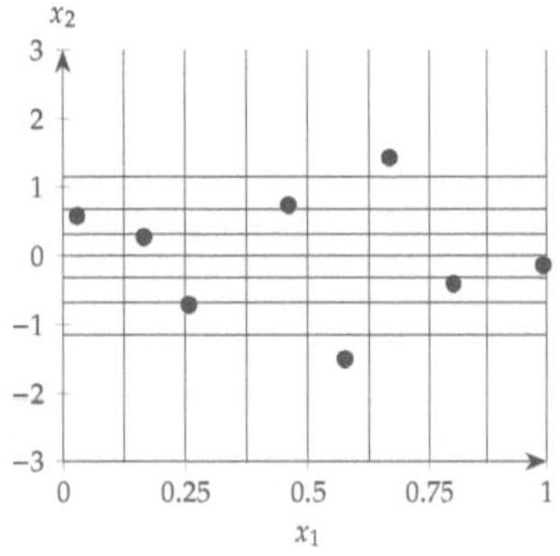

An example using Alg. 10.1 for eight points is shown in Fig. 10.8. In this example, we use a uniform distribution for x_1 and a normal distribution for x_2. There is one point in each equiprobable interval. As stated before, randomness does not necessarily ensure a good spread, but optimizing the spread is difficult because the function is highly multimodal. Instead, to encourage high spread, we could generate multiple Latin hypercube samples with this algorithm and select the one with the largest sum of the distance between points.

Fig. 10.8 An example from the LHS algorithm showing uniform distribution in x_1 and a Gaussian distribution in x_2 with eight sample points. The equiprobable bins are shown as grid lines.

10.2.2 Low-Discrepancy Sequences

Low-discrepancy sequences generate deterministic sequences of points that are well spread. Each new point added in the sequence maintains low discrepancy—*discrepancy* refers to the variation in point density throughout the domain. Hence, a low-discrepancy set is close to even

10.2 SAMPLING

density (i.e., well spread). These sequences are called *quasi-random* because they often serve as suitable replacements for applications that use random sequences, but they are not random or even pseudo-random.

An advantage of low-discrepancy sequences over LHS is that most of the approaches do not require selecting all the samples beforehand. These methods generate deterministic sequences; in other words, we generate the same sequence of points whether we choose them beforehand or add more later. This property is particularly advantageous in iterative procedures. We may choose an initial sampling plan and add more well-spread points to the sample later. This is not necessarily an advantage for the methods of this chapter because the optimization drives the selection of new points rather than continuing to seek spread-out samples. However, this feature is useful for other applications, such as quadrature, Monte Carlo simulations, and other problems where an iterative sampling process is used to determine statistical convergence (see Section 12.3). Low-discrepancy sequences add more points that are well spread without having to throw out the existing samples. Even in non-iterative procedures, these sampling strategies can be a useful alternative.

Several of these sequences are built on generalizations of the one-dimensional *van der Corput sequence* to more than one dimension. Such sequences are defined by representing an integer i in a given integer base b (the van der Corput sequence is always base 2):

$$i = a_0 + a_1 b + a_2 b^2 + \ldots + a_r b^r \text{ where } a \in [0, b-1]. \tag{10.1}$$

If the base is 2, this is just a standard binary sequence (Section 7.6.1). After determining the relevant coefficients (a_j), the ith element of the sequence is

$$\phi_b(i) = \frac{a_0}{b} + \frac{a_1}{b^2} + \frac{a_2}{b^3} + \ldots + \frac{a_r}{b^{r+1}}. \tag{10.2}$$

An algorithm to generate an element in this sequence, also known as a *radical inverse function* for base b, is given in Alg. 10.2.

For base 2, the sequence is as follows:

$$\frac{1}{2}, \frac{1}{4}, \frac{3}{4}, \frac{1}{8}, \frac{5}{8}, \frac{3}{8}, \frac{7}{8}, \frac{1}{16}, \frac{9}{16}, \ldots \tag{10.3}$$

The interval is divided in half, and then each subinterval is also halved, with new points spreading out across the domain (see Fig. 10.9).

Similarly, for base 3, the interval is split into thirds, then each subinterval is split into thirds, and so on:

$$\frac{1}{3}, \frac{2}{3}, \frac{1}{9}, \frac{4}{9}, \frac{7}{9}, \frac{2}{9}, \frac{5}{9}, \frac{8}{9}, \frac{1}{27}, \ldots \tag{10.4}$$

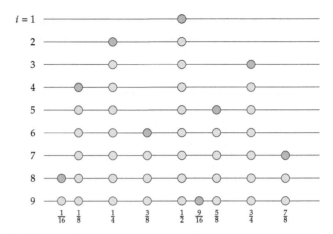

Fig. 10.9 Van Der Corput sequence.

Algorithm 10.2 | Radical inverse function

Inputs:
  i: ith point in sequence
  b: Base (integer)
Outputs:
  ϕ: Generated point

$b_d = b$ Base used in denominator
$\phi = 0$
while $i > 0$ **do**
  $a = \text{mod}(i, b)$ Coefficient
  $\phi = \phi + a/b_d$
  $b_d = b_d \cdot b$ Increase exponent in denominator
  $i = \text{Int}(i/b)$ Integer division
end while

[†] A set of numbers is pairwise prime if there is no positive integer that can evenly divide any pair of them, except 1. Typically, though, we just use the first n_x prime numbers.

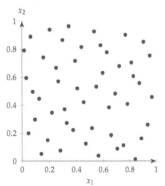

Fig. 10.10 Halton sequence with base 2 for x_1 and base 3 for x_2. First, 30 points are selected (in blue), and then 20 points are added (in red). These points would be identical to 50 points chosen at once.

Halton Sequence

A Halton sequence uses pairwise prime numbers (larger than 1) for the base of each dimension of the problem.[†] The ith point in the Halton sequence is

$$\phi(i, b_1), \phi(i, b_2), \ldots, \phi(i, b_{n_x}), \tag{10.5}$$

where the b_j set is pairwise prime. As an example in two dimensions, Fig. 10.10 shows 30 generated points of the Halton sequence where x_1 uses base 2, and x_2 uses base 3, and then a subsequent 20 generated points are added (in another color), showing the reuse of existing points.

If the dimensionality of the problem is high, then some of the base combinations lead to points that are highly correlated and thus

10.2 Sampling

undesirable for a sampling plan. For example, the left of Fig. 10.11 shows 50 generated points where x_1 uses base 17, and x_2 uses base 19. To avoid this issue, we can use a *scrambled* Halton sequence.

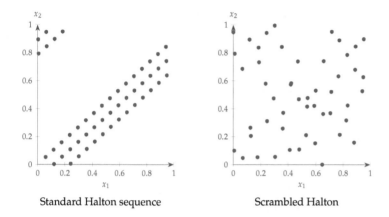

Standard Halton sequence   Scrambled Halton

Fig. 10.11 Halton sequence with base 17 for x_1 and base 19 for x_2.

Scrambling can be accomplished by generating a permutation array containing a random permutation of the integers $p = [0, 1, \ldots, b-1]$. Then, rather than using the integers a directly in Eq. 10.2, we use the entries of a as the indices of the permutation array. If p is the permutation array, we have:

$$\phi_b(i) = \frac{p_{a_0}}{b} + \frac{p_{a_1}}{b^2} + \frac{p_{a_2}}{b^3} + \ldots + \frac{p_{a_r}}{b^{r+1}}. \quad (10.6)$$

The permutation array is fixed for all digits a and for all n_p points in the domain. The right side of Fig. 10.11 shows the same example (with base 17 and base 19) but with scrambling to weaken the strong correlations.

Hammersley Sequence

The Hammersley sequence is closely related to the Halton sequence. However, it provides better spacing if we know beforehand the number of points (n_p) that we are going to use. This approach only needs $n_x - 1$ bases (still pairwise prime) because the first dimension uses regular spacing:

$$\frac{i}{n_p}, \phi(i, b_1), \phi(i, b_2), \ldots, \phi(i, b_{n_x-1}). \quad (10.7)$$

Because this sequence needs to know the number of points (n_p) beforehand, it is less useful for iterative procedures. However, the implementation is straightforward, and so it may still be a useful alternative to LHS. Figure 10.12 shows 50 points generated from a Hammersley sequence where the x_2-axis uses base 2.

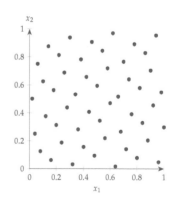

Fig. 10.12 Hammersley sequence with base 2 for the x_2-axis.

Other Sequences

A wide variety of other low-discrepancy sequences exist. The Faure sequence is similar to the Halton, but it uses the same base for all dimensions and uses permutation scrambling for each dimension instead.[167,168] Sobol sequences use base 2 sequences but with a reordering based on "direction numbers".[169] Niederreiter sequences are effectively a generalization of Sobol sequences to other bases.[170]

167. Faure, *Discrépance des suites associées à un systéme de numération (en dimension s)*. 1982.

168. Faure and Lemieux, *Generalized Halton sequences in 2008: A comparative study*, 2009.

169. Sobol, *On the distribution of points in a cube and the approximate evaluation of integrals*, 1967.

170. Niederreiter, *Low-discrepancy and low-dispersion sequences*, 1988.

10.3 Constructing a Surrogate

Once sampling is completed, we have a list of data points, often called *training data*:

$$\left(x^{(i)}, f^{(i)}\right), \tag{10.8}$$

where $x^{(i)}$ is an input vector from the sampling plan, and $f^{(i)}$ contains the corresponding outputs from evaluating the model: $f^{(i)} = f\left(x^{(i)}\right)$. We seek to construct a surrogate model from this data set. Surrogate models can be based on physics, mathematics, or a combination of the two. Incorporating known physics into a model is often desirable to improve model accuracy. However, functional relationships are unknown for many complex problems, and a data-driven mathematical model can be more effective.

Surrogate-based models can be based on interpolation or regression, as illustrated in Fig. 10.13. *Interpolation* builds a function that exactly matches the provided training data. *Regression* models do not try to match the training data points exactly; instead, they minimize the error between a smooth trend function and the training data. The nature of the training data can help decide between these two types of surrogate models. Regression is particularly useful when the data are noisy. Interpolatory models may produce undesirable oscillations when fitting the noise. In contrast, regression models can find a smooth function that is less sensitive to the noise. Interpolation is useful when the data are highly multimodal (and not noisy). This is because a regression model may smooth over variations that are actually physical, whereas an interpolatory model can accurately capture those variations.

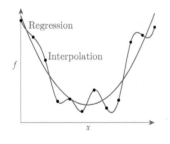

Fig. 10.13 Interpolation models match the training data at the provided points, whereas regression models minimize the error between the training data and a function with an assumed trend.

There are two main steps involved in either type of surrogate model. First, we select a set of basis functions, which represent the form for the model. Second, we determine the model parameters that provide the best fit to the provided data. Determining the model parameters is an optimization problem, which we discuss first. We discuss linear regression and nonlinear regression, which are techniques for choosing model parameters for a given set of basis functions. Next, we discuss cross validation, which is a critical technique for selecting an appropriate model form. Finally, we discuss common basis functions.

10.3.1 Linear Least Squares Regression

A linear regression model does not mean that the surrogate is linear in the input variables but rather that the model is linear in its coefficients (i.e., linear in the parameters we are estimating). For example, the following equation is a two-dimensional linear regression model, where we use $\hat{f}$ to represent our estimated model of the function f:

$$\hat{f}(x) = w_1 x_1^2 + w_2 x_1 x_2 + w_3 \exp(x_2) + w_4 x_1 + w_5. \tag{10.9}$$

This function is highly nonlinear, but it is classified as a linear regression model because the regression seeks to choose the appropriate values for the coefficients w_i (and the function is linear in w).

A general linear regression model can be expressed as

$$\hat{f} = w^\mathsf{T} \psi(x) = \sum_i w_i \psi_i(x), \tag{10.10}$$

where w is a vector of weights, and ψ is a vector of basis functions. In this section, we assume that the basis functions are provided. In general, the basis functions can be any set of functions that we choose (and typically they are nonlinear). It is usually desirable for these functions to be orthogonal.

Example 10.2 Data fitting can be posed as a linear regression model

Consider a quadratic fit: $\hat{f}(x) = ax^2 + bx + c$. This can be posed as a linear regression model (Eq. 10.10) where the coefficients we wish to estimate are $w = [a, b, c]$ and the basis functions are $\psi = [x^2, x, 1]$. For a more general n-dimensional polynomial model, the basis functions would be polynomials with terms combining the dependencies on all input variables x up to a certain order. For example, for two input variables up to second order, the basis functions would be $\psi = [1, x_1, x_2, x_1 x_2, x_1^2, x_2^2]$, and w would consist of seven coefficients.

The coefficients are chosen to minimize the error between our predicted function values $\hat{f}$ and the actual function values $f^{(i)}$. Because we want to minimize both positive and negative errors, we minimize the sum of the square of the errors (or a weighted sum of squared errors):*

*The choice of minimizing the sum of the squares rather than the sum of the absolute values or some other metric is not arbitrary. The motivation for using the sum of the squares is discussed further in the following section.

$$\underset{w}{\text{minimize}} \quad \sum_i \left(\hat{f}\left(w; x^{(i)}\right) - f^{(i)}\right)^2. \tag{10.11}$$

The solution to this optimization problem is called a *least squares solution*. If the regression model is linear, we can simplify this objective and solve the problem analytically. Recall that $\hat{f} = \psi^\mathsf{T} w$, so the objective can be written as

$$\underset{w}{\text{minimize}} \quad \sum_i \left(\psi\left(x^{(i)}\right)^\mathsf{T} w - f^{(i)}\right)^2. \tag{10.12}$$

We can express this in matrix form by defining the following:

$$\Psi = \begin{bmatrix} - & \psi\left(x^{(1)}\right)^\mathsf{T} & - \\ - & \psi\left(x^{(2)}\right)^\mathsf{T} & - \\ & \vdots & \\ - & \psi\left(x^{(n_s)}\right)^\mathsf{T} & - \end{bmatrix}. \tag{10.13}$$

The matrix Ψ is of size $(n_s \times n_w)$, where n_s is the number of samples, n_w the number of parameters in w, and $n_s \geq n_w$. This means that there should be more equations than unknowns or that we have sampled more points than the number of coefficients we need to estimate. This should make sense because our surrogate function is only an assumed form and generally not an exact fit to the actual underlying function. Thus, we need more data to create a good fit.

Then the optimization problem can be written in matrix form as:

$$\underset{w}{\text{minimize}} \quad \|\Psi w - f\|_2^2. \tag{10.14}$$

Expanding the squared norm (i.e., $\|x\|_2^2 = x^\mathsf{T} x$) gives

$$\underset{w}{\text{minimize}} \quad w^\mathsf{T} \Psi^\mathsf{T} \Psi w - 2 f^\mathsf{T} \Psi w + f^\mathsf{T} f. \tag{10.15}$$

We can omit the last term from the objective because our optimization variables are w, and the last term has no w dependence:

$$\underset{w}{\text{minimize}} \quad w^\mathsf{T} \Psi^\mathsf{T} \Psi w - 2 f^\mathsf{T} \Psi w. \tag{10.16}$$

This fits the general form for an unconstrained quadratic programming (QP) problem, as shown in Section 5.5.1:

$$\underset{x}{\text{minimize}} \quad \frac{1}{2} x^\mathsf{T} Q x + q^\mathsf{T} x, \tag{10.17}$$

where

$$Q = 2 \Psi^\mathsf{T} \Psi \tag{10.18}$$

$$q = -2 \Psi^\mathsf{T} f. \tag{10.19}$$

10.3 Constructing a Surrogate

Recall that an equality constrained QP (of which unconstrained is a subset) has an analytic solution as long as the QP is positive definite. In our case, we can show that Q is positive definite as long as Ψ is full rank:

$$x^T Q x = 2 x^T \Psi^T \Psi x = 2 \|\Psi x\|_2^2 > 0. \tag{10.20}$$

This is not surprising because the objective is a sum of squared values. Referring back to the solution in Section 5.5.1, and removing the portions associated with the constraints, the solution is

$$Q x = -q. \tag{10.21}$$

In our case, this becomes

$$2 \Psi^T \Psi w = 2 \Psi^T f. \tag{10.22}$$

After simplifying, we have an analytic solution for the weights:

$$w = (\Psi^T \Psi)^{-1} \Psi^T f. \tag{10.23}$$

We sometimes express the linear relationship in Eq. 10.12 as $\Psi w = f$, although the case where there are more equations than unknowns does not typically have a solution (the problem is *overdetermined*). Instead, we seek the solution that minimizes the error $\|\Psi w - f\|^2$, that is, Eq. 10.23. The quantity $\Psi^\dagger = (\Psi^T \Psi)^{-1} \Psi^T$ is called the *pseudoinverse* of Ψ (or more specifically, the Moore–Penrose pseudoinverse), and thus we can write Eq. 10.23 in the more compact form

$$w = \Psi^\dagger f. \tag{10.24}$$

This allows for a similar form to solving a linear system of equations where an inverse would be used instead. In solving both a linear system and the linear least-squares equation (Eq. 10.23), we do not explicitly invert a matrix. For linear least squares, a QR factorization is commonly used for improved numerical conditioning as compared to solving Eq. 10.23 directly.

Tip 10.2 Least squares is not the same as a linear system solution

In MATLAB or Julia, the backslash operator is overloaded, so you can solve an overdetermined system of equations $Ax = b$ with `x = A\b`, but keep in mind that for an A of size $(m \times n)$, where $m > n$, this syntax performs a least-squares solution, not a linear system solution as it would for a full rank $(n \times n)$ system. The overloading of this operator is generally not used in other languages; for example, in Python, rather than using `numpy.linalg.solve`, you would use `numpy.linalg.lstsq`.

Example 10.3 Linear regression

Consider the quadratic fit discussed in Ex. 10.2. We are provided the data points, x and f, shown as circles in Fig. 10.14. From these data, we construct the matrix Ψ for our basis functions as follows:

$$\Psi = \begin{bmatrix} x^{(1)2} & x^{(1)} & 1 \\ x^{(2)2} & x^{(2)} & 1 \\ & \vdots & \\ x^{(n_s)2} & x^{(n_s)} & 1 \end{bmatrix}.$$

We can then solve for the coefficients w using the linear least squares solution (Eq. 10.23). Substituting the coefficients and respective basis functions into Eq. 10.10, we obtain the surrogate model,

$$\hat{f}(x) = w_1 x^2 + w_2 x + w_3,$$

which is also plotted in Fig. 10.14 as a solid line.

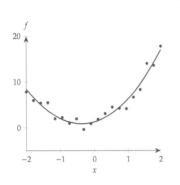

Fig. 10.14 Linear least squares example with a quadratic fit on a one-dimensional function.

A common variation of this approach is to use *regularized* least squares, which adds a term in the objective. The new objective is

$$\underset{w}{\text{minimize}} \quad \|\Psi w - f\|_2^2 + \mu \|w\|_2^2, \tag{10.25}$$

where μ is a weight assigned to the second term. This second term attempts to reduce the magnitudes of the entries in w while balancing the fit in the first term. This approach is particularly beneficial if the data contain strong outliers or are particularly noisy. The rationale for this approach is that we may want to accept a higher error (quantified by the first term) in exchange for smaller values for the coefficients. This generally leads to simpler, more generalizable models (e.g., by reducing the influence of some terms). A related extension uses a second term of the form $\|w - w_0\|_2^2$. The idea is that we want a good fit, while maintaining parameters that are close to some known nominal values w_0.

A regularized least squares problem can be solved with the same linear least squares approach. We can write the previous problem using concatenated matrices and vectors:

$$\underset{w}{\text{minimize}} \quad \left\| \begin{bmatrix} \Psi \\ \sqrt{\mu} I \end{bmatrix} w - \begin{bmatrix} f \\ 0 \end{bmatrix} \right\|_2^2. \tag{10.26}$$

This is of the same linear form as before ($\|Aw - b\|^2$), so we can reuse the solution (Eq. 10.23):

$$\begin{aligned} w^* &= (A^\mathsf{T} A)^{-1} A^\mathsf{T} b \\ &= (\Psi^\mathsf{T} \Psi + \mu I)^{-1} \Psi^\mathsf{T} f. \end{aligned} \tag{10.27}$$

10.3 Constructing a Surrogate

For linear least squares (with or without regularization), we have seen that the optimization problem of determining the appropriate coefficients can be found analytically. We can also add linear constraints to the problem (equality or inequality), and the optimization remains a QP. In that case, the problem is still convex. Although it does not generally have an analytic solution, we can still quickly find the global optimum. This topic is discussed in Section 11.3.

10.3.2 Maximum Likelihood Interpretation

This section presents an alternative motivation for the sum of squared error approach used in the previous section. It is somewhat of a diversion from the present discussion, but it will be helpful in several results later in this chapter. In the previous section, we assumed linear models of the form

$$\hat{f}^{(i)} = w^\mathsf{T} x^{(i)}, \tag{10.28}$$

where we use x for simplicity in writing instead of $\psi(x^{(i)})$. The derivation remains the same for any arbitrary function of x. The function $\hat{f}$ is just a model, so we could say that it is equal to our actual observations $f^{(i)}$ plus an error term:

$$f^{(i)} = w^\mathsf{T} x^{(i)} + \varepsilon^{(i)}, \tag{10.29}$$

where ε captures the error associated with the ith data point. We assume that the error is normally distributed with mean zero and a standard deviation of σ:

$$p\left(\varepsilon^{(i)}\right) = \frac{1}{\sigma\sqrt{2\pi}} \exp\left(-\frac{\varepsilon^{(i)2}}{2\sigma^2}\right). \tag{10.30}$$

The use of Gaussian uncertainty can be motivated by the central limit theorem, which states that for large sample sizes, the sum of random variables tends toward a Gaussian distribution regardless of the original distribution associated with each variable. In other words, the sample distribution of the sample mean is approximately Gaussian. Because we assume the error terms to be the sum of various random, independent perturbations, then by the central limit theorem, we expect the errors to be normally distributed.

We now substitute Eq. 10.29 into Eq. 10.30 to show the probability of f conditioned on x and parameterized by w:

$$p\left(f^{(i)}|x^{(i)}; w\right) = \frac{1}{\sigma\sqrt{2\pi}} \exp\left(-\frac{(f^{(i)} - w^\mathsf{T} x^{(i)})^2}{2\sigma^2}\right). \tag{10.31}$$

Once we include all the data points $x^{(i)}$, we would like to compute the probability of observing f conditioned on the inputs x for a given set of parameters in w. We call this the *likelihood function* $L(w)$. In this case, assuming all errors are independent, the total probability for observing the outputs is the product of the probability of observing each output:

$$L(w) = \prod_{i=1}^{n_s} \frac{1}{\sigma\sqrt{2\pi}} \exp\left(-\frac{\left(f^{(i)} - w^\mathsf{T} x^{(i)}\right)^2}{2\sigma^2}\right). \tag{10.32}$$

Now we can pose this as an optimization problem where we wish to find the parameters w that maximize the likelihood function; in other words, we maximize the probability that our model is consistent with the observed data. Because the objective is a product of multiple terms, it is helpful to take the logarithm of the objective. Maximizing L or maximizing $\ell = \ln(L)$ does not change the solution to the problem but makes it easier to solve. We call this the *log likelihood* function:

$$\ell(w) = \ln\left(\prod_{i=1}^{n_s} \frac{1}{\sigma\sqrt{2\pi}} \exp\left(-\frac{\left(f^{(i)} - w^\mathsf{T} x^{(i)}\right)^2}{2\sigma^2}\right)\right) \tag{10.33}$$

$$= \sum_{i=1}^{n_s} \ln\left(\frac{1}{\sigma\sqrt{2\pi}} \exp\left(-\frac{\left(f^{(i)} - w^\mathsf{T} x^{(i)}\right)^2}{2\sigma^2}\right)\right) \tag{10.34}$$

$$= n_s \ln\left(\frac{1}{\sigma\sqrt{2\pi}}\right) - \sum_{i=1}^{n_s} \frac{\left(f^{(i)} - w^\mathsf{T} x^{(i)}\right)^2}{2\sigma^2}. \tag{10.35}$$

The first term has no dependence on w, and so when optimizing $\ell(w)$; it is just a scalar term that can be removed as follows:

$$\underset{w}{\text{maximize}}\ \ell(w) \quad \Rightarrow \tag{10.36}$$

$$\underset{w}{\text{maximize}} \left(-\sum_{i=1}^{n_s} \frac{\left(f^{(i)} - w^\mathsf{T} x^{(i)}\right)^2}{2\sigma^2}\right) \quad \Rightarrow \tag{10.37}$$

$$\underset{w}{\text{minimize}} \sum_{i=1}^{n_s} \frac{(f^{(i)} - w^\mathsf{T} x^{(i)})^2}{2\sigma^2}. \tag{10.38}$$

Similarly, the denominator of the second term has no dependence on w and is just a scalar that can also be removed:

$$\underset{w}{\text{maximize}}\ \ell(w) \quad \Rightarrow \quad \underset{w}{\text{minimize}} \sum_{i=1}^{n_s} \left(f^{(i)} - w^\mathsf{T} x^{(i)}\right)^2. \tag{10.39}$$

Thus, maximizing the log likelihood function (maximizing the probability of observing the data) is equivalent to minimizing the sum of squared errors (the least-squares formulation). This derivation provides another motivation for using the sum of squared errors in regression.

10.3.3 Nonlinear Least Squares Regression

A surrogate model can be nonlinear in the coefficients. For example, building on the simple function shown earlier in Eq. 10.9 we can add the coefficients w_6 and w_7 as follows:

$$\hat{f}(x) = w_1 x_1^{w_6} + w_2 x_1 x_2 + w_3 \exp(w_7 x_2) + w_4 x_1 + w_5. \qquad (10.40)$$

The addition of w_6 and w_7 makes this function nonlinear in the coefficients. We can still estimate these parameters, but not analytically.

Equation 10.11 is still relevant; we still seek to minimize the sum of the squared errors:

$$\underset{w}{\text{minimize}} \sum_i \left(\hat{f}\left(w; x^{(i)}\right) - f^{(i)} \right)^2. \qquad (10.41)$$

For general nonlinear regression models, we cannot write a more specific form for $\hat{f}$ as we could for the linear case, so we leave the objective as it is.

This is a *nonlinear least-squares problem*. The optimization problem is unconstrained, so any of the methods from Chapter 4 apply. We could also easily add constraints, for example, bounds on parameters, known relationships between parameters, and so forth, and use the methods from Chapter 5.

In contrast to the linear case, we need to provide a starting point, our best guess for the parameters, and we may need to deal with scaling, noise, multimodality, or any of the other potential challenges of general nonlinear optimization. Still, this is a relatively straightforward problem within the broader realm of engineering optimization problems.

Although the methods of Chapter 4 can be used if the problem remains unconstrained, there are more specialized methods available that take advantage of the specific structure of the problem. One popular approach to solving the nonlinear least-squares problem is the *Levenberg–Marquardt algorithm*, which we discuss in this section.

As a stepping stone towards the Levenberg–Marquardt algorithm, we first derive the *Gauss–Newton algorithm*, which is a modification of Newton's method (Section 3.8) for solving nonlinear least-squares problems. One way to think of this algorithm is as an iterative linearization of the residual. Once it is linearized, we can apply the same methods we derived for linear least squares. We linearize the residual $r = \hat{f}(w) - f$ at iteration k as

$$r(w) \approx r(w_k) + J_r \Delta w, \qquad (10.42)$$

where Δw is the step and the Jacobian is

$$J_{r_{ij}} = \frac{\partial r_i}{\partial w_j}. \tag{10.43}$$

After the linearization, the objective becomes

$$\text{minimize} \quad \|J_r \Delta w + r\|_2^2. \tag{10.44}$$

This is now the same form as linear least squares (Eq. 10.14), so we can reuse its solution (Eq. 10.23) to solve for the step

$$\Delta w = -\left(J_r^\mathsf{T} J_r\right)^{-1} J_r^\mathsf{T} r. \tag{10.45}$$

We now have an update formula for the coefficients at each iteration:

$$w_{k+1} = w_k - \left(J_r^\mathsf{T} J_r\right)^{-1} J_r^\mathsf{T} r. \tag{10.46}$$

An alternative derivation for this formula is to consider a Newton step for an unconstrained optimizer. The objective is $e = \sum_i r_i^2$, and the formula for a Newton step (Section 4.4.3) is

$$w_{k+1} = w_k - H_e^{-1} \nabla e. \tag{10.47}$$

The gradient is

$$\nabla e_j = 2 r_i \frac{\partial r_i}{\partial w_j}, \tag{10.48}$$

or in matrix form:

$$\nabla e = 2 J_r^\mathsf{T} r. \tag{10.49}$$

The Hessian in index notation is

$$H_{e_{jk}} = 2 \frac{\partial r_i}{\partial w_j} \frac{\partial r_i}{\partial w_k} + 2 r_i \frac{\partial^2 r_i}{\partial w_j \partial w_k}. \tag{10.50}$$

We can write it in matrix form as follows:

$$H_e = 2 J_r^\mathsf{T} J_r + 2 r H_r. \tag{10.51}$$

If we neglect the second term in the Hessian, then the Newton update is:

$$\begin{aligned} w_{k+1} &= w_k - \frac{1}{2} \left(J_r^\mathsf{T} J_r\right)^{-1} 2 J_r^\mathsf{T} r \\ &= w_k - \left(J_r^\mathsf{T} J_r\right)^{-1} J_r^\mathsf{T} r, \end{aligned} \tag{10.52}$$

which is the same update as before.

Thus, another interpretation of this method is that a Gauss–Newton step is a modified Newton step where the second derivatives of the

10.3 Constructing a Surrogate

residual are neglected (and thus, a quasi-Newton approach to estimate second derivatives is not needed). This method is particularly effective near convergence because as $r \to 0$ (i.e., as we approach the solution to our residual minimization), the neglected term also approaches zero. The appeal of this approach is that we can often obtain an accurate prediction for the Hessian using only the first derivatives because of the known structure of the objective.

When the second term is not small, then the Gauss–Newton step may be too inaccurate. We could use a line search, but the Levenberg–Marquardt algorithm utilizes a different strategy. The idea is to regularize the problem as discussed in the previous section or, in other words, provide the ability to dampen the steps as needed. Each linearized subproblem becomes

$$\underset{\Delta w}{\text{minimize}} \quad \|J_r \Delta w + r\|_2^2 + \mu \|\Delta w\|_2^2. \tag{10.53}$$

Recall that the solution to this problem (see Eq. 10.27) is

$$\Delta w = -\left(J_r^T J_r + \mu I\right)^{-1} J_r^T r. \tag{10.54}$$

If $\mu = 0$, then we retain the Gauss–Newton step. Conversely, as μ becomes large, so that the $J_r^T J_r$ is negligible, the step becomes

$$\Delta w = -\frac{1}{\mu} J_r^T r. \tag{10.55}$$

This is precisely the steepest-descent direction for our objective (see Eq. 10.49), although with a small magnitude because μ is large. The parameter μ provides some control for directions ranging between Gauss–Newton and steepest descent.

The Levenberg–Marquardt algorithm has been revised to improve the scaling for components of the gradient that are small. The second minimization term weights all parameters equally. The scaling can be improved by multiplying by a diagonal matrix in the regularization as follows:

$$\underset{\Delta w}{\text{minimize}} \quad \|J_r \Delta w + r\|_2^2 + \mu \|D \Delta w\|_2^2, \tag{10.56}$$

where D is defined as

$$D^2 = \text{diag}\left(J_r^T J_r\right). \tag{10.57}$$

This matrix scales the objective by the diagonal elements of the Hessian. Thus, when μ is large, and the direction tends toward the steepest descent, the components of the gradient are scaled by the curvature, which reduces the amount of zigzagging. The solution to the minimization problem of Eq. 10.56 is

$$\Delta w = - \left(J_r^T J_r + \mu \operatorname{diag}\left(J_r^T J_r \right) \right)^{-1} J_r^T r . \qquad (10.58)$$

Finally, we describe one of the possible heuristics for selecting and updating the damping parameter μ. After a successful step (a sufficient reduction in the objective), μ is increased by a factor of ρ. Conversely, an unsuccessful step is rejected, and μ is reduced by a factor ($\mu = \mu/\rho$).

Rather than returning a scalar objective ($\sum_i r_i^2$), the user function should return a vector of the residuals because that vector is needed in the update steps (Eq. 10.58). A potential convergence metric is a tolerance on objective value changes between subsequent iterations. The full procedure is described in Alg. 10.3.

Algorithm 10.3 Levenberg–Marquardt algorithm for solving a nonlinear least squares problem

Inputs:
- x_0: Starting point
- μ_0: Initial damping parameter
- ρ: Damping parameter factor

Outputs:
- x^*: Optimal solution

$k = 0$
$x = x_0$
$\mu = \mu_0$
$r, J = \text{residual}(x)$
$e = \|r\|_2^2$ *Residual error*
while $|\Delta| > \tau$ **do**
   $s = - \left(J^T J + \mu \operatorname{diag}(J^T J) \right)^{-1} J^T r$ *Evaluate step*
   $r_s, J_s = \text{residual}(x + s)$
   $e_s = \|r_s\|_2^2$
   $\Delta = e_s - e$ *Change in residual error*
   if $\Delta < 0$ **then**   *Objective decreased; accept step*
      $x = x + s$
      $r, J, e = r_s, J_s, e_s$
      $\mu = \mu/\rho$
   else   *Reject step*
      $\mu = \mu \cdot \rho$ *Increase damping*
   end if
   $k = k + 1$
end while

10.3 Constructing a Surrogate

Example 10.4 Rosenbrock as a nonlinear least-squares problem

The Rosenbrock function is a sum of squared terms, so it can be posed as a nonlinear least squares problem:

$$r(x) = \begin{bmatrix} (1 - x_1) \\ 10(x_2 - x_1^2) \end{bmatrix}.$$

In the following example, we use the same starting point as Ex. 4.18 ($x_0 = [-1.2, -1]$), an initial damping parameter of $\mu = 0.01$, an update factor of $\rho = 10$, and a tolerance of $\tau = 10^{-6}$ (change in sum of squared errors). The iteration path is shown on the left of Fig. 10.15, and the convergence of the sum of squared errors is shown on the right side.

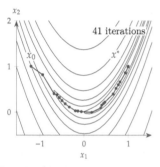
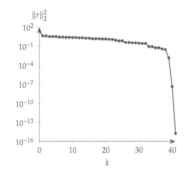

Iteration history

Convergence of the sum of squared residuals

Fig. 10.15 Levenberg–Marquardt algorithm applied to the minimization of the Rosenbrock function.

10.3.4 Cross Validation

The other important consideration for developing a surrogate model is the choice of the basis functions in ψ. In some instances, we may know something about the model behavior and thus what type of basis functions should be used, but generally, the best way to determine the basis functions is through cross validation. Cross validation is also helpful in characterizing error, even if we already have a chosen set of basis functions. One of the reasons we use cross validation is to prevent *overfitting*. Overfitting occurs when we have too many degrees of freedom and closely fit a given set of data, but the resulting model has a poor predictive ability. In other words, we are fitting noise. The following example illustrates this idea with a one-dimensional function.

> **Example 10.5** The dangers of overfitting
>
> Consider the set of training data (Fig. 10.16, left), which we use to create a surrogate function. This is a one-dimensional problem so that it can be easily visualized. In general, however, visualization is limited, and determining the right basis functions to use can be difficult. If we use a polynomial basis, we might attempt to determine the appropriate order by trying each case (e.g., quadratic, cubic, quartic) and measuring the error in our fit (Fig. 10.16, center).
>
>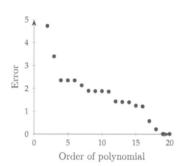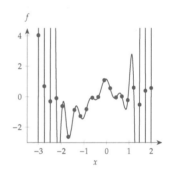
>
> Training data | The error in fitting the data decreases with the order of the polynomial | A 19th-order polynomial fit to the data has low error but poor predictive ability
>
> **Fig. 10.16** Fitting different order polynomials to data.
>
> It seems as if the higher the order of the polynomial, the lower the error. For example, a 20th-order polynomial reduces the error to almost zero. The problem is that although the error is low on this set of data, the predictive capability of such a model for other data points is poor. For example, the right side of Fig. 10.16 shows a 19th-order polynomial fit to the data. The model passes right through the points, but it does not work well for many of the points that are not part of the training set (which is the whole purpose of the surrogate).
>
> The opposite of overfitting is *underfitting*, which is also a potential issue. When underfitting, we do not have enough degrees of freedom to create a useful model (e.g., imagine using a linear fit for the previous example).

The solution to the overfitting problem highlighted in Ex. 10.5 is cross validation. *Cross validation* means that we use one set of data for training (creating the model) and a different set of data for assessing its predictive error. There are many different ways to perform cross validation; we describe two. *Simple cross validation* is illustrated in Fig. 10.17 and consists of the following steps:

1. Randomly split your data into a training set and a validation set (e.g., a 70–30 split).

10.3 Constructing a Surrogate

2. Train each candidate model (the different options for ψ) using only the training set, but evaluate the error with the validation set. The error on previously unseen data is called the *generalization error* (e_g in Fig. 10.17).
3. Choose the model with the lowest generalization error, and optionally retrain that model using all of the data.

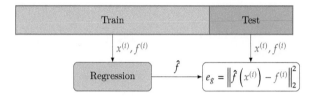

Fig. 10.17 Simple cross-validation process.

An alternative option that is more involved but uses the data more effectively is called *k-fold cross validation*. It is particularly advantageous when we have a small data set where we cannot afford to leave much out. This procedure is illustrated in Fig. 10.18 and consists of the following steps:

1. Randomly split your data into n sets (e.g., $n = 10$).
2. Train each candidate model using the data from all sets except one (e.g., 9 of the 10 sets) and use the remaining set for validation. Repeat for all n possible validation sets and average the performance.
3. Choose the model with the lowest average generalization error. Optionally, retrain with all the data.

The extreme version of this process, when training data are very limited, is *leave-one-out* cross validation (i.e., each testing subset consists of one data point).

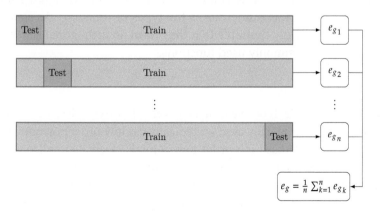

Fig. 10.18 Diagram of k-fold cross-validation process.

Example 10.6 Cross validation helps to avoid overfitting

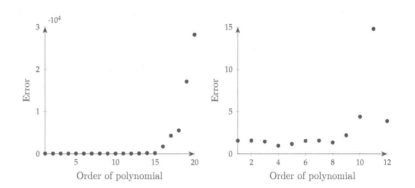

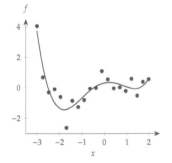

Fig. 10.19 Error from k-fold validation.

This example continues from Ex. 10.5. First, we perform k-fold cross validation using 10 divisions. The average error across the divisions using the training data is shown in Fig. 10.19 (with a smaller y-axis scale on the right).

The error increases dramatically as the polynomial order increases. Zooming in on the flat region, we see a range of options with similar errors. Among the similar solutions, we generally prefer the simplest model. In this case, a fourth-order polynomial seems reasonable. A fourth-order polynomial is compared against the data in Fig. 10.20. This model has a much better predictive ability.

Fig. 10.20 A fourth-order polynomial fit to the data.

10.3.5 Common Basis Functions

Although cross validation can help us find the lowest generalization error among a provided set of basis functions, we still need to determine what sets of options to consider. This selection is crucial because our model is only as good as the available options, but increasing the number of options increases computational time. The possibilities for basis functions are as numerous as the types of function. As stated before, it is generally desirable that they form an orthogonal set. We focus on a few commonly used functions.

Polynomials

Polynomials, of which we have already seen a few examples, are useful in many applications. However, we typically use low-order polynomials for regression because high-order polynomials rarely generalize well. Polynomials can be particularly effective in cases where a knowledge of the physics suggests them to be an appropriate choice (e.g., drag varies quadratically with speed) Because a lot of structure is already

10.3 Constructing a Surrogate

built into the model form, fewer data points are needed to create a reasonable model (e.g., a quadratic function in n dimensions needs at least $n(n + 1)/2 + n + 1$ points, so this amounts to 6 points in two dimensions, 10 points in three dimensions, and so on).

Radial Basis Functions

Another common type of basis function is a *radial basis function*. Radial basis functions are functions that depend on the distance from some center point and can be written as follows:

$$\psi^{(i)} = \psi\left(\left\|x - c^{(i)}\right\|\right) = \psi\left(r^{(i)}\right), \quad (10.59)$$

where c is the center point, and r is the radius about the center point. Although the center points can be placed anywhere, we usually choose the sampling data as centering points:

$$\psi^{(i)} = \psi\left(\left\|x - x^{(i)}\right\|\right). \quad (10.60)$$

This is often a useful choice because it captures the idea that our ability to predict function behavior is related to how close we are to known function values (in other words, nearby points are more highly correlated). This form naturally lends itself to interpolation, although regularization can be added to allow for regression. Polynomials are often combined with radial basis functions because the polynomial can capture global function behavior, while the radial basis functions can introduce modifications to capture local behavior.

One popular radial basis function is the *Gaussian basis*:

$$\psi^{(i)}(x) = \exp\left(-\sum_j \theta_j \left|x - x_j^{(i)}\right|^2\right), \quad (10.61)$$

where θ_j are the model parameters. One of the forms of kriging discussed in the following section can be viewed as a radial basis function model with a Gaussian basis.

Tip 10.3 Surrogate modeling toolbox

The surrogate modeling toolbox (SMT)[†] is a useful package for surrogate modeling, with a particular focus on providing derivatives for use in gradient-based optimization.[171] SMT includes surrogate modeling techniques that utilize gradients as training data to enhance accuracy and scalability with the number of inputs.[172]

[†] https://smt.readthedocs.io/

171. Bouhlel et al., *A Python surrogate modeling framework with derivatives*, 2019.

172. Bouhlel and Martins, *Gradient-enhanced kriging for high-dimensional problems*, 2019.

10.4 Kriging

Kriging is a popular surrogate modeling technique that can build approximations of highly nonlinear engineering simulations. We may not have a simple parametric form for such simulations that we can use with regression and expect a good fit. Instead of tuning the parameters of a functional form that describes *what the function is*, kriging tunes the parameters of a statistical model that describes *how the function behaves*.

The kriging statistical model that approximates f consists of two terms: a function $\mu(x)$ that is meant to capture some of the function behavior and a random variable $Z(x)$. Thus, we can write the kriging model as

$$F(x) = \mu(x) + Z(x) \quad \text{where} \quad Z(x) \sim \mathcal{N}(0, \sigma^2). \tag{10.62}$$

When we evaluate the function we want to approximate at point x, we get a scalar value $f(x)$. In contrast, when we evaluate the stochastic process (Eq. 10.62) at x, we get a random variable $F(x)$ that has a normal distribution with mean μ and variance σ^2. Although we wrote μ as a function of x, most kriging models consider this to be constant because the random variable term alone is effective in capturing the function behavior. For the rest of this section, we discuss the case with constant μ, which is called *ordinary kriging*. Kriging is also referred to as *Gaussian process* interpolation, or more generally in the regression case discussed later in this section, as Gaussian process regression.

The power of the statistical model lies in how it treats the *correlation* between the random variables. Although we do not know the exact form of the error term $Z(x)$, we can still make some reasonable assumptions about it. Consider two points in a sampling plan, $x^{(i)}$ and $x^{(j)}$, and the corresponding terms, $Z\left(x^{(i)}\right)$ and $Z\left(x^{(j)}\right)$. Intuitively, we expect $Z\left(x^{(i)}\right)$ to be close to $Z\left(x^{(j)}\right)$ whenever $x^{(i)}$ is close to $x^{(j)}$. Therefore, it seems reasonable to assume that the correlation between $Z\left(x^{(i)}\right)$ and $Z\left(x^{(j)}\right)$ is a function of the distance between the two points.

In kriging, we assume that this correlation is given by a *kernel* function $K\left(x^{(i)}, x^{(j)}\right)$:

$$K\left(x^{(i)}, x^{(j)}\right) = \text{corr}\left(Z\left(x^{(i)}\right), Z\left(x^{(j)}\right)\right) \tag{10.63}$$

As a matrix, the kernel is represented as $K_{ij} = K\left(x^{(i)}, x^{(j)}\right)$. Various kernel functions are used with kriging.* The most commonly used kernel function is

*Kernel functions must be symmetric and positive definite because a covariance matrix is always symmetric and positive definite.

$$K\left(x^{(i)}, x^{(j)}\right) = \exp\left(-\sum_{l=1}^{n_d} \theta_l \left|x_l^{(i)} - x_l^{(j)}\right|^{p_l}\right), \tag{10.64}$$

10.4 Kriging

where $\theta_l \geq 0$, $0 \leq p_l \leq 2$, and n_d is the number of dimensions (i.e., the length of the vector x). If every $p_l = 2$, this becomes a *Gaussian kernel*.

Let us examine how the statistical model F defined in Eq. 10.62 captures the typical behavior of the function f. The parameter μ captures the typical value, and σ^2 captures the expected variance. The kernel (or correlation) function (Eq. 10.64) implicitly models *continuous* functions. If f is continuous, we know that, as $|x^{(i)} - x^{(j)}| \to 0$, then $|f(x^{(i)}) - f(x^{(j)})| \to 0$. This is captured in the kernel function because as $|x^{(i)} - x^{(j)}| \to 0$, the correlation approaches 1. The parameter θ_l captures how active the function f is in the lth coordinate direction. A unit difference in variable l ($|x_l^{(i)} - x_l^{(j)}| = 1$) has a more significant impact on the correlation when θ_l is large. The exponent p_l describes the smoothness of the function in the lth coordinate direction. Values of p_l close to 2 produce smooth functions, whereas values closer to zero produce functions with more variation.

Kriging surrogate modeling involves two main steps. The first step consists of using the data to estimate the statistical model parameters μ, σ^2, $\theta_1, \ldots, \theta_{n_d}$, and $p_1, \ldots, p_{n_d}$. The second step consists of making predictions using the statistical model and these estimated parameter values.

The parameter estimation uses the same maximum likelihood approach from Section 10.3.2, but now it is more complicated. Let us denote the random variable as $F^{(i)} \equiv F(x^{(i)})$ and the vector of random variables as $F = [F^{(1)}, \ldots, F^{(n_s)}]$, where n_s is the number of samples. Similarly, $f^{(i)} \equiv f(x^{(i)})$ and the vector of observed function values is $f \equiv [f^{(1)}, \ldots, f^{(n_s)}]$. Using this notation, we can say that the vector F is jointly normally distributed. This is also known as a *multivariate Gaussian distribution*.[†] The probability density function (PDF) (the likelihood that $F = f$) is

[†]There are other ways to derive kriging that do not require making assumptions on the random variable distribution type.

$$p(f) = \frac{1}{(2\pi)^{n_s/2} |\Sigma|^{1/2}} \exp\left[-\frac{1}{2}(f - e\mu)^\mathsf{T} \Sigma^{-1} (f - e\mu)\right], \quad (10.65)$$

where e is a vector of 1s with size n_s, and $|\Sigma|$ is the determinant of the covariance,

$$\Sigma_{ij} = \sigma^2 K\left(x^{(i)}, x^{(j)}\right). \quad (10.66)$$

The covariance between two elements $F^{(i)}$ and $F^{(j)}$ of F is related to correlation by the following definition:[‡]

[‡]Covariance and correlation are briefly reviewed in Appendix A.9.

$$\Sigma_{ij} = \text{cov}\left(F^{(i)}, F^{(j)}\right) = \sigma^2 \text{corr}\left(F^{(i)}, F^{(j)}\right) = \sigma^2 K\left(x^{(i)}, x^{(j)}\right). \quad (10.67)$$

We assume stationarity of the second moment, that is, the variance σ^2 is constant in the domain.

The statistical model parameters $\theta_1, \ldots, \theta_{n_d}$ and $p_1, \ldots, p_{n_d}$ enter the likelihood (Eq. 10.65) via their effect on the kernel K (Eq. 10.64) and hence on the covariance matrix Σ (Eq. 10.66).

We estimate the parameters using the maximum log likelihood approach from Section 10.3.2; that is, we maximize the probability of observing our data f conditioned on the parameters μ and Σ. Using a PDF (Eq. 10.65) where μ is constant and the covariance is $\Sigma = \sigma^2 K$, yields the following likelihood function:

$$L(\mu, \sigma, \theta, p) = \frac{1}{(2\pi)^{n_s/2} \sigma^{n_s} |K|^{1/2}} \exp\left[-\frac{(f - e\mu)^\mathsf{T} K^{-1}(f - e\mu)}{2\sigma^2}\right].$$

We now need to find the parameters μ, σ, θ_i, and p_i that maximize this likelihood function, that is, maximize the probability of our observations f. As before, we take the logarithm to form the log likelihood function:

$$\ell(\mu, \sigma, \theta, p) = -\frac{n_s}{2} \ln(2\pi) - \frac{n_s}{2} \ln(\sigma^2)$$
$$- \frac{1}{2} \ln |K| - \frac{(f - e\mu)^\mathsf{T} K^{-1}(f - e\mu)}{2\sigma^2}. \quad (10.68)$$

We can maximize part of this term analytically by taking derivatives with respect to μ and σ, setting them equal to zero, and solving for their optimal values to obtain:

$$\mu^* = \frac{e^\mathsf{T} K^{-1} f}{e^\mathsf{T} K^{-1} e} \quad (10.69)$$

$$\sigma^{*2} = \frac{(f - e\mu^*)^\mathsf{T} K^{-1}(f - e\mu^*)}{n_s}. \quad (10.70)$$

We now substitute these values back into the log likelihood function (Eq. 10.68), which yields

$$\ell(\theta, p) = -\frac{n_s}{2} \ln(\sigma^{*2}) - \frac{1}{2} \ln |K|. \quad (10.71)$$

This function, also called the *concentrated likelihood function*, only depends on the kernel K, which depends on θ and p.

We cannot solve for optimal values of θ and p analytically. Instead, we rely on numerical optimization to maximize Eq. 10.71. Because θ can vary across a broad range, it is often better to search using logarithmic scaling. Once we solve that optimization problem, we compute the mean and variance in Eqs. 10.69 and 10.70.

Now that we have a fitted model, we can make predictions at new points where we have not sampled. We do this by substituting x_p into a formula called the *kriging predictor*. The formula is unique, but there are many ways to derive it. One way to derive it is to find the function value at x_p that is the most consistent with the behavior of the function captured by the fitted kriging model.

Let f_p be our guess for the value of the function at x_p. One way to assess the consistency of our guess is to add (x_p, f_p) as an artificial point to our training data (so that we now have $n_s + 1$ points) and estimate the likelihood using the parameters from our fitted kriging model. The likelihood of this augmented data can now be thought of as a function of f_p: high values correspond to guessed values of f_p that are consistent with function behavior captured by the fitted kriging model. Therefore, the value of f_p that maximizes the likelihood of this augmented data set is a natural way to predict the value of the function. This is an optimization problem with a closed-form solution, and the corresponding formula is the kriging predictor.

Now we outline the derivation of the kriging predictor.§ With the augmented point, our function values are $\bar{f} = [f, f_p]$, where f is the n_s-vector of function values from the original training data. Then, the correlation matrix with the additional data point is

§ Jones[173] provides the complete derivation; here we show only a few key steps.

173. Jones, *A taxonomy of global optimization methods based on response surfaces*, 2001.

$$\bar{K} = \begin{bmatrix} K & k \\ k^\mathsf{T} & 1 \end{bmatrix}, \qquad (10.72)$$

where k is the correlation of the new point with the training data given by

$$k = \begin{bmatrix} \operatorname{corr}\left(F\left(x^{(1)}\right), F(x_p)\right) = K\left(x^{(1)}, x_p\right) \\ \vdots \\ \operatorname{corr}\left(F\left(x^{(n_s)}\right), F(x_p)\right) = K\left(x^{(n_s)}, x_p\right) \end{bmatrix}. \qquad (10.73)$$

The 1 in the bottom right of the augmented correlation matrix (Eq. 10.72) is because the correlation of the new variable $F(x_p)$ with itself is 1. The log likelihood function with these new augmented vectors and the previously determined parameters is as follows (see Eq. 10.68):

$$\ell(f_p) = -\frac{n_s}{2}\ln(2\pi) - \frac{n_s}{2}\ln(\sigma^{*2}) - \frac{1}{2}\ln|\bar{K}| - \frac{(\bar{f} - e\mu^*)^\mathsf{T} \bar{K}^{-1}(\bar{f} - e\mu^*)}{2\sigma^{*2}}.$$

We want to maximize this function with respect to f_p. Because only the last term depends on f_p (it is a part of $\bar{f}$) we can omit the other terms and formulate the following:

$$\underset{f_p}{\text{maximize}}\ \ell(f_p) = -\frac{(\bar{f} - ev)^\mathsf{T} \bar{K}^{-1}(\bar{f} - e\mu^*)}{2\sigma^{*2}}. \qquad (10.74)$$

This problem can be solved analytically, yielding the mean value of the kriging prediction,

$$f_p = \mu^* + k^\mathsf{T} K^{-1}(f - e\mu^*). \tag{10.75}$$

The mean square error of the kriging prediction (that is, the expected squared value of the error) is given by¶

$$\sigma_p^2 = \sigma^{*2}\left[1 - k^\mathsf{T} K^{-1} k + \frac{(1 - k^\mathsf{T} K^{-1} e)^2}{e^\mathsf{T} K^{-1} e}\right]. \tag{10.76}$$

¶ The formula for mean squared error does not come from the augmented likelihood approach, but is a byproduct of showing that the kriging predictor is the "best linear unbiased predictor" for the assumed statistical model.[174]

174. Sacks et al., *Design and analysis of computer experiments*, 1989.

One attractive feature of kriging models is that they are interpolatory and thus match the training data exactly. To see how this is true, if x_p is the same as one of the training data points, $x^{(i)}$, then k is just ith column of K. Hence, $K^{-1}k$ is a vector e_i, with all zeros except for 1 in the ith element. In the prediction (Eq. 10.75), $k^\mathsf{T} K^{-1} = e_i^\mathsf{T}$ and so the last term is $f^{(i)} - \mu^*$, which means that $f_p = f^{(i)}$.

In the mean square error (Eq. 10.76), $k^\mathsf{T} K^{-1} k$ is the same as $k^\mathsf{T} e_i$. This is the ith element of k, which is 1. Therefore, the first two terms in the brackets in Eq. 10.76 cancel, and the last term is zero, yielding $\sigma_p^2 = 0$. This is expected; if we already sampled the point, the uncertainty about its function value should be zero.

When describing a fitted kriging model, we often refer to the standard error as the square root of this quantity (i.e., $\sqrt{\sigma_p^2}$). The standard error is directly related to the *confidence interval* (e.g., ±1 standard error corresponds to a 68 percent confidence interval).

Example 10.7 One-dimensional kriging model

In this example, we consider the decaying sinusoid:

$$f(x) = \exp(-0.1x)\sin(x).$$

We assume, however, that this function is unknown, and we sample at the following points:

$$x = [0.5, 2, 2.5, 9, 10].$$

We can fit a kriging model to this data by following the procedure in this section. This includes solving the optimization problem of Eq. 10.71 using a gradient-based method with exact derivatives. We fix $p = 2$ and search for θ in the range $[10^{-3}, 10^2]$ with the exponent as the optimization variable.

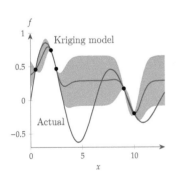

Fig. 10.21 Kriging model showing the training data (dots), the kriging predictor (blue line) and the confidence interval corresponding to ±1 standard error (shaded areas), compared to the actual function (gray line).

The resulting interpolation is shown in Fig. 10.21, where we plot the mean line. The shaded area represents the uncertainty corresponding to ±1 standard error. The uncertainty goes to zero at the known data points and is largest when far from known data points.

10.4 KRIGING

If we can provide the gradients of the function at the training data points (in addition to the function values), we can use that information to build a more accurate kriging model. This approach is called *gradient-enhanced kriging* (GEK). The methodology is the same as before, except we add more observed outputs (i.e., in addition to the function values at the sampled points, we add their gradients). In addition to considering the correlation between the function values at different sampled points, the kernel matrix K needs to be expanded to consider correlations between function values and gradients, gradients and function values, and among gradient components.

We can use still use equation (Eq. 10.75) for the GEK predictor and equation (Eq. 10.76) for the mean square error if we plug in "expanded versions" of the outputs f, the vector k, the matrix K, and the vector of 1s, e.

We expand the output vector to include not just the function values at the sampled points but also their gradients:

$$f_{\text{GEK}} \equiv \begin{bmatrix} f_1 \\ \vdots \\ f_{n_s} \\ \nabla f_1 \\ \vdots \\ \nabla f_{n_s} \end{bmatrix}. \tag{10.77}$$

This vector is of length $n_s + n_s n_d$, where n_d is the dimension of x. The gradients are usually provided at the same x locations as the function samples, but that is not required.

Recall that the term $e\mu^*$ in Eq. 10.75 for the kriging predictor represents the expected value of the random variables $F^{(1)}, \ldots, F^{(n_s)}$. Now that we have expanded the outputs to include the gradients at the sampled points, the mean vector needs to be expanded to include the expected values of $\nabla F^{(i)}$, which are all zero. We can still use $e\mu^*$ in the formula for the predictor if we use the following definition:

$$e_{\text{GEK}} \equiv [1, \ldots, 1, 0, \ldots, 0], \tag{10.78}$$

where 1 occurs for the first n_s entries, and 0 for the remaining $n_s n_d$ entries.

The additional correlations (between function values and derivatives and between the derivatives themselves) are as follows:

$$\text{corr}\left(F\left(x^{(i)}\right), F\left(x^{(j)}\right)\right) = K_{ij}$$

$$\text{corr}\left(F\left(x^{(i)}\right), \frac{\partial F\left(x^{(j)}\right)}{\partial x_l}\right) = \frac{\partial K_{ij}}{\partial x_l^{(j)}}$$

$$\text{corr}\left(\frac{\partial F\left(x^{(i)}\right)}{\partial x_l}, F\left(x^{(j)}\right)\right) = \frac{\partial K_{ij}}{\partial x_l^{(i)}} \quad (10.79)$$

$$\text{corr}\left(\frac{\partial F\left(x^{(i)}\right)}{\partial x_l}, \frac{\partial F\left(x^{(j)}\right)}{\partial x_k}\right) = \frac{\partial^2 K_{ij}}{\partial x_l^{(i)} \partial x_k^{(j)}}.$$

Here, we use l and k to represent a component of a vector, and we use $K_{ij} \equiv K\left(x^{(i)}, x^{(j)}\right)$ as shorthand. For our particular kernel choice (Eq. 10.64), these correlations become the following:

$$K_{ij} = \exp\left(-\sum_{k=1}^{n_d} \theta_l \left(x_l^{(i)} - x_l^{(j)}\right)^2\right)$$

$$\frac{\partial K_{ij}}{\partial x_l^{(j)}} = 2\theta_l \left(x_l^{(i)} - x_l^{(j)}\right) K_{ij} \quad (10.80)$$

$$\frac{\partial K_{ij}}{\partial x_l^{(i)}} = -\frac{\partial K_{ij}}{\partial x_l^{(j)}}$$

$$\frac{\partial^2 K_{ij}}{\partial x_l^{(i)} \partial x_k^{(j)}} = \begin{cases} -4\theta_l \theta_k \left(x_k^{(i)} - x_k^{(j)}\right)\left(x_l^{(i)} - x_l^{(j)}\right) K_{ij} & l \neq k \\ -4\theta_l^2 \left(x_l^{(i)} - x_l^{(j)}\right)^2 K_{ij} + 2\theta_l K_{ij} & l = k, \end{cases}$$

where we used $p = 2$. Putting this all together yields the expanded correlation matrix:

$$K_{\text{GEK}} \equiv \begin{bmatrix} K & J_K \\ J_K^T & H_K \end{bmatrix}, \quad (10.81)$$

where the $(n_s \times n_s n_d)$ block representing the first derivatives is

$$J_K = \begin{bmatrix} \frac{\partial K_{11}}{\partial x^{(1)}}^T & \cdots & \frac{\partial K_{1n_s}}{\partial x^{(n_s)}}^T \\ \vdots & \ddots & \vdots \\ \frac{\partial K_{n_s 1}}{\partial x^{(1)}}^T & \cdots & \frac{\partial K_{n_s n_s}}{\partial x^{(n_s)}}^T \end{bmatrix} \quad (10.82)$$

and the $(n_s n_d \times n_s n_d)$ matrix of second derivatives is

$$H_K = \begin{bmatrix} \frac{\partial^2 K_{11}}{\partial x^{(1)} \partial x^{(1)}} & \cdots & \frac{\partial^2 K_{1n_s}}{\partial x^{(1)} \partial x^{(n_s)}} \\ \vdots & \ddots & \vdots \\ \frac{\partial^2 K_{n_s 1}}{\partial x^{(n_s)} \partial x^{(1)}} & \cdots & \frac{\partial^2 K_{n_s n_s}}{\partial x^{(n_s)} \partial x^{(n_s)}} \end{bmatrix}. \quad (10.83)$$

10.4 Kriging

We can still get the estimates μ^* and σ^{*2} with Eqs. 10.69 and 10.70 using the expanded versions of K, e, f and replacing n_s in Eq. 10.76 with $n_s(n_d + 1)$, which is the new length of the outputs.

The predictor equations (Eqs. 10.75 and 10.76) also apply with the expanded matrices and vectors. However, we also need to expand k in these computations to include the correlations between the gradients at the sampled points with the gradient at the point x where we make a prediction. Thus, the expanded k is:

$$k_{\text{GEK}} \equiv \begin{bmatrix} k \\ \text{corr}\left(\dfrac{\partial F\left(x^{(1)}\right)}{\partial x^{(1)}}, F(x_p)\right) = \dfrac{\partial K\left(x^{(1)}, x_p\right)}{\partial x^{(1)}} \\ \vdots \\ \text{corr}\left(\dfrac{\partial F\left(x^{(n_s)}\right)}{\partial x^{(n_s)}}, F(x_p)\right) = \dfrac{\partial K\left(x^{(n_s)}, x_p\right)}{\partial x^{(n_s)}} \end{bmatrix}. \quad (10.84)$$

Example 10.8 Gradient-enhanced kriging

We repeat Ex. 10.7 but this time include the gradients (Fig. 10.22). The standard error reduces dramatically between points. The additional information contained in the derivatives significantly helps in creating a more accurate fit.

Example 10.9 Two-dimensional kriging

The Jones function (Appendix D.1.4) is shown on the left in Fig. 10.23. Using GEK with only 10 training points from a Hammersley sequence (shown as the dots), created the surrogate model on the right. A reasonable representation of this multimodal space can be captured even with a small number of samples.

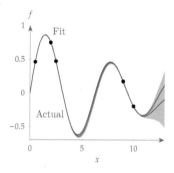

Fig. 10.22 A GEK fit to the input data (circles) and a shaded confidence interval.

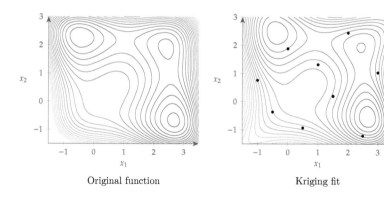

Fig. 10.23 Kriging fit to the multimodal Jones function.

One difficulty with GEK is that the kernel matrix quickly grows in size as the dimension of the problem increases, the number of samples increases, or both. Various approaches have been proposed to improve the scaling with higher dimensions, such as a weighted sum of smaller correlation matrices[175] or a partial least squares approach.[172]

175. Han et al., *Weighted gradient-enhanced kriging for high-dimensional surrogate modeling and design optimization*, 2017.

172. Bouhlel and Martins, *Gradient-enhanced kriging for high-dimensional problems*, 2019.

The version of kriging in this section is interpolatory. For noisy data, a regression approach can be used by modifying the correlation matrix as follows:

$$K_{\text{reg}} \equiv K + \tau I, \tag{10.85}$$

with $\tau > 0$. This adds a positive constant along the diagonal, so the model no longer correlates perfectly with the provided points. The parameter τ is then an additional parameter to estimate in the maximum likelihood optimization. Even for interpolatory models, this term is often still added to the covariance matrix with a small constant value of τ (near machine precision) to ensure that the correlation matrix is invertible. This section focused on the most common choices when using kriging, but many other versions exist.[176]

176. Forrester et al., *Engineering Design via Surrogate Modelling: A Practical Guide*, 2008.

10.5 Deep Neural Networks

Like kriging, deep neural nets can be used to approximate highly nonlinear simulations where we do not need to provide a parametric form. Neural networks follow the same basic steps described for other surrogate models but with a unique model leading to specialized approaches for derivative computation and optimization strategy. Neural networks loosely mimic the brain, which consists of a vast network of neurons. In neural networks, each neuron is a node that represents a simple function. A network defines chains of these simple functions to obtain composite functions that are much more complex. For example, three simple functions, $f^{(1)}, f^{(2)}$, and $f^{(3)}$, may be chained into the composite function (or network):

$$f(x) = f^{(3)}\left(f^{(2)}\left(f^{(1)}(x)\right)\right). \tag{10.86}$$

Even though each function may be simple, the composite function can express complex behavior. Most neural networks are *feedforward* networks, meaning that information flows from inputs x to outputs f. *Recurrent* neural networks include feedback connections.

Figure 10.24 shows a diagram of a neural network. Each node represents a neuron. The neurons are connected between consecutive layers, forming a dense network. The first layer is the *input layer*, the last one is the *output layer*, and the middle ones are the *hidden layers*.

10.5 Deep Neural Networks

The total number of layers is called the network's *depth*. *Deep neural networks* have many layers, enabling the modeling of complex behavior.

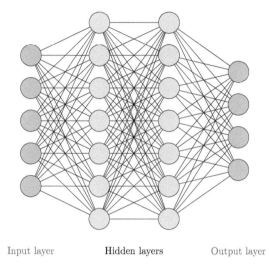

Fig. 10.24 Deep neural network with two hidden layers.

The first and last layers can be viewed as the inputs and outputs of a surrogate model. Each neuron in the hidden layer represents a function. This means that the output from a neuron is a number, and thus the output from a whole layer can be represented as a vector x. We represent the vector of values for layer k by $x^{(k)}$, and the value for the ith neuron in layer k by $x_i^{(k)}$.

Consider a neuron in layer k. This neuron is connected to many neurons from the previous layer $k-1$ (see the first part of Fig. 10.25). We need to choose a functional form for each neuron in the layer that takes in the values from the previous layer as inputs. Chaining together linear functions would yield another linear function. Therefore, some layers must use nonlinear functions.

The most common choice for hidden layers is a layer of linear functions followed by a layer of functions that create nonlinearity. A neuron in the linear layer produces the following intermediate variable:

$$z = \sum_{j=1}^{n} w_j x_j^{(k-1)} + b. \tag{10.87}$$

In vector form:

$$z = w^\mathsf{T} x^{(k-1)} + b. \tag{10.88}$$

The first term is a weighted sum of the values from the neurons in the previous layer. The w vector contains the weights. The term b is the *bias*, which is an offset that scales the significance of the overall output.

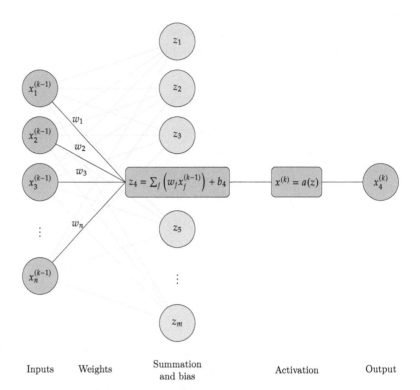

Fig. 10.25 Typical functional form for a neuron in the neural net.

These two terms are analogous to the weights used in the previous section but with the constant term separated for convenience. The second column of Fig. 10.25 illustrates the linear (summation and bias) layer.

Next, we pass z through an *activation function*, which we call $a(z)$. Historically, one of the most common activation functions has been the sigmoid function:

$$a(z) = \frac{1}{1 + e^{-z}}. \qquad (10.89)$$

This function is shown in the top plot of Fig. 10.26. The sigmoid function produces values between 0 and 1, so large negative inputs result in insignificant outputs (close to 0), and large positive inputs produce outputs close to 1.

Most modern neural nets use a rectified linear unit (ReLU) as the activation function:

$$a(z) = \max(0, z). \qquad (10.90)$$

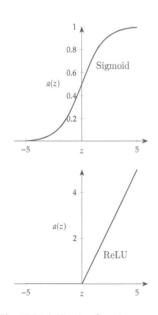

Fig. 10.26 Activation functions.

This function is shown in the bottom plot of Fig. 10.26. The ReLU has been found to be far more effective than the sigmoid function in producing accurate neural nets. This activation function eliminates negative inputs. Thus, the bias term can be thought of as a threshold

10.5 Deep Neural Networks

establishing what constitutes a significant value. The final two columns of Fig. 10.25 illustrate the activation step.

Combining the linear function with the activation function produces the output for the ith neuron:

$$x_i^{(k)} = a\left(w^\mathsf{T} x^{(k-1)} + b_i\right). \tag{10.91}$$

To compute the outputs for all the neurons in this layer, the weights w for one neuron form one row in a matrix of weights W and we can write:

$$\begin{bmatrix} x_1^{(k)} \\ \vdots \\ x_i^{(k)} \\ \vdots \\ x_{n_k}^{(k)} \end{bmatrix} = a\left(\begin{bmatrix} W_{1,1} & \cdots & W_{1,j} & \cdots & W_{1,n_{k-1}} \\ \vdots & & \vdots & & \vdots \\ W_{i,1} & \cdots & W_{i,j} & \cdots & W_{i,n_{k-1}} \\ \vdots & & \vdots & & \vdots \\ W_{n_k,1} & \cdots & W_{n_k,j} & \cdots & W_{n_k,n_{k-1}} \end{bmatrix} \begin{bmatrix} x_1^{(k-1)} \\ \vdots \\ x_j^{(k-1)} \\ \vdots \\ x_{n_{k-1}}^{(k-1)} \end{bmatrix} + \begin{bmatrix} b_1 \\ \vdots \\ b_i \\ \vdots \\ b_{n_k} \end{bmatrix}\right) \tag{10.92}$$

or

$$x^{(k)} = a\left(W x^{(k-1)} + b\right). \tag{10.93}$$

The activation function is applied separately for each row. The following equation is more explicit (where w_i is the ith row of W):

$$x_i^{(k)} = a\left(w_i^\mathsf{T} x_i^{(k-1)} + b_i\right). \tag{10.94}$$

This neural net is now parameterized by a number of weights. Like other surrogate models, we need to determine the optimal value for these parameters (i.e., train the network) using training data. In the example of Fig. 10.24, there is a layer of 5 neurons, 7 neurons, 7 neurons, and then 4 neurons, and so there would be $5 \times 7 + 7 \times 7 + 7 \times 4$ weights and $7 + 7 + 4$ bias terms, giving a total of 130 variables. This represents a small neural net because there are few inputs and few outputs. Large neural nets can have millions of variables. We need to optimize those variables to minimize a cost function.

As before, we use a maximum likelihood estimate where we optimize the parameters θ (weights and biases in this case) to maximize the probability of observing the output data y conditioned on our inputs x. As shown in Section 10.3.2, this results in a sum of squared errors function:

$$\underset{\theta}{\text{minimize}} \sum_{i=1}^{n} \left(\hat{f}\left(\theta; x^{(i)}\right) - f^{(i)}\right)^2. \tag{10.95}$$

We now have the objective and variables in place to train the neural net. As with the other models discussed in this chapter, it is critical to set aside some data for cross validation.

Because the optimization problem (Eq. 10.95) often has a large number of parameters θ, we generally use a gradient-based optimization algorithm (however the algorithms of Chapter 4 are modified as we will discuss shortly). To solve Eq. 10.95 using gradient-based optimization, we require the derivatives of the objective function with respect to the weighs θ. Because the objective is a scalar and the number of weights is large, reverse-mode algorithmic differentiation (AD) (see Section 6.6) is ideal to compute the required derivatives.

Reverse-mode AD is known in the machine learning community as *backpropagation*.* Whereas general-purpose reverse-mode AD operates at the code level, backpropagation usually operates on larger sets of operations and data structures defined in machine learning libraries. Although less general, this approach can increase efficiency and stability. The ReLU activation function (Fig. 10.26, bottom) is not differentiable at $z = 0$, but in practice, this is generally not problematic—primarily because these methods typically rely on inexact gradients anyway, as discussed next.

*The machine learning community independently developed backpropagation before becoming aware of the connection to reverse-mode AD.[58]

58. Baydin et al., *Automatic differentiation in machine learning: A survey*, 2018.

The objective function in Eq. 10.95 consists of a sum of subfunctions, each of which depends on a single data point $(x^{(i)}, f^{(i)})$. Objective functions vary across machine learning applications, but most have this same form:

$$\underset{\theta}{\text{minimize}}\, f(\theta), \qquad (10.96)$$

where

$$f(\theta) = \sum_{i=1}^{n} \ell\left(\theta; x^{(i)}, f^{(i)}\right) = \sum_{i=1}^{n} \ell_i(\theta). \qquad (10.97)$$

As previously mentioned, the challenge with these problems is that we often have large training sets where n may be in the billions. That means that computing the objective can be costly, and computing the gradient can be even more costly.

If we divide the objective by n (which does not change the solution), the objective function becomes an approximation of the expected value (see Appendix A.9):

$$f(\theta) = \frac{1}{n} \sum_{i=1}^{n} \ell_i(\theta) = \mathbb{E}(\ell(\theta)) \qquad (10.98)$$

From probability theory, we know that we can estimate an expected value from a smaller set of random samples. For the application of estimating a gradient, we call this subset of random samples a *minibatch*

10.5 Deep Neural Networks

$S = \{x^{(1)} \ldots x^{(m)}\}$, where m is usually between 1 and a few hundred. The entries $x^{(1)}, \ldots, x^{(m)}$ do not correspond to the first n entries but are drawn randomly from a uniform probability distribution (Fig. 10.27). Using the minibatch, we can estimate the gradient as the sum of the subfunction gradients at different training points:

$$\nabla_\theta f(\theta) \approx \frac{1}{m} \sum_{i \in S} \nabla_\theta \ell\left(\theta; x^{(i)}, f^{(i)}\right). \quad (10.99)$$

Thus, we divide the training data into these minibatches and use a new minibatch to estimate the gradients at each iteration in the optimization.

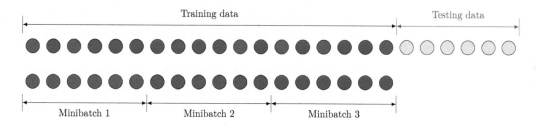

Fig. 10.27 Minibatches are randomly drawn from the training data.

This approach works well for these specific problems because of the unique form for the objective (Eq. 10.98). As an example, for one million training samples, a single gradient evaluation would require evaluating all one million training samples. Alternatively, for a similar cost, a minibatch approach can update the optimization variables a million times using the gradient estimated from one training sample at a time. This latter process usually converges much faster, mainly because we are only fitting parameters against limited data in these problems, so we generally do not need to find the exact minimum.

Typically, this gradient is used with steepest descent methods (Section 4.4.1), more typically referred to as *gradient descent* in the machine learning communities. As discussed in Chapter 4, steepest descent is not the most effective optimization algorithm. However, steepest descent with the minibatch updates, called *stochastic gradient descent*, has been found to work well in machine learning applications. This suitability is primarily because (1) many machine learning optimizations are performed repeatedly, (2) the true objective is difficult to formalize, and (3) finding the absolute minimum is not as important as finding a good enough solution quickly. One key difference in stochastic gradient descent relative to the steepest descent method is that we do not perform a line search. Instead, the step size (called the *learning rate* in machine learning applications) is a preselected value that is usually decreased between major optimization iterations.

Stochastic minibatching is easily applied to first-order methods and has thus driven the development of improvements on stochastic gradient descent, such as momentum, Adagrad, and Adam.[177] Although some of these methods may seem somewhat ad hoc, there is mathematical rigor to many of them.[178] Batching makes the gradients noisy, so second-order methods are generally not pursued. However, ongoing research is exploring stochastic batch approaches that might effectively leverage the benefits of second-order methods.

177. Ruder, *An overview of gradient descent optimization algorithms*, 2016.

178. Goh, *Why momentum really works*, 2017.

10.6 Optimization and Infill

Once a surrogate model has been built, optimization may be performed using the surrogate function values. That is, instead of minimizing the expensive function $f(x)$, we minimize the model $\hat{f}(x)$, as previously illustrated in Fig. 10.1.

The surrogate model may be static, but more commonly, it is updated between optimization iterations by adding new training data and rebuilding the model.

The process by which we select new data points is called *infill*. There are two main approaches to infill: prediction-based exploitation and error-based exploration. Typically, only one infill point is chosen at a time. The assumption is that evaluating the model is computationally expensive, but rebuilding and evaluating the surrogate is cheap.

10.6.1 Exploitation

For models that do not provide uncertainty estimates, the only real option is exploitation. A prediction-based exploitation infill strategy adds an infill point wherever the surrogate predicts the optimum. The reasoning behind this approach is that in SBO, we do not necessarily care about having a globally accurate surrogate; instead, we only care about having an accurate surrogate near the optimum.

The most logical point to sample is thus the optimum predicted by the surrogate. Likely, the location predicted by the surrogate will not be at the true optimum. However, evaluating this point adds valuable information in the region of interest.

We rebuild the surrogate and re-optimize, repeating the process until convergence. This approach usually results in the quickest convergence to an optimum, which is desirable when the actual function is expensive to evaluate. The downside is that we may converge prematurely to an inferior local optimum for problems with multiple local optima.

Even though the approach is called *exploitation*, the optimizer used on the surrogate can be a global search method (gradient-based or gradient-free), although it is usually a local search method. If uncertainty is present, using the mean value of the surrogate as the infill criteria results in essentially an exploitation strategy.

The algorithm is outlined in Alg. 10.4. Convergence could be based on a maximum number of iterations or a tolerance for the objective function's fractional change.

Algorithm 10.4 Exploitation-driven surrogate-based optimization

Inputs:
    n_s: Number of initial samples
    $\underline{x}, \overline{x}$: Variable lower and upper bounds
    τ: Convergence tolerance

Outputs:
    x^*: Best point identified
    f^*: Corresponding function value

$x^{(i)} = \text{sample}(n_s, n_d)$ Sample
$f^{(i)} = f\left(x^{(i)}\right)$ Evaluate function
$k = 0$
while $k < k_{\max}$ and $\left(\hat{f}^* - f_{\text{new}}\right)/\hat{f}^* < \tau$ **do**
    $\hat{f} = \text{surrogate}\left(x^{(i)}, f^{(i)}\right)$ Construct surrogate model
    $x^*, \hat{f}^* = \min \hat{f}(x)$ Perform optimization on the surrogate function
    $f_{\text{new}} = f(x^*)$ Evaluate true function at predicted optimum
    $x^{(i)} = x^{(i)} \cup x^*$ Append new point to training data
    $f^{(i)} = f^{(i)} \cup f_{\text{new}}$ Append corresponding function value
    $k = k + 1$
end while

10.6.2 Efficient Global Optimization

An alternative approach to infill uses error-based exploration. This approach requires using kriging (Section 10.4) or another surrogate approach that predicts not just function values but also error estimates. Although many infill metrics exist within this category, we focus on a popular one called *expected improvement*, and the associated algorithm, *efficient global optimization* (EGO).[145]

As stated previously, sampling where the mean is low is an exploitation strategy, but we do not necessarily want to sample where the uncertainty is high. That may lead to wasteful function calls in regions of the design space where the surrogate model is inaccurate but which are far from any optimum. In effect, this strategy would be

145. Jones et al., *Efficient global optimization of expensive black-box functions*, 1998.

like a larger sampling plan aiming to reduce error everywhere in the surrogate. Instead, we want to sample where we have the maximum probability of finding a better point.

Let the best solution we have found so far be $f^* = f(x^*)$. The improvement for any new test point x is then given by

$$I(x) = \max\left(f^* - f(x), 0\right). \tag{10.100}$$

If $f(x) \geq f^*$, there is no improvement, but if $f(x) < f^*$, the improvement is positive. However, $f(x)$ is not a deterministic value in this model but rather a probability distribution. Thus, the expected improvement is the expected value (or mean) of the improvement:

$$EI(x) = \mathbb{E}\left(\max(f^* - f(x), 0)\right). \tag{10.101}$$

The expected value for a kriging model can be found analytically as:

$$EI(x) = (f^* - \mu_f(x))\Phi\left(\frac{f^* - \mu_f(x)}{\sigma_f(x)}\right) + \sigma_f(x)\phi\left(\frac{f^* - \mu_f(x)}{\sigma_f(x)}\right), \tag{10.102}$$

where Φ and ϕ are the CDF and PDF, respectively, for the standard normal distribution, and μ_f and σ_f are the mean and standard error functions produced from kriging (Eqs. 10.75 and 10.76).

The algorithm is similar to that of the previous section (Alg. 10.4), but instead of choosing the minimum of the surrogate, the selected infill point is the point with the greatest expected improvement. The corresponding algorithm is detailed in Alg. 10.5.

Algorithm 10.5 Efficient global optimization

Inputs:
  n_s: Number of initial samples
  $\underline{x}, \overline{x}$: Lower and upper bounds
  τ: Minimum expected improvement

Outputs:
  x^*: Best point identified
  f^*: Corresponding function value

$x^{(i)} = \text{sample}(n_s, n_d)$ Sample
$f^{(i)} = f(x^{(i)})$ Evaluate function
$f^* = \min\{f^{(i)}\}$ Best point so far; also update corresponding x^*
$k = 0$
while $k < k_{\max}$ and $f_{ei} > \tau$ **do**

10.6 Optimization and Infill

$\mu(x), \sigma(x) = \text{GP}(x^{(i)}, f^{(i)})$ Construct Gaussian process surrogate model
$x_k, f_{ei} = \max EI(x)$ Maximize expected improvement
$f_k = f(x_k)$ Evaluate true function at predicted optimum
$f^* = \min\{f^*, f_k\}$ Update best point and x^* if necessary
$x^{(i)} \leftarrow [x^{(i)}, x_k]$ Add new point to training data
$f^{(i)} \leftarrow [f^{(i)}, f_k]$
$k = k + 1$
end while

Example 10.10   Expected improvement

Consider the same one-dimensional function of Ex. 10.7 using kriging (without gradients), where the data points and fit are shown again in Fig. 10.28. The best point we have found so far is denoted in the figure as x^*, f^*. For a Gaussian process model, the fit also provides a 1-standard-error region, represented by the shaded region in Ex. 10.7.

Now imagine we want to evaluate this function at some new test point, $x_{\text{test}} = 3.25$. In Fig. 10.28, the full probability distribution for the objective at x_{test} is shown in red. This probability distribution occurs at a fixed value of x, so we can visualize it in a dimension coming out of the page. The integral of the shaded red region is the probability of improvement over the best point. The expected value is similar to the probability of improvement. However, rather than returning a probability, it returns the expected magnitude of improvement. That magnitude may be more helpful in defining stopping criteria than quantifying a probability; that is, if the amount of improvement is negligible, it does not matter that the associated probability is high.

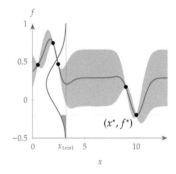

Fig. 10.28 At a given test point ($x_{\text{test}} = 3.25$), we highlight the probability distribution and the expected improvement in the shaded red region.

Now, let us evaluate the expected improvement not just at x_{test}, but across the domain. The result is shown by the red function in the top left of Fig. 10.29. The highest peak suggests that we expect the largest improvement close to our best known point at this first iteration. We also see significant potential for improvement in the middle region of high uncertainty. The expected improvement metric does not simply capture regions with high uncertainty but rather regions that are likely to lead to improvement (which may also have high uncertainty). On the left side of the figure, for example, we anticipate zero expected improvement. For our next sample, we would choose the location with the greatest expected improvement, rebuild the surrogate model, and repeat.

A few select iterations in the convergence process are shown in the remaining panes of Fig. 10.29. On the top right, after the first promising valley is well explored, the middle region becomes the most likely location of potential improvements. Eventually, the potential improvements are minor, below our convergence threshold, and we terminate (bottom right).

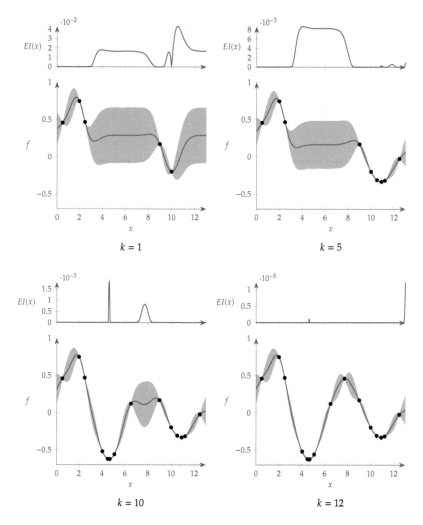

Fig. 10.29 Expected improvement evaluated across the domain.

10.7 Summary

Surrogate-based optimization can be an effective approach to optimization problems where models are expensive to evaluate or noisy. The first step in building a surrogate model is sampling, in which we select the points that are evaluated to obtain the training data. Full factorial searches are too expensive for even a modest number of variables, and random sampling does not provide good coverage, so we need techniques that provide good coverage with a small number of samples. Popular techniques for this kind of sampling include LHS and low-discrepancy sequences.

The next step is surrogate selection and construction. For a given choice of basis functions, regression is used to select optimal model parameters. Cross validation is a critical component of this process. We want good predictive capability, which means that the models work well on data that the model has not been trained against. Model selection often involves trade-offs of more rigid models that do not need as much training data versus more flexible models that require more training data. Polynomials are often used for regression problems because a relatively small number of samples can be used to capture model behavior. Radial basis functions are more often used for interpolation because they can handle multimodal behavior but may require more training data.

Kriging and deep neural nets are two options that model more complex and multimodal design spaces. When using these models, special considerations are needed for efficiency, such as using symmetric matrix factorizations and gradients for kriging and using backpropagation and stochastic gradients for deep neural nets.

The last step of the process is infill, where points are sampled during optimization to update the surrogate model. Some approaches are exploitation-based, where we perform optimization using the surrogate and then use the optimal solution to update the model. Other approaches are exploration-based, where we sample not just at the deterministic optimum but also at points where the expected improvement is high. Exploration-based approaches require surrogate models that provide uncertainty estimates, such as kriging models.

Problems

10.1 Answer *true* or *false* and justify your answer.

a. You should use surrogate-based optimization when a problem has an expensive simulation and many design variables because it is immune to the "curse of dimensionality".

b. Latin hypercube sampling is a random process that is more efficient than pure random sampling.

c. LHS seeks to minimize the distance between the samples, with the constraint that the projection on each axis must follow a chosen probability distribution.

d. Polynomial regressions are not considered to be surrogate models because they are too simple and do not consider any of the model physics.

e. There can be some overlap between the training points and cross-validation points, as long as that overlap is small.

f. Cross validation is a required step in selecting basis functions for SBO.

g. In addition to modeling the function values, kriging surrogate models also provide an estimate of the uncertainty in the values.

h. A prediction-based exploitation infill strategy adds an infill point wherever the surrogate predicts the largest error.

i. Maximizing the expected improvement maximizes the probability of finding a better function value.

j. Neural networks require many nodes with a variety of sophisticated activation functions to represent challenging nonlinear models.

k. Backpropagation is the computation of the derivatives of the neural net error with respect to the activation function weights using reverse-mode AD.

10.2 *Latin hypercube sampling*. Implement an LHS sampling algorithm and plot 20 points across two dimensions with uniform projection in both dimensions. Overlay the grid to check that one point occurs in each bin.

10.3 *Inversion sampling*. Use inversion sampling with Latin hypercube sampling to create and plot 100 points across two dimensions. Each dimension should follow a normal distribution with zero

10.7 Summary

mean and a standard deviation of 1 (cross-terms in covariance matrix are 0).

10.4 *Linear regression.* Use the following training data sampled at x with the resulting function value f (also tabulated on the resources website):

$$x = [-2.0000, -1.7895, -1.5789, -1.3684, -1.1579,$$
$$-0.9474, -0.7368, -0.5263, -0.3158, -0.1053,$$
$$0.1053, 0.3158, 0.5263, 0.7368, 0.9474,$$
$$1.1579, 1.3684, 1.5789, 1.7895, 2.0000]$$

$$f = [7.7859, 5.9142, 5.3145, 5.4135, 1.9367,$$
$$2.1692, 0.9295, 1.8957, -0.4215, 0.8553,$$
$$1.7963, 3.0314, 4.4279, 4.1884, 4.0957,$$
$$6.5956, 8.2930, 13.9876, 13.5700, 17.7481].$$

Use linear regression to determine the coefficients for a polynomial basis of $[x^2, x, 1]$ to predict $f(x)$. Plot your fit against the training data and report the coefficients for the polynomial bases.

10.5 *Cross validation.* Use the following training data sampled at x with resulting function value f (also tabulated on resources website):

$$x = [-3.0, -2.6053, -2.2105, -1.8158, -1.4211,$$
$$-1.0263, -0.6316, -0.2368, 0.1579, 0.5526,$$
$$0.9474, 1.3421, 1.7368, 2.1316, 2.5263,$$
$$2.9211, 3.3158, 3.7105, 4.1053, 4.5]$$

$$f = [43.1611, 28.1231, 12.9397, 3.7628, -2.5457,$$
$$-4.267, 2.8101, -0.6364, 1.1996, -0.9666,$$
$$-2.7332, -6.7556, -9.4515, -7.0741, -7.6989,$$
$$-8.4743, -7.9017, -2.0284, 11.9544, 33.7997].$$

a. Create a polynomial surrogate model using the set of polynomial basis functions x^i for $i = 0$ to n. Plot the error in the surrogate model while increasing n (the maximum order of the polynomial model) from 1 to 20.

b. Plot the polynomial fit for $n = 16$ against the data and comment on its suitability.

c. Re-create the error plot versus polynomial order using k-fold cross validation with 10 divisions. Be sure to limit the y-axes to the area of interest.

d. Plot the polynomial fit against the data for a polynomial order that produces low error under cross validation, and report the coefficients for the polynomial. Justify your selection.

10.6 *Nonlinear least squares.* Implement a Levenberg–Marquardt algorithm and demonstrate its performance on the Rosenbrock function from three different starting points.

10.7 *Kriging.* Implement kriging (without gradients) and demonstrate its fit on the following one-dimensional function:

$$y = \exp(-x)\cos(5x),$$

where $x \in [0, 2.5]$, using the following five sample points: $x = [0, 0.2, 1.0, 1.2, 2.2]$.

10.8 *Efficient global optimization.* Use EGO with the function from the previous problem, showing the iteration history until the expected improvement reduces below 0.001.

Convex Optimization 11

General nonlinear optimization problems are difficult to solve. Depending on the particular optimization algorithm, they may require tuning parameters, providing derivatives, adjusting scaling, and trying multiple starting points. Convex optimization problems do not have any of those issues and are thus easier to solve. The challenge is that these problems must meet strict requirements. Even for candidate problems with the potential to be convex, significant experience is usually needed to recognize and utilize techniques that reformulate the problems into an appropriate form.

> By the end of this chapter you should be able to:
> 1. Understand the benefits and limitations of convex optimization.
> 2. Identify and solve linear and quadratic optimization problems.
> 3. Formulate and solve convex optimization problems.
> 4. Identify and solve geometric programming problems.

11.1 Introduction

Convex optimization problems have desirable characteristics that make them more predictable and easier to solve. Because a convex problem has provably only one optimum, convex optimization methods always converge to the global minimum. Solving convex problems is straightforward and does not require a starting point, parameter tuning, or derivatives, and such problems scale well up to millions of design variables.[179]

All we need to solve a convex problem is to set it up appropriately; there is no need to worry about convergence, local optima, or noisy functions. Some convex problems are so straightforward that they are not recognized as an optimization problem and are just thought

179. Diamond and Boyd, *Convex optimization with abstract linear operators*, 2015.

of as a function or operation. A familiar example of the latter is the linear-least-squares problem (described previously in Section 10.3.1 and revisited in a subsequent section).

Although these are desirable properties, the catch is that convex problems must satisfy strict requirements. Namely, the objective and all inequality constraints must be convex functions, and the equality constraints must be *affine*.*

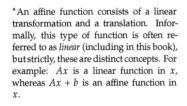

*An affine function consists of a linear transformation and a translation. Informally, this type of function is often referred to as *linear* (including in this book), but strictly, these are distinct concepts. For example: Ax is a linear function in x, whereas $Ax + b$ is an affine function in x.

A function f is convex if

$$f\left((1-\eta)x_1 + \eta x_2\right) \leq (1-\eta)f(x_1) + \eta f(x_2) \tag{11.1}$$

for all x_1 and x_2 in the domain, where $0 \leq \eta \leq 1$. This requirement is illustrated in Fig. 11.1 for the one-dimensional case. The right-hand side of the inequality is just the equation of a line from $f(x_1)$ to $f(x_2)$ (the blue line), whereas the left-hand side is the function $f(x)$ evaluated at all points between x_1 and x_2 (the black curve). The inequality says that the function must always be below a line joining any two points in the domain. Stated informally, a convex function looks something like a bowl.

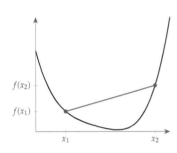

Fig. 11.1 Convex function definition in the one-dimensional case: The function (black line) must be below a line that connects any two points in the domain (blue line).

Unfortunately, even these strict requirements are not enough. In general, we cannot identify a given problem as convex or take advantage of its structure to solve it efficiently and must therefore treat it as a general nonlinear problem. There are two approaches to taking advantage of convexity. The first one is to directly formulate the problem in a known convex form, such as a linear program or a quadratic program (discussed later in this chapter). The second option is to use *disciplined convex optimization*, a specific set of rules and mathematical functions that we can use to build up a convex problem. By following these rules, we can automatically translate the problem into an efficiently solvable form.

Although both of these approaches are straightforward to apply, they also expose the main weakness of these methods: we need to express the objective and inequality constraints using only these elementary functions and operations. In most cases, this requirement means that the model must be simplified. Often, a problem is not directly expressed in a convex form, and a combination of experience and creativity is needed to reformulate the problem in an equivalent manner that is convex.

Simplifying models usually results in a fidelity reduction. This is less problematic for optimization problems intended to be solved repeatedly, such as in optimal control and machine learning, which are domains in which convex optimization is heavily used. In these cases, simplification by local linearization, for example, is less problematic because the linearization can be updated in the next time step. However, this fidelity reduction is problematic for design applications.

In design scenarios, the optimization is performed once, and the design cannot continue to be updated after it is created. For this reason, convex optimization is less frequently used for design applications, except for some limited uses in geometric programming, a topic discussed in more detail in Section 11.6.

This chapter just introduces convex optimization and is not a replacement for more comprehensive textbooks on the topic.[†] We focus on understanding what convex optimization is useful for and describing the most widely used forms.

The known categories of convex optimization problems include linear programming, quadratic programming, second-order cone programming, semidefinite programming, cone programming, and graph form programming. Each of these categories is a subset of the next (Fig. 11.2).[‡]

We focus on the first three because they are the most widely used, including in other chapters in this book. The latter three forms are less frequently formulated directly. Instead, users apply elementary functions and operations and the rules specified by disciplined convex programming, and a software tool transforms the problem into a suitable conic form that can be solved. Section 11.5 describes this procedure.

After covering the three main categories of convex optimization problems, we discuss geometric programming. Geometric programming problems are not convex, but with a change of variables, they can be transformed into an equivalent convex form, thus extending the types of problems that can be solved with convex optimization.

[†]Boyd and Vandenberghe[86] is the most cited textbook on convex optimization.

86. Boyd and Vandenberghe, *Convex Optimization*, 2004.

[‡]Several references exist with examples for those categories that we do not discuss in detail.[180–181]

180. Lobo et al., *Applications of second-order cone programming*, 1998.

181. Parikh and Boyd, *Block splitting for distributed optimization*, 2013.

182. Vandenberghe and Boyd, *Semidefinite programming*, 1996.

183. Vandenberghe and Boyd, *Applications of semidefinite programming*, 1999.

11.2 Linear Programming

A *linear program* (LP) is an optimization problem with a linear objective and linear constraints and can be written as

$$\begin{aligned}
\underset{x}{\text{minimize}} \quad & f^\mathsf{T} x \\
\text{subject to} \quad & Ax + b = 0 \\
& Cx + d \leq 0,
\end{aligned} \quad (11.2)$$

where f, b, and d are vectors and A and C are matrices. All LPs are convex.

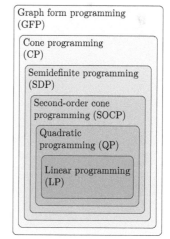

Fig. 11.2 Relationship between various convex optimization problems.

Example 11.1 Formulating a linear programming problem

Suppose we are shopping and want to find how best to meet our nutritional needs for the lowest cost. We enumerate all the food options and use the variable x_j to represent how much of food j we purchase. The parameter c_j is the cost of a unit amount of food j. The parameter N_{ij} is the amount of nutrient i contained in a unit amount of food j. We need to make sure we have at least r_i of nutrient i to meet our dietary requirements. We can now formulate the cost objective as

$$\underset{x}{\text{minimize}} \sum_j c_j x_j = c^\mathsf{T} x.$$

To meet the nutritional requirement of nutrient i, we need to satisfy

$$\sum_j N_{ij} x_j \geq r_i \Rightarrow Nx \geq r.$$

Finally, we cannot purchase a negative amount of food, so $x \geq 0$. The objective and all of the constraints are linear in x, so this is an LP (where $f \equiv c$, $C \equiv -N$, $d \equiv r$ in Eq. 11.2). We do not need to artificially restrict which foods we include in our initial list of possibilities. The formulation allows the optimizer to select a given food item x_i to be zero (i.e., do not purchase any of that food item), according to what is optimal.

As a concrete example, consider a simplified version (and a reductionist view of nutrition) with 10 food options and three nutrients with the amounts listed in the following table.

Food	Cost	Nutrient 1	Nutrient 2	Nutrient 3
A	0.46	0.56	0.29	0.48
B	0.54	0.84	0.98	0.55
C	0.40	0.23	0.36	0.78
D	0.39	0.48	0.14	0.59
E	0.49	0.05	0.26	0.79
F	0.03	0.69	0.41	0.84
G	0.66	0.87	0.87	0.01
H	0.26	0.85	0.97	0.77
I	0.05	0.88	0.13	0.13
J	0.60	0.62	0.69	0.10

If the amount of each food is x, the cost column is c, and the nutrient columns are $n_1, n_2,$ and n_3, we can formulate the LP as

$$\underset{x}{\text{minimize}} \quad c^\mathsf{T} x$$
$$\text{subject to} \quad 5 \leq n_1^\mathsf{T} x \leq 8$$
$$7 \leq n_2^\mathsf{T} x$$
$$1 \leq n_3^\mathsf{T} x \leq 10$$
$$x \leq 4.$$

The last constraint ensures that we do not overeat any one item and get tired of it. LP solvers are widely available, and because the inputs of an LP are just a table of numbers some solvers do not even require a programming language. The solution for this problem is

$$x = [0, 1.43, 0, 0, 0, 4.00, 0, 4.00, 0.73, 0],$$

suggesting that our optimal diet consists of items B, F, H, and I in the proportions shown here. The solution reached the upper limit for nutrient 1 and the lower limit for nutrient 2.

LPs frequently occur with allocation or assignment problems, such as choosing an optimal portfolio of stocks, deciding what mix of products to build, deciding what tasks should be assigned to each worker, or determining which goods to ship to which locations. These types of problems frequently occur in domains such as operations research, finance, supply chain management, and transportation.*

*See Section 2.3 for a brief historical background on the development of LP and its applications.

A common consideration with LPs is whether or not the variables should be discrete. In Ex. 11.1, x_i is a continuous variable, and purchasing fractional amounts of food may or may not be possible, depending on the type of food. Suppose we were performing an optimal stock allocation. In that case, we can purchase fractional amounts of stock. However, if we were optimizing how much of each product to manufacture, it might not be feasible to build 32.4 products. In these cases, we need to restrict the variables to be integers using integer constraints. These types of problems require discrete optimization algorithms, which are covered in Chapter 8. Specifically, we discussed a mixed-integer LP in (Section 8.3).

11.3 Quadratic Programming

A *quadratic program* (QP) has a quadratic objective and linear constraints. Quadratic programming was introduced in Section 5.5 in the context of sequential quadratic programming. A general QP can be expressed as follows:

$$\begin{aligned} \underset{x}{\text{minimize}} \quad & \frac{1}{2}x^T Q x + f^T x \\ \text{subject to} \quad & Ax + b = 0 \\ & Cx + d \leq 0. \end{aligned} \quad (11.3)$$

A QP is only convex if the matrix Q is positive semidefinite. If $Q = 0$, a QP reduces to an LP.

One of the most common QP examples is least squares regression, which was discussed previously in Section 10.3.1 and is used in many applications such as data fitting.

The linear least-squares problem has an analytic solution if A has full rank, so the machinery of a QP is not necessary. However, we can add constraints in QP form to solve *constrained least squares* problems, which do not have analytic solutions in general.

Example 11.2 A constrained least squares QP

The left pane of Fig. 11.3 shows some example data that are both noisy and biased relative to the true (but unknown) underlying curve, represented as a dashed line. Given the data points, we would like to estimate the underlying functional relationship. We assume that the relationship is cubic and write it as

$$y(x) = a_1 x^3 + a_2 x^2 + a_3 x + a_4.$$

We need to estimate the coefficients $a_1, \ldots, a_4$. As discussed previously, this can be posed as a QP problem or, even more simply, as an analytic problem. The middle pane of Fig. 11.3 shows the resulting least squares fit.

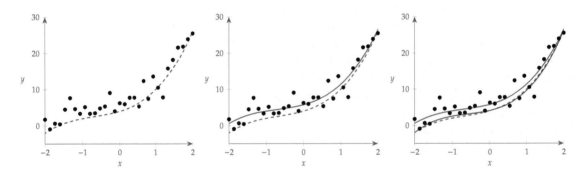

Fig. 11.3 True function on the left, least squares in the middle, and constrained least squares on the right.

Suppose that we know the upper bound of the function value based on measurements or additional data at a few locations. In this example, assume that we know that $f(-2) \leq -2$, $f(0) \leq 4$, and $f(2) \leq 26$. These requirements can be posed as linear constraints:

$$\begin{bmatrix} (-2)^3 & (-2)^2 & -2 & 1 \\ 0 & 0 & 0 & 1 \\ 2^3 & 2^2 & 2 & 1 \end{bmatrix} \begin{bmatrix} a_1 \\ a_2 \\ a_3 \\ a_4 \end{bmatrix} \leq \begin{bmatrix} -2 \\ 4 \\ 26 \end{bmatrix}.$$

After adding these linear constraints and retaining a quadratic objective (the sum of the squared error), the resulting problem is still a QP. The resulting solution is shown in the right pane of Fig. 11.3, which results in a much more accurate fit.

> **Example 11.3** Linear-quadratic regulator (LQR) controller
>
> Another common example of a QP occurs in optimal control. Consider the following discrete-time linear dynamic system:
>
> $$x_{t+1} = Ax_t + Bu_t,$$
>
> where x_t is the deviation from a desired state at time t (e.g., the positions and velocities of an aircraft), and u_t represents the control inputs that we want to optimize (e.g., control surface deflections). This dynamic equation can be used as a set of linear constraints in an optimization problem, but we must decide on an objective.
>
> We would like to have small x_t because that would mean reducing the error in our desired state quickly, but we would also like to have small u_t because small control inputs require less energy. These are competing objectives, where a small control input will take longer to minimize error in a state, and vice versa.
>
> One way to express this objective is as a quadratic function,
>
> $$\underset{x,u}{\text{minimize}} \quad \frac{1}{2} \sum_{t=0}^{n} \left(x_t^T Q x_t + u_t^T R u_t \right),$$
>
> where the weights in Q and R reflect our preferences on how important it is to have a small state error versus small control inputs.* This function has a form similar to kinetic energy, and the LQR problem could be thought of as determining the control inputs that minimize the energy expended, subject to the vehicle dynamics. This choice of the objective function was intentional because the problem is a convex QP (as long as we choose positive weights). Because it is convex, this problem can be solved reliably and efficiently, which are necessary conditions for a robust control law.

*This is an example of a multiobjective function, which is explained in Chapter 9.

11.4 Second-Order Cone Programming

A *second-order cone program* (SOCP) has a linear objective and a second-order cone constraint:

$$\begin{aligned}
\underset{x}{\text{minimize}} \quad & f^T x \\
\text{subject to} \quad & \|A_i x + b_i\|_2 \leq c_i^T x + d_i \quad (11.4) \\
& Gx + h = 0.
\end{aligned}$$

If $A_i = 0$, then this form reduces to an LP.

One useful subset of SOCP is a *quadratically constrained quadratic program* (QCQP). A QCQP is the same as a QP but has quadratic inequality constraints instead of linear ones, that is,

$$\begin{aligned}
\underset{x}{\text{minimize}} \quad & \tfrac{1}{2} x^\mathsf{T} Q x + f^\mathsf{T} x \\
\text{subject to} \quad & A x + b = 0 \\
& \tfrac{1}{2} x^\mathsf{T} R_i x + c_i^\mathsf{T} x + d_i \leq 0 \text{ for } i = 1, \ldots, m,
\end{aligned} \quad (11.5)$$

where Q and R must be positive semidefinite for the QCQP to be convex. A QCQP reduces to a QP if $R = 0$. We formulated QCQPs when solving trust-region problems in Section 4.5. However, for trust-region problems, only an approximate solution method is typically used.

Every QCQP can be expressed as an SOCP (although not vice versa). The QCQP in Eq. 11.5 can be written in the equivalent form,

$$\begin{aligned}
\underset{x,\beta}{\text{minimize}} \quad & \beta \\
\text{subject to} \quad & \|Fx + g\|_2 \leq \beta \\
& Ax + b = 0 \\
& \|G_i x + h_i\|_2 \leq 0.
\end{aligned} \quad (11.6)$$

If we square both sides of the first and last constraints, this formulation is exactly equivalent to the QCQP where $Q = 2F^\mathsf{T} F$, $f = 2F^\mathsf{T} g$, $R_i = 2G_i^\mathsf{T} G_i$, $c_i = 2G_i^\mathsf{T} h_i$, and $d_i = h_i^\mathsf{T} h_i$. The matrices F and G_i are the square roots of the matrices Q and R_i, respectively (divided by 2), and would be computed from a factorization.

11.5 Disciplined Convex Optimization

Disciplined convex optimization builds convex problems using a specific set of rules and mathematical functions. By following this set of rules, the problem can be translated automatically into a conic form that we can efficiently solve using convex optimization algorithms.[46] Table 11.1 shows several examples of convex functions that can be used to build convex problems. Notice that not all functions are continuously differentiable because this is not a requirement of convexity.

[46] Grant et al., *Disciplined convex programming*, 2006.

A disciplined convex problem can be formulated using any of these functions for the objective and inequality constraints. We can also use various operations that preserve convexity to build up more complex functions. Some of the more common operations are as follows:

- Multiplying a convex function by a positive constant
- Adding convex functions

11.5 Disciplined Convex Optimization

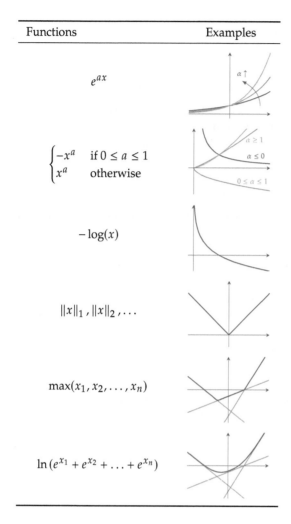

Table 11.1 Examples of convex functions.

- Composing a convex function with an affine function (i.e., if $f(x)$ is convex, then $f(Ax + b)$ is also convex)
- Taking the maximum of two convex functions

Although these functions and operations greatly expand the types of convex problems that we can solve beyond LPs and QPs, they are still restrictive within the broader scope of nonlinear optimization. Still, for objectives and constraints that require only simple mathematical expressions, there is the possibility that the problem can be posed as a disciplined convex optimization problem.

The original expression of a problem is often not convex but can be made convex through a transformation to a mathematically equivalent problem. These transformation techniques include implementing a change of variables, adding slack variables, or expressing the objective in a different form. Successfully recognizing and applying these techniques is a skill requiring experience.

Tip 11.1 Software for disciplined convex programming

CVX and its variants are free popular tools for disciplined convex programming with interfaces for multiple programming languages.*

*https://stanford.edu/~boyd/software.html

Example 11.4 A supervised learning classification problem

A *classification* problem seeks to determine a decision boundary between two sets of data. For example, given a large set of engineering parts, each associated with a label identifying whether it was defective or not, we would like to determine an optimal set of parameters that allow us to predict whether a new part will be defective or not. First, we have to decide on a set of *features*, or properties that we use to characterize each data point. For an engineering part, for example, these features might include dimensions, weights and moments of inertia, or surface finish.

If the data are separable, we could find a hyperplane,

$$f(x) = a^\mathsf{T} x + \beta,$$

that separates the two data sets, or in other words, a function that classifies the objects. For example, if we call one data set y_i, for $i = 1 \ldots n_y$, and the other z_i, for $i = 1 \ldots n_z$, we need to satisfy the following constraints:

$$\begin{aligned} a^\mathsf{T} y_i + \beta &\geq \varepsilon \\ a^\mathsf{T} z_i + \beta &\leq -\varepsilon, \end{aligned} \quad (11.7)$$

for some small tolerance ε. In general, there are an infinite number of separating hyperplanes, so we seek the one that maximizes the distance between the points. However, such a problem is not yet well defined because we can multiply a and β in the previous equations by an arbitrary constant to achieve any separation we want, so we need to normalize or fix some reference dimension (only the ratio of the parameters matters in defining the hyperplane, not their absolute magnitudes). We define the optimization problem as follows:

$$\begin{aligned} \text{maximize} \quad & \gamma \\ \text{by varying} \quad & \gamma, a, \beta \\ \text{subject to} \quad & a^\mathsf{T} y_i + \beta \geq \gamma \text{ for } i = 1 \ldots n_y \\ & a^\mathsf{T} z_j + \beta \leq -\gamma \text{ for } j = 1, \ldots, n_z \\ & \|a\| \leq 1. \end{aligned}$$

The last constraint provides a normalization to prevent the problem from being unbounded. This norm constraint is always active ($\|a\| = 1$), but we express it as an inequality so that the problem remains convex (recall that equality

11.5 Disciplined Convex Optimization

constraints must be affine, but inequality constraints can be any convex function). The objective and inequality constraints are all convex functions, so we can solve it in a disciplined convex programming environment. Alternatively, in this case, we could employ a change of variables to put the problem in QP form if desired.

An example is shown in Fig. 11.4 for data with two features for easy visualization. The middle line shows the separating hyperplane and the outer lines are a distance of γ away, just passing through a data point from each set.

If the data are not completely separable, we need to modify our approach. Even if the data are separable, outliers may undesirably pull the hyperplane so that points are closer to the boundary than is necessary. To address these issues, we need to relax the constraints. As discussed, Eq. 11.7 can always be multiplied by an arbitrary constant. Therefore, we can equivalently express the constraints as follows:

$$a^\mathsf{T} y_i + \beta \geq 1$$
$$a^\mathsf{T} z_j + \beta \leq -1.$$

To relax these constraints, we add nonnegative slack variables, u_i and v_j:

$$a^\mathsf{T} y_i + \beta \geq 1 - u_i$$
$$a^\mathsf{T} z_j + \beta \leq -(1 - v_j),$$

where we seek to minimize the sum of the entries in u and v. If they sum to 0, we have the original constraints for a completely separable function. However, recall that we are interested in not just creating separation but also in maximizing the distance to the classification boundary. To accomplish this, we use a regularization approach where our two objectives include maximizing the distance from the boundary and maximizing the sum of the classification margins. The width between the two planes $a^\mathsf{T} x + \beta = 1$ and $a^\mathsf{T} x + \beta = -1$ is $2/\|a\|$. Therefore, to maximize the separation distance, we minimize $\|a\|$. The optimization problem is defined as follows:[†]

$$\begin{aligned}
\text{minimize} \quad & \|a\| + \omega \left(\sum_i u_i + \sum_j v_j \right) \\
\text{by varying} \quad & a, \beta, u, v \\
\text{subject to} \quad & a^\mathsf{T} y_i + \beta \geq (1 - u_i), \quad i = 1, \ldots, n_y \\
& a^\mathsf{T} z_j + \beta \leq -(1 - v_j), \quad j = 1, \ldots, n_z \\
& u \geq 0 \\
& v \geq 0.
\end{aligned}$$

Here, ω is a user-chosen weight reflecting a preference for the trade-offs in separation margin and stricter classification. The problem is still convex, and an example is shown in Fig. 11.5 with a weight of $\omega = 1$.

The methodology can handle nonlinear classifiers by using a different form with kernel functions like those discussed in Section 10.4.

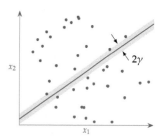

Fig. 11.4 Two separable data sets are shown as points with two different colors. A classification boundary with maximum width is shown.

[†]In the machine learning community, this optimization problem is known as a *support vector machine*. This problem is an example of *supervised learning* because classification labels were provided. Classification can be done without labels but requires a different approach under the umbrella of *unsupervised learning*.

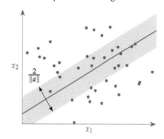

Fig. 11.5 A classification boundary is shown for nonseparable data using a regularization approach.

11.6 Geometric Programming

A *geometric program* (GP) is not convex but can be transformed into an equivalent convex problem. GPs are formulated using monomials and posynomials. A *monomial* is a function of the following form:

$$f(x) = c x_1^{a_1} x_2^{a_2} \cdots x_m^{a_m}, \tag{11.8}$$

where $c > 0$, and all $x_i > 0$. A *posynomial* is a sum of monomials:

$$f(x) = \sum_{j=1}^{n} c_j x_1^{a_{1j}} x_2^{a_{2j}} \cdots x_m^{a_{mj}}, \tag{11.9}$$

where all $c_j > 0$.

Example 11.5 Monomials and posynomials in engineering models

Monomials and posynomials appear in many engineering models. For example, the calculation of lift from the definition of the lift coefficient is a monomial:

$$L = C_L \frac{1}{2} \rho V^2 S.$$

Total incompressible drag, a sum of parasitic and induced drag, is a posynomial:

$$D = C_{D_p} q S + \frac{C_L^2}{\pi A Re} q S.$$

A GP in standard form is written as follows:

$$\begin{aligned} \underset{x}{\text{minimize}} \quad & f_0(x) \\ \text{subject to} \quad & f_i(x) \leq 1 \\ & h_i(x) = 1, \end{aligned} \tag{11.10}$$

where f_i are posynomials, and h_i are monomials. This problem does not fit into any of the convex optimization problems defined in the previous section, and it is not convex. This formulation is useful because we can convert it into an equivalent convex optimization problem.

First, we take the logarithm of the objective and of both sides of the constraints:

$$\begin{aligned} \underset{x}{\text{minimize}} \quad & \ln f_0(x) \\ \text{subject to} \quad & \ln f_i(x) \leq 0 \\ & \ln h_i(x) = 0. \end{aligned} \tag{11.11}$$

11.6 Geometric Programming

Let us examine the equality constraints further. Recall that h_i is a monomial, so writing one of the constraints explicitly results in the following form:
$$\ln\left(c x_1^{a_1} x_2^{a_2} \ldots x_m^{a_m}\right) = 0. \tag{11.12}$$

Using the properties of logarithms, this can be expanded to the equivalent expression:
$$\ln c + a_1 \ln x_1 + a_2 \ln x_2 + \ldots + a_m \ln x_m = 0. \tag{11.13}$$

Introducing the change of variables $y_i = \ln x_i$ results in the following equality constraint:
$$a_1 y_1 + a_2 y_2 + \ldots + a_m y_m + \ln c = 0, \quad a^\mathsf{T} y + \ln c = 0, \tag{11.14}$$

which is an affine constraint in y.

The objective and inequality constraints are more complex because they are posynomials. The expression $\ln f_i$ written in terms of a posynomial results in the following:
$$\ln\left(\sum_{j=1}^{n} c_j x_1^{a_{1j}} x_2^{a_{2j}} \ldots x_m^{a_{mj}}\right). \tag{11.15}$$

Because this is a sum of products, we cannot use the logarithm to expand each term. However, we still introduce the same change of variables (expressed as $x_i = e^{y_i}$):

$$\begin{aligned}
\ln f_i &= \ln\left(\sum_{j=1}^{n} c_j \exp(y_1 a_{1j}) \exp(y_2 a_{2j}) \ldots \exp(y_m a_{mj})\right) \\
&= \ln\left(\sum_{j=1}^{n} c_j \exp(y_1 a_{1j} + y_2 a_{2j} + y_m a_{mj})\right) \quad (11.16) \\
&= \ln\left(\sum_{j=1}^{n} \exp\left(a_j^\mathsf{T} y + b_j\right)\right), \quad \text{where} \quad b_j = \ln c_j.
\end{aligned}$$

This is a log-sum-exp of an affine function. As mentioned in the previous section, log-sum-exp is convex, and a convex function composed of an affine function is a convex function. Thus, the objective and inequality constraints are convex in y. Because the equality constraints are also affine, we have a convex optimization problem obtained through a change of variables.

*Based on an example from Boyd et al.[184]

184. Boyd et al., *A tutorial on geometric programming*, 2007.

> **Example 11.6** Maximizing volume of a box as a geometric program
>
> Suppose we want to maximize the volume of a box with a constraint on the total surface area (i.e., the material used), and a constraint on the aspect ratio of the base of the box.* We parameterize the box by its height x_h, width x_w, and depth x_d:
>
> $$\begin{aligned} \text{maximize} \quad & x_h x_w x_d \\ \text{by varying} \quad & x_h, x_w, x_d \\ \text{subject to} \quad & 2(x_h x_w + x_h x_d + x_w x_d) \leq A \\ & \alpha_l \leq \frac{x_w}{x_d} \leq \alpha_h. \end{aligned}$$
>
> We can express this problem in GP form (Eq. 11.10):
>
> $$\begin{aligned} \text{minimize} \quad & x_h^{-1} x_w^{-1} x_d^{-1} \\ \text{by varying} \quad & x_h, x_w, x_d \\ \text{subject to} \quad & \frac{2}{A} x_h x_w + \frac{2}{A} x_h x_d + \frac{2}{A} x_w x_d \leq 1 \\ & \frac{1}{\alpha_h} x_w x_d^{-1} \leq 1 \\ & \alpha_l x_d x_w^{-1} \leq 1. \end{aligned}$$
>
> We can now plug this into a GP solver. For this example, we use the following parameters: $\alpha_l = 2$, $\alpha_h = 8$, $A = 100$. The solution is $x_d = 2.887$, $x_h = 3.849$, $x_w = 5.774$, with a total volume of 64.16.

Unfortunately, many other functions do not fit this form (e.g., design variables that can be positive or negative, terms with negative coefficients, trigonometric functions, logarithms, and exponents). GP modelers use various techniques to extend usability, including using a Taylor series across a restricted domain, fitting functions to posynomials,[185] and rearranging expressions to other equivalent forms, including implicit relationships. Creativity and some sacrifice in fidelity are usually needed to create a corresponding GP from a general nonlinear programming problem. However, if the sacrifice in fidelity is not too great, there is a significant advantage because the formulation comes with all the benefits of convexity—guaranteed convergence, global optimality, efficiency, no parameter tuning, and limited scaling issues.

185. Hoburg et al., *Data fitting with geometric-programming-compatible softmax functions*, 2016.

One extension to geometric programming is signomial programming. A signomial program has the same form, except that the coefficients c_i can be positive or negative (the design variables x_i must still be strictly positive). Unfortunately, this problem cannot be transformed into a convex one, so a global optimum is no longer guaranteed. Still, a signomial program can usually be solved using a sequence of geometric programs, so it is much more efficient than solving the general nonlinear problem. Signomial programs have been used to extend the range of design problems that can be solved using geometric programming techniques.[186,187]

> **Tip 11.2 Software for geometric programming**
>
> GPkit[†] is a freely available software package for posing and solving geometric programming (and signomial programming) models.

186. Kirschen et al., *Application of signomial programming to aircraft design*, 2018.

187. York et al., *Turbofan engine sizing and tradeoff analysis via signomial programming*, 2018.

[†]https://gpkit.readthedocs.io

11.7 Summary

Convex optimization problems are highly desirable because they do not require parameter tuning, starting points, or derivatives and converge reliably and rapidly to the global optimum. The trade-off is that the form of the objective and constraints must meet stringent requirements. These requirements often necessitate simplifying the physics models and implementing clever reformulations. The reduction in model fidelity is acceptable in domains where optimizations are performed repeatedly in time (e.g., controls, machine learning) or for high-level conceptual design studies. Linear programming and quadratic programming, in particular, are widely used across many domains and form the basis of many of the gradient-based algorithms used to solve general nonconvex problems.

Problems

11.1 Answer *true* or *false* and justify your answer.

a. The optimum found through convex optimization is guaranteed to be the global optimum.

b. Cone programming problems are a special case of quadratic programming problems.

c. It is sometimes possible to obtain distinct feasible regions in linear optimization.

d. A quadratic problem is a problem with a quadratic objective and quadratic constraints.

e. A quadratic problem is only convex if the Hessian of the objective function is positive definite.

f. Solving a quadratic problem is easy because the solution can be obtained analytically.

g. Least squares regression is a type of quadratic programming problem.

h. Second-order cone programming problems feature a linear objective and a second-order cone constraint.

i. Disciplined convex optimization builds convex problems by using convex differentiable functions.

j. It is possible to transform some nonconvex problems into convex ones by using a change of variables, adding slack variables, or reformulating the objective function.

k. A geometric program is not convex but can be transformed into an equivalent convex program.

l. Convex optimization algorithms work well as long as a good starting point is provided.

11.2 Solve the following using a convex solver (not a general nonlinear solver):

$$\begin{aligned}
\text{minimize} \quad & x_1^2 + 3x_2^2 \\
\text{subject to} \quad & x_1 + 4x_2 \geq 2 \\
& 3x_1 + 2x_2 \geq 5 \\
& x_1 \geq 0, x_2 \geq 0.
\end{aligned}$$

11.3 The following foods are available to you at your nearest grocer:

Food	Cost	Nutrient 1	Nutrient 2	Nutrient 3
A	7.68	0.16	1.41	2.40
B	9.41	0.47	0.58	3.95
C	6.74	0.87	0.56	1.78
D	3.95	0.62	1.59	4.50
E	3.13	0.29	0.42	2.65
F	6.63	0.46	1.84	0.16
G	5.86	0.28	1.23	4.50
H	0.52	0.25	1.61	4.70
I	2.69	0.28	1.11	3.11
J	1.09	0.26	1.88	1.74

Minimize the amount you spend while making sure you get at least 5 units of nutrient 1, between 8 and 20 units of nutrient 2, and between 5 and 30 units of nutrient 3. Also be sure not to buy more than 4 units of any one food item, just for variety. Determine the optimal amount of each item to purchase and the total cost.

11.4 Consider the aircraft wing design problem described in Appendix D.1.6. Modify or approximate the model as needed to formulate it as a GP. Solve the new formulation using a GP solver.

If you want to make it more challenging, *do not read the hints that follow*. All equations except the Gaussian efficiency curve are compatible with GP. However, you may need additional optimization variables and constraints. For example, you could add L and v to a set of variables and impose

$$L = \frac{1}{2}\rho v^2 bc C_L$$

as an equality constraint. This is equivalent to a GP-compatible monomial constraint

$$\frac{\rho v^2 bc C_L}{2L} = 1.$$

The efficiency curve can be approximated by a posynomial function. For example, assuming that the optimal speed is $v^* \approx 18 \, \text{m/s}$, you may use

$$4\left(\frac{\eta}{\eta_{\max}}\right)^{10} + 16 = v,$$

which is only valid if $\eta \in [0, \eta_{\max}]$ and $v \in [16, 20]$ m/s.

12 Optimization Under Uncertainty

Uncertainty is always present in engineering design. Manufacturing processes create deviations from the specifications, operating conditions vary from the ideal, and some parameters are inherently variable. Optimization with deterministic inputs can lead to poorly performing designs. *Optimization under uncertainty* (OUU) is the optimization of systems in the presence of random parameters or design variables. The objective is to produce robust and reliable designs. A design is *robust* when the objective function is less sensitive to inherent variability. A design is *reliable* when it is less prone to violating a constraint when accounting for the variability.*

This chapter discusses how uncertainty can be used in the objective function to obtain robust designs and how it can be used in constraints to get reliable designs. We introduce methods that propagate input uncertainties through a computational model to produce output statistics.

We assume familiarity with basic statistics concepts such as expected value, variance, probability density functions (PDFs), cumulative distribution functions (CDFs), and some common probability distributions. A brief review of these topics is provided in Appendix A.9 if needed.

*Although we maintain a distinction in this book, some of the literature includes both of these concepts under the umbrella of "robust optimization".

By the end of this chapter you should be able to:

1. Define robustness and reliability in the context of optimization under uncertainty.

2. Describe and use several strategies for both robust optimization and reliability.

3. Understand the pros and cons for the following forward-propagation methods: first-order perturbation methods, direct quadrature, Monte Carlo methods, and polynomial chaos.

4. Use forward-propagation methods in optimization.

12.1 Robust Design

We call a design *robust* if its performance is less sensitive to inherent variability. In optimization, "performance" is directly associated with the objective function. Satisfying the design constraints is a requirement, but adding a margin to a constraint does not increase performance in the standard optimization formulation. Thus, for a robust design, the objective function is less sensitive to variations in the random design variables and parameters. We can achieve this by formulating an objective function that considers such variations and reflects uncertainty.

A common example of robust design is considering the performance of an engineering device at different operating conditions. If we had deterministic operating conditions, it would make sense to maximize the performance for those conditions. For example, suppose we knew the exact wind speeds and wind directions a sailboat would experience in a race. In that case, we could optimize the hull and sail design to minimize the time around the course. Unfortunately, if variability does exist, the sailboat designed for deterministic conditions will likely perform poorly in off-design conditions. A better strategy considers the uncertainty in the operating conditions and maximizes the expected performance across a range of conditions. A robust design achieves good performance even with uncertain wind speeds and directions.

There are many options for formulating robust design optimization problems. The most common OUU objective is to minimize the *expected value* of the objective function (min $\mu_f(x)$). This yields robust designs because the average performance under variability is considered.

Consider the function shown on the left in Fig. 12.1. If x is deterministic, minimizing this function yields the global minimum on the right. Now consider what happens when x is uncertain. "Uncertain" means that x is no longer a deterministic input. Instead, it is a random variable with some probability distribution. For example, $x = 0.5$ represents a random variable with a mean of $\mu_x = 0.5$. We can compute the average value of the objective μ_f at each x from the expected value of a function (Eq. A.65):

$$\mu_f(x) = \int_{-\infty}^{\infty} f(z)p(z)\,dz, \quad \text{where} \quad p(z) \sim \mathcal{N}(x, \sigma_x), \tag{12.1}$$

and z is a dummy variable for integration. Repeating this integral at each x value gives the expected value as a function of x.

Figure 12.1 shows the expected value of the objective for three different standard deviations. The probability distribution of x for

12.1 Robust Design

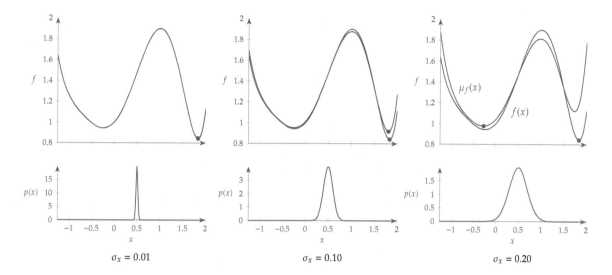

$\sigma_x = 0.01$ $\qquad$ $\sigma_x = 0.10$ $\qquad$ $\sigma_x = 0.20$

Fig. 12.1 The global minimum of the expected value μ_f can shift depending on the standard deviation of x, σ_x. The bottom row of figures shows the normal probability distributions at $x = 0.5$.

a mean value of $x = 0.5$ and three different standard deviations is shown on the bottom row the figure. For a small variance ($\sigma_x = 0.01$), the expected value function $\mu_f(x)$ is indistinguishable from the deterministic function $f(x)$, and the global minimum is the same for both functions. However, for $\sigma_x = 0.2$, the minimum of the expected value function is different from that of the deterministic function. Therefore, the minimum on the right is not as robust as the one on the left. The minimum one on the right is a narrow valley, so the expected value increases rapidly with increased variance. The opposite is true for the minimum on the left. Because it is in a broad valley, the expected value is less sensitive to variability in x. Thus, a design whose performance changes rapidly with respect to variability is not robust.

Of course, the mean is just one possible statistical output metric. Variance, or standard deviation (σ_f), is another common metric. However, directly minimizing the variance is less common because although low variability is often desirable, such an objective has no incentive to improve mean performance and so usually performs poorly. These two metrics represent a trade-off between risk (variance) and reward (mean). The compromise between these two metrics can be quantified through multiobjective optimization (see Chapter 9), which would result in a Pareto front with the notional behavior illustrated in Fig. 12.2. Because both multiobjective optimization and uncertainty quantification are costly, the overall cost of producing such a Pareto front might be prohibitive. Therefore, we might instead seek to minimize the expected value while constraining the variance to a value that the designer can tolerate. Another option is to minimize the mean plus weighted standard deviations.

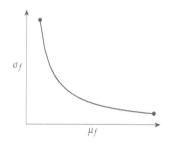

Fig. 12.2 When designing for robustness, there is an inherent trade-off between risk (represented by the variance, σ_f) and reward (represented by the expected value, μ_f).

Many other relevant statistical objectives do not involve statistical moments like mean or variance. Examples include minimizing the 95th percentile of the distribution or employing a reliability metric, $\Pr(f(x) > f_{\text{crit}})$, that minimizes the probability that the objective exceeds some critical value.

Example 12.1 Robust airfoil optimization

Consider an airfoil optimization, where the profile shape of a wing is optimized to minimize the drag coefficient while constraining the lift coefficient to be equal to a target value. Figure 12.3 shows how the drag coefficient of an RAE 2822 airfoil varies with the Mach number (the airplane speed) in blue, as evaluated by a Navier–Stokes flow solver.* This is a typical drag rise curve, where increasing the Mach number leads to stronger shock waves and an associated increase in wave drag.

Now let us optimize the airfoil shape so that we can fly faster without a large increase in drag. Minimizing the drag of this airfoil at Mach 0.71 results in the red drag curve shown in Fig. 12.3. The drag is much lower at Mach 0.71 (as requested!), but any deviation from the target Mach number causes significant drag penalties. In other words, the design is not robust.

One way to improve the design is to use *multipoint optimization*, where we minimize a weighted sum of the drag coefficient evaluated at different Mach numbers. In this case, we use Mach = 0.68, 0.71, 0.725. Compared with the single-point design, the multipoint design has a higher drag at Mach 0.71 but a lower drag at the other Mach numbers, as shown in Fig. 12.3. Thus, a trade-off in peak performance was required to achieve enhanced robustness.

A multipoint optimization is a simplified example of OUU. Effectively, we have treated the Mach number as a random parameter with a given probability at three discrete values. We then minimized the expected value of the drag. This simple change significantly increased the robustness of the design.

---

*For more details on this type of problem and on the aerodynamic shape optimization framework that produced these results, see Martins.[127]

127. Martins, *Perspectives on aerodynamic design optimization*, 2020.

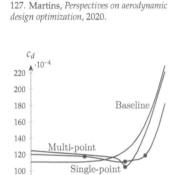

Fig. 12.3 Single-point optimization performs the best at the target speed but poorly away from the condition. Multipoint optimization is more robust to changes in speed.

†See other wind farm OUU problems with coupled farm and turbine optimization,[188] multiobjective trade-offs in mean and variance,[189] and more involved uncertainty quantification techniques discussed later in this chapter.[190]

188. Stanley and Ning, *Coupled wind turbine design and layout optimization with non-homogeneous wind turbines*, 2019.

189. Gagakuma et al., *Reducing wind farm power variance from wind direction using wind farm layout optimization*, 2021.

190. Padrón et al., *Polynomial chaos to efficiently compute the annual energy production in wind farm layout optimization*, 2019.

Example 12.2 Robust wind farm layout optimization

Wind farm layout optimization is another example of OUU but has a more involved probability distribution than the multipoint formulation.† The positions of wind turbines on a wind farm have a substantial impact on overall performance because their wakes interfere. The primary goal of wind farm layout optimization is to position the wind turbines to reduce interference and thus maximize power production. In this example, we optimized the position of nine turbines subject to the constraints that the turbines must stay within a specified boundary and must not be too close to any other turbine.

One of the primary challenges of wind farm layout optimization is that the wind is uncertain and highly variable. To keep this example simple, we assume that wind speed is constant, and only the wind direction is an uncertain

12.1 ROBUST DESIGN

parameter. Figure 12.4 shows a PDF of the wind direction for an actual wind farm, known as a *wind rose*, which is commonly visualized as shown in the plot on the right. The predominant wind directions are from the west and the south. Because of the variable nature of the wind, it would be challenging to intuit the optimal layout.

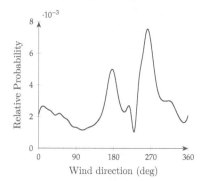
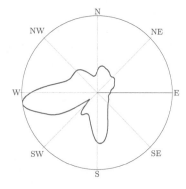

Fig. 12.4 Probability density function of wind direction (left) and corresponding wind rose (right).

We solve this problem using two approaches. The first approach is to solve the problem deterministically (i.e., ignore the variability). This is usually done by using mean values for uncertain parameters, often assuming that the variability is Gaussian or at least symmetric. The wind direction is periodic and asymmetric, so we optimize using the most probable wind direction (261°).

The second approach is to treat this as an OUU problem. Instead of maximizing the power for one direction, we maximize the expected value of the power for all directions. This is straightforward to compute from the definition of expected value because this is a one-dimensional function. Section 12.3 explains other ways to perform forward propagation.

Figure 12.5 shows the power as a function of wind direction for both cases. The deterministic approach results in higher power production when the wind comes from the west (and 180° from that), but that power reduces considerably for other directions. In contrast, the OUU result is less sensitive to changes in wind direction. The expected value of power is 58.6 MW for the deterministic case and 66.1 MW for the OUU case, an improvement of over 12 percent.[‡]

[‡] Instead of using expected power directly, wind turbine designers use annual energy production, which is the expected power multiplied by utilization time.

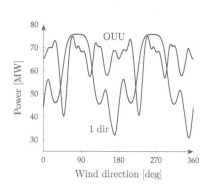
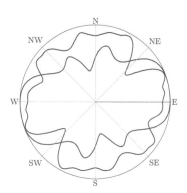

Fig. 12.5 Wind farm power as a function of wind direction for two optimization approaches: deterministic optimization using the most probable direction and OUU.

We can also analyze the trade-off in the optimal layouts. The left side of Fig. 12.6 shows the optimal layout using the deterministic formulation, with the wind coming from the predominant direction (the direction we optimized for). The wakes are shown in blue, and the boundaries are depicted with a dashed line. The optimization spaced the wind turbines out so that there is minimal wake interference. However, the performance degrades significantly when the wind changes direction. The right side of Fig. 12.6 shows the same layout but with the wind coming from the second-most-probable direction. In this case, many of the turbines are operating in the wake of another turbine and produce much less power.

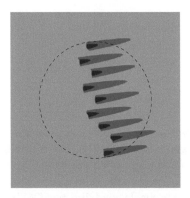
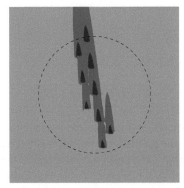

Fig. 12.6 Deterministic cases with the primary wind direction (left) and the secondary wind direction (right).

In contrast, the robust layout is shown in Fig. 12.7, with the predominant wind direction on the left and the second-most-probable direction on the right. In both cases, the wake effects are relatively minor. The turbines are not ideally placed for the predominant direction, but trading the performance for that one direction yields better overall performance when considering other wind directions.

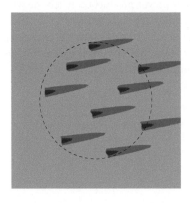
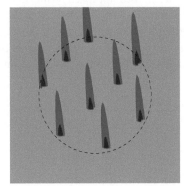

Fig. 12.7 OUU cases with the primary wind direction (left) and the secondary wind direction (right).

12.2 Reliable Design

We call a design *reliable* when it is less prone to failure under variability. In other words, the constraints have a lower probability of being violated under variations in the random design variables and parameters. In a robust design, we consider the effect of uncertainty on the objective function. In reliable design, we consider that effect on the constraints.

A common example of reliability is structural safety. Consider Ex. 3.9, where we formulated a mass minimization subject to stress constraints. In such structural optimization problems, many of the stress constraints are active at the optimum. Constraining the stress to be equal to or below the yield stress value as if this value were deterministic is probably not a good idea because variations in the material properties or manufacturing could result in structural failure. Instead, we might want to include this variability so that we can reduce the probability of failure.

To generate a reliable design, we want the probability of satisfying the constraints to exceed some preselected reliability level. Thus, we change deterministic inequality constraints $g(x) \leq 0$ to ensure that the probability of constraint satisfaction exceeds a specified reliability level r, that is,

$$\Pr(g(x) \leq 0) \geq r. \tag{12.2}$$

For example, if we set $r_i = 0.999$, then constraint i must be satisfied with a probability of 99.9 percent. Thus, we can explicitly set the reliability level that we wish to achieve, with associated trade-offs in the level of performance for the objective function.

Example 12.3 Reliability with the Barnes function

Consider the Barnes problem shown on the left side of Fig. 12.8. The three red lines are the three nonlinear constraints of the problem, and the red regions highlight regions of infeasibility. With deterministic inputs, the optimal value is on the constraint line. An uncertainty ellipse shown around the optimal point highlights the fact that the solution is not reliable. Any variability in the inputs can cause one or more constraints to be violated.

Conversely, the right side of Fig. 12.8 shows a reliable optimum, with the same uncertainty ellipse. In this case, it is much more probable that the design will satisfy all constraints under the input variations. However, as noted in the introduction, increased reliability presents a performance trade-off, with a corresponding increase in the objective function. The higher the reliability we seek, the more we need to give up on performance.

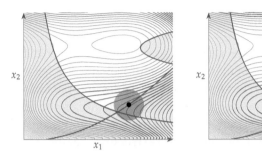

Fig. 12.8 The deterministic optimum design is on the constraint line (left), and the constraint might be violated if there is variability. The reliable design optimum (right) satisfies the constraints despite the variability.

In some engineering disciplines, increasing reliability is handled simply through safety factors. These safety factors are deterministic but are usually derived through statistical means.

Example 12.4 Relating safety factors to reliability

If we were constraining the stress (σ) in a structure to be less than the material's yield stress (σ_y), we would not want to use a constraint of the following form:

$$\sigma(x) \leq \sigma_y.$$

This would be dangerous because we know there is inherent variability in the loads and uncertainty in the yield stress of the material. Instead, we often use a simple safety factor and enforce the following constraint:

$$\sigma(x) \leq \eta \sigma_y,$$

where η is a total safety factor that accounts for safety factors from loads, materials, and failure modes. Of course, not all applications have standards-driven safety factors already determined. The statistical approach discussed in this chapter is useful in these situations to obtain reliable designs.

12.3 Forward Propagation

In the previous sections, we have assumed that we know the statistics (e.g., mean and standard deviation) of the *outputs* of interest (objectives and constraints). However, we generally do not have that information. Instead, we might only know the PDFs of the *inputs*.* *Forward-propagation* methods propagate input uncertainties through a numerical model to compute output statistics.

*Even characterizing input uncertainty might not be straightforward, but for forward propagation, we assume this information is provided.

Uncertainty quantification is a large field unto itself, and we only provide an introduction to it in this chapter. We introduce four well-known nonintrusive methods for forward propagation: first-order perturbation methods, direct quadrature, Monte Carlo methods, and polynomial chaos.

12.3.1 First-Order Perturbation Method

Perturbation methods are based on a local Taylor series expansion of the functional output. In the following, f represents an output of interest, and x represents all the random variables (not necessarily all the variables that f depends on). A first-order Taylor series approximation of f about the mean of x is given by

$$f(x) \approx f(\mu_x) + \sum_{i=1}^{n} \frac{\partial f}{\partial x_i}(x_i - \mu_{x_i}), \qquad (12.3)$$

where n is the dimensionality of x. We can estimate the average value of f by taking the expected value of both sides and using the linearity of expectation as follows:

$$\begin{aligned}\mu_f &= \mathbb{E}(f(x)) \\ &\approx \mathbb{E}(f(\mu_x)) + \sum_i \mathbb{E}\left(\frac{\partial f}{\partial x_i}(x_i - \mu_{x_i})\right) \\ &= f(\mu_x) + \sum_i \frac{\partial f}{\partial x_i}\left(\mathbb{E}(x_i) - \mu_{x_i}\right) \\ &= f(\mu_x) + \sum_i \frac{\partial f}{\partial x_i}\left(\mu_{x_i} - \mu_{x_i}\right). \end{aligned} \qquad (12.4)$$

The last first-order term is zero, so we can write

$$\mu_f = f(\mu_x). \qquad (12.5)$$

That is, when considering only first-order terms, *the mean of the function is the function evaluated at the mean of the input.*

The variance of f is given by

$$\begin{aligned}\sigma_f^2 &= \mathbb{E}(f(x)^2) - (\mathbb{E}(f(x)))^2 \\ &\approx \mathbb{E}\Bigg[f(\mu_x)^2 + 2f(\mu_x)\sum_i \frac{\partial f}{\partial x_i}(x_i - \mu_{x_i}) + \\ &\qquad \sum_i \sum_j \frac{\partial f}{\partial x_i}\frac{\partial f}{\partial x_j}(x_i - \mu_{x_i})(x_j - \mu_{x_j})\Bigg] - f(\mu_x)^2 \\ &= \sum_i \sum_j \frac{\partial f}{\partial x_i}\frac{\partial f}{\partial x_j}\mathbb{E}\left[(x_i - \mu_{x_i})(x_j - \mu_{x_j})\right]. \end{aligned} \qquad (12.6)$$

The expectation term in this equation is the covariance matrix $\Sigma(x_i, x_j)$, so we can write this in matrix notation as

$$\sigma_f^2 = (\nabla_x f)^\mathsf{T} \Sigma (\nabla_x f). \qquad (12.7)$$

We often assume that each random input variable is mutually independent. This is true for the design variables for a well-posed optimization problem, but the parameters may or may not be independent.

When the parameters are independent (this assumption is often made even if not strictly true), the covariance matrix is diagonal, and the variance estimation simplifies to

$$\sigma_f^2 = \sum_{i=1}^{n} \left(\frac{\partial f}{\partial x_i} \sigma_{x_i} \right)^2 . \tag{12.8}$$

These equations are frequently used to propagate errors from experimental measurements. Major limitations of this approach are that (1) it relies on a linearization (first-order Taylor series), which has limited accuracy;[†] (2) it assumes that all uncertain parameters are uncorrelated, which is true for design variables but is not necessarily true for parameters (this assumption can be relaxed by providing the covariances); and (3) it implicitly assumes symmetry in the input distributions because we neglect all higher-order moments (e.g., skewness, kurtosis) and is, therefore, less applicable for problems that are highly asymmetric, such as the wind farm example (Ex. 12.2).

We have *not* assumed that the input or output distributions are normal probability distributions (i.e., Gaussian). However, we can only estimate the mean and variance with a first-order series and not the higher-order moments.

The equation for the variance (Eq. 12.8) is straightforward, but the derivative terms can be challenging when using gradient-based optimization. The first-order derivatives in Eq. 12.7 can be computed using any of the methods from Chapter 6. If they are computed efficiently using a method appropriate to the problem, the forward propagation is efficient as well. However, second-order derivatives are required to use gradient-based optimization (assuming some of the design variables are also random variables). That is because the uncertain objectives and constraints now contain derivatives, and we need derivatives of those functions. Because computing accurate second derivatives is costly, these methods are used less often than the other techniques discussed in this chapter.

We can use a simpler approach if we ignore variability in the objective and focus only on the variability in the constraints (reliability-based optimization). In this case, we can approximate the effect of the uncertainty by pulling it outside of the optimization iterations. We demonstrate one such approach, where we make the additional assumption that each constraint is normally distributed.[192]

[†]Higher-order Taylor series can also be used,[191] but they are less common because of their increased complexity.

191. Cacuci, *Sensitivity & Uncertainty Analysis*, 2003.

192. Parkinson et al., *A general approach for robust optimal design*, 1993.

12.3 Forward Propagation

If $g(x)$ is normally distributed, we can rewrite the probabilistic constraint (Eq. 12.2) as

$$g(x) + z\sigma_g \leq 0, \tag{12.9}$$

where z is chosen for the desired reliability level r. For example, $z = 2$ implies a reliability level of 97.72 percent (one-sided tail of the normal distribution). In many cases, an output distribution is reasonably approximated as normal, but this method tends to be less effective for cases with nonnormal output.

With multiple active constraints, we must be careful to appropriately choose the reliability level for each constraint such that the overall reliability is in the desired range. We often simplify the problem by assuming that the constraints are uncorrelated. Thus, the total reliability is the product of the reliabilities of each constraint.

This simplified approach has the following steps:

1. Compute the deterministic optimum.
2. Estimate the standard deviation of each constraint σ_g using Eq. 12.8.
3. Adjust the constraints to $g(x) + z\sigma_g \leq 0$ for some desired reliability level and re-optimize.
4. Repeat steps 1–3 as needed.

This method is easy to use, and although approximate, the magnitude of error is usually appropriate for the conceptual design phase. If the errors are unacceptable, the standard deviation can be computed inside the optimization. The major limitation of this method is that it only applies to reliability-based optimization.

Example 12.5 Iterative reliability-based optimization

Consider the following problem:

$$\begin{aligned}
\text{minimize} \quad & f = x_1^2 + 2x_2^2 + 3x_3^2 \\
\text{by varying} \quad & x_1, x_2, x_3 \\
\text{subject to} \quad & g_1 = -2x_1 - x_2 - 2x_3 + 6 \leq 0 \\
& g_2 = -5x_1 + x_2 + 3x_3 + 10 \leq 0.
\end{aligned}$$

All the design variables are random variables with standard deviations $\sigma_{x_1} = \sigma_{x_2} = 0.033$, and $\sigma_{x_3} = 0.0167$. We seek a reliable optimum, where each constraint has a target reliability of 99.865 percent.

First, we compute the deterministic optimum, which is

$$x^* = [2.3\overline{15}, 0.3\overline{75}, 0.4\overline{60}], \quad f^* = 6.4\overline{48}.$$

We compute the standard deviation of each constraint, using Eq. 12.8, about the deterministic optimum, yielding $\sigma_{g_1} = 0.081, \sigma_{g_2} = 0.176$. Using an inverse CDF function (discussed in Section 10.2.1) shows that a CDF of 0.99865 corresponds to a z-score of 3. We then re-optimize with the new reliability constraints to obtain the solution:

$$x^* = [2.462, 0.3836, 0.4673], \quad f^* = 7.013.$$

In this case, we sacrificed approximately 9 percent in the objective value to obtain a more reliable design. Because there are two constraints, and each had a target reliability of 99.865 percent, the estimated overall reliability (assuming independence of constraints) is 99.865 percent × 99.865 percent = 99.73 percent.

To check these results, we use Monte Carlo simulations (explained in Section 12.3.3) with 100,000 samples to produce the output histograms shown in Fig. 12.9. The deterministic optimum fails often ($\|g(x)\|_\infty > 0$), so its reliability is a surprisingly poor 34.6 percent. The reliable optimum shifts the distribution to the left, yielding a reliability of 99.75 percent, which is close to our design target.

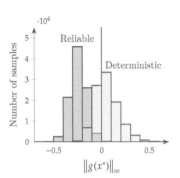

Fig. 12.9 Histogram of maximum constraint violation across 100,000 samples for both the deterministic and reliability-based optimization.

12.3.2 Direct Quadrature

Another approach to estimating statistical outputs of interest is to apply numerical integration (also known as *quadrature*) directly to their definitions. For example:

$$\mu_f = \int f(x) p(x) \, dx \qquad (12.10)$$

$$\sigma_f^2 = \int f(x)^2 p(x) \, dx - \mu_f^2. \qquad (12.11)$$

Discretizing x using n points, we get the summation

$$\int f(x) \, dx \approx \sum_{i=1}^{n} f(x_i) w_i. \qquad (12.12)$$

The quadrature strategy determines the evaluation nodes (x_i) and the corresponding weights (w_i).

The most common quadratures originate from composite Newton–Cotes formulas: the composite midpoint, trapezoidal, and Simpson's rules. These methods use equally spaced nodes, a specification that can be relaxed but still results in a predetermined set of fixed nodes. To reach a specified level of accuracy, it is often desirable to use *nesting*. In this strategy, a refined mesh (smaller spacing between nodes) reuses nodes from the coarser spacing. For example, a simple nesting strategy is to add a new node between all existing nodes. Thus, the accuracy of the integral can be improved up to a specified tolerance while reusing previous function evaluations.

Although straightforward to apply, the Newton–Cotes formulas are usually much less efficient than *Gaussian quadrature*, at least for smooth, nonperiodic functions. Efficiency is highly desirable because the output functions must be called many times for forward propagation, as well as throughout the optimization. The Newton–Cotes formulas are based on fitting polynomials: constant (midpoint), linear (trapezoidal), and quadratic (Simpson's). The weights are adjusted between the different methods, but the nodes are fixed. Gaussian quadrature includes the nodes as degrees of freedom selected by the quadrature strategy. The method approximates the integrand as a polynomial and then efficiently evaluates the integral for the polynomial exactly. Because some of the concepts from Gaussian quadrature are used later in this chapter, we review them here.

An n-point Gaussian quadrature has $2n$ degrees of freedom (n node positions and n corresponding weights), so it can be used to exactly integrate any polynomial up to order $2n - 1$ if the weights and nodes are appropriately chosen. For example, a 2-point Gaussian quadrature can exactly integrate all polynomials up to order 3. To illustrate, consider an integral over the bounds -1 to 1 (we will later see that these bounds can be used as a general representation of any finite bounds through a change of variables):

$$\int_{-1}^{1} f(x)\,dx \approx w_1 f(x_1) + w_2 f(x_2). \tag{12.13}$$

We want this model to be exact for all polynomials up to order 3. If the actual function were a constant ($f(x) = a$), then the integral equation would result in the following:

$$2a = a(w_1 + w_2). \tag{12.14}$$

Repeating this process for polynomials of order 1, 2, and 3 yields four equations and four unknowns:

$$\begin{aligned} 2 &= w_1 + w_2 \\ 0 &= w_1 x_1 + w_2 x_2 \\ \frac{2}{3} &= w_1 x_1^2 + w_2 x_2^2 \\ 0 &= w_1 x_1^3 + w_2 x_2^3. \end{aligned} \tag{12.15}$$

Solving these equations yields $w_1 = w_2 = 1, x_1 = -x_2 = 1/\sqrt{3}$. Thus, we have the weights and node positions that integrate a cubic (or lower-order) polynomial exactly using just two function evaluations, that is,

$$\int_{-1}^{1} f(x)\,dx = f\left(-\frac{1}{\sqrt{3}}\right) + f\left(\frac{1}{\sqrt{3}}\right). \qquad (12.16)$$

More generally, this means that if we can reasonably approximate a general function with a cubic polynomial over the interval, we can provide a good estimate for its integral efficiently.

We would like to extend this procedure to any number of points without the cumbersome approach just applied. The derivation is lengthy (particularly for the weights), so it is not repeated here, other than to explain some of the requirements and the results. The derivation of Gaussian quadrature requires *orthogonal polynomials*. Two vectors are orthogonal if their dot product is zero. The definition is similar for functions, but because functions have an infinite dimension, we require an integral instead of a summation. Thus, two functions f and g are orthogonal over an interval a to b if their inner product is zero. Different definitions can be used for the inner product. The simplest definition is as follows:

$$\int_{a}^{b} f(x)g(x)\,dx = 0. \qquad (12.17)$$

For the Gaussian quadrature derivation, we need a set of polynomials that are not only orthogonal to each other but also to any polynomial of lower order. For the previous inner product, it turns out that Legendre polynomials (L_n is a Legendre polynomial of order n) possess the desired properties:

$$\int_{-1}^{1} x^{k} L_n(x)\,dx = 0, \text{ for any } k < n. \qquad (12.18)$$

Legendre polynomials can be generated by the recurrence relationship,

$$L_{n+1}(x) = \frac{(2n+1)}{(n+1)} x L_n(x) - \frac{n}{(n+1)} L_{n-1}(x), \qquad (12.19)$$

where $L_0 = 1$, and $L_1 = x$. Figure 12.10 shows a plot of the first few Legendre polynomials.

From the Gaussian quadrature derivation, we find that we can integrate any polynomial of order $2n - 1$ exactly by choosing the node positions x_i as the roots of the Legendre polynomial L_n, with the corresponding weights given by

$$w_i = \frac{2}{(1 - x_i^2)\,[L'_n(x_i)]^2}. \qquad (12.20)$$

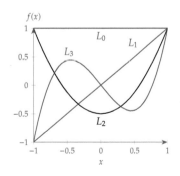

Fig. 12.10 The first few Legendre polynomials.

12.3 Forward Propagation

Legendre polynomials are defined over the interval $[-1, 1]$, but we can reformulate them for an arbitrary interval $[a, b]$ through a change of variables:

$$x = \left(\frac{b-a}{2}\right)z + \left(\frac{b+a}{2}\right), \qquad (12.21)$$

where $z \in [-1, 1]$.

Using the change of variables, we can write

$$\int_a^b f(x)\,dx = \int_{-1}^1 f\left(\frac{(b-a)}{2}z + \frac{b+a}{2}\right)\left(\frac{b-a}{2}\right)dz. \qquad (12.22)$$

Now, applying a quadrature rule, we can approximate the integral as

$$\int_a^b f(x)\,dx \approx \left(\frac{b-a}{2}\right)\sum_{i=1}^m w_i f\left(\frac{(b-a)}{2}z_i + \frac{b+a}{2}\right), \qquad (12.23)$$

where the node locations and respective weights come from the Legendre polynomials.

Recall that what we are after in this section is not just any generic integral but, rather, metrics such as the expected value,

$$\mu_f = \int f(x)p(x)\,dx. \qquad (12.24)$$

As compared to our original integral (Eq. 12.12), we have an additional function $p(x)$, referred to as a *weight function*. Thus, we extend the definition of orthogonal polynomials (Eq. 12.17) to orthogonality with respect to the weight $p(x)$, also known as a *weighted inner product*:

$$\langle f, g \rangle = \int_a^b f(x)g(x)p(x)\,dx = 0. \qquad (12.25)$$

For our purposes, the weight function is $p(x)$, or it is related to it through a change of variables.

Orthogonal polynomials for various weight functions are listed in Table 12.1. The weight function in the table does not always correspond exactly to the typically used PDF ($p(x)$), so a change of variables (like Eq. 12.22) might be needed. The formula described previously is known as *Gauss–Legendre quadrature*, whereas the variants listed in Table 12.1 are called *Gauss–Hermite*, and so on. Formulas and tables with node locations and corresponding weight values exist for most standard probability distributions. For any given weight function, we can generate orthogonal polynomials,[193] and we can generate orthogonal polynomials for general distributions (e.g., ones that were empirically derived).

[193] Golub and Welsch, *Calculation of Gauss quadrature rules*, 1969.

Table 12.1 Orthogonal polynomials that correspond to some common probability distributions.

Prob. dist.	Weight function	Polynomial	Support range
Uniform	1	Legendre	$[-1, 1]$
Normal	e^{-x^2}	Hermite	$(-\infty, \infty)$
Exponential	e^{-x}	Laguerre	$[0, \infty)$
Beta	$(1-x)^\alpha (1+x)^\beta$	Jacobi	$(-1, 1)$
Gamma	$x^\alpha e^{-x}$	Generalized Laguerre	$[0, \infty)$

We now provide more details on Gauss–Hermite quadrature because normal distributions are common. The Hermite polynomials (H_n) follow the recurrence relationship,

$$H_{n+1}(x) = xH_n(x) - nH_{n-1}(x), \tag{12.26}$$

where $H_0(x) = 1$, and $H_1(x) = x$. The first few polynomials are plotted in Fig. 12.11. For Gauss–Hermite quadrature, the nodes are positioned at the roots of $H_n(x)$, and their weights are

$$w_i = \frac{\sqrt{\pi} n!}{n^2 \left(H_{n-1}(\sqrt{2}x_i)\right)^2}. \tag{12.27}$$

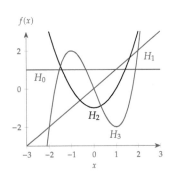

Fig. 12.11 The first few Hermite polynomials.

A coordinate transformation is needed because the standard normal distribution differs slightly from the weight function in Table 12.1. For example, if we are seeking an expected value, with x normally distributed, then the integral is given by

$$\mu_f = \int_{-\infty}^{\infty} f(x) \frac{1}{\sigma\sqrt{2\pi}} \exp\left(-\frac{1}{2}\left(\frac{x-\mu}{\sigma}\right)^2\right) dx. \tag{12.28}$$

We use the change of variables,

$$z = \frac{x-\mu}{\sqrt{2}\sigma}. \tag{12.29}$$

Then, the resulting integral becomes

$$\mu_f = \frac{1}{\sqrt{\pi}} \int_{-\infty}^{\infty} f\left(\mu + \sqrt{2}\sigma z\right) \exp\left(-z^2\right) dz. \tag{12.30}$$

This is now in the appropriate form, so the quadrature rule (using the Hermite nodes and weights) is

$$\mu_f = \frac{1}{\sqrt{\pi}} \sum_{i=1}^{n} w_i f\left(\mu + \sqrt{2}\sigma z_i\right). \tag{12.31}$$

12.3 FORWARD PROPAGATION

Example 12.6 Gauss–Hermite quadrature

Suppose we want to compute the expected value μ_f for the one-dimensional function $f(x) = \cos(x^2)$ at $x = 2$, assuming that x is normally distributed as $x \sim \mathcal{N}(2, 0.2)$.

Let us use Gauss–Hermite quadrature with an increasing number of nodes. We plot the absolute value of the error, $|\varepsilon|$, relative to the exact result ($\mu_f = -0.466842330417276$) versus the number of quadrature points in Fig. 12.12. The Gauss–Hermite quadrature converges quickly; with only six points, we reduce the error to around 10^{-6}. Trapezoidal integration, by comparison, requires over 35 function evaluations for a similar error.

In this problem, we could have taken advantage of symmetry, but we are only interested in the trend (for a smooth function, trapezoidal integration generally converges at least quadratically, whereas Gaussian quadrature converges exponentially).

The first-order method of the previous section predicts $\mu_f = -0.6536$, which is not an acceptable approximation because of the nonlinearity of f.

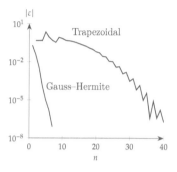

Fig. 12.12 Error in the integral as a function of the number of nodes.

Gaussian quadrature does not naturally lead to nesting, which, as previously mentioned, can increase the accuracy by adding points to a given quadrature. However, methods such as Gauss–Konrod quadrature adapt Gaussian quadrature to utilize nesting. Although Gaussian quadrature is often used to compute one-dimensional integrals efficiently, it is not always the best method. For non-smooth functions, trapezoidal integration is usually preferable because polynomials are ill-suited for capturing discontinuities. Additionally, for periodic functions such as the one shown in Fig. 12.4, the trapezoidal rule is better than Gaussian quadrature, exhibiting exponential convergence.[194,195] This is most easily seen by using a Fourier series expansion.[196]

Clenshaw–Curtis quadrature applies this idea to a general function by employing a change of variables ($x = \cos\theta$) to create a periodic function that can then be efficiently integrated with the trapezoidal rule. Clenshaw–Curtis quadrature also has the advantage that nesting is straightforward and thus desirable for higher-dimensional functions, as discussed next.

The direct quadrature methods discussed so far focused on integration in one dimension, but most problems have more than one random variable. Extending numerical integration to multiple dimensions (also known as *cubature*) is much more challenging. The most obvious extension for multidimensional quadrature is a full grid tensor product. This type of grid is created by discretizing each dimension and then evaluating at every combination of nodes. Mathematically,

194. Wilhelmsen, *Optimal quadrature for periodic analytic functions*, 1978.

195. Trefethen and Weideman, *The exponentially convergent trapezoidal rule*, 2014.

196. Johnson, *Notes on the convergence of trapezoidal-rule quadrature*, 2010.

the quadrature formula can be written as

$$\int f(x)\,\mathrm{d}x_1\,\mathrm{d}x_2\ldots\mathrm{d}x_n \approx$$

$$\sum_i \sum_j \cdots \sum_n f(x_i, x_j, \ldots, x_n) w_i w_j \ldots w_n. \quad (12.32)$$

Although conceptually straightforward, this approach is subject to the *curse of dimensionality*.[‡] The number of points we need to evaluate grows exponentially with the number of input dimensions.

[‡]This is the same issue as with the full factorial sampling used to construct surrogate models in Section 10.2.

One approach to dealing with exponential growth is to use a sparse grid method.[197] The basic idea is to neglect higher-order cross terms. For example, assume that we have a two-dimensional problem and that both variables used a fifth-degree polynomial in the quadrature strategy. The cross terms would include terms up to the 10th order. Although we can integrate these high-order polynomials exactly, their contributions become negligible beyond a specific order. We specify a maximum degree that we want to include and remove all higher-order terms from the evaluation. This method significantly reduces the number of evaluation nodes, with minimal loss in accuracy.

197. Smolyak, *Quadrature and interpolation formulas for tensor products of certain classes of functions*, 1963.

Example 12.7 Sparse grid methods for quadrature

Figure 12.13 compares a two-dimension full tensor grid using the Clenshaw–Curtis exponential rule (left) with a level 5 sparse grid using the same quadrature strategy (right).

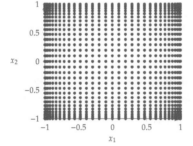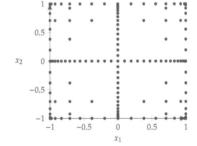

Fig. 12.13 Comparison between a two-dimensional full tensor grid (left) and a level 5 sparse grid using the Clenshaw–Curtis exponential rule (right).

For a problem with dimension d and n sample points in each dimension, the entire tensor grid has a computational complexity of $O(n^d)$. In contrast, the sparse grid method has a complexity of $O(n(\log n)^{d-1})$ with comparable accuracy. This scaling alleviates the curse of dimensionality to some extent. However, the number of evaluation points is still strongly dependent on problem dimensionality, making it intractable in high dimensions.

12.3.3 Monte Carlo Simulation

Monte Carlo simulation is a sampling-based procedure that computes statistics and output distributions. Sampling methods approximate the integrals mentioned in the previous section by using the law of large numbers. The concept is that output probability distributions can be approximated by running the simulation many times with randomly sampled inputs from the corresponding probability distributions. There are three steps:

1. *Random sampling.* Sample n points x_i from the input probability distributions using a random number generator.
2. *Numerical experimentation.* Evaluate the outputs at these points, $f_i = f(x_i)$.
3. *Statistical analysis.* Compute statistics on the discrete output distribution f_i.

For example, the discrete form of the mean is

$$\mu_f = \frac{1}{n} \sum_{i=1}^{n} f_i,, \qquad (12.33)$$

and the unbiased estimate of the variance is computed as

$$\sigma_f^2 = \frac{1}{n-1} \left(\sum_{i=1}^{n} (f_i^2) - n\mu_f^2 \right). \qquad (12.34)$$

We can also estimate $\Pr(g(x) \leq 0)$ by counting how many times the constraint was satisfied and dividing by n. If we evaluate enough samples, our output statistics converge to the actual values by the law of large numbers. Therein also lies this method's disadvantage: it requires a *large number* of samples.

Monte Carlo simulation has three main advantages. First, the convergence rate is independent of the number of inputs. Whether we have 3 or 300 random input variables, the convergence rate is similar because we randomize all input variables for each sample. This is an advantage over direct quadrature for high-dimensional problems because, unlike quadrature, Monte Carlo does not suffer from the curse of dimensionality. Second, the algorithm is easy to parallelize because all of the function evaluations are independent. Third, in addition to statistics like the mean and variance, Monte Carlo generates the output probability distributions. This is a unique advantage compared with first-order perturbation and direct quadrature, which provide summary statistics but not distributions.

Example 12.8 Monte Carlo applied to a one-dimensional function

Consider the one-dimensional function from Fig. 12.1:

$$f(x) = \exp\left(\frac{x}{2}\right) + 0.2x^6 - 0.2x^5 - 0.75x^4 + x^2.$$

We compute the expected value function at each x location using Monte Carlo simulation, for $\sigma = 0.2$. Using different numbers of samples, we obtain the expected value functions plotted in Fig. 12.14. For 100 samples, the noise in the expected value is visible. The noise decreases as the number of samples increases. For 100,000 samples, the noise is barely noticeable in the plot.

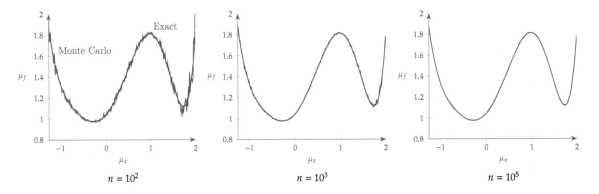

Fig. 12.14 Monte Carlo requires a large number of samples for an accurate prediction of the expected value.

The major disadvantage of the Monte Carlo method is that even though the convergence rate does not depend on the number of inputs, the convergence rate is slow—$O(1/\sqrt{n})$. This means that every additional digit of accuracy requires about 100 times more samples. It is also hard to know which value of n to use a priori. Usually, we need to determine an appropriate value for n through convergence testing (trying larger n values until the statistics converge).

One approach to achieving converged statistics with fewer iterations is to use Latin hypercube sampling (LHS) or low-discrepancy sequences, as discussed in Section 10.2. Both methods allow us to approximate the input distributions with fewer samples. Low-discrepancy sequences are particularly well suited for this application because convergence testing is iterative. When combined with low-discrepancy sequences, the method is called *quasi-Monte Carlo,* and the scaling improves to

12.3 Forward Propagation

$O(1/n)$. Thus, each additional digit of accuracy requires 10 times as many samples. Even with better sampling methods, many simulations are usually required, which can be prohibitive if used as part of an OUU problem.

Example 12.9 Forward propagation with Monte Carlo

Consider a problem with the following objective and constraint:

$$f(x) = x_1^2 + 2x_2^2 + 3x_3^2$$
$$g(x) = x_1 + x_2 + x_3 - 3.5 \leq 0.$$

Suppose that the current optimization iteration is $x = [1, 1, 1]$. We assume that the first variable is deterministic, whereas the latter two variables have uncertainty under a normal distribution with the following standard deviations: $\sigma_2 = 0.06$ and $\sigma_3 = 0.2$. We would like to compute the output statistics for f (mean, variance, and a histogram) and compute the reliability of the constraint at this current iteration.

We do not know how many samples we need to get reasonably converged statistics, so we need to perform a convergence study. For a given number of samples, we generate random numbers normally distributed with mean x_i and standard deviation σ_i. Then we evaluate the functions and compute the mean (Eq. 12.33), variance (Eq. 12.34), and reliability of the outputs.

Figure 12.15 shows the convergence of the mean and standard deviation using a random sampling curve, LHS (Section 10.2.1), and quasi-Monte Carlo (using Halton sequence sampling from Section 10.2.2). The latter two methods converge much more quickly than random sampling. LHS performs better for few samples in this case, but generating the convergence data requires more function evaluations than quasi-Monte Carlo because an all-new set of sample points is generated for each n (instead of being incrementally generated as in the Halton sequence for quasi-Monte Carlo). That cost is less problematic for optimization applications because the convergence testing is only done at the preprocessing stage. Once a number of samples n is chosen for convergence, n is fixed throughout the optimization.

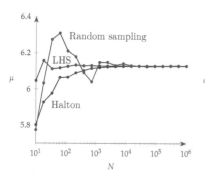
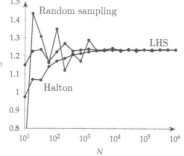

Fig. 12.15 Convergence of the mean (left) and standard deviation (right) versus the number of samples using Monte Carlo.

From the data, we conclude that we need about $n = 10^4$ samples to have well-converged statistics. Using $n = 10^4$ yields $\mu = 6.127, \sigma = 1.235$, and $r = 0.9914$. The random sampling of these results varies between simulations (except for the Halton sequence in quasi-Monte Carlo, which is deterministic).

The production of an output histogram is a key benefit of this method. The histogram of the objective function is shown in Fig. 12.16. Notice that it is not normally distributed in this case.

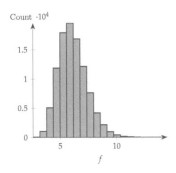

Fig. 12.16 Histogram of objective function for 10,000 samples.

§Polynomial chaos is not chaotic and does not actually need polynomials. The name *polynomial chaos* came about because it was initially derived for use in a physical theory of chaos.[198]

198. Wiener, *The homogeneous chaos*, 1938.

12.3.4 Polynomial Chaos

Polynomial chaos (also known as *spectral expansions*) is a class of forward-propagation methods that take advantage of the inherent smoothness of the outputs of interest using polynomial approximations.§

The method extends the ideas of Gaussian quadrature to estimate the output function, from which the output distribution and other summary statistics can be efficiently generated. In addition to using orthogonal polynomials to evaluate integrals, we use them to approximate the output function. As in Gaussian quadrature, the polynomials are orthogonal with respect to a specified probability distribution (see Eq. 12.25 and Table 12.1). A general function that depends on uncertain variables x can be represented as a sum of basis functions ψ_i (which are usually polynomials) with weights α_i,

$$f(x) = \sum_{i=0}^{\infty} \alpha_i \psi_i(x). \tag{12.35}$$

In practice, we truncate the series after $n + 1$ terms and use

$$f(x) \approx \sum_{i=0}^{n} \alpha_i \psi_i(x). \tag{12.36}$$

The required number of terms n for a given input dimension d and polynomial order o is

$$n + 1 = \frac{(d + o)!}{d! o!}. \tag{12.37}$$

This approach amounts to a truncated generalized Fourier series.

By definition, we choose the first basis function to be $\psi_0 = 1$. This means that the first term in the series is a constant (polynomial of order 0). Because the basis functions are orthogonal, we know that

$$\langle \psi_i, \psi_j \rangle = 0 \text{ if } i \neq j. \tag{12.38}$$

12.3 Forward Propagation

Polynomial chaos consists of three main steps:

1. Select an orthogonal polynomial basis.
2. Compute coefficients to fit the desired function.
3. Compute statistics on the function of interest.

These three steps are described in the following sections. We begin with the last step because it provides insight for the first two.

Compute Statistics

Using the polynomial approximation (Eq. 12.36) in the definition of the mean, we obtain

$$\mu_f = \int_{-\infty}^{\infty} \sum_i \alpha_i \psi_i(x) p(x) \, dx. \tag{12.39}$$

The coefficients α_i are constants that can be taken out of the integral, so we can write

$$\mu_f = \sum_i \alpha_i \int \psi_i(x) p(x) \, dx$$

$$= \alpha_0 \int \psi_0(x) p(x) \, dx + \alpha_1 \int \psi_1(x) p(x) \, dx + \alpha_2 \int \psi_2(x) p(x) \, dx + \ldots.$$

We can multiply all terms by ψ_0 without changing anything because $\psi_0 = 1$, so we can rewrite this expression in terms of the inner product as

$$\mu_f = \alpha_0 \int p(x) \, dx + \alpha_1 \langle \psi_0, \psi_1 \rangle + \alpha_2 \langle \psi_0, \psi_2 \rangle + \ldots. \tag{12.40}$$

Because the polynomials are orthogonal, all the terms except the first are zero (see Eq. 12.38). From the definition of a PDF (Eq. A.63), we know that the first term is 1. Thus, the mean of the function is simply the zeroth coefficient,

$$\mu_f = \alpha_0. \tag{12.41}$$

We can derive a formula for the variance using a similar approach. Substituting the polynomial representation (Eq. 12.36) into the definition of variance and using the same techniques used in deriving the mean, we obtain

$$\sigma_f^2 = \int \left(\sum_i \alpha_i \psi_i(x) \right)^2 p(x) \, dx - \alpha_0^2$$

$$= \sum_i \alpha_i^2 \int \psi_i(x)^2 p(x) \, dx - \alpha_0^2$$

$$\begin{aligned}
\sigma_f^2 &= \alpha_0^2 \int \psi_0^2 p(x)\,dx + \sum_{i=1}^{n} \alpha_i^2 \int \psi_i(x)^2 p(x)\,dx - \alpha_0^2 \\
&= \alpha_0^2 + \sum_{i=1}^{n} \alpha_i^2 \int \psi_i(x)^2 p(x)\,dx - \alpha_0^2 \\
&= \sum_{i=1}^{n} \alpha_i^2 \int \psi_i(x)^2 p(x)\,dx \\
&= \sum_{i=1}^{n} \alpha_i^2 \langle \psi_i^2 \rangle.
\end{aligned} \qquad (12.42)$$

That last step is just the definition of the weighted inner product (Eq. 12.25), providing the variance in terms of the coefficients and polynomials:

$$\sigma_f^2 = \sum_{i=1}^{n} \alpha_i^2 \langle \psi_i^2 \rangle. \qquad (12.43)$$

The inner product $\langle \psi_i^2 \rangle = \langle \psi_i, \psi_i \rangle$ can often be computed analytically. For example, using Hermite polynomials with a normal distribution yields

$$\langle H_n^2 \rangle = n!. \qquad (12.44)$$

For cases without analytic solutions, Gaussian quadrature of this inner product is still straightforward and exact because it only includes polynomials.

For multiple uncertain variables, the formulas are the same, but we use multidimensional basis polynomials. Denoting these multidimensional basis polynomials as Ψ_i, we can write

$$\mu_f = \alpha_0 \qquad (12.45)$$

$$\sigma_f^2 = \sum_{i=1}^{n} \alpha_i^2 \langle \Psi_i^2 \rangle. \qquad (12.46)$$

The multidimensional basis polynomials are defined by products of one-dimensional polynomials, as detailed in the next section. Polynomial chaos computes the mean and variance using these equations and our definition of the inner product. Other statistics can be estimated by sampling the polynomial expansion. Because we now have a simple polynomial representation that no longer requires evaluating the original (potentially expensive) function f, we can use sampling procedures (e.g., Monte Carlo) to create output distributions without incurring high costs. Of course, we have to evaluate the function f to generate the coefficients, as we will discuss later.

12.3 Forward Propagation

Selecting an Orthogonal Polynomial Basis

As discussed in Section 12.3.2, we already know appropriate orthogonal polynomials for many continuous probability distributions (see Table 12.1[¶]). We also have methods to generate other exponentially convergent polynomial sets for any given empirical distribution.[199]

[¶]Other polynomials can be used, but these polynomials are optimal because they yield exponential convergence.

[199]. Eldred et al., *Evaluation of non-intrusive approaches for Wiener–Askey generalized polynomial chaos*, 2008.

The multidimensional basis functions we need are defined by tensor products. For example, if we had two variables from a uniform probability distribution (and thus Legendre bases), then the polynomials up through the second-order terms would be as follows:

$$\Psi_0(x) = \psi_0(x_1)\psi_0(x_2) = 1$$
$$\Psi_1(x) = \psi_1(x_1)\psi_0(x_2) = x_1$$
$$\Psi_2(x) = \psi_0(x_1)\psi_1(x_2) = x_2$$
$$\Psi_3(x) = \psi_1(x_1)\psi_1(x_2) = x_1 x_2$$
$$\Psi_4(x) = \psi_2(x_1)\psi_0(x_2) = \frac{1}{2}\left(3x_1^2 - 1\right)$$
$$\Psi_5(x) = \psi_0(x_1)\psi_2(x_2) = \frac{1}{2}\left(3x_2^2 - 1\right).$$

The $\psi_1(x_1)\psi_2(x_2)$ term, for example, does not appear in this list because it is a third-order polynomial, and we truncated the series after the second-order terms. We should expect this number of basis functions because Eq. 12.37 with $d = 2$ and $o = 2$ yields $n = 6$.

Determine Coefficients

Now that we have selected an orthogonal polynomial basis, $\psi_i(x)$, we need to determine the coefficients α_i in Eq. 12.36. We discuss two approaches for determining the coefficients. The first approach is quadrature, which is also known as *spectral projection*. The second is with regression, which is also known as *stochastic collocation*.

Let us start with the quadrature approach. Beginning with the polynomial approximation

$$f(x) = \sum_i \alpha_i \psi_i(x), \qquad (12.47)$$

we take the inner product of both sides with respect to ψ_j,

$$\langle f(x), \psi_j \rangle = \sum_i \alpha_i \langle \psi_i, \psi_j \rangle. \qquad (12.48)$$

Using the orthogonality property of the basis functions (Eq. 12.38), all the terms in the summation are zero except for

$$\langle f(x), \psi_i \rangle = \alpha_i \langle \psi_i^2 \rangle. \qquad (12.49)$$

Thus, we can find each coefficient by

$$\alpha_i = \frac{1}{\langle \psi_i^2 \rangle} \int f(x)\psi_i(x)p(x)\,\mathrm{d}x, \quad (12.50)$$

where we replaced the inner product with the definition given by Eq. 12.17.

As expected, the zeroth coefficient corresponds to the definition of the mean,

$$\alpha_0 = \int f(x)p(x)\,\mathrm{d}x. \quad (12.51)$$

These coefficients can be obtained through multidimensional quadrature (see Section 12.3.2) or Monte Carlo simulation (Section 12.3.3), which means that this approach inherits the same limitations of the chosen quadrature approach. However, the process can be more efficient if the selected basis functions are good approximations of the distributions. These integrals are usually evaluated using Gaussian quadrature (e.g., Gauss–Hermite quadrature if $p(x)$ is a normal distribution).

Suppose all we are interested in is the mean (Eqs. 12.41 and 12.51). In that case, the polynomial chaos approach amounts to just Gaussian quadrature. However, if we want to compute other statistical properties or produce an output PDF, the additional effort of obtaining the higher-order coefficients produces a polynomial approximation of $f(x)$ that we can then sample to predict other quantities of interest.

It may appear that to estimate $f(x)$ (Eq. 12.36), we need to know $f(x)$ (Eq. 12.50). The distinction is that we just need to be able to evaluate $f(x)$ at some predefined quadrature points, which in turn gives a polynomial approximation for any x.

The second approach to determining the coefficients is regression. Equation 12.36 is linear, so we can estimate the coefficients using least squares (although an underdetermined system with regularization can be used as well). If we evaluate the function m times, where $x^{(i)}$ is the ith sample, the resulting linear system is as follows:

$$\begin{bmatrix} \psi_0\left(x^{(1)}\right) & \cdots & \psi_n\left(x^{(1)}\right) \\ \vdots & & \vdots \\ \psi_0\left(x^{(m)}\right) & \cdots & \psi_n\left(x^{(m)}\right) \end{bmatrix} \begin{bmatrix} \alpha_0 \\ \vdots \\ \alpha_n \end{bmatrix} = \begin{bmatrix} f\left(x^{(1)}\right) \\ \vdots \\ f\left(x^{(m)}\right) \end{bmatrix}. \quad (12.52)$$

∥ There are software packages that facilitate the use of polynomial chaos methods.[200,201]

200. Adams et al., *Dakota, a multilevel parallel object-oriented framework for design optimization, parameter estimation, uncertainty quantification, and sensitivity analysis: Version 6.14 user's manual*, 2021.

201. Feinberg and Langtangen, *Chaospy: An open source tool for designing methods of uncertainty quantification*, 2015.

As a rule of thumb, the number of sample points m should be at least twice as large as the number of unknowns, $n + 1$. The sampling points, also known as the *collocation points*, typically correspond to the nodes in the corresponding quadrature strategy or utilize random sequences.∥

12.3 Forward Propagation

> **Example 12.10** Forward propagation with polynomial chaos

Consider the following objective function:

$$f(x) = 3 + \cos(3x_1) + \exp(-2x_2),$$

where the current iteration is at $x = [1, 1]$, and we assume that both design variables are normally distributed with the following standard deviations: $\sigma = [0.06, 0.2]$.

We approximate the function with fourth-order Hermite polynomials. Using Eq. 12.37, we see that there are 15 basis functions from the various combinations of $H_i H_j$:

$$\Psi_0 = H_0(x_1)H_0(x_2)$$
$$\Psi_1 = H_0(x_1)H_1(x_2) = x_2$$
$$\Psi_2 = H_0(x_1)H_2(x_2) = x_2^2 - 1$$
$$\Psi_3 = H_0(x_1)H_3(x_2) = x_2^3 - 3x_2$$
$$\Psi_4 = H_0(x_1)H_4(x_2) = x_2^4 - 6x_2^2 + 3$$
$$\Psi_5 = H_1(x_1)H_0(x_2) = x_1$$
$$\Psi_6 = H_1(x_1)H_1(x_2) = x_1 x_2$$
$$\Psi_7 = H_1(x_1)H_2(x_2) = x_1 x_2^2 - x_1$$
$$\Psi_8 = H_1(x_1)H_3(x_2) = x_1 x_2^3 - 3x_1 x_2$$
$$\Psi_9 = H_2(x_1)H_0(x_2) = x_1^2 - 1$$
$$\Psi_{10} = H_2(x_1)H_1(x_2) = x_1^2 x_2 - x_2$$
$$\Psi_{11} = H_2(x_1)H_2(x_2) = x_1^2 x_2^2 - x_1^2 - x_2^2 + 1$$
$$\Psi_{12} = H_3(x_1)H_0(x_2) = x_1^3 - 3x_1$$
$$\Psi_{13} = H_3(x_1)H_1(x_2) = x_1^3 x_2 - 3x_1 x_2$$
$$\Psi_{14} = H_4(x_1)H_0(x_2) = x_1^4 - 6x_1^2 + 3.$$

The integrals for the basis functions (Hermite polynomials) have analytic solutions:

$$\langle \Psi_k^2 \rangle = \langle (H_m H_n)^2 \rangle = m! n! .$$

We now compute the following double integrals to obtain the coefficients using Gaussian quadrature:

$$\alpha_k = \frac{1}{\langle \Psi_k^2 \rangle} \int_{-\infty}^{\infty} \int_{-\infty}^{\infty} f(x) \Psi_k(x) p(x) \, dx_1 \, dx_2$$

We must be careful with variable definitions because the inputs are not standard normal distributions. The function f is defined over the unnormalized variable x, whereas our basis functions are defined over a standard normal distribution: $y = (x - \mu)/\sigma$. The probability distribution in this case is a bivariate, uncorrelated, normal distribution:

$$\alpha_k = \frac{1}{\langle \Psi_k^2 \rangle} \int_{-\infty}^{\infty} \int_{-\infty}^{\infty} f(x) \Psi_k \left(\frac{x-\mu}{\sigma} \right) \times$$

$$\frac{1}{2\pi\sigma_1\sigma_2} \exp\left(-\left(\frac{x_1-\mu_1}{\sqrt{2}\sigma_1}\right)^2 \right) \exp\left(-\left(\frac{x_2-\mu_2}{\sqrt{2}\sigma_2}\right)^2 \right) dx_1\, dx_2.$$

To put this in the proper form for Gauss–Hermite quadrature, we use the change of variable $z = (x - \mu)/(\sqrt{2}\sigma)$, as follows:

$$\alpha_k = \frac{1}{\langle \Psi_k^2 \rangle} \frac{1}{\pi} \int_{-\infty}^{\infty} \int_{-\infty}^{\infty} f\left(\sqrt{2}\sigma z + \mu \right) \Psi_k \left(\sqrt{2}z \right) e^{-z_1^2} e^{-z_2^2}\, dz_1\, dz_2.$$

Applying Gauss–Hermite quadrature, the integral is approximated by

$$\alpha_k \approx \frac{1}{\pi \langle \Psi_k^2 \rangle} \sum_{i=1}^{n_i} \sum_{j=1}^{n_j} w_i w_j f(X_{ij}) \Psi_k\left(\sqrt{2} Z_{ij}\right),$$

where n_i and n_j determine the number of quadrature nodes we choose to include, and X_{ij} is the tensor product

$$X = \left(\sqrt{2}\sigma_1 z_1 + \mu_1 \right) \otimes \left(\sqrt{2}\sigma_2 z_2 + \mu_2 \right),$$

and $Z = z_1 \otimes z_2$.

In this case, we choose a full tensor product mesh of the fifth order in both dimensions. The nodes and weights are given by

$$z_1 = z_2 = [-2.02018, -0.95857, 0.0, 0.95857, 2.02018]$$
$$w_1 = w_2 = [0.01995, 0.39362, 0.94531, 0.39362, 0.01995]$$

and visualized as a tensor product of evaluation points in Fig. 12.17. The nonzero coefficients (within a tolerance of approximately 10^{-4}) are as follows:

$$\alpha_0 = 2.1725$$
$$\alpha_1 = -0.0586$$
$$\alpha_2 = 0.0117$$
$$\alpha_3 = -0.00156$$
$$\alpha_5 = -0.0250$$
$$\alpha_9 = 0.01578.$$

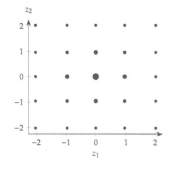

Fig. 12.17 Evaluation nodes with area proportional to weight.

We can now easily compute the mean and standard deviation as

$$\mu_f = \alpha_0 = 2.1725$$

$$\sigma_f = \sqrt{\sum_{i=1}^{n} \alpha_i^2 \langle \Psi_i^2 \rangle} = 0.06966.$$

In this case, we are able to accurately estimate the mean and standard deviation with only 25 function evaluations. In contrast, applying Monte Carlo to this same problem, with LHS, requires about 10,000 function calls to estimate the mean and over 100,000 function calls to estimate the standard deviation (with less accuracy).

Although direct quadrature would work equally well if all we wanted was the mean and standard deviation, polynomial chaos gives us a polynomial approximation of our function near μ_x:

$$\tilde{f}(x) = \sum_i \alpha_i \Psi_i(x).$$

This fourth-order polynomial is compared to the original function in Fig. 12.18, where the dot represents the mean of x.

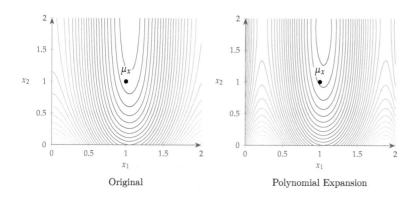

Fig. 12.18 Original function on left, polynomial expansion about μ_x on right.

The primary benefit of this new function is that it is very inexpensive to evaluate (and the original function is often expensive), so we can use sampling procedures to compute other statistics, such as percentiles or reliability levels, or simply to visualize the output PDF, as shown in Fig. 12.19.

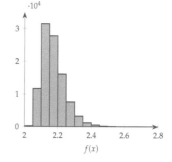

Fig. 12.19 Output histogram produced by sampling the polynomial expansion.

12.4 Summary

Engineering problems are subject to variation under uncertainty. OUU deals with optimization problems where the design variables or other parameters have uncertain variability. Robust design optimization seeks designs that are less sensitive to inherent variability in the objective function. Common OUU objectives include minimizing the mean or standard deviation or performing multiobjective trade-offs between the mean performance and standard deviation. Reliable design optimization seeks designs with a reduced probability of failure, considering the variability in the constraint values. To quantify robustness and reliability, we need a forward-propagation procedure that propagates the probability distributions of the inputs (either design variables or parameters that are fixed during optimization) to the statistics or probability distributions of the outputs (objective and constraint functions). Four classes of forward propagation methods were discussed in this chapter.*

*This list is not exhaustive. For example, the methods discussed in this chapter are nonintrusive. Intrusive polynomial chaos uses expansions inside governing equations. Like intrusive methods for derivative computation (Chapter 6), intrusive methods for forward propagation require more implementation effort but are more accurate and efficient.

Perturbation methods use a Taylor series expansion of the output functions to estimate the mean and variance. These methods can be efficient for a range of problem sizes, especially if accurate derivatives are available. Their main weaknesses are that they require derivatives (and hence second derivatives when using a gradient-based optimization), only work well with symmetric input probability distributions, and only provide the mean and variance (for first-order methods).

Direct quadrature uses numerical quadrature to evaluate the summary statistics. This process is straightforward and effective. Its primary weakness is that it is limited to low-dimensional problems (number of random inputs). Sparse grids enable these methods to handle a higher number of dimensions, but the scaling is still lacking.

Monte Carlo methods approximate the summary statistics and output distributions using random sampling and the law of large numbers. These methods are straightforward to use and are independent of the problem dimension. Their major weakness is that they are inefficient. However, because the alternatives are intractable for a large number of random inputs, Monte Carlo is an appropriate choice for many high-dimensional problems.

Polynomial chaos represents uncertain variables as a sum of orthogonal basis functions. This method is often a more efficient way to characterize both statistical moments and output distributions. However, the methodology is usually limited to a small number of dimensions because the number of required basis functions grows exponentially.

Problems

12.1 Answer *true* or *false* and justify your answer.

   a. The greater the reliability, the less likely the design is to have a worse objective function value.

   b. Reliability can be handled in a deterministic way using safety factors, which ensure that the optimum has some margin before the original constraint is violated.

   c. Forward propagation computes the PDFs of the outputs and inputs for a given numerical model.

   d. The computational cost of direct quadrature scales exponentially with the number of random variables, whereas the cost of Monte Carlo is independent of the number of random variables.

   e. Monte Carlo methods approximate PDFs using random sampling and converges slowly.

   f. The first-order perturbation method computes the PDFs using local Taylor series expansions.

   g. Because the first-order perturbation method requires first-order derivatives to compute the uncertainty metrics, OUU using the first-order perturbation method requires second-order derivatives.

   h. Polynomial chaos is a forward-propagation technique that uses polynomial approximations with random coefficients to model the input uncertainties.

   i. The number of basis functions required by polynomial chaos grows exponentially with the number of uncertain input variables.

12.2 Consider the following problem:

$$\text{minimize} \quad f = x_1^2 + x_2^4 + x_2 \exp(x_3)$$
$$\text{subject to} \quad x_1^2 + x_2^2 + x_3^3 \geq 10$$
$$x_1 x_2 + x_2 x_3 \geq 5.$$

Assume that all design variables are random variables with the following standard deviations: $\sigma_{x_1} = 0.1, \sigma_{x_2} = 0.2, \sigma_{x_3} = 0.05$. Use the iterative reliability-based optimization procedure to find a reliable optimum with an overall reliability of 99.9 percent. How much did the objective decrease relative to the deterministic optimum? Check your reliability level with Monte Carlo simulation.

12.3 Using Gaussian quadrature, find the mean and variance of the function $\exp(\cos(x))$ at $x = 1$, assuming x is normally distributed with a standard deviation of 0.1. Determine how many evaluation points are needed to converge to 5 decimal places. Compare your results to trapezoidal integration.

12.4 Repeat the previous problem, but assume a uniform distribution with a half-width of 0.1.

12.5 Consider the function in Ex. 12.10. Solve the same problem, but use Monte Carlo sampling instead. Compare the output histogram and how many function calls are required to achieve well-converged results for the mean and variance.

12.6 Repeat Ex. 12.10 using polynomial chaos, except with a uniform distribution in both dimensions, where the standard deviations from the example correspond to the half-width of a uniform distribution.

12.7 *Robust optimization of a wind farm.* We want to find the optimal turbine layout for a wind farm to minimize the cost of energy (COE). We will consider a very simplified wind farm with only three wind turbines. The first turbine will be fixed at $(0, 0)$, and the x-positions of the back two turbines will be fixed with 4-diameter spacing between them. The only thing we can change is the y-position of the two back turbines, as shown in Fig. 12.20 (all dimensions in this problem are in terms of rotor diameters). In other words, we just have two design variables: y_2 and y_3.

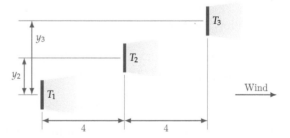

Fig. 12.20 Wind farm layout.

We further simplify by assuming the wind always comes from the west, as shown in the figure, and is always at a constant speed. The wake model has a few parameters that define things like its

12.4 Summary

spread angle and decay rate. We will refer to these parameters as $\alpha, \beta,$ and δ (knowing exactly what each parameter corresponds to is not important for our purposes). The supplementary resources repository contains code for this problem.

a. Run the optimization deterministically, assuming that the three wake parameters are $\alpha = 0.1$, $\beta = 9$, and $\delta = 5$. Because there are several possible similar solutions, we add the following constraints: $y_i \geq 0$ (bound) and $y_3 \geq y_2$ (linear). Do not use $[0, 0]$ as the starting point for the optimization because that occurs right at a flat spot in the wake (a fixed point), so you might not make any progress. Report the optimal spacing that you find.

b. Now assume that the wake parameters are uncertain variables under some probability distribution. Specifically, we have the following information for the three parameters:

- α is governed by a Weibull distribution with a scale parameter of 0.1 and a shape parameter of 1.
- β is given by a normal distribution with a mean and standard deviation of $\mu=9$, $\sigma=1$.
- δ is given by a normal distribution with a mean and standard deviation of $\mu=5$, $\sigma=0.4$.

Note that the mean for all of these distributions corresponds to the deterministic value we used previously.

Using a Monte Carlo method, run an OUU minimizing the 95th percentile for COE.

c. Once you have completed both optimizations, perform a cross analysis by filling out the four numbers in the table that follows.

	Deterministic COE	95th percentile COE
Deterministic layout	[ ]	[ ]
OUU layout	[ ]	[ ]

Take the two optimal designs that you found, and then compare each on the two objectives (deterministic and 95th percentile). The first row corresponds to the performance of the optimal deterministic layout. Evaluate the performance of this layout using the deterministic value for COE and the 95th percentile that accounts for uncertainty. Repeat for the optimal solution for the OUU case. Discuss your findings.

Multidisciplinary Design Optimization 13

As mentioned in Chapter 1, most engineering systems are multidisciplinary, motivating the development of multidisciplinary design optimization (MDO). The analysis of multidisciplinary systems requires coupled models and coupled solvers. We prefer the term *component* instead of *discipline* or *model* because it is more general. However, we use these terms interchangeably depending on the context. When components in a system represent different physics, the term *multiphysics* is commonly used.

All the optimization methods covered so far apply to multidisciplinary problems if we view the coupled multidisciplinary analysis as a single analysis that computes the objective and constraint functions by solving the coupled model for a given set of design variables. However, there are additional considerations in the solution, derivative computation, and optimization of coupled systems.

In this chapter, we build on Chapter 3 by introducing models and solvers for coupled systems. We also expand the derivative computation methods of Chapter 6 to handle such systems. Finally, we introduce various MDO *architectures*, which are different options for formulating and solving MDO problems.

> By the end of this chapter you should be able to:
>
> 1. Describe when and why you might want to use MDO.
> 2. Read and create XDSM diagrams.
> 3. Compute derivatives of coupled models.
> 4. Understand the differences between monolithic and distributed architectures.

13.1 The Need for MDO

In Chapter 1, we mentioned that MDO increases the system performance, decreases the design time, reduces the total cost, and reduces the uncertainty at a given point in time (recall Fig. 1.3). Although these benefits still apply when modeling and optimizing a single discipline or component, broadening the modeling and optimization to the whole system brings on additional benefits.

Even without performing any optimization, constructing a multidisciplinary (coupled) model that considers the whole engineering system is beneficial. Such a model should ideally consider all the interactions between the system components. In addition to modeling physical phenomena, the model should also include other relevant considerations, such as economics and human factors. The benefit of such a model is that it better reflects the actual state and performance of the system when deployed in the real world, as opposed to an isolated component with assumed boundary conditions. Using such a model, designers can quantify the actual impact of proposed changes on the whole system.

When considering optimization, the main benefit of MDO is that optimizing the design variables for the various components *simultaneously* leads to a better system than when optimizing the design variables for each component separately. Currently, many engineering systems are designed and optimized sequentially, which leads to suboptimal designs. This approach is often used in industry, where engineers are grouped by discipline, physical subsystem, or both. This might be perceived as the only choice when the engineering system is too complex and the number of engineers too large to coordinate a simultaneous design involving all groups.

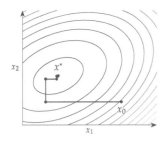

Fig. 13.1 Sequential optimization is analogous to coordinate descent.

Sequential optimization is analogous to coordinate descent, which consists of optimizing each variable sequentially, as shown in Fig. 13.1. Instead of optimizing one variable at a time, sequential optimization optimizes distinct sets of variables at a time, but the principle remains the same. This approach tends to work for unconstrained problems, although the convergence rate is limited to being linear.

One issue with sequential optimization is that it might converge to a suboptimal point for a constrained problem. An example of such a case is shown in Fig. 13.2, where sequential optimization gets stuck at the constraint because it cannot decrease the objective while remaining feasible by only moving in one of the directions. In this case, the optimization must consider both variables simultaneously to find a feasible descent direction.

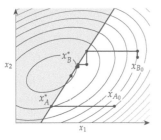

Fig. 13.2 Sequential optimization can fail to find the constrained optimum because the optimization with respect to a set of variables might not see a feasible descent direction that otherwise exists when considering all variables simultaneously.

Another issue is that when there are variables that affect multiple disciplines (called *shared design variables*), we must make a choice about which discipline handles those variables. If we let each discipline optimize the same shared variable, the optimizations likely yield different values for those variables each time, in which case they will not converge. On the other hand, if we let one discipline handle a shared variable, it will likely converge to a value that violates one or more constraints from the other disciplines.

By considering the various components and optimizing a multidisciplinary performance metric with respect to as many design variables as possible simultaneously, MDO automatically finds the best trade-off between the components—this is the key principle of MDO. Suboptimal designs also result from decisions at the system level that involve power struggles between designers. In contrast, MDO provides the right trade-offs because mathematics does not care about politics.

Example 13.1 MDO applied to wing design

Consider a multidisciplinary model of an aircraft wing, where the aerodynamics and structures disciplines are coupled to solve an aerostructural analysis and design optimization problem. For a given flow condition, the aerodynamic solver computes the forces on the wing for a given wing shape, whereas the structural solver computes the wing displacement for a given set of applied forces. Thus, these two models are coupled as shown in Fig. 13.3. For a steady flow condition, there is only one wing shape and a corresponding set of forces that satisfies both disciplinary models simultaneously.

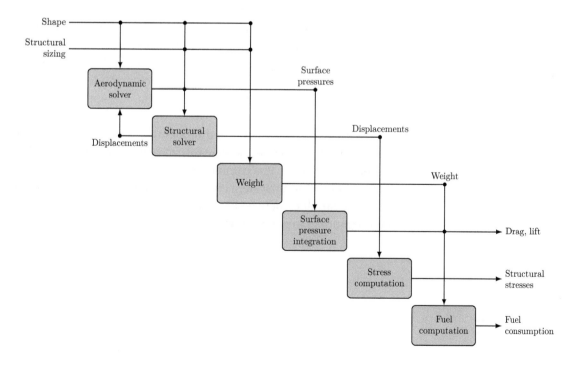

Fig. 13.3 Multidisciplinary numerical model for an aircraft wing.

In the absence of a coupled model, aerodynamicists may have to assume a fixed wing shape at the flight conditions of interest. Similarly, structural designers may assume fixed loads in their structural analysis. However, solving the coupled model is necessary to get the actual flying shape of the wing and the corresponding performance metrics.

One possible design optimization problem based on these models would be to minimize the drag by changing the wing shape and the structural sizing while satisfying a lift constraint and structural stress constraints. Optimizing the wing shape and structural sizing simultaneously yields the best possible result because it finds feasible descent directions that would not be available with sequential optimization. Wing shape variables, such as wingspan, are shared design variables because they affect both the aerodynamics and the structure. They cannot be optimized by considering aerodynamics or structures separately.

13.2 Coupled Models

As mentioned in Chapter 3, a model is a set of equations that we solve to predict the state of the engineering system and compute the objective and constraint function values. More generally, we can have a *coupled model*, which consists of multiple models (or components) that depend on each other's state variables.

The same steps for formulating a design optimization problem (Section 1.2) apply in the formulation of MDO problems. The main difference in MDO problems is that the objective and constraints are computed by the coupled model. Once such a model is in place, the design optimization problem statement (Eq. 1.4) applies, with no changes needed.

A generic example of a coupled model with three components is illustrated in Fig. 13.4. Here, the states of each component affect all other components. However, it is common for a component to depend only on a subset of the other system components. Furthermore, we might distinguish variables between internal state variables and coupling variables (more in this in Section 13.2.2).

Fig. 13.4 Coupled model composed of three numerical models. This coupled model would replace the single model in Fig. 3.21.

Mathematically, a coupled model is no more than a larger set of equations to be solved, where all the governing equation residuals (r), the corresponding state variables (u), and all the design variables (x) are concatenated into single vectors. Then, we can still just write the whole multidisciplinary model as $r(x, u) = 0$.

However, it is often necessary or advantageous to partition the system into smaller components for three main reasons. First, specialized solvers are often already in place for a given set of governing equations, which may be more efficient at solving their set of equations than a general-purpose solver. In addition, some of these solvers might be black boxes that do not provide an interface for using alternative solvers. Second, there is an incentive for building the multidisciplinary system in a modular way. For example, a component might be useful on its own and should therefore be usable outside the multidisciplinary system.

A modular approach also facilitates the extension of the multidisciplinary system and makes it easy to replace the model of a given discipline with an alternative one. Finally, the overall system of equations may be more efficiently solved if it is partitioned in a way that exploits the system structure. These reasons motivate an implementation of coupled models that is flexible enough to handle a mixture of different types of models and solvers for each component.

Tip 13.1 Beware of loss of precision when coupling components

Precision can be lost when coupling components, leading to a loss of precision in the overall coupled system solution. Ideally, the various components would be coupled through memory, that is, a component can provide a pointer to or a copy of the variable or array to the other components. If the type (e.g., double-precision float) is maintained, then there would be no loss in precision.

However, the number type might not be maintained in some conversions, so it is crucial to be aware of this possibility and mitigate it. One common issue is that components need to be coupled through file input and output. Codes do not usually write all the available digits to the file, causing a loss in precision. Casting a read variable to another type might also introduce errors. Find the level of numerical error (Tip 3.2) and mitigate these issues as much as possible.

We start the remainder of this section by defining components in more detail (Section 13.2.1). We explain how the coupling variables relate to the state variables (Section 13.2.2) and coupled system formulation (Section 13.2.3). Then, we discuss the coupled system structure (Section 13.2.4). Finally, we explain methods for solving coupled systems (Section 13.2.5), including a hierarchical approach that can handle a mixture of models and solvers (Section 13.2.6).

13.2.1 Components

In Section 3.3, we explained how all models can ultimately be written as a system of residuals, $r(x, u) = 0$. When the system is large or includes submodels, it might be natural to *partition* the system into *components*. We prefer to use the more general term *components* instead of *disciplines* to refer to the submodels resulting from the partitioning because the partitioning of the overall model is not necessarily by discipline (e.g., aerodynamics, structures). A system model might also be partitioned by physical system components (e.g., wing, fuselage, or an aircraft in a fleet) or by different conditions applied to the same model (e.g., aerodynamic simulations at different flight conditions).

The partitioning can also be performed within a given discipline for the same reasons cited previously. In theory, the system model equations in $r(x, u) = 0$ can be partitioned in any way, but only some partitions are advantageous or make sense. We denote a partitioning into n components as

$$r(u) = 0 \equiv \begin{cases} r_1(u_1; u_2, \ldots, u_i, \ldots, u_n) = 0 \\ \quad \vdots \\ r_i(u_i; u_1, \ldots, u_{i-1}, u_{i+1}, \ldots, u_n) = 0 \\ \quad \vdots \\ r_n(u_n; u_1, \ldots, u_i, \ldots, u_{n-1}) = 0 \end{cases} \quad . \tag{13.1}$$

Each r_i and u_i are *vectors* corresponding to the residuals and states of component i. The semicolon denotes that we solve each component i by driving its residuals (r_i) to zero by varying only its states (u_i) while keeping the states from all other components constant. We assume this is possible, but this is not guaranteed in general. We have omitted the dependency on x in Eq. 13.1 because, for now, we just want to find the state variables that solve the governing equations for a fixed design.

Components can be either *implicit* or *explicit*, a concept we introduced in Section 3.3. To solve an implicit component i, we need an algorithm for driving the equation residuals, $r_i(u_1, \ldots, u_i, \ldots, u_n)$, to zero by varying the states u_i while the other states (u_j for all $j \neq i$) remain fixed. This algorithm could involve a matrix factorization for a linear system or a Newton solver for a nonlinear system.

An explicit component is much easier to solve because that components' states are explicit functions of other components' states. The states of an explicit component can be computed without factorization or iteration. Suppose that the states of a component i are given by the explicit function $u_i = f(u_j)$ for all $j \neq i$. As previously explained in Section 3.3, we can convert an explicit equation to the residual form by

13.2 Coupled Models

moving the function on the right-hand side to the left-hand side. Then, we obtain set of residuals,

$$r_i(u_1, \ldots, u_n) = u_i - f(u_j) \quad \text{for all} \quad j \neq i. \tag{13.2}$$

Therefore, there is no loss of generality when using the residual notation in Eq. 13.1.

Most disciplines involve a mix of implicit and explicit components because, as mentioned in Section 3.3 and shown in Fig. 3.21, the state variables are implicitly defined, whereas the objective function and constraints are usually explicit functions of the state variables. In addition, a discipline usually includes functions that convert inputs and outputs, as discussed in Section 13.2.3.

As we will see in Section 13.2.6, the partitioning of a model can be hierarchical, where components are gathered in multiple *groups*. These groups can be nested to form a hierarchy with multiple levels. Again, this might be motivated by efficiency, modularity, or both.

Example 13.2 Residuals of the coupled aerostructural problem

Let us formulate models for the aerostructural problem described in Ex. 13.1.* A possible model for the aerodynamics is a vortex-lattice model given by the linear system

$$A\Gamma = v,$$

where A is the matrix of aerodynamic influence coefficients, and v is a vector of boundary conditions, both of which depend on the wing shape. The state Γ is a vector that represents the circulation (vortex strength) at each spanwise position on the wing, as shown on the left-hand side of Fig. 13.5. The lift and drag scalars can be computed explicitly for a given Γ, so we write these dependencies as $L = L(\Gamma)$ and $D = D(\Gamma)$, omitting the detailed explicit expressions for conciseness.

*This description omits many details for brevity. Jasa et al.[202] describe the aerostructural model in more detail and cite other references on the background theory.

202. Jasa et al., *Open-source coupled aerostructural optimization using Python*, 2018.

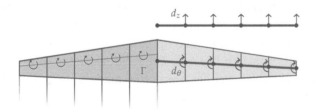

Fig. 13.5 Aerostructural wing model showing the aerodynamic state variables (circulations Γ) on the left and structural state variables (displacements d_z and rotations d_θ) on the right.

A possible model for the structures is a cantilevered beam modeled with Euler–Bernoulli elements,

$$Kd = q, \tag{13.3}$$

where K is the stiffness matrix, which depends on the beam shape and sizing. The right-hand-side vector q represents the applied forces at the spanwise position on the beam. The states d are the displacements and rotations at each node, as shown on the right-hand side of Fig. 13.5. The weight does not depend on the states, and it is an explicit function of the beam sizing and shape, so it does not involve the structural model (Eq. 13.3). The stresses are an explicit function of the displacements, so we can write $\sigma = \sigma(d)$, where σ is a vector whose size is the number of elements.

When we couple these two models, A and v depend on the wing displacements d, and q depends on Γ. We can write all the implicit and explicit equations as residuals:

$$r_1 = A(d)\Gamma - v(d)$$
$$r_2 = Kd - q(\Gamma).$$

The states of this system are as follows:

$$u = \begin{bmatrix} u_1 \\ u_2 \end{bmatrix} \equiv \begin{bmatrix} \Gamma \\ d \end{bmatrix}.$$

This coupled system is illustrated in Fig. 13.6.

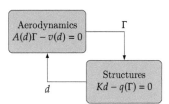

Fig. 13.6 The aerostructural model couples aerodynamics and structures through a displacement and force transfer.

13.2.2 Models and Coupling Variables

In MDO, the *coupling variables* are variables that need to be passed from the model of one discipline to the others because of interdependencies in the system. Thus, the coupling variables are the inputs and outputs of each model. Sometimes, the coupling variables are just the state variables of one model (or a subset of these) that get passed to another model, but often we need to convert between the coupling variables and other variables within the model.

We represent the coupling variables by a vector $\hat{u}_i$, where the subscript i denotes the model that computes these variables. In other words, $\hat{u}_i$ contains the outputs of model i. A model i can take any coupling variable vector $\hat{u}_{j \neq i}$ as one of its inputs, where the subscript indicates that j can be the output from any model except its own. Figure 13.7 shows the inputs and outputs for a model. The model solves for the set of its state variables, u_i. The residuals in the solver depend on the input variables coming from other models. In general, this is not a direct dependency, so the model may require an explicit function (P_i) that converts the inputs ($\hat{u}_{j \neq i}$) to the required parameters p_i. These parameters remain fixed when the model solves its implicit equations for u_i.

After the model solves for its state variables (u_i), there may be another explicit function (Q_i) that converts these states to output variables ($\hat{u}_i$) for the other models. The function (Q_i) typically reduces

13.2 Coupled Models

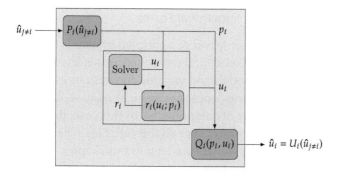

Fig. 13.7 In the general case, a model may require conversions of inputs and outputs distinct from the states that the solver computes.

the number of output variables relative to the number of internal states, sometimes by orders of magnitude.

The model shown in Fig. 13.7 can be viewed as an implicit function that computes its outputs as a function of all the inputs, so we can write $\hat{u}_i = U_i(\hat{u}_{j \neq i})$. The model contains three components: two explicit and one implicit. We can convert the explicit components to residual equations using Eq. 13.2 and express the model as three sets of residuals as shown in Fig. 13.8. The result is a group of three components that we can represent as $r(u) = 0$. This conversion and grouping hint at a powerful concept that we will use later, which is *hierarchy*, where components can be grouped using multiple levels.

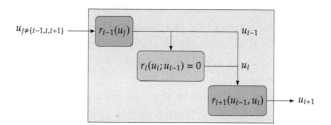

Fig. 13.8 The conversion of inputs and outputs can be represented as explicit components with corresponding state variables. Using this form, any model can be entirely expressed as $r(u) = 0$. The inputs could be any subset of u except for those handled in the component (u_{i-1}, u_i, and u_{i+1}).

Example 13.3 Conversion of inputs and outputs in aerostructural problems

Consider the structural model from Ex. 13.2. We wrote $q(\Gamma)$ to represent the dependency of the external forces on the aerodynamic model circulations to keep the notation simple, but in reality, there should be a separate explicit component that converts Γ into q. The circulation translates to a lift force at each spanwise position, which in turn needs to be distributed consistently to the nodes of each beam element. Also, the displacements given by the structural model (translations and rotations of each node) must be converted into a twist distribution on the wing, which affects the right-hand side of the aerodynamic model, $\theta(d)$. Both of these conversions are explicit functions.

13.2.3 Residual and Functional Forms

The *system-level* representation of a coupled system is determined by the variables that are "seen" and controlled at this level.

Representing all models and variable conversions as $r(u) = 0$ leads to the *residual form* of the coupled system, already written in Eq. 13.1, where n is the number of components. In this case, the system level has direct access and control over all the variables. This residual form is desirable because, as we will see later in this chapter, it enables us to formulate efficient ways to solve coupled systems and compute their derivatives.

The *functional form* is an alternate system-level representation of the coupled system that considers only the coupling variables and expresses them as implicit functions of the others. We can write this form as

$$\hat{u} = U(\hat{u}) \Leftrightarrow \begin{cases} \hat{u}_1 = U_1(\hat{u}_2, \ldots, \hat{u}_m) \\ \quad \vdots \\ \hat{u}_i = U_i(\hat{u}_1, \ldots, \hat{u}_{i-1}, \hat{u}_{i+1}, \ldots, \hat{u}_m) \\ \quad \vdots \\ \hat{u}_m = U_m(\hat{u}_1, \ldots, \hat{u}_{m-1}) \end{cases}, \qquad (13.4)$$

where m is the number of models and $m \leq n$. If a model U_i is a black box and we have no access to the residuals and the conversion functions, this is the only form we can use. In this case, the system-level solver only iterates the coupling variables $\hat{u}$ and relies on each model i to solve or compute its outputs $\hat{u}_i$.

These two forms are shown in Fig. 13.9 for a generic example with three models (or disciplines). The left of this figure shows the residual form, where each model is represented as residuals and states, as in Fig. 13.8. This leads to a system with nine sets of residuals and corresponding state variables. The number of state variables in each of these sets is not specified but could be any number.

The functional form of these three models is shown on the right of Fig. 13.9. In the case where the model is a black box, the residuals and conversion functions shown in Fig. 13.7 are hidden, and the system level can only access the coupling variables. In this case, each black-box is considered to be a component, as shown in the right of Fig. 13.9.

In an even more general case, these two views can be mixed in a coupled system. The models in residual form expose residuals and states, in which case, the model potentially has multiple components at the system level. The models in functional form only expose inputs and outputs; in that case, the model is just a single component.

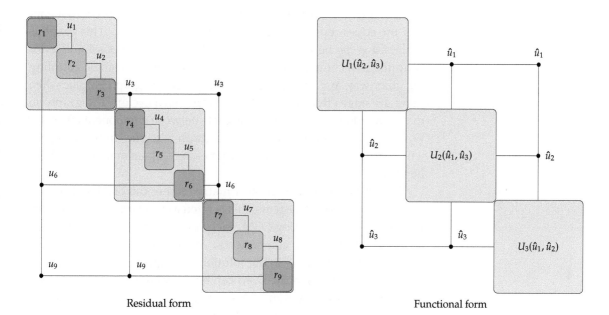

Fig. 13.9 Two system-level views of coupled system with three solvers. In the residual form, all components and their states are exposed (left); in the functional (black-box) form, only inputs and outputs for each solver are visible (right), where $\hat{u}_1 \equiv u_3$, $\hat{u}_2 \equiv u_6$, and $\hat{u}_3 \equiv u_9$.

13.2.4 Coupled System Structure

To show how multidisciplinary systems are coupled, we use a design structure matrix (DSM), which is sometimes referred to as a *dependency structure matrix* or an N^2 *matrix*. An example of the DSM for a hypothetical system is shown on the left in Fig. 13.10. In this matrix, the diagonal elements represent the components, and the off-diagonal entries denote coupling variables. A given coupling variable is computed by the component in its row and is passed to the component in its column.[†] As shown in the DSM on the left side of Fig. 13.10, there are generally off-diagonal entries both above and below the diagonal, where the entries above feed forward, whereas entries below feed backward.

[†] In some of the DSM literature, this definition is reversed, where "row" and "column" are interchanged, resulting in a transposed matrix.

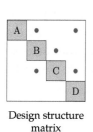

Design structure matrix

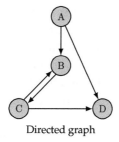

Directed graph

Fig. 13.10 Different ways to represent the dependencies of a hypothetical coupled system.

The mathematical representation of these dependencies is given by a graph (Fig. 13.10, right), where the graph nodes are the components, and the edges represent the information dependency. This graph is a *directed graph* because, in general, there are three possibilities for coupling two components: single coupling one way, single coupling the other way, and two-way coupling. A directed graph is *cyclic* when there are edges that form a closed loop (i.e., a cycle). The graph on the right of Fig. 13.10 has a single cycle between components B and C. When there are no closed loops, the graph is *acyclic*. In this case, the whole system can be solved by solving each component in turn without iterating.

The DSM can be viewed as a matrix where the blank entries are zeros. For real-world systems, this is often a *sparse matrix*. This means that in the corresponding DSM, each component depends only on a subset of all the other components. We can take advantage of the structure of this sparsity in the solution of coupled systems.

The components in the DSM can be reordered without changing the solution of the system. This is analogous to reordering sparse matrices to make linear systems easier to solve. In one extreme case, reordering could achieve a DSM with no entries below the diagonal. In that case, we would have only feedforward connections, which means all dependencies could be resolved in one forward pass (as we will see in Ex. 13.4). This is analogous to having a linear system where the matrix is lower triangular, in which case the linear solution can be obtained with forward substitution.

The sparsity of the DSM can be exploited using ideas from sparse linear algebra. For example, reducing the bandwidth of the matrix (i.e., moving nonzero elements closer to the diagonal) can also be helpful. This can be achieved using algorithms such as Cuthill–McKee,[203] reverse Cuthill–McKee (RCM), and approximate minimum degree (AMD) ordering.[204]‡

203. Cuthill and McKee, *Reducing the bandwidth of sparse symmetric matrices*, 1969.

204. Amestoy et al., *An approximate minimum degree ordering algorithm*, 1996.

‡Although these methods were designed for symmetric matrices, they are still useful for non-symmetric ones. Several numerical libraries include these methods.

205. Lambe and Martins, *Extensions to the design structure matrix for the description of multidisciplinary design, analysis, and optimization processes*, 2012.

We now introduce an extended version of the DSM, called *XDSM*,[205] which we use later in this chapter to show the *process* in addition to the data dependencies. Figure 13.11 shows the XDSM for the same four-component system. When showing only the data dependencies, the only difference relative to DSM is that the coupling variables are labeled explicitly, and the data paths are drawn. In the next section, we add the process to the XDSM.

13.2.5 Solving Coupled Numerical Models

The solution of coupled systems, also known as *multidisciplinary analysis* (MDA), requires concepts beyond the solvers reviewed in Section 3.6

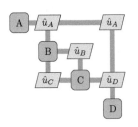

Fig. 13.11 XDSM showing data dependencies for the four-component coupled system of Fig. 13.10.

because it usually involves multiple levels of solvers.

When using the residual form described in Section 13.2.3, any solver (such as a Newton solver) can be used to solve for the state of all components (the entire vector u) simultaneously to satisfy $r(u) = 0$ for the coupled system (Eq. 13.1). This is a monolithic solution approach.

When using the functional form, we do not have access to the internal states of each model and must rely on the model's solvers to compute the coupling variables. The model solver is responsible for computing its output variables for a given set of coupling variables from other models, that is,

$$\hat{u}_i = U_i(\hat{u}_{j \neq i}). \tag{13.5}$$

In some cases, we have access to the model's internal states, but we may want to use a dedicated solver for that model anyway.

Because each model, in general, depends on the outputs of all other models, we have a coupled dependency that requires a solver to resolve. This means that the functional form requires two levels: one for the model solvers and another for the system-level solver. At the system level, we only deal with the coupling variables ($\hat{u}$), and the internal states (u) are hidden.

The rest of this section presents several system-level solvers. We will refer to each model as a component even though it is a group of components in general.

Tip 13.2 Avoid coupling components with file input and output

The coupling variables are often passed between components through files. This is undesirable because of a potential loss in precision (see Tip 13.1) and because it can substantially slow down the coupled solution.

Instead of using files, pass the coupling variable data through memory whenever possible. You can do this between codes written in different languages by wrapping each code using a common language. When using files is unavoidable, be aware of these issues and mitigate them as much as possible.

Nonlinear Block Jacobi

The most straightforward way to solve coupled numerical models (systems of components) is through a fixed-point iteration, which is analogous to the fixed-point iteration methods mentioned in Section 3.6 and detailed in Appendix B.4.1. The difference here is that instead of updating one state at a time, we update a vector of coupling variables at each iteration corresponding to a subset of the coupling variables in the overall coupled system. Obtaining this vector of coupling variables generally involves the solution of a nonlinear system. Therefore, these are called *nonlinear block* variants of the linear fixed-point iteration methods.

The nonlinear block Jacobi method requires an initial guess for all coupling variables to start with and calls for the solution of all components given those guesses. Once all components have been solved, the coupling variables are updated based on the new values computed by the components, and all components are solved again. This iterative process continues until the coupling variables do not change in subsequent iterations. Because each component takes the coupling variable values from the previous iteration, which have already been computed, all components can be solved in parallel without communication. This algorithm is formalized in Alg. 13.1. When applied to a system of components, we call it the *block Jacobi method*, where *block* refers to each component.

The nonlinear block Jacobi method is also illustrated using an XDSM in Fig. 13.12 for three components. The only input is the initial guess for the coupling variables, $\hat{u}^{(0)}$.§ The MDA block (step 0) is responsible for iterating the system-level analysis loop and for checking if the system has converged. The process line is shown as a thin black line to distinguish it from the data dependency connections (thick gray lines) and follows the sequence of numbered steps. The analyses for each component are all numbered the same (step 1) because they can be done in parallel. Each component returns the coupling variables it computes to the MDA iterator, closing the loop between step 2 and step 1 (denoted as "2 → 1").

§In this chapter, we use a superscript for the iteration number instead of subscript to avoid a clash with the component index.

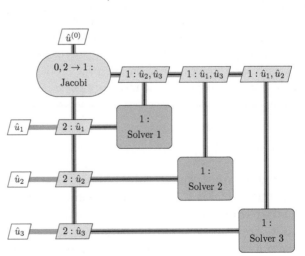

Fig. 13.12 Nonlinear block Jacobi solver for a three-component coupled system.

13.2 COUPLED MODELS

Algorithm 13.1 Nonlinear block Jacobi algorithm

Inputs:
$\hat{u}^{(0)} = \left[\hat{u}_1^{(0)}, \ldots, \hat{u}_m^{(0)}\right]$: Initial guess for coupling variables
Outputs:
$\hat{u} = [\hat{u}_1, \ldots, \hat{u}_m]$: System-level states

---

$k = 0$
while $\left\|\hat{u}^{(k)} - \hat{u}^{(k-1)}\right\|_2 > \varepsilon$ or $k = 0$ **do**   Do not check convergence for first iteration
    for all $i \in \{1, \ldots, m\}$ **do**   Can be done in parallel
        $\hat{u}_i^{(k+1)} \leftarrow$ solve $r_i\left(\hat{u}_i^{(k+1)}; \hat{u}_j^{(k)}\right) = 0, \; j \neq i$ Solve for component i's states
                                          using the states from the previous iteration of other components
    end for
    $k = k + 1$
end while

The block Jacobi solver (Alg. 13.1) can also be used when one or more components are linear solvers. This is useful for computing the derivatives of the coupled system using implicit analytics methods because that involves solving a coupled linear system with the same structure as the coupled model (see Section 13.3.3).

Nonlinear Block Gauss–Seidel

The nonlinear block Gauss–Seidel algorithm is similar to its Jacobi counterpart. The only difference is that when solving each component, we use the latest coupling variables available instead of just using the coupling variables from the previous iteration. We cycle through each component $i = 1, \ldots, m$ in order. When computing $\hat{u}_i$ by solving component i, we use the latest available states from the other components. Figure 13.13 illustrates this process.

Both Gauss–Seidel and Jacobi converge linearly, but Gauss–Seidel tends to converge more quickly because each equation uses the latest information available. However, unlike Jacobi, the components can no longer be solved in parallel.

The convergence of nonlinear block Gauss–Seidel can be improved by using a relaxation. Suppose that $\hat{u}_{\text{temp}}$ is the state of component i resulting from the solving of that component given the states of all other components, as we would normally do for each block in the Gauss–Seidel or Jacobi method. If we used this, the step would be

$$\Delta \hat{u}_i^{(k)} = \hat{u}_{\text{temp}} - \hat{u}_i^{(k)}. \tag{13.6}$$

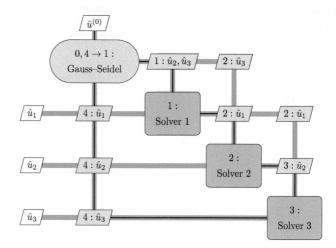

Fig. 13.13 Nonlinear block Gauss–Seidel solver for the three-discipline coupled system of Fig. 13.9.

Instead of using that step, relaxation updates the variables as

$$\hat{u}_i^{(k)} = \hat{u}_{\text{temp}} + \theta^{(k)} \Delta \hat{u}_i^{(k)}, \tag{13.7}$$

where $\theta^{(k)}$ is the relaxation factor, and $\Delta \hat{u}_i^{(k)}$ is the previous update for component i. The relaxation factor, θ, could be a fixed value, which would normally be less than 1 to dampen oscillations and avoid divergence.

Aitken's method[206] improves on the fixed relaxation approach by adapting the θ. The relaxation factor at each iteration changes based on the last two updates according to

206. Irons and Tuck, *A version of the Aitken accelerator for computer iteration*, 1969.

$$\theta^{(k)} = \theta^{(k-1)} \left(1 - \frac{\left(\Delta \hat{u}^{(k)} - \Delta \hat{u}^{(k-1)}\right)^\mathsf{T} \Delta \hat{u}^{(k)}}{\left\|\Delta \hat{u}^{(k)} - \Delta \hat{u}^{(k-1)}\right\|^2} \right). \tag{13.8}$$

Aitken's method usually accelerates convergence and has been shown to work well for nonlinear block Gauss–Seidel with multidisciplinary systems.[207] It is advisable to override the value of the relaxation factor given by Eq. 13.8 to keep it between 0.25 and 2.[208]

207. Kenway et al., *Scalable parallel approach for high-fidelity steady-state aeroelastic analysis and derivative computations*, 2014.

208. Chauhan et al., *An automated selection algorithm for nonlinear solvers in MDO*, 2018.

The steps for the full Gauss–Seidel algorithm with Aitken acceleration are listed in Alg. 13.2. Similar to the block Jacobi solver, the block Gauss–Seidel solver can also be used when one or more components are linear solvers. Aitken acceleration can be used in the linear case without modification and it is still useful.

The order in which the components are solved makes a significant difference in the efficiency of the Gauss–Seidel method. In the best possible scenario, the components can be reordered such that there are no entries in the lower diagonal of the DSM, which means that each

13.2 COUPLED MODELS

component depends only on previously solved components, and there are therefore no feedback dependencies (see Ex. 13.4). In this case, the block Gauss–Seidel method would converge to the solution in one forward sweep.

In the more general case, even though we might not eliminate the lower diagonal entries completely, minimizing these entries by reordering results in better convergence. This reordering can also mean the difference between convergence and nonconvergence.

Algorithm 13.2 Nonlinear block Gauss–Seidel algorithm with Aitken acceleration

Inputs:
$\hat{u}^{(0)} = \left[\hat{u}_1^{(0)}, \ldots, \hat{u}_m^{(0)}\right]$: Initial guess for coupling variables
$\theta^{(0)}$: Initial relaxation factor for Aitken acceleration
Outputs:
$\hat{u} = [\hat{u}_1, \ldots, \hat{u}_m]$: System-level states

---

$k = 0$
while $\left\|\hat{u}^{(k)} - \hat{u}^{(k-1)}\right\|_2 > \varepsilon$ or $k = 0$ **do**   Do not check convergence for first iteration
  for $i = 1, m$ **do**
    $\hat{u}_{\text{temp}} \leftarrow$ solve $r_i\left(\hat{u}_i^{(k+1)}; \hat{u}_1^{(k+1)}, \ldots, \hat{u}_{i-1}^{(k+1)}, \hat{u}_{i+1}^{(k)}, \ldots, \hat{u}_m^{(k)}\right) = 0$
    Solve for component i's states using the latest states from other components
    $\Delta \hat{u}_i^{(k)} = \hat{u}_{\text{temp}} - \hat{u}_i^{(k)}$ Compute step
    if $k > 0$ **then**
      $\theta^{(k)} = \theta^{(k-1)}\left(1 - \frac{\left(\Delta \hat{u}^{(k)} - \Delta \hat{u}^{(k-1)}\right)^T \Delta \hat{u}^{(k)}}{\|\Delta \hat{u}^{(k)} - \Delta \hat{u}^{(k-1)}\|^2}\right)$ Update the relaxation factor
    end if
    $\hat{u}_i^{(k+1)} = \hat{u}_i^{(k)} + \theta^{(k)} \Delta \hat{u}_i^{(k)}$ Update component i's states
  end for
  $k = k + 1$
end while

Example 13.4 Making Gauss–Seidel converge in one pass by reordering components

Consider the coupled system of six components with the dependencies shown on the left in Fig. 13.14. This system includes both feedforward and feedback dependencies and would normally require an iterative solver. In this case, however, we can reorder the components as shown on the right in Fig. 13.14 to eliminate the feedback loops. Then, we only need to solve the sequence of components E → C → A → D → F → B once to get a converged coupled solution.

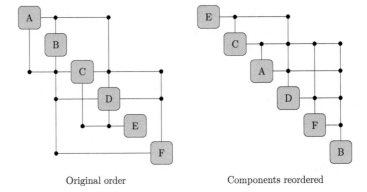

Fig. 13.14 The solution of the components of the system shown on the left can be reordered to get the equivalent system shown on the right. This new system has no feedback loops and can therefore be solved in one pass of a Gauss–Seidel solver.

Original order

Components reordered

Newton's Method

As mentioned previously, Newton's method can be applied to the residual form illustrated in Fig. 13.9 and expressed in Eq. 13.1. Recall that in this form, we have n components and the coupling variables are part of the state variables. In this case, Newton's method is as described in Section 3.8.

Concatenating the residuals and state variables for all components and applying Newton's method yields the coupled block Newton system,

$$\underbrace{\begin{bmatrix} \dfrac{\partial r_1}{\partial u_1} & \dfrac{\partial r_1}{\partial u_2} & \cdots & \dfrac{\partial r_1}{\partial u_n} \\ \dfrac{\partial r_2}{\partial u_1} & \dfrac{\partial r_2}{\partial u_2} & \cdots & \dfrac{\partial r_2}{\partial u_n} \\ \vdots & \vdots & \ddots & \vdots \\ \dfrac{\partial r_n}{\partial u_1} & \dfrac{\partial r_n}{\partial u_2} & \cdots & \dfrac{\partial r_n}{\partial u_n} \end{bmatrix}}_{\dfrac{\partial r}{\partial u}} \underbrace{\begin{bmatrix} \Delta u_1 \\ \Delta u_2 \\ \vdots \\ \Delta u_n \end{bmatrix}}_{\Delta u} = -\underbrace{\begin{bmatrix} r_1 \\ r_2 \\ \vdots \\ r_n \end{bmatrix}}_{r}. \tag{13.9}$$

We can solve this linear system to compute the Newton step for all components' state variables u simultaneously, and then iterate to satisfy $r(u) = 0$ for the complete system. This is the *monolithic Newton* approach illustrated on the left panel of Fig. 13.15. As with any Newton method, a globalization strategy (such as a line search) is required to increase the likelihood of successful convergence when starting far from the solution (see Section 4.2). Even with such a strategy, Newton's method does not necessarily converge robustly.

A variation on this monolithic Newton approach uses two-level solver *hierarchy*, as illustrated on the middle panel of Fig. 13.15. The system-level solver is the same as in the monolithic approach, but each

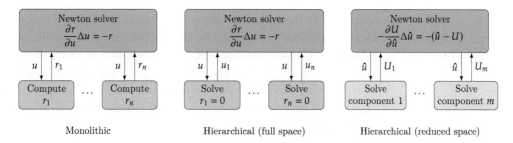

component is solved first using the latest states. The Newton step for each component i is given by

$$\frac{\partial r_i}{\partial u_i} \Delta u_i = -r_i \left(u_i; u_{j \neq i}\right), \quad (13.10)$$

where u_j represents the states from other components (i.e., $j \neq i$), which are fixed at this level. Each component is solved before taking a step in the entire state vector (Eq. 13.9). The procedure is given in Alg. 13.3 and illustrated in Fig. 13.16. We call this the *full-space hierarchical Newton* approach because the system-level solver iterates the entire state vector. Solving each component before taking each step in the full space Newton iteration acts as a preconditioner. In general, the monolithic approach is more efficient, and the hierarchical approach is more robust, but these characteristics are case-dependent.

Fig. 13.15 There are three options for solving a coupled system with Newton's method. The monolithic approach (left) solves for all state variables simultaneously. The block approach (middle) solves the same system as the monolithic approach, but solves each component for its states at each iteration. The black box approach (right) applies Newton's method to the coupling variables.

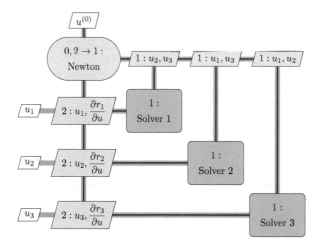

Fig. 13.16 Full-space hierarchical Newton solver for a three-component coupled system.

Newton's method can also be applied to the functional form illustrated in Fig. 13.9 to solve only for the coupling variables. We call this the *reduced-space hierarchical Newton* approach because the system-level solver iterates only in the space of the coupling variables, which is smaller than the full space of the state variables. Using this approach, each component's solver can be a black box, as in the nonlinear block Jacobi and Gauss–Seidel solvers. This approach is illustrated on the right panel of Fig. 13.15. The reduced-space approach is mathematically equivalent and follows the same iteration path as the full-space approach if each component solver in the reduced-space approach is converged well enough.[132]

132. Gray et al., *OpenMDAO: An open-source framework for multidisciplinary design, analysis, and optimization*, 2019.

Algorithm 13.3 Full-space hierarchical Newton

Inputs:
$u^{(0)} = \left[u_1^{(0)}, \ldots, u_n^{(0)}\right]$: Initial guess for coupling variables
Outputs:
$u = [u_1, \ldots, u_n]$: System-level states

$k = 1$	Iteration counter for full-space iteration
while $\|r\|_2 > \varepsilon$ **do**	Check residual norm for all components
for all $i \in \{1, \ldots, n\}$ **do**	Can be done in parallel; k is constant in this loop
while $\|r_i\|_2 > \varepsilon$ **do**	Check residual norm for component i
Compute $r_i\left(u_i^{(k)}; u_{j \neq i}^{(k-1)}\right)$	States for other components are fixed
Compute $\dfrac{\partial r_i}{\partial u_i}$	Jacobian block for component i for current state
Solve $\dfrac{\partial r_i}{\partial u_i} \Delta u_i = -r_i$	Solve for Newton step for ith component
$u_i^{(k)} = u_i^{(k)} + \Delta u_i$	Update state variables for component i
end while	
end for	
Compute $r\left(u^{(k)}\right)$	Full residual vector for current states
Compute $\dfrac{\partial r}{\partial u}$	Full Jacobian for current states
Solve $\dfrac{\partial r}{\partial u} \Delta u = -r$	Coupled Newton system (Eq. 13.9)
$u^{(k+1)} = u^{(k)} + \Delta u$	Update full state variable vector
$k = k + 1$	
end while	

To apply the reduced-space Newton's method, we express the functional form (Eq. 13.4) as residuals by using the same technique we used to convert an explicit function to the residual form (Eq. 13.2). This yields

$$\hat{r}_i(\hat{u}) = \hat{u}_i - U_i(\hat{u}_{j \neq i}), \tag{13.11}$$

where $\hat{u}_i$ represents the guesses for the coupling variables, and U_i represents the actual computed values. For a system of nonlinear

13.2 Coupled Models

residual equations, the Newton step in the coupling variables, $\Delta \hat{u} = \hat{u}^{(k+1)} - \hat{u}^{(k)}$, can be found by solving the linear system

$$\left.\frac{\partial \hat{r}}{\partial \hat{u}}\right|_{\hat{u}=\hat{u}^{(k)}} \Delta \hat{u} = -\hat{r}\left(\hat{u}^{(k)}\right), \quad (13.12)$$

where we need the partial derivatives of all the residuals with respect to the coupling variables to form the Jacobian matrix $\partial \hat{r}/\partial \hat{u}$. The Jacobian can be found by differentiating Eq. 13.11 with respect to the coupling variables. Then, expanding the concatenated residuals and coupling variable vectors yields

$$\begin{bmatrix} I & -\frac{\partial U_1}{\partial \hat{u}_2} & \cdots & -\frac{\partial U_1}{\partial \hat{u}_m} \\ -\frac{\partial U_2}{\partial \hat{u}_1} & I & \cdots & -\frac{\partial U_2}{\partial \hat{u}_m} \\ \vdots & \vdots & \ddots & \vdots \\ -\frac{\partial U_m}{\partial \hat{u}_1} & -\frac{\partial U_m}{\partial \hat{u}_2} & \cdots & I \end{bmatrix} \begin{bmatrix} \Delta \hat{u}_1 \\ \Delta \hat{u}_2 \\ \vdots \\ \Delta \hat{u}_m \end{bmatrix} = - \begin{bmatrix} \hat{u}_1 - U_1(\hat{u}_2, \ldots, \hat{u}_m) \\ \hat{u}_2 - U_2(\hat{u}_1, \hat{u}_3, \ldots, \hat{u}_m) \\ \vdots \\ \hat{u}_m - U_m(\hat{u}_1, \ldots, \hat{u}_{m-1}) \end{bmatrix}.$$
(13.13)

The residuals in the right-hand side of this equation are evaluated at the current iteration.

The derivatives in the block Jacobian matrix are also computed at the current iteration. Each row i represents the derivatives of the (potentially implicit) function that computes the outputs of component i with respect to all the inputs of that component. The Jacobian matrix in Eq. 13.13 has the same structure as the DSM (but transposed) and is often sparse. These derivatives can be computed using the methods from Chapter 6. These are partial derivatives in the sense that they do not take into account the coupled system. However, they must take into account the respective model and can be computed using implicit analytic methods when the model is implicit.

This Newton solver is shown in Fig. 13.17 and detailed in Alg. 13.4. Each component corresponds to a set of rows in the block Newton system (Eq. 13.13). To compute each set of rows, the corresponding component must be solved, and the derivatives of its outputs with respect to its inputs must be computed as well. Each set can be computed in parallel, but once the system is assembled, a step in the coupling variables is computed by solving the full system (Eq. 13.13).

These coupled Newton methods have similar advantages and disadvantages to the plain Newton method. The main advantage is that it converges quadratically once it is close enough to the solution (if the problem is well-conditioned). The main disadvantage is that it might

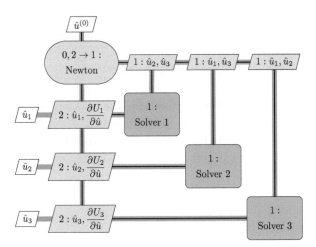

Fig. 13.17 Reduced-space hierarchical Newton solver for a three-component coupled system.

not converge at all, depending on the initial guess. One disadvantage specific to the coupled Newton methods is that it requires formulating and solving the coupled linear system (Eq. 13.13) at each iteration.

Algorithm 13.4 Reduced-space hierarchical Newton

Inputs:
$\hat{u}^{(0)} = \left[\hat{u}_1^{(0)}, \ldots, \hat{u}_m^{(0)}\right]$: Initial guess for coupling variables
Outputs:
$\hat{u} = [\hat{u}_1, \ldots, \hat{u}_m]$: System-level states

---

$k = 0$	
while $\|\hat{r}\|_2 > \varepsilon$ **do**	Check residual norm for all components
for all $i \in \{1, \ldots, m\}$ **do**	Can be done in parallel
$U_i \leftarrow$ compute $U_i\left(\hat{u}_{j \neq i}^{(k)}\right)$	Solve component i and compute its outputs
end for	
$\hat{r} = \hat{u}^{(k)} - U$	Compute all coupling variable residuals
Compute $\dfrac{\partial U}{\partial \hat{u}}$	Jacobian of coupling variables for current state
Solve $\dfrac{\partial \hat{r}}{\partial \hat{u}} \Delta \hat{u} = -\hat{r}$	Coupled Newton system (Eq. 13.13)
$\hat{u}^{(k+1)} = \hat{u}^{(k)} + \Delta \hat{u}$	Update all coupling variables
$k = k + 1$	
end while	

If the Jacobian $\partial r / \partial u$ is not readily available, Broyden's method can approximate the Jacobian inverse ($\tilde{J}^{-1}$) by starting with a guess (say,

13.2 Coupled Models

$\tilde{J}_0^{-1} = I$) and then using the update

$$\tilde{J}^{-1(k+1)} = \tilde{J}^{-1(k)} + \frac{\left(\Delta u^{(k)} - \tilde{J}^{-1(k)} \Delta r^{(k)}\right) \Delta u^{(k)\mathsf{T}}}{\Delta r^{(k)\mathsf{T}} \Delta r^{(k)}}, \quad (13.14)$$

where $\Delta u^{(k)}$ is the last step in the states and $\Delta r^{(k)}$ is the difference between the two latest residual vectors. Because the inverse is provided explicitly, we can find the update by performing the multiplication,

$$\Delta u^{(k)} = -\tilde{J}^{-1} r^{(k)}. \quad (13.15)$$

Broyden's method is analogous to the quasi-Newton methods of Section 4.4.4 and is derived in Appendix C.1.

Example 13.5 Aerostructural solver comparison

We now apply the coupled solution methods presented in this section to the implicit parts of the aerostructural model, which are the two first residuals from Ex. 13.2,

$$r = \begin{bmatrix} r_1 \\ r_2 \end{bmatrix} = \begin{bmatrix} A(d)\Gamma - v(d) \\ Kd - q(\Gamma) \end{bmatrix},$$

and the variables are the circulations and displacements,

$$u = \begin{bmatrix} u_1 \\ u_2 \end{bmatrix} = \begin{bmatrix} \Gamma \\ d \end{bmatrix}.$$

In this case, the linear systems defined by r_1 and r_2 are small enough to be solved using a direct method, such as LU factorization. Thus, we can solve r_1 for Γ, for a given d, and solve r_2 for d, for a given Γ. Also, no conversions are involved, so the set of coupling variables is equivalent to the set of state variables ($\hat{u} = u$).

Using the nonlinear block Jacobi method (Alg. 13.1), we start with an initial guess (e.g., $\Gamma = 0$, $d = 0$) and solve $r_1 = 0$ and $r_2 = 0$ separately for the new values of Γ and d, respectively. Then we use these new values of Γ and d to solve $r_1 = 0$ and $r_2 = 0$ again, and so on until convergence.

Nonlinear block Gauss–Seidel (Alg. 13.2) is similar, but we need to solve the two components in sequence. We can start by solving $r_1 = 0$ for Γ with $d = 0$. Then we use the Γ obtained from this solution in r_2 and solve for a new d. We now have a new d to use in r_1 to solve for a new Γ, and so on.

The Jacobian for the Newton system (Eq. 13.9) is

$$\frac{\partial r}{\partial u} = \begin{bmatrix} \dfrac{\partial r_1}{\partial u_1} & \dfrac{\partial r_1}{\partial u_2} \\ \dfrac{\partial r_2}{\partial u_1} & \dfrac{\partial r_2}{\partial u_2} \end{bmatrix} = \begin{bmatrix} A & \dfrac{\partial A}{\partial d}\Gamma - \dfrac{\partial v}{\partial d} \\ -\dfrac{\partial q}{\partial \Gamma} & K \end{bmatrix}.$$

We already have the block diagonal matrices in this Jacobian from the governing equations, but we need to compute the off-diagonal partial derivative blocks, which can be done analytically or with algorithmic differentiation (AD).

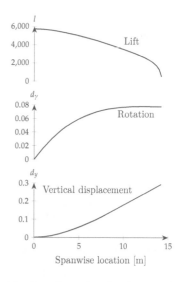

Fig. 13.18 Spanwise distribution of the lift, wing rotation (d_θ), and vertical displacement (d_z) for the coupled aerostructural solution.

The solution is shown in Fig. 13.18, where we plot the variation of lift, vertical displacement, and rotation along the span. The vertical displacements are a subset of d, and the rotations are a conversion of a subset of d representing the rotations of the wing section at each spanwise location. The lift is the vertical force at each spanwise location, which is proportional to Γ times the wing chord at that location.

The monolithic Newton approach does not converge in this case. We apply the full-space hierarchical approach (Alg. 13.3), which converges more reliably. In this case, the reduced-space approach is not used because there is no distinction between coupling variables and state variables.

In Fig. 13.19, we compare the convergence of the methods introduced in this section.¶ The Jacobi method has the poorest convergence rate and oscillates. The Gauss–Seidel method is much better, and it is even better with Aitken acceleration. Newton has the highest convergence rate, as expected. Broyden performs about as well as Gauss–Seidel in this case.

¶These results and subsequent results based on the same example were obtained using OpenAeroStruct,²⁰² which was developed using OpenMDAO. The description in these examples is simplified for didactic purposes; check the paper and code for more details.

202. Jasa et al., *Open-source coupled aerostructural optimization using Python*, 2018.

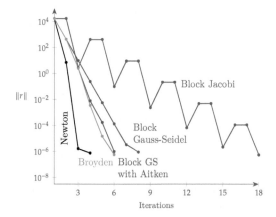

Fig. 13.19 Convergence of each solver for aerostructural system.

13.2.6 Hierarchical Solvers for Coupled Systems

The coupled solvers we discussed so far already use a two-level hierarchy because they require a solver for each component and a second level that solves the group of components. This hierarchy can be extended to three and more levels by making groups of groups.

Modular analysis and unified derivatives (MAUD) is a mathematical framework developed for this purpose. Using MAUD, we can mix residual and functional forms and seamlessly handle implicit and explicit components.∥

The hierarchy of solvers can be represented as a tree data structure, where the nodes are the solvers and the leaves are the components, as

∥MAUD was developed by Hwang and Martins⁴⁴ when they realized that the unified derivatives equation (UDE) provides the mathematical foundation for a framework of parallel hierarchical solvers through a small set of user-defined functions. MAUD can also compute the derivatives of coupled systems, as we will see in Section 13.3.3.

44. Hwang and Martins, *A computational architecture for coupling heterogeneous numerical models and computing coupled derivatives*, 2018.

shown in Fig. 13.20 for a system of six components and five solvers. The root node ultimately solves the complete system, and each solver is responsible for a subsystem and thus handles a subset of the variables.

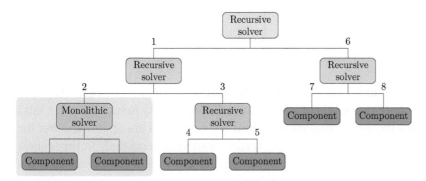

Fig. 13.20 A system of components can be organized in a solver hierarchy.

There are two possible types of solvers: monolithic and recursive. Monolithic solvers can only have components as children and handle all their variables simultaneously using the residual form. Of the methods we introduced in the previous section, only monolithic and full-space Newton (and Broyden) can do this for nonlinear systems. Linear systems can be solved in a monolithic fashion using a direct solver or an iterative linear solver, such as a Krylov subspace method. Recursive solvers, as the name implies, visit all the child nodes in turn. If a child node turns out to be another recursive solver, it does the same until a component is reached. The block Jacobi and Gauss–Seidel methods can be used as recursive solvers for nonlinear and linear systems. The reduced-space Newton and Broyden methods can also be recursive solvers. For the hypothetical system shown in Fig. 13.20, the numbers show the order in which each solver and component would be called.

The hierarchy of solvers should be chosen to exploit the system structure. MAUD also facilitates parallel computation when subsystems are uncoupled, which provides further opportunities to exploit the structure of the problem. Figs. 13.21 and 13.22 show several possibilities.

The three systems in Fig. 13.21 show three different coupling modes. In the first mode, the two components are independent of each other and can be solved in parallel using any solvers appropriate for each of the components. In the serial case, component 2 depends on 1, but not the other way around. Therefore, we can converge to the coupled solution using one block Gauss–Seidel iteration. If the dependency were reversed (feedback but no feedforward), the order of the two components would be switched. Finally, the fully coupled case requires an iterative solution using any of the methods from Section 13.2.5. MAUD is designed to handle these three coupling modes.

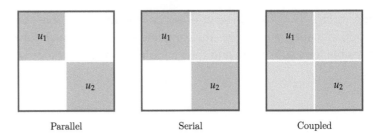

Fig. 13.21 There are three main possibilities involving two components.

Figure 13.22 shows three possibilities for a four-component system where two levels of solvers can be used. In the first one (on the left), we require a coupled solver for components 1 and 2 and another for components 3 and 4, but no further solving is needed. In the second (Fig. 13.22, middle), components 1 and 2 as well as components 3 and 4 can be solved serially, but these two groups require a coupled solution. For the two levels to converge, the serial and coupled solutions are called repeatedly until the two solvers agree with each other. The third possibility (Fig. 13.22, right) has two systems that have two independent components, which can each be solved in parallel, but the overall system is coupled. With MAUD, we can set up any of these sequences of solvers through the solver hierarchy tree, as illustrated in Fig. 13.20.

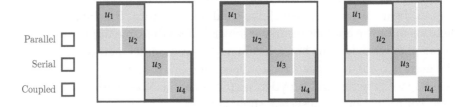

Fig. 13.22 Three examples of a system of four components with a two-level solver hierarchy.

To solve the system from Ex. 13.3 using hierarchical solvers, we can use the hierarchy shown in Fig. 13.23. We form three groups with three components each. Each group includes the input and output conversion components (which are explicit) and one implicit component (which requires its own solver). Serial solvers can be used to handle the input and output conversion components. A coupled solver is required to solve the entire coupled system, but the coupling between the groups is restricted to the corresponding outputs (components 3, 6, and 9).

Alternatively, we could apply a coupled solver to the functional representation (Fig. 13.9, right). This would also use two levels of solvers: a solver within each group and a system-level solver for the coupling of the three groups. However, the system-level solver would handle coupling variables rather than the residuals of each component.

13.3 Coupled Derivatives Computation

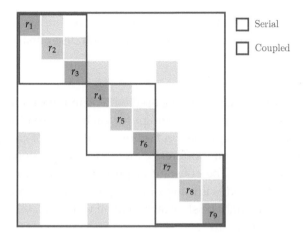

Fig. 13.23 For the case of Fig. 13.9, we can use a serial evaluation within each of the three groups and require a coupled solver to handle the coupling between the three groups.

Tip 13.3 Framework for implementing coupled system solvers

The development of coupled solvers is often done for a specific set of models from scratch, which requires substantial effort. OpenMDAO is an open-source framework that facilitates such efforts by implementing MAUD.[132] All the solvers introduced in this chapter are available in OpenMDAO. This framework also makes it easier to compute the derivatives of the coupled system, as we will see in the next section. Users can assemble systems of mixed explicit and implicit components.

132. Gray et al., *OpenMDAO: An open-source framework for multidisciplinary design, analysis, and optimization*, 2019.

For implicit components, they must give OpenMDAO access to the residual computations and the corresponding state variables. For explicit components, OpenMDAO only needs access to the inputs and the outputs, so it supports black-box models.

OpenMDAO is usually more efficient when the user provides access to the residuals and state variables instead of treating models as black boxes. A hierarchy of multiple solvers can be set up in OpenMDAO, as illustrated in Fig. 13.20. OpenMDAO also provides the necessary interfaces for user-defined solvers. Finally, OpenMDAO encourages coupling through memory, which is beneficial for numerical precision (see Tip 13.1) and computational efficiency (see Tip 13.2).

13.3 Coupled Derivatives Computation

The gradient-based optimization algorithms from Chapters 4 and 5 require the derivatives of the objective and constraints with respect to the design variables. Any of the methods for computing derivatives from Chapter 6 can be used to compute the derivatives of coupled models, but some modifications are required. The main difference is that in MDO, the computation of the functions of interest (objective and constraints) requires the solution of the multidisciplinary model.

13.3.1 Finite Differences

The finite-difference method can be used with no modification, as long as an MDA is converged well enough for each perturbation in the design variables. As explained in Section 6.4, the cost of computing derivatives with the finite-difference method is proportional to the number of variables. The constant of proportionality can increase significantly compared with that of a single discipline because the MDA convergence might be slow (especially if using a block Jacobi or Gauss–Seidel iteration).

The accuracy of finite-difference derivatives depends directly on the accuracy of the functions of interest. When the functions are computed from the solution of a coupled system, their accuracy depends both on the accuracy of each component and the accuracy of the MDA. To address the latter, the MDA should be converged well enough.

13.3.2 Complex Step and AD

The complex-step method and forward-mode AD can also be used for a coupled system, but some modifications are required. The complex-step method requires all components to be able to take complex inputs and compute the corresponding complex outputs. Similarly, AD requires inputs and outputs that include derivative information. For a given MDA, if one of these methods is applied to each component and the coupling includes the derivative information, we can compute the derivatives of the coupled system. The propagation of the forward-mode seed (or the complex step) is illustrated in Fig. 13.24 for a system of two components.

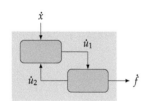

Fig. 13.24 Forward mode of AD for a system of two components.

When using AD, manual coupling is required if the components and the coupling are programmed in different languages. The complex-step method can be more straightforward to implement than AD for cases where the models are implemented in different languages, and all the languages support complex arithmetic. Although both of these methods produce accurate derivatives for each component, the accuracy of the derivatives for the coupled system could be compromised by a low level of convergence of the MDA.

The reverse mode of AD for coupled systems would be more involved: after an initial MDA, we would run a reverse MDA to compute the derivatives, as illustrated in Fig. 13.25.

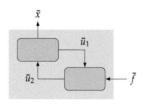

Fig. 13.25 Reverse mode of AD for a system of two components.

13.3.3 Implicit Analytic Methods

The implicit analytic methods from Section 6.7 (both direct and adjoint) can also be extended to compute the derivatives of coupled systems. All the equations derived for a single component in Section 6.7 are valid for coupled systems if we concatenate the residuals and the state variables. Furthermore, we can mix explicit and implicit components using concepts introduced in the UDE. Finally, when using the MAUD approach, the coupled derivative computation can be done using the same hierarchy of solvers.

Coupled Derivatives of Residual Representation

In Eq. 13.1, we denoted the coupled system as a series of concatenated residuals, $r_i(u) = 0$, and variables u_i corresponding to each component $i = 1, \ldots, n$ as

$$r(u) \equiv \begin{bmatrix} r_1(u) \\ \vdots \\ r_n(u) \end{bmatrix}, \quad u \equiv \begin{bmatrix} u_1 \\ \vdots \\ u_n \end{bmatrix}, \quad (13.16)$$

where the residual for each component, r_i, could depend on all states u. To derive the coupled version of the direct and adjoint methods, we apply them to the concatenated vectors. Thus, the coupled version of the linear system for the direct method (Eq. 6.43) is

$$\begin{bmatrix} \frac{\partial r_1}{\partial u_1} & \cdots & \frac{\partial r_1}{\partial u_n} \\ \vdots & \ddots & \vdots \\ \frac{\partial r_n}{\partial u_1} & \cdots & \frac{\partial r_n}{\partial u_n} \end{bmatrix} \begin{bmatrix} \phi_1 \\ \vdots \\ \phi_n \end{bmatrix} = \begin{bmatrix} \frac{\partial r_1}{\partial x} \\ \vdots \\ \frac{\partial r_n}{\partial x} \end{bmatrix}, \quad (13.17)$$

where ϕ_i represents the derivatives of the states from component i with respect to the design variables. Once we have solved for ϕ, we can use the coupled equivalent of the total derivative equation (Eq. 6.44) to compute the derivatives:

$$\frac{df}{dx} = \frac{\partial f}{\partial x} - \begin{bmatrix} \frac{\partial f}{\partial u_1} & \cdots & \frac{\partial f}{\partial u_n} \end{bmatrix} \begin{bmatrix} \phi_1 \\ \vdots \\ \phi_n \end{bmatrix}. \quad (13.18)$$

Similarly, the adjoint equations (Eq. 6.46) can be written for a coupled system using the same concatenated state and residual vectors. The coupled adjoint equations involve a corresponding concatenated adjoint vector and can be written as

$$\begin{bmatrix} \dfrac{\partial r_1}{\partial u_1}^\mathsf{T} & \cdots & \dfrac{\partial r_n}{\partial u_1}^\mathsf{T} \\ \vdots & \ddots & \vdots \\ \dfrac{\partial r_1}{\partial u_n}^\mathsf{T} & \cdots & \dfrac{\partial r_n}{\partial u_n}^\mathsf{T} \end{bmatrix} \begin{bmatrix} \psi_1 \\ \vdots \\ \psi_n \end{bmatrix} = \begin{bmatrix} \dfrac{\partial f}{\partial u_1}^\mathsf{T} \\ \vdots \\ \dfrac{\partial f}{\partial u_n}^\mathsf{T} \end{bmatrix}. \qquad (13.19)$$

After solving this equations for the coupled-adjoint vector, we can use the coupled version of the total derivative equation (Eq. 6.47) to compute the desired derivatives as

$$\frac{\mathrm{d}f}{\mathrm{d}x} = \frac{\partial f}{\partial x} - \begin{bmatrix} \psi_1^\mathsf{T} \cdots \psi_n^\mathsf{T} \end{bmatrix} \begin{bmatrix} \dfrac{\partial r_1}{\partial x} \\ \vdots \\ \dfrac{\partial r_n}{\partial x} \end{bmatrix}. \qquad (13.20)$$

Like the adjoint method from Section 6.7, the coupled adjoint is a powerful approach for computing gradients with respect to many design variables.*

*The coupled-adjoint approach has been implemented for aerostructural problems governed by coupled PDEs[207] and demonstrated in wing design optimization.[209]

207. Kenway et al., *Scalable parallel approach for high-fidelity steady-state aeroelastic analysis and derivative computations*, 2014.

209. Kenway and Martins, *Multipoint high-fidelity aerostructural optimization of a transport aircraft configuration*, 2014.

The required partial derivatives are the derivatives of the residuals or outputs of each component with respect to the state variables or inputs of all other components. In practice, the block structure of these partial derivative matrices is sparse, and the matrices themselves are sparse. This sparsity can be exploited using graph coloring to drastically reduce the computation effort of computing Jacobians at the system or component level, as explained in Section 6.8.

Figure 13.26 shows the structure of the Jacobians in Eq. 13.17 and Eq. 13.19 for the three-group case from Fig. 13.23. The sparsity structure of the Jacobian is the *transpose* of the DSM structure. Because the Jacobian in Eq. 13.19 is transposed, the Jacobian in the adjoint equation has the same structure as the DSM.

The structure of the linear system can be exploited in the same way as for the nonlinear system solution using hierarchical solvers: serial solvers within each group and a coupled solver for the three groups. The block Jacobi and Gauss–Seidel methods from Section 13.2.5 are applicable to coupled linear components, so these methods can be re-used to solve this coupled linear system for the total coupled derivatives.

The partial derivatives in the coupled Jacobian, the right-hand side of the linear systems (Eqs. 13.17 and 13.19), and the total derivatives equations (Eqs. 13.18 and 13.20) can be computed with any of the methods from Chapter 6. The nature of these derivatives is the same

13.3 Coupled Derivatives Computation

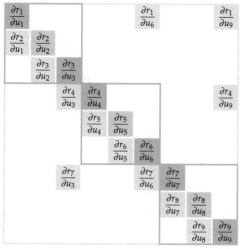
Direct Jacobian

Adjoint Jacobian

Fig. 13.26 Jacobian structure for residual form of the coupled direct (left) and adjoint (right) equations for the three-group system of Fig. 13.23. The structure of the transpose of the Jacobian is the same as that of the DSM.

as we have seen previously for implicit analytic methods (Section 6.7). They do not require the solution of the equation and are typically cheap to compute. Ideally, the components would already have analytic derivatives of their outputs with respect to their inputs, which are all the derivatives needed at the system level.

The partial derivatives can also be computed using the finite-difference or complex-step methods. Even though these are not efficient for cases with many inputs, it might still be more efficient to compute the partial derivatives with these methods and then solve the coupled derivative equations instead of performing a finite difference of the coupled system, as described in Section 13.3.1. The reason is that computing the partial derivatives avoids having to reconverge the coupled system for every input perturbation. In addition, the coupled system derivatives should be more accurate when finite differences are used only to compute the partial derivatives.

Coupled Derivatives of Functional Representation

Variants of the coupled direct and adjoint methods can also be derived for the functional form of the system-level representation (Eq. 13.4), by using the residuals defined for the system-level Newton solver (Eq. 13.11),

$$\hat{r}_i(\hat{u}) = \hat{u}_i - U_i(\hat{u}_{j \neq i}) = 0, \quad i = 1, \ldots, m. \quad (13.21)$$

Recall that driving these residuals to zero relies on a solver for each component to solve for each component's states and another solver to solve for the coupling variables $\hat{u}$.

Using this new residual definition and the coupling variables, we can derive the functional form of the coupled direct method as

$$\begin{bmatrix} I & -\dfrac{\partial U_1}{\partial \hat{u}_2} & \cdots & -\dfrac{\partial U_1}{\partial \hat{u}_m} \\ -\dfrac{\partial U_2}{\partial \hat{u}_1} & I & \cdots & -\dfrac{\partial U_2}{\partial \hat{u}_m} \\ \vdots & \vdots & \ddots & \vdots \\ -\dfrac{\partial U_m}{\partial \hat{u}_1} & -\dfrac{\partial U_m}{\partial \hat{u}_2} & \cdots & I \end{bmatrix} \begin{bmatrix} \hat{\phi}_1 \\ \hat{\phi}_2 \\ \vdots \\ \hat{\phi}_m \end{bmatrix} = \begin{bmatrix} \dfrac{\partial \hat{U}_1}{\partial x} \\ \dfrac{\partial \hat{U}_2}{\partial x} \\ \vdots \\ \dfrac{\partial \hat{U}_m}{\partial x} \end{bmatrix}, \qquad (13.22)$$

where the Jacobian is identical to the one we derived for the coupled Newton step (Eq. 13.13). Here, $\hat{\phi}_i$ represents the derivatives of the coupling variables from component i with respect to the design variables. The solution can then be used in the following equation to compute the total derivatives:

$$\dfrac{df}{dx} = \dfrac{\partial f}{\partial x} - \begin{bmatrix} \dfrac{\partial f}{\partial \hat{u}_1} & \cdots & \dfrac{\partial f}{\partial \hat{u}_m} \end{bmatrix} \begin{bmatrix} \hat{\phi}_1 \\ \vdots \\ \hat{\phi}_m \end{bmatrix}. \qquad (13.23)$$

Similarly, the functional version of the coupled adjoint equations can be derived as

$$\begin{bmatrix} I & -\dfrac{\partial U_2}{\partial \hat{u}_1}^T & \cdots & -\dfrac{\partial U_m}{\partial \hat{u}_1}^T \\ -\dfrac{\partial U_1}{\partial \hat{u}_2}^T & I & \cdots & -\dfrac{\partial U_m}{\partial \hat{u}_2}^T \\ \vdots & \vdots & \ddots & \vdots \\ -\dfrac{\partial U_1}{\partial \hat{u}_m}^T & -\dfrac{\partial U_2}{\partial \hat{u}_m}^T & \cdots & I \end{bmatrix} \begin{bmatrix} \hat{\psi}_1 \\ \hat{\psi}_2 \\ \vdots \\ \hat{\psi}_m \end{bmatrix} = \begin{bmatrix} \dfrac{\partial f}{\partial \hat{u}_1}^T \\ \dfrac{\partial f}{\partial \hat{u}_2}^T \\ \vdots \\ \dfrac{\partial f}{\partial \hat{u}_m}^T \end{bmatrix}. \qquad (13.24)$$

After solving for the coupled-adjoint vector using the previous equation, we can use the total derivative equation to compute the desired derivatives:

$$\dfrac{df}{dx} = \dfrac{\partial f}{\partial x} - \begin{bmatrix} \hat{\psi}_1^T & \cdots & \hat{\psi}_m^T \end{bmatrix} \begin{bmatrix} \dfrac{\partial \hat{r}_1}{\partial x} \\ \vdots \\ \dfrac{\partial \hat{r}_m}{\partial x} \end{bmatrix}. \qquad (13.25)$$

Because the coupling variables ($\hat{u}$) are usually a reduction of the internal state variables (u), the linear systems in Eqs. 13.22 and 13.24 are usually much smaller than that of the residual counterparts (Eqs. 13.17

and 13.19). However, unlike the partial derivatives in the residual form, the partial derivatives in the functional form Jacobian need to account for the solution of the corresponding component. When viewed at the component level, these derivatives are actually total derivatives of the component. When the component is an implicit set of equations, computing these derivatives with finite-differencing would require solving the component's equations for each variable perturbation. Alternatively, an implicit analytic method (from Section 6.7) could be applied to the component to compute these derivatives.

Figure 13.27 shows the Jacobian structure in the functional form of the coupled direct method (Eq. 13.22) for the case of Fig. 13.23. The dimension of this Jacobian is smaller than that of the residual form. Recall from Fig. 13.9 that U_1 corresponds to r_3, U_2 corresponds to r_6, and U_3 corresponds to r_9. Thus, the total size of this Jacobian corresponds to the sum of the sizes of components 3, 6, and 9, as opposed to the sum of the sizes of all nine components for the residual form. However, as mentioned previously, partial derivatives for the functional form are more expensive to compute because they need to account for an implicit solver in each of the three groups.

$$\begin{bmatrix} I & -\frac{\partial U_1}{\partial \hat{u}_2} & -\frac{\partial U_1}{\partial \hat{u}_3} \\ -\frac{\partial U_2}{\partial \hat{u}_1} & I & -\frac{\partial U_2}{\partial \hat{u}_3} \\ -\frac{\partial U_3}{\partial \hat{u}_1} & -\frac{\partial U_3}{\partial \hat{u}_2} & I \end{bmatrix}$$

Fig. 13.27 Jacobian of coupled derivatives for the functional form of Fig. 13.23.

UDE for Coupled Systems

As in the single-component case in Section 6.9, the coupled direct and adjoint equations derived in this section can be obtained from the UDE with the appropriate definitions of residuals and variables. The components corresponding to each block in these equations can also be implicit or explicit, which provides the flexibility to represent systems of heterogeneous components.

MAUD implements the linear systems from these coupled direct and adjoint equations using the UDE. The overall linear system inherits the hierarchical structure defined for the nonlinear solvers. Instead of nonlinear solvers, we use linear solvers, such as a direct solver and Krylov (both monolithic). As mentioned in Section 13.2.5, the nonlinear block Jacobi and Gauss–Seidel (both recursive) can be reused to solve coupled linear systems. Components can be expressed using residual or functional forms, making it possible to include black-box components.

The example originally used in Chapter 6 to demonstrate how to compute derivatives with the UDE (Ex. 6.15) can be viewed as a coupled derivative computation where each equation is a component. Example 13.6 demonstrates the UDE approach to computing derivatives by building on the wing design problem presented in Ex. 13.2.

> **Tip 13.4** Implementing coupled derivative computation
>
> Obtaining derivatives for each component of a multidisciplinary model and assembling them to compute the coupled derivatives usually requires a high implementation effort. In addition to implementing hierarchical coupled solvers (as mentioned in Tip 13.3), the OpenMDAO framework also implements the MAUD approach to computing coupled derivatives. The linear system mirrors the hierarchy set up for nonlinear coupled solvers.[132] Ideally, users provide the partial derivatives for each component using accurate and efficient methods. However, if derivatives are not available, OpenMDAO can automatically compute them using finite differences or the complex-step method. OpenMDAO also facilitates efficient derivative computation for sparse Jacobians using the graph coloring techniques introduced in Section 6.8.

132. Gray et al., *OpenMDAO: An open-source framework for multidisciplinary design, analysis, and optimization*, 2019.

> **Example 13.6** Aerostructural derivatives
>
> Let us now consider a wing design optimization problem based on the aerostructural model considered in Ex. 13.1.[†] The design variables are as follows:
>
> α: Angle of attack. This controls the amount of lift produced by the airplane.
>
> b: Wingspan. This is a shared variable because it directly affects both the aerodynamic and structural models.
>
> θ: Twist distribution along the wingspan, represented by a vector. This controls the relative lift loading in the spanwise direction, which affects the drag and the load distribution on the structure. It affects the aerodynamic model but not the structural model (because it is idealized as a beam).
>
> t: Thickness distribution of beam along the wingspan, represented by a vector. This directly affects the weight and the stiffness. It does not affect the aerodynamic model.
>
> The objective is to minimize the fuel required for a given range R, which can be written as a function of drag, lift, and weight, as follows:
>
> $$f = W\left[\exp\left(\frac{RcD}{VL}\right) - 1\right]. \quad (13.26)$$
>
> The empty weight W only depends on t and b, and the dependence is explicit (it does not require solving the aerodynamic or structural models). The drag D and lift L depend on all variables once we account for the coupled system of equations. The remaining variables are fixed: R is the required range, V is the airplane's cruise speed, and c is the specific fuel consumption of the airplane's engines. We also need to constrain the stresses in the structure, σ, which are an explicit function of the displacements (see Ex. 6.12).
>
> To solve this optimization problem using gradient-based optimization, we need the *coupled* derivatives of f and σ with respect to α, b, θ, and t. Computing the derivatives of the aerodynamic and structural models separately is not

[†] As in Ex. 13.5, these results were obtained using OpenAeroStruct, and the description and equations are simplified for brevity.

13.3 Coupled Derivatives Computation

sufficient. For example, a perturbation on the twist changes the loads, which then changes the wing displacements, which requires solving the aerodynamic model again. Coupled derivatives take this effect into account.

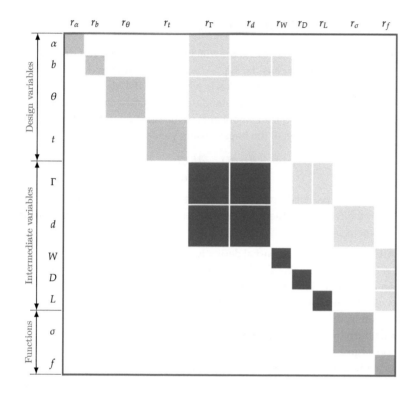

Fig. 13.28 The DSM of the aerostructural problem shows the structure of the reverse UDE.

We show the DSM for the system in Fig. 13.28. Because the DSM has the same sparsity structure as the transpose of the Jacobian, this diagram reflects the structure of the reverse UDE. The blocks that pertain to the design variables have unit diagonals because they are independent variables, but they directly affect the solver blocks. The blocks responsible for solving for Γ and d are the only ones with feedback coupling. The part of the UDE pertaining to Γ and d is the Jacobian of residuals for the aerodynamic and structural components, which we already derived in Ex. 13.5 to apply Newton's method on the coupled system. The functions of interest are all explicit components and only depend directly on the design variables or the state variables. For example, the weight W depends only on t; drag and lift depend only on the converged Γ; σ depends on the displacements; and finally, the fuel burn f just depends on drag, lift, and weight. This whole coupled chain of derivatives is computed by solving the linear system shown in Fig. 13.28.

For brevity, we only discuss the derivatives required to compute the derivative of fuel burn with respect to span, but the other partial derivatives would follow the same rationale.

- $\partial r/\partial u$ is identical to what we derived when solving the coupled aerostructural system in Ex. 13.5.

- $\partial r/\partial x$ has two components, which we can obtain by differentiating the residuals:

$$\frac{\partial}{\partial b}(A\Gamma - v) = \frac{\partial A}{\partial b}\Gamma - \frac{\partial v}{\partial b}, \quad \frac{\partial}{\partial b}(Kd - q) = \frac{\partial K}{\partial b}d.$$

- $\partial f/\partial x = \partial f/\partial b = 0$ because the fuel burn does not depend directly on the span if we just consider Eq. 13.26. However, it does depend on the span through W, D, and L. This is where the UDE description is more general and clearer than the standard direct and adjoint formulation. By defining the explicit components of the function in the bottom-right corner, the solution of the linear system yields the chain rule

$$\frac{df}{db} = \frac{\partial f}{\partial D}\frac{dD}{db} + \frac{\partial f}{\partial L}\frac{dL}{db} + \frac{\partial f}{\partial W}\frac{dW}{db},$$

where the partial derivatives can be obtained by differentiating Eq. 13.26 symbolically, and the total derivatives are part of the coupled linear system solution.

After computing all the partial derivative terms, we solve either the forward or reverse UDE system. For the derivative with respect to span, neither method has an advantage. However, for the derivatives of fuel burn with respect to the twist and thickness variables, the reverse mode is much more efficient. In this example, $df/db = -11.0$ kg/m, so each additional meter of span reduced the fuel burn by 11 kg. If we compute this same derivative without coupling (by converging the aerostructural model but not considering the off-diagonal terms in the aerostructural Jacobian), we obtain $df/db = -17.7$ kg/m, which is significantly different. The derivatives of the fuel burn with respect to the twist distribution and the thickness distribution along the wingspan are plotted in Fig. 13.29, where we can see the difference between coupled and uncoupled derivatives.

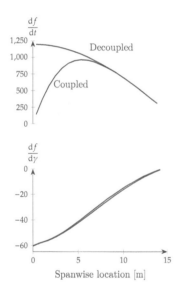

Fig. 13.29 Derivatives of the fuel burn with respect to the spanwise distribution of twist and thickness variables. The coupled derivatives differ from the uncoupled derivatives, especially for the derivatives with respect to structural thicknesses near the wing root.

13.4 Monolithic MDO Architectures

So far in this chapter, we have extended the models and solvers from Chapter 3 and derivative computation methods from Chapter 6 to coupled systems. We now discuss the options to optimize coupled systems, which are given by various MDO architectures.

Monolithic MDO architectures cast the design problem as a single optimization. The only difference between the different monolithic architectures is the set of design variables that the optimizer is responsible for, which affects the constraint formulation and how the governing equations are solved.

13.4.1 Multidisciplinary Feasible

The multidisciplinary design feasible (MDF) architecture is the architecture that is most similar to a single-discipline problem and usually the most intuitive for engineers. The design variables, objective, and constraints are the same as we would expect for a single-discipline problem. The only difference is that the computation of the objective and constraints requires solving a coupled system instead of a single system of governing equations. Therefore, all the optimization algorithms covered in the previous chapters can be applied without modification when using the MDF architecture. This approach is also called a *reduced-space* approach because the optimizer does not handle the space of the state and coupling variables. Instead, it relies on a solver to find the state variables that satisfy the governing equations for the current design (see Eq. 3.32).

The resulting optimization problem is as follows:*

$$
\begin{aligned}
&\text{minimize} && f(x; \hat{u}^*) \\
&\text{by varying} && x \\
&\text{subject to} && g(x, \hat{u}^*) \leq 0 \\
&\text{while solving} && \hat{r}(\hat{u}; x) = 0 \\
&\text{for} && \hat{u}.
\end{aligned}
\qquad (13.27)
$$

*The quantities after the semicolon in the variable dependence correspond to variables that remain fixed in the current context. For simplicity, we omit the design equality constraints ($h = 0$) without loss of generality.

At each optimization iteration, the optimizer has a multidisciplinary feasible point $\hat{u}^*$ found through the MDA. For a design given by the optimizer (x), the MDA finds the internal component states (u) and the coupling variables ($\hat{u}$). To denote the MDA solution, we use the residuals of the functional form, where the residuals for component i are[†]

$$\hat{r}_i(\hat{u}, u_i) = \hat{u}_i - U_i(u_i, \hat{u}_{j \neq i}) = 0. \qquad (13.28)$$

[†]These are identical to the residuals of the system-level Newton solver (Eq. 13.11).

Each component is assumed to solve for its state variables u_i internally. The MDA finds the coupling variables by solving the coupled system of components $i = 1, \ldots, m$ using one of the methods from Section 13.2.5.

Then, the objective and constraints can be computed based on the current design variables and coupling variables. Figure 13.30 shows an XDSM for MDF with three components. Here we use a nonlinear block Gauss–Seidel method (see Alg. 13.2) to converge the MDA, but any other method from Section 13.2.5 could be used.

One advantage of MDF is that the system-level states are physically compatible if an optimization stops prematurely. This is advantageous in an engineering design context when time is limited, and we are not as concerned with finding an optimal design in the strict mathematical

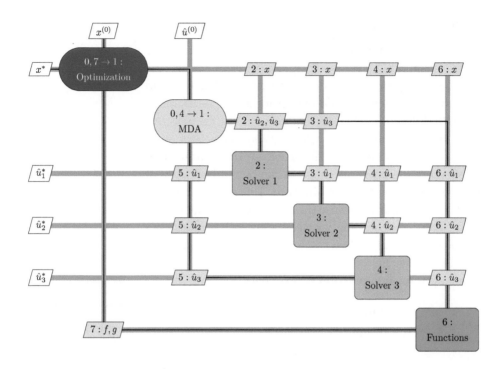

Fig. 13.30 The MDF architecture relies on an MDA to solve for the coupling and state variables at each optimization iteration. In this case, the MDA uses the block Gauss–Seidel method.

sense as we are with finding an improved design. However, it is not guaranteed that the design constraints are satisfied if the optimization is terminated early; that depends on whether the optimization algorithm maintains a feasible design point or not.

The main disadvantage of MDF is that it solves an MDA for each optimization iteration, which requires its own algorithm outside of the optimization. Implementing an MDA algorithm can be time-consuming if one is not already in place.

As mentioned in Tip 13.3, a MAUD-based framework such as OpenMDAO can facilitate this. MAUD naturally implements the MDF architecture because it focuses on solving the MDA (Section 13.2.5) and on computing the derivatives corresponding to the MDA (Section 13.3.3).‡

When using a gradient-based optimizer, gradient computations are also challenging for MDF because coupled derivatives are required. Finite-difference derivative approximations are easy to implement, but their poor scalability and accuracy are compounded by the MDA, as explained in Section 13.3. Ideally, we would use one of the analytic coupled derivative computation methods of Section 13.3, which require a substantial implementation effort. Again, OpenMDAO was developed to facilitate coupled derivative computation (see Tip 13.4).

‡The first application of MAUD was the design optimization of a satellite and its orbit dynamics. The problem consisted of over 25,000 design variables and over 2 million state variables[210]

210. Hwang et al., *Large-scale multidisciplinary optimization of a small satellite's design and operation*, 2014.

13.4 Monolithic MDO Architectures

Example 13.7 Aerostructural optimization using MDF

Continuing the wing aerostructural problem from Ex. 13.6, we are finally ready to optimize the wing. The MDF formulation is as follows:

$$\begin{aligned}
\text{minimize} \quad & f \\
\text{by varying} \quad & \alpha, b, \theta, t \\
\text{subject to} \quad & L - W = 0 \\
& 2.5|\sigma| - \sigma_{\text{yield}} \leq 0 \\
\text{while solving} \quad & A(d)\Gamma - v(d, \alpha) = 0 \\
& Kd - q(\Gamma) = 0 \\
\text{for} \quad & \Gamma, d.
\end{aligned}$$

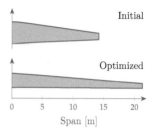

Fig. 13.31 The optimization reduces the fuel burn by increasing the span.

The structural stresses are constrained to be less than the yield stress of the material by a safety factor (2.5 in this case). In Ex. 13.5, we set up the MDA for the aerostructural problem, and in Ex. 13.6, we set up the coupled derivative computations needed to solve this problem using gradient-based optimization.

Solving this optimization resulted in the larger span wing shown in Fig. 13.31. This larger span increases the structural weight, but decreases drag. Although the increase in weight would typically increase the fuel burn, the drag decrease more than compensates for this adverse effect, and the fuel burn ultimately decreases up to this value of span. Beyond this optimal span value, the weight penalty would start to dominate, resulting in a fuel burn increase.

The twist and thickness distributions are shown in Fig. 13.32. The wing twist directly controls the spanwise lift loading. The baseline wing had no twist, which resulted in the loading shown in Fig. 13.33. In this figure, the gray line represents a hypothetical elliptical lift distribution, which results in the theoretical minimum for induced drag. The loading distributions for the level flight (1 g) and maneuver conditions (2.5 g) are indistinguishable. The optimization increases the twist in the midspan and drastically decreases it toward the tip. This twist distribution differentiates the loading at the two conditions: it makes the loading at level flight closer to the elliptical ideal while shifting the loading at the maneuver condition toward the wing root.

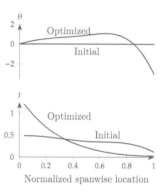

Fig. 13.32 Twist and thickness distributions for the baseline and optimized wings.

The thickness distribution also changes significantly, as shown in Fig. 13.32. The optimization tailors the thickness by adding more thickness in the spar near the root, where the moments are larger, and thins out the wing much more toward the tip, where the loads decrease. This more radical thickness distribution is enabled by the tailoring of the spanwise lift loading discussed previously.

These trades make sense because, at the level flight condition, the optimizer is concerned with minimizing drag, whereas, at the maneuver condition, the optimizer just wants to satisfy the stress constraint for a given total lift.

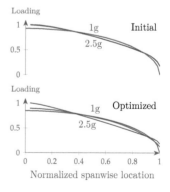

Fig. 13.33 Lift loading for the baseline and optimized wings.

Example 13.8 Aerostructural sequential optimization

In Section 13.1, we argued that sequential optimization does not, in general, converge to the true optimum for constrained problems. We now demonstrate this for a modified version of the wing aerostructural design optimization problem from Ex. 13.7. One major modification was to reduce the problem to two design variables to visualize the optimization path: one structural variable corresponding to a constant spar thickness and one twist variable corresponding to the wing tip twist, which controls the slope of a linear twist distribution. The simultaneous optimization of these two variables using the MDF architecture from Ex. 13.7 yields the path labeled "MDO" in Fig. 13.34.

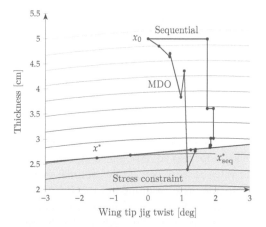

Fig. 13.34 Sequential optimization gets stuck at the stress constraint, whereas simultaneous optimization of the aerodynamic and structural variable finds the true multidisciplinary optimum.

To perform sequential optimization for the wing design problem of Ex. 13.1, we could start by optimizing the aerodynamics by solving the following problem:

$$\begin{aligned} \text{minimize} \quad & f \\ \text{by varying} \quad & \alpha, \theta \\ \text{subject to} \quad & L - W = 0 \,. \end{aligned}$$

Here, W is constant because the structural thicknesses t are fixed, but L is a function of the aerodynamic design variables and states. We cannot include the span b because it is a shared variable, as explained in Section 13.1. Otherwise, this optimization would tend to increase b indefinitely to reduce the lift-induced drag. Because f is a function of D and L, and L is constant because $L = W$, we could replace the objective with D.

Once the aerodynamic optimization has converged, the twist distribution and the forces are fixed, and we then optimize the structure by minimizing weight subject to stress constraints by solving the following problem:

$$\begin{aligned} \text{minimize} \quad & f \\ \text{by varying} \quad & t \\ \text{subject to} \quad & 2.5|\sigma| - \sigma_{\text{yield}} \leq 0 \,. \end{aligned}$$

Because the drag and lift are constant, the objective could be replaced by W. Again, we cannot include the span in this problem because it would decrease indefinitely to reduce the weight and internal loads due to bending.

These two optimizations are repeated until convergence. As shown in Fig. 13.34, sequential optimization only changes one variable at a time, and it converges to a point on the constraint with about 3.5° more twist than the true optimum of the MDO. When including more variables, these differences are likely to be even larger.

13.4.2 Individual Discipline Feasible

The individual discipline feasible (IDF) architecture adds independent copies of the coupling variables to allow component solvers to run independently and possibly in parallel. These copies are known as *target variables* and are controlled by the optimizer, whereas the actual coupling variables are computed by the corresponding component. Target variables are denoted by a superscript t, so the coupling variables produced by discipline i are denoted as $\hat{u}_i^t$. These variables represent the current guesses for the coupling variables that are independent of the corresponding actual coupling variables computed by each component. To ensure the eventual consistency between the target coupling variables and the actual coupling variables at the optimum, we define a set of *consistency constraints*, $h_i^c = \hat{u}_i^t - \hat{u}_i$, which we add to the optimization problem formulation.

The optimization problem for the IDF architecture is

$$\begin{aligned}
& \text{minimize} && f(x; \hat{u}) \\
& \text{by varying} && x, \hat{u}^t \\
& \text{subject to} && g(x; \hat{u}) \leq 0 \\
& && h_i^c = \hat{u}_i^t - \hat{u}_i = 0 && i = 1, \ldots, m \\
& \text{while solving} && r_i\left(\hat{u}_i; x, \hat{u}_{j \neq i}^t\right) = 0 && i = 1, \ldots, m \\
& \text{for} && \hat{u}.
\end{aligned} \quad (13.29)$$

Each component i is solved independently to compute the corresponding output coupling variables $\hat{u}_i$, where the inputs $\hat{u}_{j \neq i}^t$ are given by the optimizer. Thus, each component drives its residuals to zero to compute

$$\hat{u}_i = U_i\left(x, \hat{u}_{j \neq i}^t\right). \quad (13.30)$$

The consistency constraint quantifies the difference between the target coupling variables guessed by the optimizer and the actual coupling variables computed by the components. The optimizer iterates the target coupling variables simultaneously with the design variables to find a multidisciplinary feasible point that is also an optimum. At each iteration, the objective and constraints are computed using the latest available coupling variables. Figure 13.35 shows the XDSM for IDF.

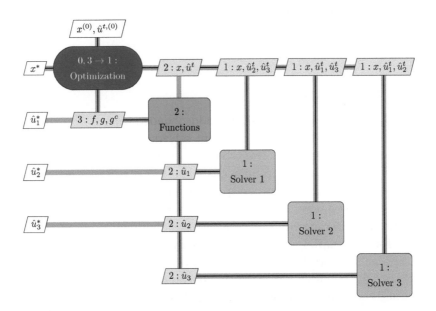

Fig. 13.35 The IDF architecture breaks up the MDA by letting the optimizer solve for the coupling variables that satisfy interdisciplinary feasibility.

One advantage of IDF is that each component can be solved in parallel because they do not depend on each other directly. Another advantage is that if gradient-based optimization is used to solve the problem, the optimizer is typically more robust and has a better convergence rate than the fixed-point iteration algorithms of Section 13.2.5.

The main disadvantage of IDF is that the optimizer must handle more variables and constraints compared with the MDF architecture. If the number of coupling variables is large, the size of the resulting optimization problem may be too large to solve efficiently. This problem can be mitigated by careful selection of the components or by aggregating the coupling variables to reduce their dimensionality.

Unlike MDF, IDF does not guarantee a multidisciplinary feasible state at every design optimization iteration. Multidisciplinary feasibility is only guaranteed at the end of the optimization through the satisfaction of the consistency constraints. This is a disadvantage because if the optimization stops prematurely or we run out of time, we do not have a valid state for the coupled system.

13.4 Monolithic MDO Architectures

Example 13.9 Aerostructural optimization using IDF

For the IDF architecture, we need to make copies of the coupling variables (Γ^t and d^t) and add the corresponding consistency constraints, as highlighted in the following problem statement:

$$\begin{aligned}
\text{minimize} \quad & f \\
\text{by varying} \quad & \alpha, b, \theta, t, \Gamma^t, d^t \\
\text{subject to} \quad & L = W \\
& 2.5|\sigma| - \sigma_{\text{yield}} \leq 0 \\
& \Gamma^t - \Gamma = 0 \\
& d^t - d = 0 \\
\text{while solving} \quad & A\left(d^t\right)\Gamma - \theta\left(d^t, \alpha\right) = 0 \\
& Kd - q\left(\Gamma^t\right) = 0 \\
\text{for} \quad & \Gamma, d.
\end{aligned}$$

The aerodynamic and structural models are solved independently. The aerodynamic solver finds Γ for the d^t given by the optimizer, and the structural solver finds d for the given Γ^t.

When using gradient-based optimization, we do not require coupled derivatives, but we do need the derivatives of each model with respect to both state variables. The derivatives of the consistency constraints are just a unit matrix when taken with respect to the variable copies and are zero otherwise.

13.4.3 Simultaneous Analysis and Design

Simultaneous analysis and design (SAND) extends the idea of IDF by moving not only the coupling variables to the optimization problem but also all component states. The SAND architecture requires exposing all the components in the form of the system-level view previously introduced in Fig. 13.9. The residuals of the analysis become constraints for which the optimizer is responsible.§

This means that component solvers are no longer needed, and the optimizer becomes responsible for simultaneously solving the components for their states, the interdisciplinary compatibility for the coupling variables, and the design optimization problem for the design variables. All that is required from the model is the computation of residuals. Because the optimizer is controlling all these variables, SAND is also known as a *full-space* approach. SAND can be stated as

§When the residual equations arise from discretized PDEs, we have what is called *PDE-constrained optimization*.[211]

211. Biegler et al., *Large-Scale PDE-Constrained Optimization*, 2003.

follows:

$$
\begin{aligned}
\text{minimize} \quad & f(x, \hat{u}, u) \\
\text{by varying} \quad & x, \hat{u}, u \\
\text{subject to} \quad & g(x, \hat{u}) \le 0 \\
& r(x, \hat{u}, u) = 0.
\end{aligned}
\tag{13.31}
$$

Here, we use the representation shown in Fig. 13.7, so there are two sets of explicit functions that convert the input coupling variables of the component. The SAND architecture is also applicable to single components, in which case there are no coupling variables. The XDSM for SAND is shown in Fig. 13.36.

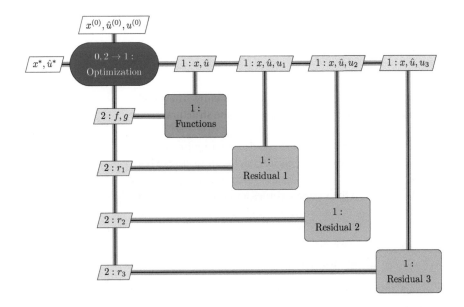

Fig. 13.36 The SAND architecture lets the optimizer solve for all variables (design, coupling, and state variables), and component solvers are no longer needed.

Because it solves for all variables simultaneously, the SAND architecture can be the most efficient way to get to the optimal solution. In practice, however, it is unlikely that this is advantageous when efficient component solvers are available.

The resulting optimization problem is the largest of all MDO architectures and requires an optimizer that scales well with the number of variables. Therefore, a gradient-based optimization algorithm is likely required, in which case the derivative computation must also be considered. Fortunately, SAND does not require derivatives of the coupled system or even total derivatives that account for the component solution; only partial derivatives of residuals are needed.

SAND is an intrusive approach because it requires access to residuals.

These might not be available if components are provided as black boxes. Rather than computing coupling variables $\hat{u}_i$ and state variables u_i by converging the residuals to zero, each component i just computes the current residuals r_i for the current values of the coupling variables $\hat{u}$ and the component states u_i.

Example 13.10 Aerostructural optimization using SAND

For the SAND approach, we do away completely with the solvers and let the optimizer find the states. The resulting problem is as follows:

$$\begin{aligned}
\text{minimize} \quad & f \\
\text{by varying} \quad & a, b, \theta, t, \Gamma, d \\
\text{subject to} \quad & L = W \\
& 2.5|\sigma| - \sigma_{\text{yield}} \leq 0 \\
& A\Gamma - \theta = 0 \\
& Kd - q = 0.
\end{aligned}$$

Instead of being solved separately, the models are now solved by the optimizer.

When using gradient-based optimization, the required derivatives are just partial derivatives of the residuals (the same partial derivatives we would use for an implicit analytic method).

13.5 Distributed MDO Architectures

The monolithic MDO architectures we have covered so far form and solve a single optimization problem. Distributed architectures decompose this single optimization problem into a set of smaller optimization problems, or *disciplinary subproblems*, which are then coordinated by a *system-level subproblem*. One key requirement for these architectures is that they must be mathematically equivalent to the original monolithic problem to converge to the same solution.

There are two primary motivations for distributed architectures. The first one is the possibility of decomposing the problem to reduce the computational time. The second motivation is to mimic the structure of large engineering design teams, where disciplinary groups have the autonomy to design their subsystems so that MDO is more readily adopted in industry. Overall, distributed MDO architectures have fallen short on both of these expectations. Unless a problem has a special structure, there is no distributed architecture that converges as rapidly as a monolithic one. In practice, distributed architectures have not been used much recently.

There are two main types of distributed architectures: those that enforce multidisciplinary feasibility via an MDA somewhere in the process and those that enforce multidisciplinary feasibility in some other way (using constraints or penalties at the system level). This is analogous to MDF and IDF, respectively, so we name these types *distributed MDF* and *distributed IDF*.*

In MDO problems, it can be helpful to distinguish between design variables that affect only one component directly (called *local design variables*) and design variables that affect two or more components directly (called *shared design variables*). We denote the vector of design variables local to component i by x_i and the shared variables by x_0. The full vector of design variables is given by concatenating the shared and local design variables into a single vector $x = \left[x_0^\mathsf{T}, x_1^\mathsf{T}, \ldots, x_m^\mathsf{T}\right]$, where m is the number of components.

If a constraint can be computed using a single component and satisfied by varying only the local design variables for that component, it is a *local constraint*; otherwise, it is nonlocal. Similarly, for the design variables, we concatenate the constraints as $g = \left[g_0^\mathsf{T}, g_1^\mathsf{T}, \ldots, g_m^\mathsf{T}\right]$. The same distinction could be applied to the objective function, but we do not usually do this.

The MDO problem representation we use here is shown in Fig. 13.37 for a general three-component system. We use the functional form introduced in Section 13.2.3, where the states in each component are hidden. In this form, the system level only has access to the outputs of each solver, which are the coupling coupling variables and functions of interest.

*Martins and Lambe[41] present a more comprehensive description of all MDO architectures, including references to known applications of each architecture.

41. Martins and Lambe, *Multidisciplinary design optimization: A survey of architectures*, 2013.

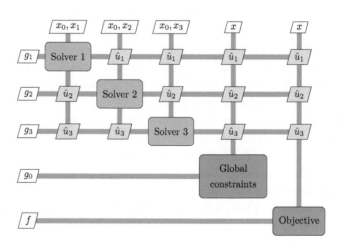

Fig. 13.37 MDO problem nomenclature and dependencies.

The set of constraints is also split into shared constraints and local ones. Local constraints are computed by the corresponding component and depend only on the variables available in that component. Shared constraints depend on more than one set of coupling variables. These dependencies are also shown in Fig. 13.37.

13.5.1 Collaborative Optimization

The collaborative optimization (CO) architecture is inspired by how disciplinary teams work to design complex engineered systems.[212] This is a distributed IDF architecture, where the disciplinary optimization problems are formulated to be independent of each other by using target values of the coupling and shared design variables. These target values are then shared with all disciplines during every iteration of the solution procedure. The complete independence of disciplinary subproblems combined with the simplicity of the data-sharing protocol makes this architecture attractive for problems with a small amount of shared data.

212. Braun and Kroo, *Development and application of the collaborative optimization architecture in a multidisciplinary design environment*, 1997.

The system-level subproblem modifies the original problem as follows: (1) local constraints are removed, (2) target coupling variables, $\hat{u}^t$, are added as design variables, and (3) a *consistency constraint* is added. This optimization problem can be written as follows:

$$
\begin{aligned}
& \text{minimize} && f\left(x_0, x_1^t, \ldots, x_m^t, \hat{u}^t\right) \\
& \text{by varying} && x_0, x_1^t, \ldots, x_m^t, \hat{u}^t \\
& \text{subject to} && g_0\left(x_0, x_1^t, \ldots, x_m^t, \hat{u}^t\right) \le 0 \\
& && J_i^* = \left\| x_{0i}^t - x_0 \right\|_2^2 + \left\| x_i^t - x_i \right\|^2 \\
& && \quad + \left\| \hat{u}_i^t - \hat{u}_i\left(x_{0i}^t, x_i, \hat{u}_{j \ne i}^t\right) \right\|^2 = 0 \quad \text{for } i = 1, \ldots, m,
\end{aligned}
\tag{13.32}
$$

where x_{0i}^t are copies of the shared design variables that are passed to discipline i, and x_i^t are copies of the local design variables passed to the system subproblem.

The constraint function J_i^* is a measure of the inconsistency between the values requested by the system-level subproblem and the results from the discipline i subproblem. The disciplinary subproblems do not include the original objective function. Instead, the objective of each subproblem is to minimize the inconsistency function.

For each discipline i, the subproblem is as follows:

$$\begin{aligned}
\text{minimize} \quad & J_i\left(x_{0i}^t, x_i; \hat{u}_i\right) \\
\text{by varying} \quad & x_{0i}^t, x_i \\
\text{subject to} \quad & g_i\left(x_{0i}^t, x_i; \hat{u}_i\right) \leq 0 \\
\text{while solving} \quad & r_i\left(\hat{u}_i; x_{0i}^t, x_i, \hat{u}_{j\neq i}^t\right) = 0 \\
\text{for} \quad & \hat{u}_i.
\end{aligned} \quad (13.33)$$

These subproblems are independent of each other and can be solved in parallel. Thus, the system-level subproblem is responsible for minimizing the design objective, whereas the discipline subproblems minimize system inconsistency while satisfying local constraints.

The CO problem statement has been shown to be mathematically equivalent to the original MDO problem.[212] There are two versions of the CO architecture: CO_1 and CO_2. Here, we only present the CO_2 version. The XDSM for CO is shown in Fig. 13.38 and the procedure is detailed in Alg. 13.5.

212. Braun and Kroo, *Development and application of the collaborative optimization architecture in a multidisciplinary design environment*, 1997.

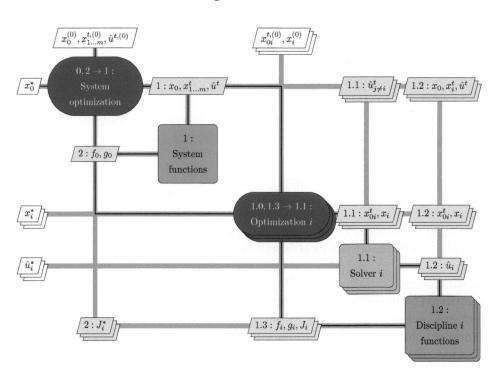

Fig. 13.38 Diagram for the CO architecture.

CO has the organizational advantage of having entirely separate disciplinary subproblems. This is desirable when designers in each discipline want to maintain some autonomy. However, the CO for-

13.5 Distributed MDO Architectures

mulation suffers from numerical ill-conditioning. This is because the constraint gradients of the system problem at an optimal solution are all zero vectors, which violates the constraint qualification requirement for the Karush–Kuhn–Tucker (KKT) conditions (see Section 5.3.1). This ill-conditioning slows down convergence when using a gradient-based optimization algorithm or prevents convergence altogether.

Algorithm 13.5 Collaborative optimization

Inputs:
  x: Initial design variables
Outputs:
  x^*: Optimal variables
  f^*: Optimal objective value
  g^*: Optimal constraint values

0: Initiate system optimization iteration
repeat
  1: Compute system subproblem objectives and constraints
  for Each discipline i (in parallel) **do**
    1.0: Initiate disciplinary subproblem optimization
    repeat
      1.1: Evaluate disciplinary analysis
      1.2: Compute disciplinary subproblem objective and constraints
      1.3: Compute new disciplinary subproblem design point and J_i
    until 1.3 → 1.1: Optimization i has converged
  end for
  2: Compute a new system subproblem design point
until 2 → 1: System optimization has converged

Example 13.11 Aerostructural optimization using CO

To apply CO to the wing aerostructural design optimization problem (Ex. 13.1), we need to set up a system-level optimization problem and two discipline-level optimization subproblems.

The system-level optimization problem is formulated as follows:

$$\begin{aligned}
\text{minimize} \quad & f \\
\text{by varying} \quad & b^t, \Gamma^t, d^t, W^t \\
\text{subject to} \quad & J_1^* \leq \varepsilon \\
& J_2^* \leq \varepsilon,
\end{aligned}$$

where ε is a specified convergence tolerance. The set of variables that are copied as targets includes the shared design variable (b) and the coupling variables (Γ and d).

The aerodynamics subproblem is as follows:

$$\text{minimize} \quad J_1 \equiv \left(1 - \frac{b}{b^t}\right)^2 + \sum_{i=1}^{n_\Gamma} \left(1 - \frac{\Gamma_i}{\Gamma_i^t}\right)^2$$

$$\text{by varying} \quad b, \alpha, \theta$$
$$\text{subject to} \quad L - W^t = 0$$
$$\text{while solving} \quad A\Gamma - \theta = 0$$
$$\text{for} \quad \Gamma.$$

In this problem, the aerodynamic optimization minimizes the discrepancy between the span requested by the system-level optimization (b^t) and the span that aerodynamics is optimizing (b). The same applies to the coupling variables Γ. The aerodynamics subproblem is fully responsible for optimizing α and θ.

The structures subproblem is as follows:

$$\text{minimize} \quad J_2 \equiv \left(1 - \frac{b}{b^t}\right)^2 + \sum_{i=1}^{n_d} \left(1 - \frac{d_i}{d_i^t}\right)^2 + \left(1 - \frac{W}{W^t}\right)^2$$

$$\text{by varying} \quad b, t$$
$$\text{subject to} \quad 2.5|\sigma| - \sigma_{\text{yield}} \leq 0$$
$$\text{while solving} \quad Kd - q = 0$$
$$\text{for} \quad d.$$

Here, the structural optimization minimizes the discrepancy between the span wanted by the structures (a decrease) versus what the system level requests (which takes into account the opposite trend from aerodynamics). The structural subproblem is fully responsible for satisfying the stress constraints by changing the structural sizing t, which are local variables.

13.5.2 Analytical Target Cascading

Analytical target cascading (ATC) is a distributed IDF architecture that uses penalties in the objective function to minimize the difference between the target variables requested by the system-level optimization and the actual variables computed by each discipline.[†]

The idea of ATC is similar to the CO architecture in the previous section, except that ATC uses penalties instead of a constraint. The ATC system-level problem is as follows:

$$\text{minimize} \quad f_0(x, \hat{u}^t) + \sum_{i=1}^{m} \Phi_i \left(x_{0i}^t - x_0, \hat{u}_i^t - \hat{u}_i(x_0, x_i, \hat{u}^t)\right)$$
$$+ \Phi_0(g_0(x, \hat{u}^t)) \quad (13.34)$$
$$\text{by varying} \quad x_0, \hat{u}^t,$$

[†] ATC was originally developed as a method to handle design requirements in a system's hierarchical decomposition.[213] ATC became an MDO architecture after further developments.[214] A MATLAB implementation of ATC is available.[215]

213. Kim et al., *Analytical target cascading in automotive vehicle design*, 2003.

214. Tosserams et al., *An augmented Lagrangian relaxation for analytical target cascading using the alternating direction method of multipliers*, 2006.

215. Talgorn and Kokkolaras, *Compact implementation of non-hierarchical analytical target cascading for coordinating distributed multidisciplinary design optimization problems*, 2017.

13.5 Distributed MDO Architectures

where Φ_0 is a penalty relaxation of the shared design constraints, and Φ_i is a penalty relaxation of the discipline i consistency constraints.

Although the most common penalty functions in ATC are quadratic penalty functions, other penalty functions are possible. As mentioned in Section 5.4, penalty methods require a good selection of the penalty weight values to converge quickly and accurately enough. The ATC architecture converges to the same optimum as other MDO architectures, provided that problem is unimodal and all the penalty terms in the optimization problems approach zero.

Figure 13.39 shows the ATC architecture XDSM, where w denotes the penalty function weights used in the determination of Φ_0 and Φ_i. The details of ATC are described in Alg. 13.6.

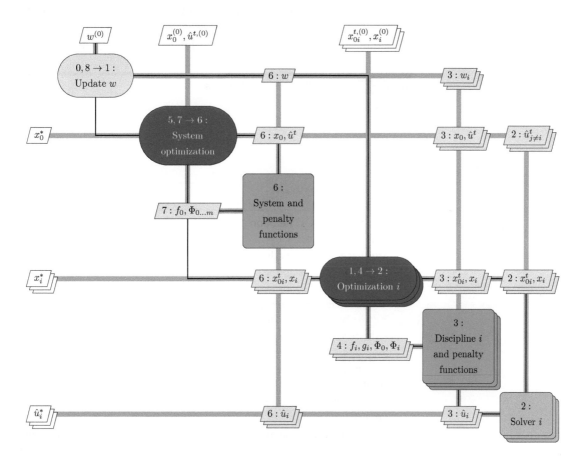

Fig. 13.39 Diagram for the ATC architecture

The ith discipline subproblem is as follows:

$$\begin{aligned}
\text{minimize} \quad & f_0\left(x_{0i}^t, x_i; \hat{u}_i, \hat{u}_{j\neq i}^t\right) + f_i\left(x_{0i}^t, x_i; \hat{u}_i\right) \\
& + \Phi_i\left(\hat{u}_i^t - \hat{u}_i, x_{0i}^t - x_0\right) \\
& + \Phi_0\left(g_0\left(x_{0i}^t, x_i; \hat{u}_i, \hat{u}_{j\neq i}^t\right)\right) \\
\text{by varying} \quad & x_{0i}^t, x_i \\
\text{subject to} \quad & g_i\left(x_{0i}^t, x_i; \hat{u}_i\right) \leq 0 \\
\text{while solving} \quad & r_i\left(\hat{u}_i; x_{0i}^t, x_i, \hat{u}_{j\neq i}^t\right) = 0 \\
\text{for} \quad & \hat{u}_i.
\end{aligned} \qquad (13.35)$$

The most common penalty functions used in ATC are quadratic penalty functions (see Section 5.4.1). Appropriate penalty weights are important for multidisciplinary consistency and convergence.

Algorithm 13.6 Analytical target cascading

Inputs:
    x: Initial design variables
Outputs:
    x^*: Optimal variables
    f^*: Optimal objective value
    c^*: Optimal constraint values

0: Initiate main ATC iteration
repeat
    for Each discipline i **do**
        1: Initiate discipline optimizer
        repeat
            2: Evaluate disciplinary analysis
            3: Compute discipline objective and constraint functions and penalty function values
            4: Update discipline design variables
        until $4 \rightarrow 2$: Discipline optimization has converged
    end for
    5: Initiate system optimizer
    repeat
        6: Compute system objective, constraints, and all penalty functions
        7: Update system design variables and coupling targets.
    until $7 \rightarrow 6$: System optimization has converged
    8: Update penalty weights
until $8 \rightarrow 1$: Penalty weights are large enough

13.5.3 Bilevel Integrated System Synthesis

Bilevel integrated system synthesis (BLISS) uses a series of linear approximations to the original design problem, with bounds on the design variable steps to prevent the design point from moving so far away that the approximations are too inaccurate.[216] This is an idea similar to that of the trust-region methods in Section 4.5. These approximations are constructed at each iteration using coupled derivatives (see Section 13.3).

216. Sobieszczanski–Sobieski et al., *Bilevel integrated system synthesis for concurrent and distributed processing*, 2003.

BLISS optimizes the local design variables within the discipline subproblems and the shared variables at the system level. The approach consists of using a series of linear approximations to the original optimization problem with limits on the design variable steps to stay within the region where the linear prediction yields the correct trend. This idea is similar to that of trust-region methods (see Section 4.5).

The system-level subproblem is formulated as follows:

$$\text{minimize} \quad (f_0^*)_0 + \left(\frac{df_0^*}{dx_0}\right)\Delta x_0$$

$$\text{by varying} \quad \Delta x_0$$

$$\text{subject to} \quad (g_0^*)_0 + \left(\frac{dg_0^*}{dx_0}\right)\Delta x_0 \le 0 \quad\quad (13.36)$$

$$(g_i^*)_0 + \left(\frac{dg_i^*}{dx_0}\right)\Delta x_0 \le 0 \quad \text{for} \quad i = 1,\ldots,m$$

$$\underline{\Delta x_0} \le \Delta x_0 \le \overline{\Delta x_0}.$$

The linearization is performed at each iteration using coupled derivative computation (see Section 13.3). The discipline i subproblem is given by the following:

$$\text{minimize} \quad (f_0)_0 + \left(\frac{df_0}{dx_i}\right)\Delta x_i$$

$$\text{by varying} \quad \Delta x_i$$

$$\text{subject to} \quad (g_0)_0 + \left(\frac{dg_0}{dx_i}\right)\Delta x_i \le 0 \quad\quad (13.37)$$

$$(g_i)_0 + \left(\frac{dg_i}{dx_i}\right)\Delta x_i \le 0$$

$$\underline{\Delta x_i} \le \Delta x_i \le \overline{\Delta x_i}.$$

The extra set of constraints in both system-level and discipline subproblems denotes the design variable bounds.

To prevent violation of the disciplinary constraints by changes in the shared design variables, post-optimality derivatives are required to solve the system-level subproblem. In this case, the post-optimality derivatives quantify the change in the optimized disciplinary constraints with respect to a change in the system design variables, which can be estimated with the Lagrange multipliers of the active constraints (see Sections 5.3.3 and 5.3.4).

Figure 13.40 shows the XDSM for BLISS, and the corresponding steps are listed in Alg. 13.7. Because BLISS uses an MDA, it is a distributed MDF architecture. As a result of the linear nature of the optimization problems, repeated interrogation of the objective and constraint functions is not necessary once we have the gradients. If the underlying problem is highly nonlinear, the algorithm may converge slowly. The variable bounds may help the convergence if these bounds are properly chosen, such as through a trust-region framework.

Algorithm 13.7 Bilevel integrated system synthesis

Inputs:
 x: Initial design variables
Outputs:
 x^*: Optimal variables
 f^*: Optimal objective value
 c^*: Optimal constraint values

0: Initiate system optimization
repeat
    1: Initiate MDA
    repeat
        2: Evaluate discipline analyses
        3: Update coupling variables
    until $3 \to 2$: MDA has converged
    4: Initiate parallel discipline optimizations
    for Each discipline i **do**
        5: Evaluate discipline analysis
        6: Compute objective and constraint function values and derivatives with respect to local design variables
        7: Compute the optimal solutions for the disciplinary subproblem
    end for
    8: Initiate system optimization
    9: Compute objective and constraint function values and derivatives with respect to shared design variables using post-optimality analysis
    10: Compute optimal solution to system subproblem
until $11 \to 1$: System optimization has converged

13.5 Distributed MDO Architectures

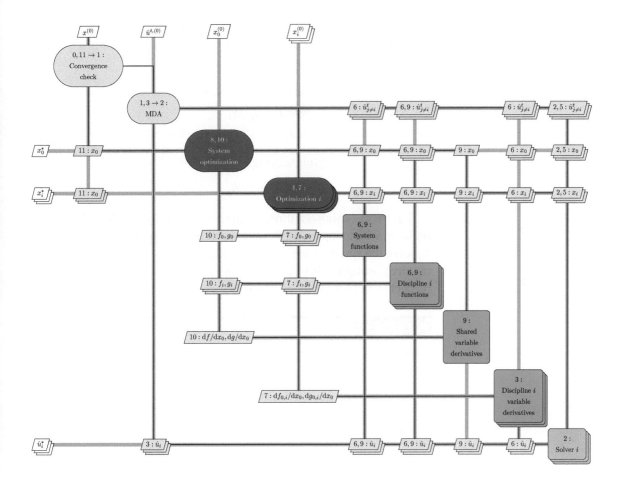

Fig. 13.40 Diagram for the BLISS architecture.

13.5.4 Asymmetric Subspace Optimization

Asymmetric subspace optimization (ASO) is a distributed MDF architecture motivated by cases where there is a large discrepancy between the cost of the disciplinary solvers. The cheaper disciplinary analyses are replaced by disciplinary design optimizations inside the overall MDA to reduce the number of more expensive disciplinary analyses.

The system-level optimization subproblem is as follows:

$$\begin{aligned}
\text{minimize} \quad & f(x; \hat{u}) \\
\text{by varying} \quad & x_0, x_k \\
\text{subject to} \quad & g_0(x; \hat{u}) \le 0 \\
& g_k(x; \hat{u}_k) \le 0 \quad \text{for all } k, \\
\text{while solving} \quad & r_k\left(\hat{u}_k; x_k, \hat{u}_{j \ne i}^t\right) = 0 \\
\text{for} \quad & \hat{u}_k .
\end{aligned} \quad (13.38)$$

The subscript k denotes disciplinary information that remains outside of the MDA. The disciplinary optimization subproblem for discipline i, which is resolved inside the MDA, is as follows:

$$\begin{aligned}
\text{minimize} \quad & f(x; \hat{u}) \\
\text{by varying} \quad & x_i \\
\text{subject to} \quad & g_i(x_0, x_i; \hat{u}_i) \leq 0 \\
\text{while solving} \quad & r_i\left(\hat{u}_i; x_i, \hat{u}^t_{j \neq i}\right) = 0 \\
\text{for} \quad & \hat{u}_i.
\end{aligned} \quad (13.39)$$

Figure 13.41 shows a three-discipline case where the third discipline replaced with an optimization subproblem. ASO is detailed in Alg. 13.8. To solve the system-level problem with a gradient-based optimizer, we require post-optimality derivatives of the objective and constraints with respect to the subproblem inputs (see Section 5.3.4).

Algorithm 13.8 ASO

Inputs:
  x: Initial design variables
Outputs:
  x^*: Optimal variables
  f^*: Optimal objective value
  c^*: Optimal constraint values

0: Initiate system optimization
repeat
  1: Initiate MDA
  repeat
    2: Evaluate analysis 1
    3: Evaluate analysis 2
    4: Initiate optimization of discipline 3
    repeat
      5: Evaluate analysis 3
      6: Compute discipline 3 objectives and constraints
      7: Update local design variables
    until 7 → 5: Discipline 3 optimization has converged
    8: Update coupling variables
  until 8 → 2 MDA has converged
  9: Compute objective and constraint function values for all disciplines 1 and 2
  10: Update design variables
until 10 → 1: System optimization has converged

13.5 Distributed MDO Architectures

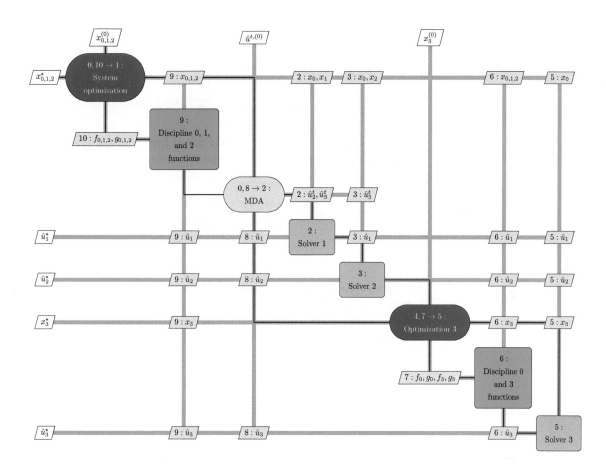

Fig. 13.41 Diagram for the ASO architecture.

For a gradient-based system-level optimizer, the gradients of the objective and constraints must take into account the suboptimization. This requires coupled post-optimality derivative computation, which increases computational and implementation time costs compared with a normal coupled derivative computation. The total optimization cost is only competitive with MDF if the discrepancy between each disciplinary solver is high enough.

Example 13.12 Aerostructural optimization using ASO

Aerostructural optimization is an example of asymmetry in the cost of the models. When the aerodynamic model consists of computational fluid dynamics, it is usually much more expensive than a finite-element structural model. If that is the case, we might be able to solve a structural sizing optimization in parallel within the time required for an aerodynamic analysis.

In this example, we formulate the system-level optimization problem as follows:

$$\begin{aligned}
\text{minimize} \quad & f \\
\text{by varying} \quad & b, \theta \\
\text{subject to} \quad & L - W^* = 0 \\
\text{while solving} \quad & A(d^*)\Gamma - \theta(d^*) = 0 \\
\text{for} \quad & \Gamma,
\end{aligned}$$

where W^* and d^* correspond to values obtained from the structural suboptimization. The suboptimization is formulated as follows:

$$\begin{aligned}
\text{minimize} \quad & f \\
\text{by varying} \quad & t \\
\text{subject to} \quad & 2.5|\sigma| - \sigma_{\text{yield}} \leq 0 \\
\text{while solving} \quad & Kd - q = 0 \\
\text{for} \quad & d.
\end{aligned}$$

Similar to the sequential optimization, we could replace f with W in the suboptimization because the other parameters in f are fixed. To solve the system-level problem with a gradient-based optimizer, we would need post-optimality derivatives of W^* with respect to span and Γ.

13.5.5 Other Distributed Architectures

There are other distributed MDF architectures in addition to BLISS and ASO: concurrent subspace optimization (CSSO) and MDO of independent subspaces (MDOIS).[41]

41. Martins and Lambe, *Multidisciplinary design optimization: A survey of architectures*, 2013.

CSSO requires surrogate models for the analyses for all disciplines. The system-level optimization subproblem is solved based on the surrogate models and is therefore fast. The discipline-level optimization subproblem uses the actual analysis from the corresponding discipline and surrogate models for all other disciplines. The solutions for each discipline subproblem are used to update the surrogate models.

MDOIS only applies when no shared variables exist. In this case, discipline subproblems are solved independently, assuming fixed coupling variables, and then an MDA is performed to update the coupling.

There are also other distributed IDF architectures. Some of these are similar to CO in that they use a multilevel approach to enforce multidisciplinary feasibility: BLISS-2000 and quasi-separable decomposition (QSD). Other architectures enforce multidisciplinary feasibility with penalties, like ATC: inexact penalty decomposition (IPD), exact penalty decomposition (EPD), and enhanced collaborative optimization (ECO).

BLISS-2000 is a variation of BLISS that uses surrogate models to represent the coupling variables for all disciplines. Each discipline subproblem minimizes the linearized objective with respect to local variables subject to local constraints. The system-level subproblem minimizes the objective with respect to the shared variables and coupling variables while enforcing consistency constraints.

When using QSD, the objective and constraint functions are assumed to depend only on the shared design variables and coupling variables. Each discipline is assigned a "budget" for a local objective, and the discipline problems maximize the margin in their local constraints and the budgeted objective. The system-level subproblem minimizes the objective and budgets of each discipline while enforcing the shared constraints and a positive margin for each discipline.

IPD and EPD apply to MDO problems with no shared objectives or constraints. They are similar to ATC in that copies of the shared variables are used for every discipline subproblem, and the consistency constraints are relaxed with a penalty function. Unlike ATC, however, the more straightforward structure of the discipline subproblems is exploited to compute post-optimality derivatives to guide the system-level optimization subproblem.

Like CO, ECO uses copies of the shared variables. The discipline subproblems minimize quadratic approximations of the objective while enforcing local constraints and linear models of the nonlocal constraints. The system-level subproblem minimizes the total violation of all consistency constraints with respect to the shared variables.

13.6 Summary

MDO architectures provide different options for solving MDO problems. An acceptable MDO architecture must be mathematically equivalent to the original problem and converge to the same optima. Sequential optimization, although intuitive, is not mathematically equivalent to the original problem and yields a design inferior to the MDO optimum.

MDO architectures are divided into two broad categories: monolithic architectures and distributed architectures. Monolithic architectures solve a single optimization problem, whereas distributed architectures solve optimization subproblems for each discipline and a system-level optimization problem. Overall, monolithic architectures exhibit a much better convergence rate than distributed architectures.[217] In the last few years, the vast majority of MDO applications have used monolithic MDO architectures. The MAUD architecture, which can implement MDF, IDF, or a hybrid of the two, successfully solves large-scale MDO problems.[132]

217. Tedford and Martins, *Benchmarking multidisciplinary design optimization algorithms*, 2010.

132. Gray et al., *OpenMDAO: An open-source framework for multidisciplinary design, analysis, and optimization*, 2019.

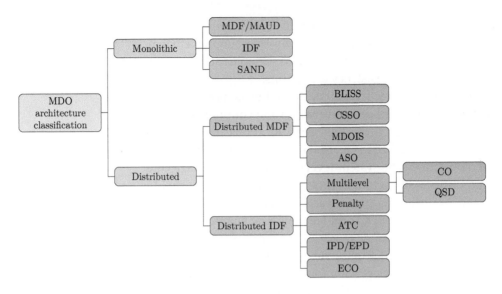

Fig. 13.42 Classification of MDO architectures.

*Martins and Lambe[41] describe all of these MDO architectures in detail.

41. Martins and Lambe, *Multidisciplinary design optimization: A survey of architectures*, 2013.

218. Golovidov et al., *Flexible implementation of approximation concepts in an MDO framework*, 1998.

219. Balabanov et al., *VisualDOC: A software system for general purpose integration and design optimization*, 2002.

The distributed architectures can be divided according to whether or not they enforce multidisciplinary feasibility (through an MDA of the whole system), as shown in Fig. 13.42. Distributed MDF architectures enforce multidisciplinary feasibility through an MDA. The distributed IDF architectures are like IDF in that no MDA is required. However, they must ensure multidisciplinary feasibility in some other way. Some do this by formulating an appropriate multilevel optimization (such as CO), and others use penalties to ensure this (such as ATC).*

Several commercial MDO frameworks are available, including Isight/SEE [218] by Dassault Systèmes, ModelCenter/CenterLink by Phoenix Integration, modeFRONTIER by Esteco, AML Suite by TechnoSoft, Optimus by Noesis Solutions, and VisualDOC by Vanderplaats Research and Development.[219] These frameworks focus on making it easy for users to couple multiple disciplines and use the optimization algorithms through graphical user interfaces. They also provide convenient wrappers to popular commercial engineering tools. Typically, these frameworks use fixed-point iteration to converge the MDA. When derivatives are needed for a gradient-based optimizer, finite-difference approximations are used rather than more accurate analytic derivatives.

Problems

13.1 Answer *true* or *false* and justify your answer.

   a. We prefer to use the term *component* instead of *discipline* because it is more general.

   b. Local design variables affect only one discipline in the MDO problem, whereas global variables affect all disciplines.

   c. All multidisciplinary models can be written in the functional form, but not all can be written in the residual form.

   d. The coupling variables are a subset of component state variables.

   e. Multidisciplinary models can be represented by directed cyclic graphs, where the nodes represent components and the edges represent coupling variables.

   f. The nonlinear block Jacobi and Gauss–Seidel methods can be used with any combination of component solvers.

   g. All the derivative computation methods from Chapter 6 can be implemented for coupled multidisciplinary systems.

   h. Implicit analytic methods for derivative computation are incompatible with the functional form of multidisciplinary models.

   i. The MAUD approach is based on the UDE.

   j. The MDF architecture has fewer design variables and more constraints than IDF.

   k. The main difference between monolithic and distributed MDO architectures is that the distributed architectures perform optimization at multiple levels.

   l. Sequential optimization is a valid MDO approach, but the main disadvantage is that it converges slowly.

13.2 Pick a multidisciplinary engineering system from the literature or formulate one based on your experience.

   a. Identify the different analyses and coupling variables.

   b. List the design variables and classify them as local or global.

   c. Identify the objective and constraint functions.

   d. Draw a diagram similar to the one in Fig. 13.37 for your system.

e. *Exploration*: Think about the objective that each discipline would have if considered separately, and discuss the trades needed to optimize the multidisciplinary objective.

13.3 Consider the DSMs that follow. For each case, what is the lowest number of feedback loops you can achieve through reordering? What hierarchy of solvers would you recommend to solve the coupled problem for each case?

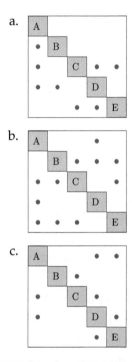

13.4 Consider the "spaghetti" diagram shown in Fig. 13.43. Draw the equivalent DSM for these dependencies. How can you exploit the structure in these dependencies? What hierarchy of solvers would you recommend to solve a coupled system with these dependencies?

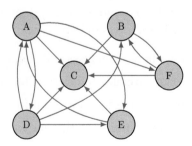

Fig. 13.43 Graph of dependencies.

13.5 Let us solve a simplified wing aerostructural problem based on simple equations for the aerodynamics and structures. We reuse the wing design problem described in Appendix D.1.6, but with a few modifications.

Suppose the lift coefficient now depends on the wing deflection:

$$C_L = C_{L_0} - C_{L,\theta}\theta,$$

where θ is the angle of deflection at the wing tip. Use $C_{L_0} = 0.4$ and $C_{L,\theta} = 0.1$ rad^{-1}. The deflection also depends on the lift. We compute θ assuming the uniform lift distribution and using the simple beam bending theory as

$$\theta = \frac{(L/b)(b/2)^3}{6EI} = \frac{Lb^2}{48EI}.$$

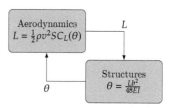

Fig. 13.44 The aerostructural model couples aerodynamics and structures through lift and wing deflection.

The Young's modulus is $E = 70$ GPa. Use the H-shaped cross-section described in Prob. 5.17 to compute the second moment of inertia, I.

We add the flight speed v to the set of design variables and handle $L = W$ as a constraint. The objective of the aerostructural optimization problem is to minimize the power with respect to $x = [b, c, v]$, subject to $L = W$.

Solve this problem using MDF, IDF, and a distributed MDO architecture. Compare the aerostructural optimal solution with the original solution from Appendix D.1.6 and discuss your results.

Mathematics Background

This appendix briefly reviews various mathematical concepts used throughout the book.

A.1 Taylor Series Expansion

Series expansions are representations of a given function in terms of a series of other (usually simpler) functions. One common series expansion is the *Taylor series*, which is expressed as a polynomial whose coefficients are based on the derivatives of the original function at a fixed point.

The Taylor series is a general tool that can be applied whenever the function has derivatives. We can use this series to estimate the value of the function near the given point, which is useful when the function is difficult to evaluate directly. The Taylor series is used to derive algorithms for finding the zeros of functions and algorithms for minimizing functions in Chapters 4 and 5.

To derive the Taylor series, we start with an infinite polynomial series about an arbitrary point, x, to approximate the value of a function at $x + \Delta x$ using

$$f(x + \Delta x) = a_0 + a_1 \Delta x + a_2 \Delta x^2 + \ldots + a_k \Delta x^k + \ldots . \quad \text{(A.1)}$$

We can make this approximation exact at $\Delta x = 0$ by setting the first coefficient to $f(x)$. To find the appropriate value for a_1, we take the first derivative to get

$$f'(x + \Delta x) = a_1 + 2a_2 \Delta x + \ldots + i a_k \Delta x^{k-1} + \ldots , \quad \text{(A.2)}$$

which means that we need $a_1 = f'(x)$ to obtain an exact derivative at x. To derive the other coefficients, we systematically take the derivative of both sides and the appropriate value of the first nonzero term (which is always constant). Identifying the pattern yields the general formula for the nth-order coefficient:

$$a_k = \frac{f^{(k)}(x)}{k!} . \quad \text{(A.3)}$$

Substituting this into the polynomial in Eq. A.1 yields the Taylor series

$$f(x + \Delta x) = \sum_{k=0}^{\infty} \frac{\Delta x^k}{k!} f^{(k)}(x). \tag{A.4}$$

The series is typically truncated to use terms up to order m,

$$f(x + \Delta x) = \sum_{k=0}^{m} \frac{\Delta x^k}{k!} f^{(k)}(x) + O\left(\Delta x^{m+1}\right), \tag{A.5}$$

which yields an approximation with a truncation error of order $O(\Delta x^{m+1})$. In optimization, it is common to use the first three terms (up to $m = 2$) to get a quadratic approximation.

Example A.1 Taylor series expansion for single variable

Consider the scalar function of a single variable, $f(x) = x - 4\cos(x)$. If we use Taylor series expansions of this function about $x = 0$, we get

$$f(\Delta x) = -4 + \Delta x + 2\Delta x^2 - \frac{1}{6}\Delta x^4 + \frac{1}{180}\Delta x^6 - \ldots.$$

Four different truncations of this series are plotted and compared to the exact function in Fig. A.1.

Fig. A.1 Taylor series expansions for one-dimensional example. The more terms we consider from the Taylor series, the better the approximation.

The Taylor series in multiple dimensions is similar to the single-variable case but more complicated. The first derivative of the function becomes a gradient vector, and the second derivatives become a Hessian matrix. Also, we need to define a direction along which we want to approximate the function because that information is not inherent like it is in a one-dimensional function. The Taylor series expansion in n dimensions along a direction p can be written as

$$f(x + \alpha p) = f(x) + \alpha \sum_{k=1}^{n} p_k \frac{\partial f}{\partial x_k} + \frac{1}{2}\alpha^2 \sum_{k=1}^{n} \sum_{l=1}^{n} p_k p_l \frac{\partial^2 f}{\partial x_k \partial x_l} + O\left(\alpha^3\right), \tag{A.6}$$

where α is a scalar that determines how far to go in the direction p. In matrix form, we can write

$$f(x + \alpha p) = f(x) + \alpha \nabla f(x)^\mathsf{T} p + \frac{1}{2}\alpha^2 p^\mathsf{T} H(x) p + O\left(\alpha^3\right), \tag{A.7}$$

where H is the Hessian matrix.

Example A.2 Taylor series expansion for two variables

Consider the following function of two variables:

$$f(x_1, x_2) = (1 - x_1)^2 + (1 - x_2)^2 + \frac{1}{2}\left(2x_2 - x_1^2\right)^2.$$

Performing a Taylor series expansion about $x = [0, -2]$, we get

$$f(x + \alpha p) = 18 + \alpha \begin{bmatrix} -2 & -14 \end{bmatrix} p + \frac{1}{2}\alpha^2 p^\mathsf{T} \begin{bmatrix} 10 & 0 \\ 0 & 6 \end{bmatrix} p.$$

The original function, the linear approximation, and the quadratic approximation are compared in Fig. A.2.

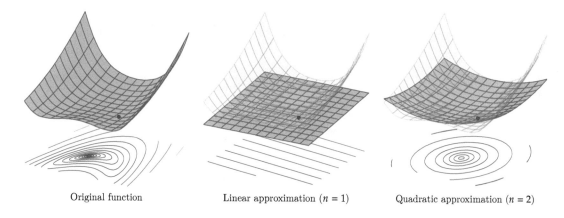

Original function   Linear approximation ($n = 1$)   Quadratic approximation ($n = 2$)

Fig. A.2 Taylor series approximations for two-dimensional example.

A.2 Chain Rule, Total Derivatives, and Differentials

The single-variable chain rule is needed for differentiating composite functions. Given a composite function, $f(g(x))$, the derivative with respect to the variable x is

$$\frac{\mathrm{d}}{\mathrm{d}x}(f(g(x))) = \frac{\mathrm{d}f}{\mathrm{d}g}\frac{\mathrm{d}g}{\mathrm{d}x}. \tag{A.8}$$

Example A.3 Single-variable chain rule

Let $f(g(x)) = \sin(x^2)$. In this case, $f(g) = \sin(g)$, and $g(x) = x^2$. The derivative with respect to x is

$$\frac{\mathrm{d}}{\mathrm{d}x}(f(g(x))) = \frac{\mathrm{d}}{\mathrm{d}g}(\sin(g))\frac{\mathrm{d}}{\mathrm{d}x}(x^2) = \cos(x^2)(2x).$$

If a function depends on more than one variable, then we need to distinguish between *partial* and *total* derivatives. For example, if $f(g(x), h(x))$, then f is a function of two variables: g and h. The application of the chain rule for this function is

$$\frac{d}{dx}(f(g(x), h(x))) = \frac{\partial f}{\partial g}\frac{dg}{dx} + \frac{\partial f}{\partial h}\frac{dh}{dx}, \quad (A.9)$$

where $\partial/\partial x$ indicates a partial derivative, and d/dx is a total derivative. When taking a partial derivative, we take the derivative with respect to only that variable, treating all other variables as constants. More generally,

$$\frac{d}{dx}(f(g_1(x), \ldots, g_n(x))) = \sum_{i=1}^{n}\left(\frac{\partial f}{\partial g_i}\frac{dg_i}{dx}\right). \quad (A.10)$$

Example A.4 Partial versus total derivatives

Consider $f(x, y(x)) = x^2 + y^2$, where $y(x) = \sin(x)$. The *partial* derivative of f with respect to x is

$$\frac{\partial f}{\partial x} = 2x,$$

whereas the *total* derivative of f with respect to x is

$$\frac{df}{dx} = \frac{\partial f}{\partial x} + \frac{\partial f}{\partial y}\frac{dy}{dx}$$
$$= 2x + 2y\cos(x)$$
$$= 2x + 2\sin(x)\cos(x).$$

Notice that the partial derivative and total derivative are quite different. For this simple case, we could also find the total derivative by direct substitution and then using an ordinary one-dimensional derivative. Substituting $y(x) = \sin(x)$ directly into the original expression for f gives

$$f(x) = x^2 + \sin^2(x)$$
$$\frac{df}{dx} = 2x + 2\sin(x)\cos(x).$$

Example A.5 Multivariable chain rule

Expanding on our single-variable example, let $g(x) = \cos(x)$, $h(x) = \sin(x)$, and $f(g, h) = g^2 h^3$. Then, $f(g(x), h(x)) = \cos^2(x)\sin^3(x)$. Applying Eq. A.9,

A.2 Chain Rule, Total Derivatives, and Differentials

we have the following:

$$\frac{d}{dx}\left(f\left(g(x), h(x)\right)\right) = \frac{\partial f}{\partial g}\frac{dg}{dx} + \frac{\partial f}{\partial h}\frac{dh}{dx}$$
$$= 2gh^3\frac{dg}{dx} + g^23h^2\frac{dh}{dx}$$
$$= -2gh^3\sin(x) + g^23h^2\cos(x)$$
$$= -2\cos(x)\sin^4(x) + 3\cos^3(x)\sin^2(x).$$

The *differential* of a function represents the linear change in that function with respect to changes in the independent variable. We introduce them here because they are helpful for finding total derivatives of multivariable equations that are implicit.

If function $y = f(x)$ is differentiable, the differential dy is

$$dy = f'(x)\,dx, \tag{A.11}$$

where dx is a nonzero real number (considered small) and dy is an approximation of the change (due to the linear term in the Taylor series). We can solve for $f'(x)$ to get $f'(x) = dy/dx$. This states that the derivative of f with respect to x is the differential of y divided by the differential of x. Strictly speaking, dy/dx here is not the derivative, although it is written in the same way. The derivative is a symbol, not a fraction. However, for our purposes, we will use these representations interchangeably and treat differentials algebraically. We also write the differentials of functions as

$$df = f'(x)\,dx. \tag{A.12}$$

Example A.6 Multivariable chain rule using differentials

We can solve Ex. A.5 using differentials as follows. Taking the definition of each function, we write their differentials,

$$df = 2gh^3\,dg + 3g^2h^2\,dh, \quad dg = -\sin(x)\,dx, \quad dh = \cos(x)\,dx.$$

Substituting g, dg, h, and dh into the differential of f we get obtain

$$df = 2\cos(x)\sin(x)^3(-\sin(x)\,dx) + 3\cos(x)^2\sin(x)^2\cos(x)\,dx.$$

Simplifying and dividing by dx yields the total derivative

$$\frac{df}{dx} = -2\cos(x)\sin^4(x) + 3\cos^3(x)\sin^2(x).$$

In Ex. A.5, there is no clear advantage in using differentials. However, differentials are more straightforward for finding total derivatives of multivariable implicit equations because there is no need to identify the independent variables. Given an equation, we just need to (1) find the differential of the equation and (2) solve for the derivative of interest. When we want quantities to remain constant, we can set the corresponding differential to zero. Differentials can be applied to vectors (say a vector x of size n), yielding a vector of differentials with the same size (dx of size n). We use this technique to derive the unified derivatives equation (UDE) in Section 6.9.

Example A.7 Total derivatives of an implicit equation

Suppose we have the equation for a circle,

$$x^2 + y^2 = r^2.$$

The differential of this equation is

$$2x\,dx + 2y\,dy = 2r\,dr.$$

Say we want to find the slope of the tangent of a circle with a fixed radius. Then, $dr = 0$, and we can solve for the derivative dy/dx as follows:

$$2x\,dx + 2y\,dy = 0 \quad \Rightarrow \quad \frac{dy}{dx} = -\frac{x}{y}.$$

Another interpretation of this derivative is that it is the first-order change in y with respect to a change in x subject to the constraint of staying on a circle (keeping a constant r). Similarly, we could find the derivative of x with respect to y as $dx/dy = -y/x$. Furthermore, we can find relationships between any derivative involving r, x, or y.

A.3 Matrix Multiplication

Consider a matrix A of size $(m \times n)$* and a matrix B of size $(n \times p)$. The two matrices can be multiplied together ($C = AB$) as follows:

$$C_{ij} = \sum_{k=1}^{n} A_{ik} B_{kj}, \quad (A.13)$$

where C is an $(m \times p)$ matrix. This multiplication is illustrated in Fig. A.3. Two matrices can be multiplied only if their inner dimensions are equal (n in this case). The remaining products discussed in this section are just special cases of matrix multiplication, but they are common enough that we discuss them separately.

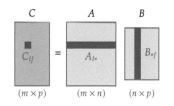

Fig. A.3 Matrix product and resulting size.

*In this notation, m is the number of rows and n is the number of columns.

A.3 Matrix Multiplication

A.3.1 Vector-Vector Products

In this book, a vector u is a column vector; thus, the row vector is represented as u^T. The product of two vectors can be performed in two ways. The more common is called an *inner product* (also known as a *dot product* or *scalar product*). The inner product is a functional, meaning that it is an operator that acts on vectors and produces a scalar. This product is illustrated in Fig. A.4. In the real vector space of n dimensions, the inner product of two vectors, u and v, whose dimensions are equal, is defined algebraically as

$$\alpha = u^\mathsf{T} v = \begin{bmatrix} u_1 & u_2 & \cdots & u_n \end{bmatrix} \begin{bmatrix} v_1 \\ v_2 \\ \vdots \\ v_n \end{bmatrix} = \sum_{i=1}^{n} u_i v_i . \tag{A.14}$$

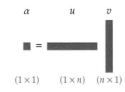

Fig. A.4 Dot (or inner) product of two vectors.

The order of multiplication is irrelevant, and therefore,

$$u^\mathsf{T} v = v^\mathsf{T} u . \tag{A.15}$$

In Euclidean space, where vectors have magnitude and direction, the inner product is defined as

$$u^\mathsf{T} v = \|u\| \, \|v\| \cos(\theta) , \tag{A.16}$$

where $\|\cdot\|$ represents the 2-norm (Eq. A.25), and θ is the angle between the two vectors.

The *outer product* takes the two vectors and multiplies them elementwise to produce a matrix, as illustrated in Fig. A.5. Unlike the inner product, the outer product does not require the vectors to be of the same length. The matrix form is as follows:

$$C = u v^\mathsf{T} = \begin{bmatrix} u_1 \\ u_2 \\ \vdots \\ u_m \end{bmatrix} \begin{bmatrix} v_1 & v_2 & \cdots & v_n \end{bmatrix}$$

$$= \begin{bmatrix} u_1 v_1 & u_1 v_2 & \cdots & u_1 v_n \\ u_2 v_1 & u_2 v_2 & \cdots & u_2 v_n \\ \vdots & \vdots & \ddots & \vdots \\ u_m v_1 & u_m v_2 & \cdots & u_m v_n \end{bmatrix} . \tag{A.17}$$

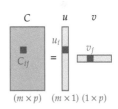

Fig. A.5 Outer product of two vectors.

The index form is as follows:

$$(u v^\mathsf{T})_{ij} = u_i v_j . \tag{A.18}$$

Outer products generate rank 1 matrices. They are used in quasi-Newton methods (Section 4.4.4 and Appendix C).

A.3.2 Matrix-Vector Products

Consider multiplying a matrix A of size $(m \times n)$ by vector u of size n. The result is a vector of size m:

$$v = Au \Rightarrow v_i = \sum_{j=1}^{n} A_{ij} u_j. \quad (A.19)$$

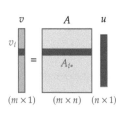

Fig. A.6 Matrix-vector product.

This multiplication is illustrated in Fig. A.6. The entries in v are dot products between the rows of A and u:

$$v = \begin{bmatrix} \text{---} A_{1*} \text{---} \\ \text{---} A_{2*} \text{---} \\ \vdots \\ \text{---} A_{m*} \text{---} \end{bmatrix} u, \quad (A.20)$$

where A_{i*} is the ith row of the matrix A. Thus, a matrix-vector product transforms a vector in n-dimensional space ($\mathbb{R}^n$) to a vector in m-dimensional space ($\mathbb{R}^m$).

A matrix-vector product can be thought of as a linear combination of the columns of A, where the u_j values are the weights:

$$v = \begin{bmatrix} | \\ A_{*1} \\ | \end{bmatrix} u_1 + \begin{bmatrix} | \\ A_{*2} \\ | \end{bmatrix} u_2 + \ldots + \begin{bmatrix} | \\ A_{*n} \\ | \end{bmatrix} u_n, \quad (A.21)$$

and A_{*j} are the columns of A.

We can also multiply by a vector on the left, instead of on the right:

$$v^\mathsf{T} = u^\mathsf{T} A. \quad (A.22)$$

In this case, a row vector is multiplied with a matrix, producing a row vector.

A.3.3 Quadratic Form (Vector-Matrix-Vector Product)

Another common product is a *quadratic form*. A quadratic form consists of a row vector, times a matrix, times a column vector, producing a scalar:

$$\alpha = u^\mathsf{T} A u = \begin{bmatrix} u_1 & u_2 & \ldots & u_n \end{bmatrix} \begin{bmatrix} A_{11} & A_{12} & \cdots & A_{1n} \\ A_{21} & A_{22} & \cdots & A_{2n} \\ \vdots & \vdots & \ddots & \vdots \\ A_{n1} & A_{n2} & \cdots & A_{nn} \end{bmatrix} \begin{bmatrix} u_1 \\ u_2 \\ \vdots \\ u_n \end{bmatrix} \quad (A.23)$$

The index form is as follows:

$$\alpha = \sum_{i=1}^{n} \sum_{j=1}^{n} u_i A_{ij} u_j. \quad (A.24)$$

In general, a vector-matrix-vector product can have a nonsquare A matrix, and the vectors would be two different sizes, but for a quadratic form, the two vectors u are identical, and thus A is square. Also, in a quadratic form, we assume that A is symmetric (even if it is not, only the symmetric part of A contributes, so effectively, it acts like a symmetric matrix).

A.4 Four Fundamental Subspaces in Linear Algebra

This section reviews how the dimensions of a matrix in a linear system relate to dimensional spaces.* These concepts are especially helpful for understanding constrained optimization (Chapter 5) and build on the review in Section 5.2.

*Strang[87] provides a comprehensive coverage of linear algebra and is credited with popularizing the concept of the "four fundamental subspaces".

87. Strang, *Linear Algebra and its Applications*, 2006.

A *vector space* is the set of all points that can be obtained by linear combinations of a given set of vectors. The vectors are said to *span* the vector space. A *basis* is a set of linearly independent vectors that generates all points in a vector space. A *subspace* is a space of lower dimension than the space that contains it (e.g., a line is a subspace of a plane).

Two vectors are *orthogonal* if the angle between them is 90 degrees. Then, their dot product is zero. A subspace S_1 is orthogonal to another subspace S_2 if every vector in S_1 is orthogonal to every vector in S_2.

Consider an $(m \times n)$ matrix A. The *rank* (r) of a matrix A is the maximum number of linearly independent row vectors of A or, equivalently, the maximum number of linearly independent column vectors. The rank can also be defined as the dimensionality of the vector space spanned by the rows or columns of A. For an $(m \times n)$ matrix, $r \leq \min(m, n)$.

Through a matrix-vector multiplication $Ax = b$, this matrix maps an n-vector x into an m-vector b. Figure A.7 shows this mapping and illustrates the four fundamental subspaces that we now explain.

The *column space* of a matrix A is the vector space spanned by the vectors in the columns of A. The dimensionality of this space is given by r, where $r \leq n$, so the column space is a subspace of n-dimensional space. The *row space* of a matrix A is the vector space spanned by the vectors in the rows of A (or equivalently, it is the column space of A^T). The dimensionality of this space is given by r, where $r \leq m$, so the row space is a subspace of n-dimensional space.

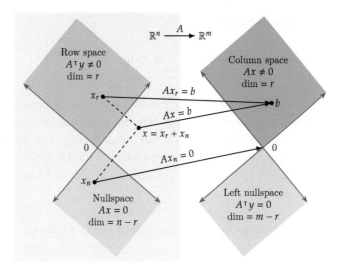

Fig. A.7 The four fundamental subspaces of linear algebra. An $(m \times n)$ matrix A maps vectors from n-space to m-space. When the vector is in the row space of the matrix, it maps to the column space of A ($x_r \to b$). When the vector is in the nullspace of A, it maps to zero ($x_n \to 0$). Combining the row space and nullspace of A, we can obtain any vector in n-dimensional space ($x = x_r + x_n$), which maps to the column space of A ($x \to b$).

The *nullspace* of a matrix A is the vector space consisting of all the vectors that are orthogonal to the rows of A. Equivalently, the nullspace of A is the vector space of all vectors x_n such that $Ax_n = 0$. Therefore, the nullspace is orthogonal to the row space of A. The dimension of the nullspace of A is $n - r$.

Combining the nullspace and row space of A adds up to the whole n-dimensional space, that is, $x = x_r + x_n$, where x_r is in the row space of A and x_n is in the nullspace of A.

The *left nullspace* of a matrix A is the vector space of all x such that $A^\mathsf{T} x = 0$. Therefore, the left nullspace is orthogonal to the column space of A. The dimension of the left nullspace of A is $m - r$. Combining the left nullspace and column space of A adds up to the whole m-dimensional space.

A.5 Vector and Matrix Norms

Norms give an idea of the magnitude of the entries in vectors and matrices. They are a generalization of the absolute value for real numbers. A norm $\|\cdot\|$ is a real-valued function with the following properties:

- $\|x\| \geq 0$ for all x.
- $\|x\| = 0$ if an only if $x = 0$.
- $\|\alpha x\| = |\alpha| \|x\|$ for all real numbers α.
- $\|x + y\| \leq \|x\| + \|y\|$ for all x and y.

Most common matrix norms also have the property that $\|xy\| \leq \|x\| \|y\|$, although this is not required in general.

A.5 Vector and Matrix Norms

We start by defining vector norms, where the vector is $x = [x_1, \ldots, x_n]$. The most familiar norm for vectors is the 2-norm, also known as the *Euclidean norm*, which corresponds to the Euclidean length of the vector:

$$\|x\|_2 = \left(\sum_{i=1}^{n} x_i^2 \right)^{\frac{1}{2}} = (x_1^2 + x_2^2 + \ldots + x_n^2)^{\frac{1}{2}}. \tag{A.25}$$

Because this norm is used so often, we often omit the subscript and just write $\|x\|$. In this book, we sometimes use the square of the 2-norm, which can be written as the dot product,

$$\|x\|_2^2 = x^\mathsf{T} x. \tag{A.26}$$

More generally, we can refer to a class of norms called *p*-norms:

$$\|x\|_p = \left(\sum_{i=1}^{n} |x_i|^p \right)^{\frac{1}{p}} = (|x_1|^p + |x_2|^p + \ldots + |x_n|^p)^{\frac{1}{p}}, \tag{A.27}$$

where $1 \leq p < \infty$. Of all the *p*-norms, three are most commonly used: the 2-norm (Eq. A.25), the 1-norm, and the ∞-norm. From the previous definition, we see that the 1-norm is the sum of the absolute values of all the entries in x:

$$\|x\|_1 = \sum_{i=1}^{n} |x_i| = |x_1| + |x_2| + \ldots + |x_n|. \tag{A.28}$$

The application of ∞ in the *p*-norm definition is perhaps less obvious, but as $p \to \infty$, the largest term in that sum dominates all of the others. Raising that quantity to the power of $1/p$ causes the exponents to cancel, leaving only the largest-magnitude component of x. Thus, the infinity norm is

$$\|x\|_\infty = \max_i |x_i|. \tag{A.29}$$

The infinity norm is commonly used in optimization convergence criteria.

The vector norms are visualized in Fig. A.8 for $n = 2$. If $x = [1, \ldots, 1]$, then

$$\|x\|_1 = n, \quad \|x\|_2 = n^{\frac{1}{2}}, \quad \|x\|_\infty = 1. \tag{A.30}$$

It is also possible to assign different weights to each vector component to form a *weighted norm*:

$$\|x\|_p = (w_1 |x_1|^p + w_2 |x_2|^p + \ldots + w_n |x_n|^p)^{\frac{1}{p}}. \tag{A.31}$$

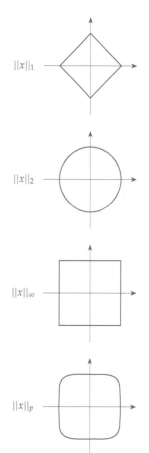

Fig. A.8 Norms for two-dimensional case.

Several norms for matrices exist. There are matrix norms similar to the vector norms that we defined previously. Namely,

$$\|A\|_1 = \max_{1 \le j \le n} \sum_{i=1}^{n} |A_{ij}|$$

$$\|A\|_2 = (\lambda_{\max}(A^\mathsf{T} A))^{\frac{1}{2}} \tag{A.32}$$

$$\|A\|_\infty = \max_{1 \le i \le n} \sum_{i=1}^{n} |A_{ij}|,$$

where $\lambda_{\max}(A^\mathsf{T} A)$ is the largest eigenvalue of $A^\mathsf{T} A$. When A is a square symmetric matrix, then

$$\|A\|_2 = |\lambda_{\max}(A)|. \tag{A.33}$$

Another matrix norm that is useful but not related to any vector norm is the Frobenius norm, which is defined as the square root of the absolute squares of its elements, that is,

$$\|A\|_F = \sqrt{\sum_{i=1}^{m} \sum_{j=1}^{n} A_{ij}^2}. \tag{A.34}$$

The Frobenius norm can be weighted by a matrix W as follows:

$$\|A\|_W = \left\|W^{\frac{1}{2}} A W^{\frac{1}{2}}\right\|_F. \tag{A.35}$$

This norm is used in the formal derivation of the Broyden–Fletcher–Goldfarb–Shanno (BFGS) update formula (see Appendix C).

A.6 Matrix Types

There are several common types of matrices that appear regularly throughout this book. We review some terminology here.

A *diagonal matrix* is a matrix where all off-diagonal terms are zero. In other words, A is diagonal if:

$$A_{ij} = 0 \text{ for all } i \ne j. \tag{A.36}$$

The *identity matrix* I is a special diagonal matrix where all diagonal components are 1.

The *transpose* of a matrix is defined as follows:

$$[A^\mathsf{T}]_{ij} = A_{ji}. \tag{A.37}$$

A.6 Matrix Types

Note that
$$(A^\mathsf{T})^\mathsf{T} = A$$
$$(A + B)^\mathsf{T} = A^\mathsf{T} + B^\mathsf{T} \quad \text{(A.38)}$$
$$(AB)^\mathsf{T} = B^\mathsf{T} A^\mathsf{T}.$$

A *symmetric matrix* is one where the matrix is equal to its transpose:
$$A^\mathsf{T} = A \implies A_{ij} = A_{ji}. \quad \text{(A.39)}$$

The *inverse* of a matrix, A^{-1}, satisfies
$$AA^{-1} = I = A^{-1}A. \quad \text{(A.40)}$$

Not all matrices are invertible. Some common properties for inverses are as follows:
$$\left(A^{-1}\right)^{-1} = A$$
$$(AB)^{-1} = B^{-1}A^{-1} \quad \text{(A.41)}$$
$$\left(A^{-1}\right)^\mathsf{T} = \left(A^\mathsf{T}\right)^{-1}.$$

A symmetric matrix A is *positive definite* if and only if
$$x^\mathsf{T} A x > 0 \quad \text{(A.42)}$$

for all nonzero vectors x. One property of positive-definite matrices is that their inverse is also positive definite.

The positive-definite condition (Eq. A.42) can be challenging to verify. Still, we can use equivalent definitions that are more practical.

For example, by choosing appropriate xs, we can derive the necessary conditions for positive definiteness:
$$\begin{aligned} A_{ii} &> 0 \quad \text{for all} \quad i \\ A_{ij} &< \sqrt{A_{ii} A_{jj}} \quad \text{for all} \quad i \neq j. \end{aligned} \quad \text{(A.43)}$$

These are necessary but not sufficient conditions. Thus, if any diagonal element is less than or equal to zero, we know that the matrix is not positive definite.

An equivalent condition to Eq. A.42 is that all the eigenvalues of A are positive. This is a sufficient condition.

Another practical condition equivalent to Eq. A.42 is that all the leading principal minors of A are positive. A leading principal minor is the determinant of a leading principal submatrix. A leading principal submatrix of order k, A_k of an $(n \times n)$ matrix A is obtained by removing the last $n - k$ rows and columns of A, as shown in Fig. A.9. Thus, to verify if A is positive definite, we start with $k = 1$, check that $A_1 > 0$ (only

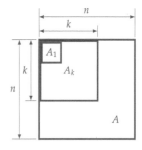

Fig. A.9 For A to be positive definite, the determinants of the submatrices $A_1, A_2, \ldots A_n$ must be greater than zero.

one element), then check that det(A_2) > 0, and so on, until det(A_n) > 0. Suppose any of the determinants in this sequence is not positive. In that case, we can stop the process and conclude that A is not positive definite.

A *positive-semidefinite* matrix satisfies

$$x^\mathsf{T} A x \geq 0 \qquad (A.44)$$

for all nonzero vectors x. In this case, the eigenvalues are nonnegative, and there is at least one that is zero. A *negative-definite* matrix satisfies

$$x^\mathsf{T} A x < 0 \qquad (A.45)$$

for all nonzero vectors x. In this case, all the eigenvalues are negative. An indefinite matrix is one that is neither positive definite nor negative definite. Then, there are at least two nonzero vectors x and y such that

$$x^\mathsf{T} A x > 0 > y^\mathsf{T} A y . \qquad (A.46)$$

A.7 Matrix Derivatives

Let us consider the derivatives of a few common cases: linear and quadratic functions. Combining the concept of partial derivatives and matrix forms of equations allows us to find the gradients of matrix functions. First, let us consider a linear function, f, defined as

$$f(x) = a^\mathsf{T} x + b = \sum_{i=1}^{n} a_i x_i + b_i , \qquad (A.47)$$

where a, x, and b are vectors of length n, and a_i, x_i, and b_i are the ith elements of a, x, and b, respectively. If we take the partial derivative of each element with respect to an arbitrary element of x, namely, x_k, we get

$$\frac{\partial}{\partial x_k} \left[\sum_{i=1}^{n} a_i x_i + b_i \right] = a_k . \qquad (A.48)$$

Thus,

$$\nabla_x (a^\mathsf{T} x + b) = a . \qquad (A.49)$$

Recall the quadratic form presented in Appendix A.3.3; we can combine that with a linear term to form a general quadratic function:

$$f(x) = x^\mathsf{T} A x + b^\mathsf{T} x + c , \qquad (A.50)$$

where x, b, and c are still vectors of length n, and A is an n-by-n symmetric matrix. In index notation, f is as follows:

$$f(x) = \sum_{i=1}^{n} \sum_{j=1}^{n} x_i a_{ij} x_j + b_i x_i + c_i. \tag{A.51}$$

For convenience, we separate the diagonal terms from the off-diagonal terms, leaving us with

$$f(x) = \sum_{i=1}^{n} \left[a_{ii} x_i^2 + b_i x_i + c_i \right] + \sum_{j \neq i} x_i a_{ij} x_j. \tag{A.52}$$

Now we take the partial derivatives with respect to x_k as before, yielding

$$\frac{\partial f}{\partial x_k} = 2 a_{kk} x_k + b_k + \sum_{j \neq i} x_j a_{jk} + \sum_{j \neq i} a_{kj} x_j. \tag{A.53}$$

We now move the diagonal terms back into the sums to get

$$\frac{\partial f}{\partial x_k} = b_k + \sum_{j=1}^{n} (x_j a_{jk} + a_{kj} x_j), \tag{A.54}$$

which we can put back into matrix form as follows:

$$\nabla_x f(x) = A^\mathsf{T} x + A x + b. \tag{A.55}$$

If A is symmetric, then $A^\mathsf{T} = A$, and thus

$$\nabla_x (x^\mathsf{T} A x + b^\mathsf{T} x + c) = 2 A x + b. \tag{A.56}$$

A.8 Eigenvalues and Eigenvectors

Given an $(n \times n)$ matrix, if there is a scalar λ and a nonzero vector v that satisfy

$$A v = \lambda v, \tag{A.57}$$

then λ is an *eigenvalue* of the matrix A, and v is an *eigenvector*. The left-hand side of Eq. A.57 is a matrix-vector product that represents a linear transformation applied to v. The right-hand side of Eq. A.57 is a scalar-vector product that represents a vector aligned with v. Therefore, the eigenvalue problem (Eq. A.57) answers the question: Which vectors, when transformed by A, remain in the same direction, and how much do their corresponding lengths change in that transformation?

The solutions of the eigenvalue problem (Eq. A.57) are given by the solutions of the scalar equation,

$$\det(A - \lambda I) = 0. \tag{A.58}$$

This equation yields a polynomial of degree n called the *characteristic equation*, whose roots are the eigenvalues of A.

If A is symmetric, it has n real eigenvalues $(\lambda_1, \ldots, \lambda_n)$ and n linearly independent eigenvectors $(v_1, \ldots, v_n)$ corresponding to those eigenvalues. It is possible to choose the eigenvectors to be orthogonal to each other (i.e., $v_i^T v_j = 0$ for $i \neq j$) and to normalize them (so that $v_i^T v_i = 1$).

We use the eigenvalue problem in Section 4.1.2, where the eigenvectors are the directions of principal curvature, and the eigenvalues quantify the curvature. Eigenvalues are also helpful in determining if a matrix is positive definite.

A.9 Random Variables

Imagine measuring the axial strength of a rod by performing a tensile test with many rods, each designed to be identical. Even with "identical" rods, every time you perform the test, you get a different result (hopefully with relatively small differences). This variation has many potential sources, including variation in the manufactured size and shape, in the composition of the material, and in the contact between the rod and testing fixture. In this example, we would call the axial strength a *random variable*, and the result from one test would be a random sample. The random variable, axial strength, is a function of several other random variables, such as bar length, bar diameter, and material Young's modulus.

One measurement does not tell us anything about how variable the axial strength is, but if we perform the test many times, we can learn a lot about its distribution. From this information, we can infer various statistical quantities, such as the *mean value* of the axial strength. The mean of some variable x that is measured n times is estimated as follows:

$$\mu_x = \frac{1}{n} \sum_{i=1}^{n} x_i . \tag{A.59}$$

This is actually a sample mean, which would differ from the population mean (the true mean if you could measure every bar). With enough samples, the sample mean approaches the population mean. In this brief review, we do not distinguish between sample and population statistics.

Another important quantity is the *variance* or *standard deviation*. This is a measure of spread, or how far away our samples are from the mean. The unbiased* estimate of the variance is

---

Unbiased means that the expected value of the sample variance is the same as the true population variance. If n were used in the denominator instead of $n-1$, then the two quantities would differ by a constant.

A.9 Random Variables

$$\sigma_x^2 = \frac{1}{n-1} \sum_{i=1}^{n} (x_i - \mu_x)^2, \quad (A.60)$$

and the standard deviation is just the square root of the variance. A small variance implies that measurements are clustered tightly around the mean, whereas a large variance means that measurements are spread out far from the mean. The variance can also be written in the following mathematically equivalent but more computationally-friendly format:

$$\sigma_x^2 = \frac{1}{n-1} \left(\sum_{i=1}^{n} (x_i^2) - n\mu_x^2 \right). \quad (A.61)$$

More generally, we might want to know what the probability is of getting a bar with a specific axial strength. In our testing, we could tabulate the frequency of each measurement in a histogram. If done enough times, it would define a smooth curve, as shown in Fig. A.10. This curve is called the *probability density function* (PDF), $p(x)$, and it tells us the *relative* probability of a certain value occurring.

More specifically, a PDF gives the probability of getting a value with a certain range:

$$\text{Prob}(a \le x \le b) = \int_a^b p(x)\,dx. \quad (A.62)$$

The total integral of the PDF must be 1 because it contains all possible outcomes (100 percent):

$$\int_{-\infty}^{\infty} p(x)\,dx = 1. \quad (A.63)$$

From the PDF, we can also measure various statistics, such as the mean value:

$$\mu_x = \mathbb{E}(x) = \int_{-\infty}^{\infty} x p(x)\,dx. \quad (A.64)$$

This quantity is also referred to as the *expected value* of x ($\mathbb{E}[x]$). The expected value of a function of a random variable, $f(x)$, is given by:[†]

$$\mu_f = \mathbb{E}\left(f(x)\right) = \int_{-\infty}^{\infty} f(x) p(x)\,dx. \quad (A.65)$$

[†] This is not a definition, but rather uses the expected value definition with a somewhat lengthy derivation.

We can also compute the *variance*, which is the expected value of the squared difference from the mean:

$$\sigma_x^2 = \mathbb{E}\left((x - \mathbb{E}(x))^2\right) = \int_{-\infty}^{\infty} (x - \mu_x)^2 p(x)\,dx, \quad (A.66)$$

or in a mathematically equivalent format:

$$\sigma_x^2 = \int_{-\infty}^{\infty} x^2 p(x)\, dx - \mu_x^2. \tag{A.67}$$

The mean and variance are the first and second moments of the distribution. In general, a distribution may require an infinite number of moments to describe it fully. Higher-order moments are generally mean centered and are normalized by the standard deviation so that the nth normalized moment is computed as follows:

$$\mathbb{E}\left(\left(\frac{x - \mu_x}{\sigma}\right)^n\right). \tag{A.68}$$

The third moment is called *skewness*, and the fourth is called *kurtosis*, although these higher-order moments are less commonly used.

The *cumulative distribution function* (CDF) is related to the PDF, which is the cumulative integral of the PDF and is defined as follows:

$$P(x) = \int_{-\infty}^{x} p(t)\, dt. \tag{A.69}$$

The capital P denotes the CDF, and the lowercase p denotes the PDF. As an example, the PDF and corresponding CDF for the axial strength are shown in Fig. A.10. The CDF always approaches 1 as $x \to \infty$.

Fig. A.10 Comparison between PDF and CDF for a simple example.

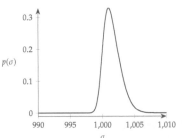
PDF for the axial strength of a rod.

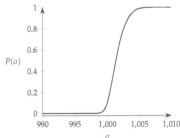
CDF for the axial strength of a rod.

We often fit a named distribution to the PDF of empirical data. One of the most popular distributions is the *normal distribution*, also known as the *Gaussian distribution*. Its PDF is as follows:

$$p(x; \mu, \sigma^2) = \frac{1}{\sigma\sqrt{2\pi}} \exp\left(\frac{-(x-\mu)^2}{2\sigma^2}\right). \tag{A.70}$$

For a normal distribution, the mean and variance are visible in the function, but these quantities are defined for any distribution. Figure A.11

A.9 Random Variables

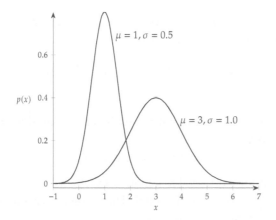

Fig. A.11 Two normal distributions. Changing the mean causes a shift along the x-axis. Increasing the standard deviation causes the PDF to spread out.

shows two normal distributions with different means and standard deviations to illustrate the effect of those parameters.

Several other popular distributions are shown in Fig. A.12: uniform distribution, Weibull distribution, lognormal distribution, and exponential distribution. These are only a few of many other possible probability distributions.

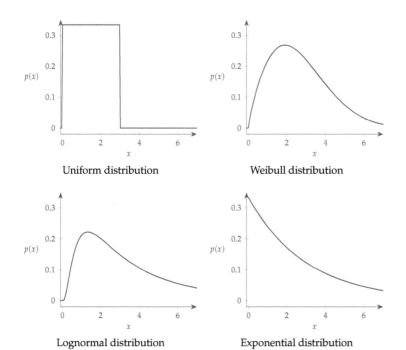

Fig. A.12 Popular probability distributions besides the normal distribution.

An extension of variance is the covariance, which measures the variability between two random variables:

$$\begin{aligned} \mathrm{cov}(x,y) &= \mathbb{E}\left((x - \mathbb{E}(x))(y - \mathbb{E}(y))\right) \\ &= \mathbb{E}(xy) - \mu_x \mu_y. \end{aligned} \quad (A.71)$$

From this definition, we see that the variance is related to covariance by the following:

$$\sigma_x^2 = \mathrm{var}(x) = \mathrm{cov}(x,x). \quad (A.72)$$

Covariance is often expressed as a matrix, in which case the variance of each variable appears on the diagonal. The correlation is the covariance divided by the standard deviations:

$$\mathrm{corr}(x,y) = \frac{\mathrm{cov}(x,y)}{\sigma_x \sigma_y}. \quad (A.73)$$

Linear Solvers B

In Section 3.6, we present an overview of solution methods for discretized systems of equations, followed by an introduction to Newton-based methods for solving nonlinear equations. Here, we review the solvers for linear systems required to solve for each step of Newton-based methods.*

*Trefethen and Bau III[220] provides a much more detailed explanation of linear solvers.

220. Trefethen and Bau III, *Numerical Linear Algebra*, 1997.

B.1 Systems of Linear Equations

If the equations are linear, they can be written as

$$Au = b, \quad (B.1)$$

where A is a square ($n \times n$) matrix, and b is a vector, and neither of these depends on u. If this system of equations has a unique solution, then the system and the matrix A are *nonsingular*. This is equivalent to saying that A has an inverse, A^{-1}. If A^{-1} does not exist, the matrix and the system are *singular*.

A matrix A is singular if its rows (or equivalently, its columns) are linearly dependent (i.e., if one of the rows can be written as a linear combination of the others).

If the matrix A is nonsingular and we know its inverse A^{-1}, the solution of the linear system (Eq. B.1) can be written as $x = A^{-1}b$. However, the numerical methods described here do not form A^{-1}. The main reason for this is that forming A^{-1} is expensive: the computational cost is proportional to n^3.

For practical problems with large n, it is typical for the matrix A to be *sparse*, that is, for most of its entries to be zeros. An entry A_{ij} represents the interaction between variables i and j. When solving differential equations on a discretized grid, for example, a given variable i only interacts with variables j in its vicinity in the grid. These interactions correspond to nonzero entries, whereas all other entries are zero. Sparse linear systems tend to have a number of nonzero terms that is proportional to n. This is in contrast with a *dense* matrix, which has n^2 nonzero entries. Solvers should take advantage of sparsity to remain efficient for large n.

We rewrite the linear system (Eq. B.1) as a set of residuals,

$$r(u) = Au - b = 0. \tag{B.2}$$

To solve this system of equations, we can use either a direct method or an iterative method. We explain these briefly in the rest of this appendix, but we do not cover more advanced techniques that take advantage of sparsity.

B.2 Conditioning

The distinction between singular and nonsingular systems blurs once we have to deal with finite-precision arithmetic. Systems that are singular in the exact sense are *ill-conditioned* when a small change in the data (entries of A or b) results in a large change in the solution. This large sensitivity of the solution to the problem parameters is an issue because the parameters themselves have finite precision. Then, any imprecision in these parameters can lead to significant errors in the solution, even if no errors are introduced in the numerical solution of the linear system.

The conditioning of a linear system can be quantified by the *condition number* of the matrix, which is defined as the scalar

$$\text{cond}(A) = \|A\| \cdot \|A^{-1}\|, \tag{B.3}$$

where any matrix norm can be used. Because $\|A\| \cdot \|A^{-1}\| \geq \|AA^{-1}\|$, we have

$$\text{cond}(A) \geq 1 \tag{B.4}$$

for all matrices. A matrix A is well-conditioned if $\text{cond}(A)$ is small and ill-conditioned if $\text{cond}(A)$ is large.

B.3 Direct Methods

The standard way to solve linear systems of equations with a computer is Gaussian elimination, which in matrix form is equivalent to *LU factorization*. This is a factorization (or decomposition) of A, such as $A = LU$, where L is a unit lower triangular matrix, and U is an upper triangular matrix, as shown in Fig. B.1.

The factorization transforms the matrix A into an upper triangular matrix U by introducing zeros below the diagonal, one column at a time, starting with the first one and progressing from left to right. This is done by subtracting multiples of each row from subsequent rows.

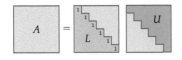

Fig. B.1 *LU* factorization.

B.3 Direct Methods

These operations can be expressed as a sequence of multiplications with lower triangular matrices L_i,

$$\underbrace{L_{n-1} \cdots L_2 L_1}_{L^{-1}} A = U. \tag{B.5}$$

After completing these operations, we have U, and we can find L by computing $L = L_1^{-1} L_2^{-1} \cdots L_{n-1}^{-1}$.

Once we have this factorization, we have $LUu = b$. Setting Uu to y, we can solve $Ly = b$ for y by forward substitution. Now we have $Uu = y$, which we can solve by back substitution for u.

Algorithm B.1 Solving $Au = b$ by LU factorization

Inputs:
 A: Nonsingular square matrix
 b: A vector

Outputs:
 u: Solution to $Au = b$

Perform forward substitution to solve $Ly = b$ for y:

$$y_1 = \frac{b_1}{L_{11}}, \quad y_i = \frac{1}{L_{ii}} \left(b_i - \sum_{j=1}^{i-1} L_{ij} y_j \right) \quad \text{for} \quad i = 2, \ldots, n$$

Perform backward substitution to solve the following $Uu = y$ for u:

$$u_n = \frac{y_n}{U_{nn}}, \quad u_i = \frac{1}{U_{ii}} \left(y_i - \sum_{j=i+1}^{n} U_{ij} u_j \right) \quad \text{for} \quad i = n-1, \ldots, 1$$

This process is not stable in general because roundoff errors are magnified in the backward substitution when diagonal elements of A have a small magnitude. This issue is resolved by *partial pivoting*, which interchanges rows to obtain more favorable diagonal elements.

Cholesky factorization is an LU factorization specialized for the case where the matrix A is symmetric and positive definite. In this case, pivoting is unnecessary because the Gaussian elimination is always stable for symmetric positive-definite matrices. The factorization can be written as

$$A = LDL^\mathsf{T}, \tag{B.6}$$

where $D = \text{diag}[U_{11}, \ldots, U_{nn}]$. This can be expressed as the matrix product

$$A = GG^\mathsf{T}, \tag{B.7}$$

where $G = LD^{1/2}$.

B.4 Iterative Methods

Although direct methods are usually more efficient and robust, iterative methods have several advantages:

- Iterative methods make it possible to trade between computational cost and precision because they can be stopped at any point and still yield an approximation of u. On the other hand, direct methods only get the solution at the end of the process with the final precision.
- Iterative methods have the advantage when a good guess for u exists. This is often the case in optimization, where the u from the previous optimization iteration can be used as the guess for the new evaluations (called a *warm start*).
- Iterative methods do not require forming and manipulating the matrix A, which can be computationally costly in terms of both time and memory. Instead, iterative methods require the computation of the residuals $r(u) = Au - b$ and, in the case of Krylov subspace methods, products of A with a given vector. Therefore, iterative methods can be more efficient than direct methods for cases where A is large and sparse. All that is needed is an efficient process to get the product of A with a given vector, as shown in Fig. B.2.

Iterative methods are divided into stationary methods (also known as *fixed-point iteration methods*) and Krylov subspace methods.

Fig. B.2 Iterative methods just require a process (which can be a black box) to compute products of A with an arbitrary vector v.

B.4.1 Jacobi, Gauss–Seidel, and SOR

Fixed-point methods generate a sequence of iterates $u_1, \ldots, u_k, \ldots$ using a function

$$u_{k+1} = G(u_k), \quad k = 0, 1, \ldots \tag{B.8}$$

starting from an initial guess u_0. The function $G(u)$ is devised such that the iterates converge to the solution u^*, which satisfies $r(u^*) = 0$. Many stationary methods can be derived by *splitting* the matrix such that $A = M - N$. Then, $Au = b$ leads to $Mu = Nu + b$, and substituting this into the linear system yields

$$u = M^{-1}(Nu + b). \tag{B.9}$$

Because $Nu = Mu - Au$, substituting this into the previous equation results in the iteration

$$u_{k+1} = u_k + M^{-1}(b - Au_k). \tag{B.10}$$

B.4 Iterative Methods

Defining the residual at iteration k as

$$r(u_k) = b - Au_k, \qquad (\text{B.11})$$

we can write

$$u_{k+1} = u_k + M^{-1} r(u_k). \qquad (\text{B.12})$$

The splitting matrix M is fixed and constructed so that it is easy to invert. The closer M^{-1} is to the inverse of A, the better the iterations work. We now introduce three stationary methods corresponding to three different splitting matrices.

The Jacobi method consists of setting M to be a diagonal matrix D, where the diagonal entries are those of A. Then,

$$u_{k+1} = u_k + D^{-1} r(u_k). \qquad (\text{B.13})$$

In component form, this can be written as

$$u_{i\,k+1} = \frac{1}{A_{ii}} \left[b_i - \sum_{j=1, j \neq i}^{n_u} A_{ij} u_{j_k} \right], \quad i = 1, \ldots, n_u. \qquad (\text{B.14})$$

Using this method, each component in u_{k+1} is independent of each other at a given iteration; they only depend on the previous iteration values, u_k, and can therefore be done in parallel.

The Gauss–Seidel method is obtained by setting M to be the lower triangular portion of A and can be written as

$$u_{k+1} = u_k + E^{-1} r(u_k), \qquad (\text{B.15})$$

where E is the lower triangular matrix. Because of the triangular matrix structure, each component in u_{k+1} is dependent on the previous elements in the vector, but the iteration can be performed in a single forward sweep. Writing this in component form yields

$$u_{i\,k+1} = \frac{1}{A_{ii}} \left[b_i - \sum_{j<i} A_{ij} u_{j_{k+1}} - \sum_{j>i} A_{ij} u_{j_k} \right], \quad i = 1, \ldots, n_u. \qquad (\text{B.16})$$

Unlike the Jacobi iterations, a Gauss–Seidel iteration cannot be performed in parallel because of the terms where $j < i$, which require the latest values. Instead, the states must be updated sequentially. However, the advantage of Gauss–Seidel is that it generally converges faster than Jacobi iterations.

The successive over-relaxation (SOR) method uses an update that is a weighted average of the Gauss–Seidel update and the previous iteration,

$$u_{k+1} = u_k + \omega\left((1-\omega)D + \omega E\right)^{-1} r(u_k), \tag{B.17}$$

where ω, the *relaxation factor*, is a scalar between 1 and 2. Setting $\omega = 1$ yields the Gauss–Seidel method. SOR in component form is as follows:

$$u_{ik+1} = (1-\omega)u_{ik} + \frac{\omega}{A_{ii}}\left[b_i - \sum_{j<i} A_{ij}u_{j\,k+1} - \sum_{j>i} A_{ij}u_{j\,k}\right], \quad i = 1,\ldots,n_u. \tag{B.18}$$

With the correct value of ω, SOR converges faster than Gauss–Seidel.

Example B.1 Iterative methods applied to a simple linear system.

Suppose we have the following linear system of two equations:

$$\begin{bmatrix} 2 & -1 \\ -2 & 3 \end{bmatrix} \begin{bmatrix} u_1 \\ u_2 \end{bmatrix} = \begin{bmatrix} 0 \\ 1 \end{bmatrix}.$$

This corresponds to the two lines shown in Fig. B.3, where the solution is at their intersection.

Applying the Jacobian iteration (Eq. B.14),

$$u_{1\,k+1} = \frac{1}{2}u_{2k}$$

$$u_{2\,k+1} = \frac{1}{3}(1 + 2u_{1k}).$$

Starting with the guess $u^{(0)} = (2,1)$, we get the iterations shown in Fig. B.3. The Gauss–Seidel iteration (Eq. B.16) is similar, where the only change is that the second equation uses the latest state from the first one:

$$u_{1\,k+1} = \frac{1}{2}u_{2k}$$

$$u_{2\,k+1} = \frac{1}{3}(1 + 2u_{1\,k+1}).$$

As expected, Gauss–Seidel converges faster than the Jacobi iteration, taking a more direct path. The SOR iteration is

$$u_{1\,k+1} = (1-\omega)u_{1k} + \frac{\omega}{2}u_{2k}$$

$$u_{2\,k+1} = (1-\omega)u_{2k} + \frac{\omega}{3}(1 + 2u_{1k}).$$

SOR converges even faster for the right values of ω. The result shown here is for $\omega = 1.2$.

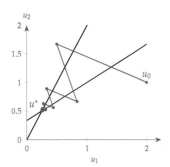

Jacobi

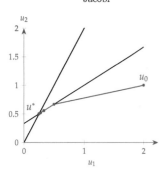

Gauss–Seidel

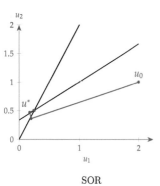

SOR

Fig. B.3 Jacobi, Gauss–Seidel, and SOR iterations.

B.4.2 Conjugate Gradient Method

The conjugate gradient method applies to linear systems where A is symmetric and positive definite. This method can be adapted to solve nonlinear minimization problems (see Section 4.4.2).

We want to solve a linear system (Eq. B.2) iteratively. This means that at a given iteration u_k, the residual is not necessarily zero and can be written as

$$r_k = Au_k - b. \tag{B.19}$$

Solving this linear system is equivalent to minimizing the quadratic function

$$f(u) = \frac{1}{2}u^\mathsf{T} Au - b^\mathsf{T} u. \tag{B.20}$$

This is because the gradient of this function is

$$\nabla f(u) = Au - b. \tag{B.21}$$

Thus, the gradient of the quadratic is the residual of the linear system,

$$r_k = \nabla f(u_k). \tag{B.22}$$

We can express the path from any starting point to a solution u^* as a sequence of n steps with directions p_k and length α_k:

$$u^* = \sum_{k=0}^{n-1} \alpha_k p_k. \tag{B.23}$$

Substituting this into the quadratic (Eq. B.20), we get

$$\begin{aligned} f(u^*) &= f\left(\sum_{k=0}^{n-1} \alpha_k p_k\right) \\ &= \frac{1}{2}\left(\sum_{k=0}^{n-1} \alpha_k p_k\right)^\mathsf{T} A \left(\sum_{k=0}^{n-1} \alpha_k p_k\right) - b^\mathsf{T}\left(\sum_{k=0}^{n-1} \alpha_k p_k\right) \\ &= \frac{1}{2}\sum_{k=0}^{n-1}\sum_{j=0}^{n-1} \alpha_k \alpha_j p_k^\mathsf{T} A p_j - \sum_{k=0}^{n-1} \alpha_k b^\mathsf{T} p_k. \end{aligned} \tag{B.24}$$

The conjugate gradient method uses a set of n vectors p_k that are *conjugate* with respect to matrix A. Such vectors have the following property:

$$p_k^\mathsf{T} A p_j = 0, \quad \text{for all} \quad k \neq j. \tag{B.25}$$

Using this conjugacy property, the double-sum term can be simplified to a single sum,

$$\frac{1}{2}\sum_{k=0}^{n-1}\sum_{j=0}^{n-1} \alpha_k \alpha_j p_k^\mathsf{T} A p_j = \frac{1}{2}\sum_{k=0}^{n-1} \alpha_k^2 p_k^\mathsf{T} A p_k. \tag{B.26}$$

Then, Eq. B.24 simplifies to

$$f(u^*) = \sum_{k=0}^{n-1} \left(\frac{1}{2}\alpha_k^2 p_k^\mathsf{T} A p_k - \alpha_k b^\mathsf{T} p_k\right). \tag{B.27}$$

Because each term in this sum involves only one direction p_k, we have reduced the original problem to a series of one-dimensional quadratic functions that can be minimized one at a time. Each one-dimensional problem corresponds to minimizing the quadratic with respect to the step length α_k. Differentiating each term and setting it to zero yields the following:

$$\alpha_k p_k^\mathsf{T} A p_k - b^\mathsf{T} p_k = 0 \Rightarrow \alpha_k = \frac{b^\mathsf{T} p_k}{p_k^\mathsf{T} A p_k}. \tag{B.28}$$

Now, the question is: How do we find this set of directions? There are many sets of directions that satisfy conjugacy. For example, the eigenvectors of A satisfy Eq. B.25.* However, it is costly to compute the eigenvectors of a matrix. We want a more convenient way to compute a sequence of conjugate vectors.

*Suppose we have two eigenvectors, v_k and v_j. Then $v_k^\mathsf{T} A v_j = v_k^\mathsf{T}(\lambda_j v_j) = \lambda_j v_k^\mathsf{T} v_j$. This dot product is zero because the eigenvectors of a symmetric matrix are mutually orthogonal.

The conjugate gradient method sets the first direction to the steepest-descent direction of the quadratic at the first point. Because the gradient of the function is the residual of the linear system (Eq. B.22), this first direction is obtained from the residual at the starting point,

$$p_1 = -r(u_0). \tag{B.29}$$

Each subsequent direction is set to a new conjugate direction using the update

$$p_{k+1} = -r_{k+1} + \beta_k p_k, \tag{B.30}$$

where β is set such that p_{k+1} and p_k are conjugate with respect to A.

We can find the expression for β by starting with the conjugacy property that we want to achieve,

$$p_{k+1}^\mathsf{T} A p_k = 0. \tag{B.31}$$

Substituting the new direction p_{k+1} with the update (Eq. B.30), we get

$$(-r_{k+1} + \beta_k p_k)^\mathsf{T} A p_k = 0. \tag{B.32}$$

Expanding the terms and solving for β, we get

$$\beta_k = \frac{r_{k+1}^\mathsf{T} A p_k}{p_k^\mathsf{T} A p_k}. \tag{B.33}$$

For each search direction p_k, we can perform an exact line search by minimizing the quadratic analytically. The directional derivative of the quadratic at a point x along the search direction p is as follows:

$$\begin{aligned}\frac{\partial f(x+\alpha p)}{\partial \alpha} &= \frac{\partial}{\partial \alpha}\left(\frac{1}{2}(x+\alpha p)^\mathsf{T} A(x+\alpha p) - b^\mathsf{T}(x+\alpha p)\right) \\ &= p^\mathsf{T} A(x+\alpha p) - p^\mathsf{T} b \\ &= p^\mathsf{T}(Ax - b) + \alpha p^\mathsf{T} A p \\ &= p^\mathsf{T} r(x) + \alpha p^\mathsf{T} A p\,.\end{aligned} \qquad (B.34)$$

By setting this derivative to zero, we can get the step size that minimizes the quadratic along the line to be

$$\alpha_k = -\frac{r_k^\mathsf{T} p_k}{p_k^\mathsf{T} A p_k}\,. \qquad (B.35)$$

The numerator can be written as a function of the residual alone. Replacing p_k with the conjugate direction update (Eq. B.30), we get

$$\begin{aligned}r_k^\mathsf{T} p_k &= r_k^\mathsf{T}\left(-r_k^\mathsf{T} + \beta_k p_{k-1}\right) \\ &= -r_k^\mathsf{T} r_k^\mathsf{T} + \beta_k r_k^\mathsf{T} p_{k-1} \\ &= -r_k^\mathsf{T} r_k\,.\end{aligned} \qquad (B.36)$$

Here we have used the property of the conjugate directions stating that the residual vector is orthogonal to all previous conjugate directions, so that $r_i^\mathsf{T} p_i$ for $i = 0, 1, \ldots, k-1$.[†] Thus, we can now write,

$$\alpha_k = -\frac{r_k^\mathsf{T} r_k}{p_k^\mathsf{T} A p_k}\,. \qquad (B.37)$$

[†] For a proof of this property, see Theorem 5.2 in Nocedal and Wright.[79]

79. Nocedal and Wright, *Numerical Optimization*, 2006.

The numerator of the expression for β (Eq. B.33) can also be written in terms of the residual alone. Using the expression for the residual (Eq. B.19) and taking the difference between two subsequent residuals, we get

$$\begin{aligned}r_{k+1} - r_k &= (A u_{k+1} - b) - (A u_k - b) = A\left(u_{k+1} - u_k\right) \\ &= A\left(u_k + \alpha_k p_k - u_k\right) \\ &= \alpha_k A p_k\,.\end{aligned} \qquad (B.38)$$

Using this result in the numerator of β in Eq. B.33, we can write

$$\begin{aligned}r_{k+1}^\mathsf{T} A p_k &= \frac{1}{\alpha_k} r_{k+1}^\mathsf{T}\left(r_{k+1} - r_k\right) \\ &= \frac{1}{\alpha_k}\left(r_{k+1}^\mathsf{T} r_{k+1} - r_{k+1}^\mathsf{T} r_k\right)\end{aligned}$$

$$r_{k+1}^\mathsf{T} A p_k = \frac{1}{\alpha_k} \left(r_{k+1}^\mathsf{T} r_{k+1} \right), \tag{B.39}$$

where we have used the property that the residual at any conjugate residual iteration is orthogonal to the residuals at all previous iterations, so $r_{k+1}^\mathsf{T} r_k = 0$.[‡]

[‡] For a proof of this property, see Theorem 5.3 in Nocedal and Wright.[79]

79. Nocedal and Wright, *Numerical Optimization*, 2006.

Now, using this new numerator and using Eq. B.37 to write the denominator as a function of the previous residual, we obtain

$$\beta_k = \frac{r_k^\mathsf{T} r_k}{r_{k-1}^\mathsf{T} r_{k-1}}. \tag{B.40}$$

We use this result in the nonlinear conjugate gradient method for function minimization in Section 4.4.2.

The linear conjugate gradient steps are listed in Alg. B.2. The advantage of this method relative to the direct method is that A does not need to be stored or given explicitly. Instead, we only need to provide a function that computes matrix-vector products with A. These products are required to compute residuals ($r = Au - b$) and the Ap term in the computation of α. Assuming a well-conditioned problem with good enough arithmetic precision, the algorithm should converge to the solution in n steps.[§]

[§] Because the linear conjugate gradient method converges in n steps, it was originally thought of as a direct method. It was initially dismissed in favor of more efficient direct methods, such as LU factorization. However, the conjugate gradient method was later reframed as an effective iterative method to obtain approximate solutions to large problems.

Algorithm B.2 Linear conjugate gradient

Inputs:
 $u^{(0)}$: Starting point
 τ: Convergence tolerance

Outputs:
 u^*: Solution of linear system

$k = 0$ — Initialize iteration counter
while $\|r_k\|_\infty > \tau$ **do**
 if $k = 0$ **then**
  $p_k = -r_k$ — First direction is steepest descent
 else
  $\beta_k = \dfrac{r_k^\mathsf{T} r_k}{r_{k-1}^\mathsf{T} r_{k-1}}$
  $p_k = -r_k + \beta_k p_{k-1}$ — Conjugate gradient direction update
 end if
 $\alpha_k = -\dfrac{r_k^\mathsf{T} r_k}{p_k^\mathsf{T} A p_k}$ — Step length
 $u_{k+1} = u_k + \alpha_k p_k$ — Update variables
 $k = k + 1$ — Increment iteration index
end while

B.4.3 Krylov Subspace Methods

Krylov subspace methods are a more general class of iterative methods.[¶] The conjugate gradient is a special case of a Krylov subspace method that applies only to symmetric positive-definite matrices. However, more general Krylov subspace methods, such as the generalized minimum residual (GMRES) method, do not have such restrictions on the matrix. Compared with stationary methods of Appendix B.4.1, Krylov methods have the advantage that they use information gathered throughout the iterations. Instead of using a fixed splitting matrix, Krylov methods effectively vary the splitting so that M is changed at each iteration according to some criteria that use the information gathered so far. For this reason, Krylov methods are usually more efficient than stationary methods.

Like stationary iteration methods, Krylov methods do not require forming or storing A. Instead, the iterations require only matrix-vector products of the form Av, where v is some vector given by the Krylov algorithm. The matrix-vector product could be given by a black box, as shown in Fig. B.2.

For the linear conjugate gradient method (Appendix B.4.2), we found conjugate directions and minimized the residual of the linear system in a sequence of these directions.

Krylov subspace methods minimize the residual in a space,

$$x_0 + \mathcal{K}_k , \tag{B.41}$$

where x_0 is the initial guess, and $\mathcal{K}_k$ is the Krylov subspace,

$$\mathcal{K}_k(A; r_0) \equiv \mathrm{span}\{r_0, Ar_0, A^2 r_0, \ldots, A^{k-1} r_0\} . \tag{B.42}$$

In other words, a Krylov subspace method seeks a solution that is a linear combination of the vectors $r_0, Ar_0, \ldots, A^{k-1} r_0$. The definition of this particular sequence is convenient because these terms can be computed recursively with the matrix-vector product black box as $r_0, A(r_0), A(A(r_0)), A(A(A(r_0))), \ldots$. Under certain conditions, it can be shown that the solution of the linear system of size n is contained in the subspace $\mathcal{K}_n$.

Krylov subspace methods (including the conjugate gradient method) converge much faster when using *preconditioning*. Instead of solving $Ax = b$, we solve

$$(M^{-1} A) x = M^{-1} b , \tag{B.43}$$

where M is the preconditioning matrix (or simply *preconditioner*). The matrix M should be similar to A and correspond to a linear system that is easier to solve. The inverse, M^{-1}, should be available explicitly, and

---

[¶] This is just an overview of Krylov subspace methods; for more details, see Trefethen and Bau III[220] or Saad.[75]

75. Saad, *Iterative Methods for Sparse Linear Systems*, 2003.

220. Trefethen and Bau III, *Numerical Linear Algebra*, 1997.

we do not need an explicit form for M. The matrix resulting from the product $M^{-1}A$ should have a smaller condition number so that the new linear system is better conditioned.

In the extreme case where $M = A$, that means we have computed the inverse of A, and we can get x explicitly. In another extreme, M could be a diagonal matrix with the diagonal elements of A, which would scale A such that the diagonal elements are 1.‖

‖ The splitting matrix M we used in the equation for the stationary methods (Appendix B.4.1) is effectively a preconditioner. An M using the diagonal entries of A corresponds to the Jacobi method (Eq. B.13).

Krylov subspace solvers require three main components: (1) an orthogonal basis for the Krylov subspace, (2) an optimal property that determines the solution within the subspace, and (3) an effective preconditioner. Various Krylov subspace methods are possible, depending on the choice for each of these three components. One of the most popular Krylov subspace methods is the GMRES.[221]**

221. Saad and Schultz, *GMRES: A generalized minimal residual algorithm for solving nonsymmetric linear systems*, 1986.

**GMRES and other Krylov subspace methods are available in most programming languages, including C/C++, Fortran, Julia, MATLAB, and Python.

Quasi-Newton Methods

C

C.1 Broyden's Method

Broyden's method is the extension of the secant method (from Section 3.8) to n dimensions.[222] It can also be viewed as the analog of the quasi-Newton methods from Section 4.4.4 for solving equations (as opposed to finding a minimum).

222. Broyden, *A class of methods for solving nonlinear simultaneous equations*, 1965.

Using the notation from Chapter 3, suppose we have a set of n equations $r(u) = [r_1, \ldots, r_n] = 0$ and n unknowns $u = [u_1, \ldots, u_n]$. Writing a Taylor series expansion of $r(u)$ and selecting the linear term of the Taylor series expansion of r yields

$$J_{k+1}(u_{k+1} - u_k) \approx r_{k+1} - r_k, \tag{C.1}$$

where J is the $(n \times n)$ Jacobian, $\partial r / \partial u$. Defining the step in u as

$$s_k = u_{k+1} - u_k, \tag{C.2}$$

and the change in the residuals as

$$y_k = r_{k+1} - r_k, \tag{C.3}$$

we can write Eq. C.1 as

$$\tilde{J}_{k+1} s_k = y_k. \tag{C.4}$$

This is the equivalent of the secant equation (Eq. 4.80). The difference is that we now approximate the Jacobian instead of the Hessian. The right-hand side is the difference between two subsequent function values (which quantifies the directional derivative along the last step) instead of the difference between gradients (which quantifies the curvature).

We seek a rank 1 update of the form

$$\tilde{J} = \tilde{J}_k + v v^\mathsf{T}, \tag{C.5}$$

where the self outer product $v v^\mathsf{T}$ yields a symmetric matrix of rank 1. Substituting this update into the required condition (Eq. C.4) yields

$$\left(\tilde{J}_k + v v^\mathsf{T}\right) s_k = y_k. \tag{C.6}$$

Post-multiplying both sides by s^T, rearranging, and dividing by $s_k^\mathsf{T} s_k$ yields

$$vv^\mathsf{T} = \frac{\left(y_k - \tilde{J}_k s_k\right) s_k^\mathsf{T}}{s_k^\mathsf{T} s_k}. \tag{C.7}$$

Substituting this result into the update (Eq. C.5), we get the Jacobian approximation update,

$$\tilde{J}_{k+1} = \tilde{J}_k + \frac{\left(y_k - \tilde{J}_k s_k\right) s_k^\mathsf{T}}{s_k^\mathsf{T} s_k}, \tag{C.8}$$

where

$$y_k = r_{k+1} - r_k \tag{C.9}$$

is the difference in the function values (as opposed to the difference in the gradients used in optimization).

This update can be inverted using the Sherman–Morrison–Woodbury formula (Appendix C.3) to get the more useful update on the inverse of the Jacobian,

$$\tilde{J}_{k+1}^{-1} = \tilde{J}_k^{-1} + \frac{\left(s_k - \tilde{J}_k^{-1} y_k\right) y_k^\mathsf{T}}{y_k^\mathsf{T} y_k}. \tag{C.10}$$

We can start with $\tilde{J}_0^{-1} = I$. Similar to the Newton step (Eq. 3.30), the step in Broyden's method is given by solving the linear system. Because the inverse is provided explicitly, we can just perform the multiplication,

$$\Delta u_k = -\tilde{J}^{-1} r_k. \tag{C.11}$$

Then we update the variables as

$$u_{k+1} = u_k + \Delta u_k. \tag{C.12}$$

C.2 Additional Quasi-Newton Approximations

In Section 4.4.4, we introduced the Broyden–Fletcher–Goldfarb–Shanno (BFGS) quasi-Newton approximation for unconstrained optimization, which was also used in Section 5.5 for constrained optimization. Here we expand on that to introduce other quasi-Newton approximations and generalize them.

To get a unique solution for the approximate Hessian update, quasi-Newton methods quantify the "closeness" of successive Hessian

C.2 ADDITIONAL QUASI-NEWTON APPROXIMATIONS

approximations by using some norm of the difference between the two matrices, leading to the following optimization problem:

$$\begin{align}
\text{minimize} \quad & \|\tilde{H} - \tilde{H}_k\| \\
\text{by varying} \quad & \tilde{H} \\
\text{subject to} \quad & \tilde{H} = \tilde{H}^\mathsf{T} \\
& \tilde{H} s_k = y_k ,
\end{align} \tag{C.13}$$

where, $y_k = \nabla f_{k+1} - \nabla f_k$, and $s_k = x_{k+1} - x_k$ (the latest step). There are several possibilities for quantifying the "closeness" between matrices and satisfying the constraints, leading to different quasi-Newton updates. With a convenient choice of matrix norm, we can solve this optimization problem analytically to obtain a formula for $\tilde{H}_{k+1}$ as a function of $\tilde{H}_k$, s_k, and y_k.

The optimization problem (Eq. C.13) does not enforce a positive-definiteness constraint. It turns out that the update formula always produces a $\tilde{H}_{k+1}$ that is positive definite, provided that $\tilde{H}_k$ is positive definite. The fact that the curvature condition (Eq. 4.81) is satisfied for each step helps with this.

C.2.1 Davidon–Fletcher–Powell Update

The Davidon–Fletcher–Powell (DFP) update can be derived using a similar approach to that used to derive the BFGS update in Section 4.4.4. However, instead of starting with the update for the Hessian, we start with the update to the Hessian inverse,

$$\tilde{V}_{k+1} = \tilde{V}_k + \alpha u u^\mathsf{T} + \beta v v^\mathsf{T} . \tag{C.14}$$

We need the inverse version of the secant equation (Eq. 4.80), which is

$$\tilde{V}_{k+1} y_k = s_k . \tag{C.15}$$

Setting $u = s_k$ and $v = \tilde{V}_k y_k$ in the update (Eq. C.14) and substituting it into the inverse version of the secant equation (Eq. C.15), we get

$$\tilde{V}_k y_k + \alpha s_k s_k^\mathsf{T} y_k + \beta \tilde{V}_k y_k y_k^\mathsf{T} \tilde{V}_k y_k = s_k . \tag{C.16}$$

We can obtain the coefficients α and β by rearranging this equation and using similar arguments to those used in the BFGS update derivation (see Section 4.4.4). The DFP update for the Hessian inverse approximation is

$$\tilde{V}_{k+1} = \tilde{V}_k + \frac{1}{s_k s_k^\mathsf{T}} y_k^\mathsf{T} s_k - \frac{1}{\tilde{V}_k y_k y_k^\mathsf{T} \tilde{V}_k} y_k^\mathsf{T} \tilde{V}_k y_k . \tag{C.17}$$

However, the DFP update was originally derived by solving the optimization problem (Eq. C.13), which minimizes a matrix norm of the update while enforcing symmetry and the secant equation. This problem can be solved analytically through the Karush–Kuhn–Tucker (KKT) conditions and a convenient matrix norm. The weighted Frobenius norm (Eq. A.35) was the norm used in this case, where the weights were based on an averaged Hessian inverse. The derivation is lengthy and is not included here. The final result is the update,

$$\tilde{H}_{k+1} = \left(I - \sigma_k s_k y_k^\mathsf{T}\right) \tilde{H}_k \left(I - \sigma_k y_k s_k^\mathsf{T}\right) + \sigma_k y_k y_k^\mathsf{T}, \tag{C.18}$$

where

$$\sigma_k = \frac{1}{y_k^\mathsf{T} s_k}. \tag{C.19}$$

This can be inverted using the Sherman–Morrison–Woodbury formula (Appendix C.3) to get the update on the inverse (Eq. C.17).

C.2.2 BFGS

The BFGS update was informally derived in Section 4.4.4. As discussed previously, obtaining an approximation of the Hessian inverse is a more efficient way to get the quasi-Newton step.

Similar to DFP, BFGS was originally formally derived by analytically solving an optimization problem. However, instead of solving the optimization problem of Eq. C.13, we solve a similar problem using the Hessian *inverse* approximation instead. This problem can be stated as

$$\begin{aligned} \text{minimize} \quad & \|\tilde{V} - \tilde{V}_k\| \\ \text{subject to} \quad & \tilde{V} y_k = s_k \\ & \tilde{V} = \tilde{V}^\mathsf{T}, \end{aligned} \tag{C.20}$$

where $\tilde{V}$ is the updated inverse Hessian that we seek, $\tilde{V}_k$ is the inverse Hessian approximation from the previous step. The first constraint is known as the *secant equation applied to the inverse*. The second constraint enforces symmetric updates. We do not explicitly specify positive definiteness. The matrix norm is again a weighted Frobenius norm (Eq. A.35), but now the weights are based on an averaged Hessian (instead of the inverse for DFP). Solving this optimization problem (Eq. C.20), the final result is

$$\tilde{V}_{k+1} = \left(I - \sigma_k s_k y_k^\mathsf{T}\right) \tilde{V}_k \left(I - \sigma_k y_k s_k^\mathsf{T}\right) + \sigma_k s_k s_k^\mathsf{T}, \tag{C.21}$$

where

$$\sigma_k = \frac{1}{y_k^\mathsf{T} s_k}. \tag{C.22}$$

This is identical to Eq. 4.88.

C.2.3 Symmetric Rank 1 Update

The symmetric rank 1 (SR1) update is a quasi-Newton update that is rank 1 as opposed to the rank 2 update of DFP and BFGS (Eq. C.14). The SR1 update can be derived formally without solving the optimization problem of Eq. C.13 because there is only one update that satisfies the secant equation.

Similar to the rank 2 update of the approximate inverse Hessian (Eq. 4.82), we construct the update,

$$\tilde{V} = \tilde{V}_k + \alpha v v^\mathsf{T}, \tag{C.23}$$

where we only need one self outer product to produce a rank 1 update (as opposed to two).

Substituting the rank 1 update (Eq. C.23) into the secant equation, we obtain

$$\tilde{V}_k y_k + \alpha v v^\mathsf{T} y_k = s_k. \tag{C.24}$$

Rearranging yields

$$\left(\alpha v^\mathsf{T} y_k\right) v = s_k - \tilde{V}_k y_k. \tag{C.25}$$

Thus, we have to make sure that v is in the direction of $y_k - H_k s_k$. The scalar α must be such that the scaling of the vectors on both sides of the equation match each other. We define a normalized v in the desired direction,

$$v = \frac{s_k - \tilde{V}_k y_k}{\left\| s_k - \tilde{V}_k y_k \right\|_2}. \tag{C.26}$$

To find the correct value for α, we substitute Eq. C.26 into Eq. C.25 to get

$$s_k - \tilde{V}_k y_k = \alpha \frac{s_k^\mathsf{T} y_k - y_k^\mathsf{T} \tilde{V}_k y_k}{\left\| s_k - \tilde{V}_k y_k \right\|_2^2} \left(s_k - \tilde{V}_k y_k \right). \tag{C.27}$$

Solving for α yields

$$\alpha = \frac{\left\| s_k - \tilde{V}_k y_k \right\|_2^2}{s_k^\mathsf{T} y_k - y_k^\mathsf{T} \tilde{V}_k y_k}. \tag{C.28}$$

Substituting Eqs. C.26 and C.28 into Eq. C.23, we get the SR1 update

$$\tilde{V} = \tilde{V}_k + \frac{1}{s_k^\mathsf{T} y_k - y_k^\mathsf{T} \tilde{V}_k y_k} \left(s_k - \tilde{V}_k y_k \right) \left(s_k - \tilde{V}_k y_k \right)^\mathsf{T}. \tag{C.29}$$

Because it is possible for the denominator in this update to be zero, the update requires safeguarding. This update is not positive definite in general because the denominator can be negative.

As in the BFGS method, the search direction at each major iteration is given by $p_k = -\tilde{V}_k \nabla f_k$ and a line search with $\alpha_{\text{init}} = 1$ determines the final step length.

C.2.4 Unification of SR1, DFP, and BFGS

The SR1, DFP, and BFGS updates for the inverse Hessian approximation can be expressed using the following more general formula:

$$\tilde{V}_{k+1} = \tilde{V}_k + \begin{bmatrix} \tilde{V}_k y_k & s_k \end{bmatrix} \begin{bmatrix} \alpha & \beta \\ \beta & \gamma \end{bmatrix} \begin{bmatrix} y_k^T \tilde{V}_k \\ s_k^T \end{bmatrix}. \tag{C.30}$$

For the SR1 method, we have

$$\begin{aligned} \alpha_{\text{SR1}} &= \frac{1}{y_k^T s_k - y_k^T \tilde{V}_k y_k} \\ \beta_{\text{SR1}} &= -\frac{1}{y_k^T s_k - y_k^T \tilde{V}_k y_k} \\ \gamma_{\text{SR1}} &= \frac{1}{y_k^T s_k - y_k^T \tilde{V}_k y_k}. \end{aligned} \tag{C.31}$$

For the DFP method, we have

$$\alpha_{\text{DFP}} = -\frac{1}{y_k \tilde{V}_k y_k}, \quad \beta_{\text{DFP}} = 0, \quad \gamma_{\text{DFP}} = \frac{1}{y_k^T s_k}. \tag{C.32}$$

For the BFGS method, we have

$$\alpha_{\text{BFGS}} = 0, \quad \beta_{\text{BFGS}} = -\frac{1}{y_k^T s_k}, \quad \gamma_{\text{BFGS}} = \frac{1}{y_k^T s_k} + \frac{y_k^T \tilde{V}_k y_k}{\left(y_k^T s_k\right)^2}. \tag{C.33}$$

C.3 Sherman–Morrison–Woodbury Formula

The formal derivations of the DFP and BFGS methods use the Sherman–Morrison–Woodbury formula (also known as the *Woodbury matrix identity*). Suppose that the inverse of a matrix is known, and then the matrix is perturbed. The Sherman–Morrison–Woodbury formula gives the inverse of the perturbed matrix without having to re-invert the perturbed matrix. We used this formula in Section 4.4.4 to derive the quasi-Newton update.

One possible perturbation is a rank 1 update of the form

$$\hat{A} = A + uv^T, \tag{C.34}$$

C.3 Sherman–Morrison–Woodbury Formula

where u and v are n-vectors. This is a rank 1 update to A because uv^T is an outer product that produces a matrix whose rank is equal to 1 (see Fig. 4.50).

If $\hat{A}$ is nonsingular, and A^{-1} is known, the Sherman–Morrison–Woodbury formula gives

$$\hat{A}^{-1} = A^{-1} - \frac{A^{-1} u v^\mathsf{T} A^{-1}}{1 + v^\mathsf{T} A^{-1} u}. \qquad (C.35)$$

This formula can be verified by multiplying Eq. C.34 and Eq. C.35, which yields the identity matrix.

This formula can be generalized for higher-rank updates as follows:

$$\hat{A} = A + UV^\mathsf{T}, \qquad (C.36)$$

where U and V are $(n \times p)$ matrices for some p between 1 and n. Then,

$$\hat{A}^{-1} = A^{-1} - A^{-1} U \left(I + V^\mathsf{T} A^{-1} U \right) V^\mathsf{T} A^{-1}. \qquad (C.37)$$

Although we need to invert a new matrix, $\left(I + V^\mathsf{T} A^{-1} U \right)$, this matrix is typically small and can be inverted analytically for $p = 2$ for the rank 2 update, for example.

Test Problems

D.1 Unconstrained Problems

D.1.1 Slanted Quadratic Function

This is a smooth two-dimensional function suitable for a first test of a gradient-based optimizer:

$$f(x_1, x_2) = x_1^2 + x_2^2 - \beta x_1 x_2, \tag{D.1}$$

where $\beta \in [0, 2)$. A β value of zero corresponds to perfectly circular contours. As β increases, the contours become increasingly slanted. For $\beta = 2$, the quadratic becomes semidefinite, and there is a line of weak minima. For $\beta > 2$, the quadratic is indefinite, and there is no minimum. An intermediate value of $\beta = 3/2$ is suitable for first tests and yields the contours shown in Fig. D.1.

Global minimum: $f(x^*) = 0$ at $x^* = (0, 0)$.

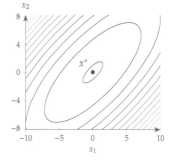

Fig. D.1 Slanted quadratic function for $\beta = 3/2$.

D.1.2 Rosenbrock Function

The two-dimensional Rosenbrock function, shown in Fig. D.2, is also known as *Rosenbrock's valley* or *banana function*. This function was introduced by Rosenbrock,[223] who used it as a benchmark problem for optimization algorithms.

The function is defined as follows:

$$f(x_1, x_2) = (1 - x_1)^2 + 100\left(x_2 - x_1^2\right)^2. \tag{D.2}$$

This became a classic benchmarking function because of its narrow turning valley. The large difference between the maximum and minimum curvatures, and the fact that the principal curvature directions change along the valley, makes it a good test for quasi-Newton methods.

The Rosenbrock function can be extended to n dimensions by defining the sum,

$$f(x) = \sum_{i=1}^{n-1}\left(100\left(x_{i+1} - x_i^2\right)^2 + (1 - x_i)^2\right). \tag{D.3}$$

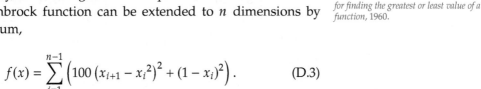

Fig. D.2 Rosenbrock function.

223. Rosenbrock, *An automatic method for finding the greatest or least value of a function*, 1960.

Global minimum: $f(x^*) = 0.0$ at $x^* = (1, 1, \ldots, 1)$.
Local minimum: For $n \geq 4$, a local minimum exists near $x = (-1, 1, \ldots, 1)$.

D.1.3 Bean Function

The "bean" function was developed in this book as a milder version of the Rosenbrock function: it has the same curved valley as the Rosenbrock function but without the extreme variations in curvature. The function, shown in Fig. D.3, is

$$f(x_1, x_2) = (1 - x_1)^2 + (1 - x_2)^2 + \frac{1}{2}(2x_2 - x_1^2)^2. \tag{D.4}$$

Global minimum: $f(x^*) = 0.09194$ at $x^* = (1.21314, 0.82414)$.

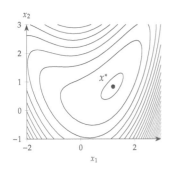

Fig. D.3 Bean function.

D.1.4 Jones Function

This is a fourth-order smooth multimodal function that is useful to test global search algorithms and also gradient-based algorithms starting from different points. There are saddle points, maxima, and minima, with one global minimum. This function, shown in Fig. D.4 along with the local and global minima, is

$$f(x_1, x_2) = x_1^4 + x_2^4 - 4x_1^3 - 3x_2^3 + 2x_1^2 + 2x_1 x_2. \tag{D.5}$$

Global minimum: $f(x^*) = -13.5320$ at $x^* = (2.6732, -0.6759)$.
Local minima: $f(x) = -9.7770$ at $x = (-0.4495, 2.2928)$.
$f(x) = -9.0312$ at $x = (2.4239, 1.9219)$.

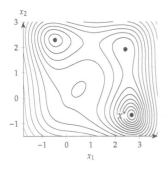

Fig. D.4 Jones multimodal function.

D.1.5 Hartmann Function

The Hartmann function is a three-dimensional smooth function with multiple local minima:

$$f(x) = -\sum_{i=1}^{4} \alpha_i \exp\left(-\sum_{j=1}^{3} A_{ij}(x_j - P_{ij})^2\right), \tag{D.6}$$

where

$$\alpha = [1.0, 1.2, 3.0, 3.2],$$

$$A = \begin{bmatrix} 3 & 10 & 30 \\ 0.1 & 10 & 35 \\ 3 & 10 & 30 \\ 0.1 & 10 & 35 \end{bmatrix}, \quad P = 10^{-4}\begin{bmatrix} 3689 & 1170 & 2673 \\ 4699 & 4387 & 7470 \\ 1091 & 8732 & 5547 \\ 381 & 5743 & 8828 \end{bmatrix}. \tag{D.7}$$

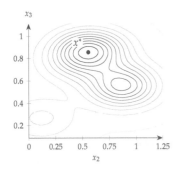

Fig. D.5 An $x_2 - x_3$ slice of Hartmann function at $x_1 = 0.1148$.

A slice of the function, at the optimal value of $x_1 = 0.1148$, is shown in Fig. D.5.

Global minimum: $f(x^*) = -3.86278$ at $x^* = (0.11480, 0.55566, 0.85254)$.

D.1.6 Aircraft Wing Design

We want to optimize the rectangular planform wing of a general aviation-sized aircraft by changing its wingspan and chord (see Ex. 1.1). In general, we would add many more design variables to a problem like this, but we are limiting it to a simple two-dimensional problem so that we can easily visualize the results.

The objective is to minimize the required power, thereby taking into account drag and propulsive efficiency, which are speed dependent. The following describes a basic performance estimation methodology for a low-speed aircraft. Implementing it may not seem like it has much to do with optimization. The physics is not important for our purposes, but practice translating equations and concepts into code is an important element of formulating optimization problems in general.

In level flight, the aircraft must generate enough lift to equal the required weight, so

$$L = W. \tag{D.8}$$

We assume that the total weight consists of a fixed aircraft and payload weight W_0 and a component of the weight that depends on the wing area S—that is,

$$W = W_0 + W_s S. \tag{D.9}$$

The wing can produce a certain lift coefficient (C_L) and so we must make the wing area (S) large enough to produce sufficient lift. Using the definition of lift coefficient, the total lift can be computed as

$$L = q C_L S, \tag{D.10}$$

where q is the dynamic pressure and

$$q = \frac{1}{2}\rho v^2. \tag{D.11}$$

If we use a rectangular wing, then the wing area can be computed from the wingspan (b) and the chord (c) as

$$S = bc. \tag{D.12}$$

The aircraft drag consists of two components: viscous drag and induced drag. The viscous drag can be approximated as

$$D_f = k C_f q S_{\text{wet}}. \tag{D.13}$$

For a fully turbulent boundary layer, the skin friction coefficient, C_f, can be approximated as

$$C_f = \frac{0.074}{Re^{0.2}}. \tag{D.14}$$

In this equation, the Reynolds number is based on the wing chord and is defined as follows:

$$Re = \frac{\rho v c}{\mu}, \quad \text{(D.15)}$$

where ρ is the air density, and μ is the air dynamic viscosity. The form factor, k, accounts for the effects of pressure drag. The wetted area, S_{wet}, is the area over which the skin friction drag acts, which is a little more than twice the planform area. We will use

$$S_{\text{wet}} = 2.05 S. \quad \text{(D.16)}$$

The induced drag is defined as

$$D_i = \frac{L^2}{q \pi b^2 e}, \quad \text{(D.17)}$$

where e is the Oswald efficiency factor. The total drag is the sum of induced and viscous drag, $D = D_i + D_f$.

Our objective function, the power required by the motor for level flight, is

$$P(b, c) = \frac{Dv}{\eta}, \quad \text{(D.18)}$$

where η is the propulsive efficiency. We assume that our electric propellers have a Gaussian efficiency curve (real efficiency curves are not Gaussian, but this is simple and will be sufficient for our purposes):

$$\eta = \eta_{\max} \exp\left(\frac{-(v - \bar{v})^2}{2\sigma^2}\right). \quad \text{(D.19)}$$

In this problem, the lift coefficient is provided. Therefore, to satisfy the lift requirement in Eq. D.8, we need to compute the velocity using Eq. D.11 and Eq. D.10 as

$$v = \sqrt{\frac{2L}{\rho C_L S}}. \quad \text{(D.20)}$$

This is the same problem that was presented in Ex. 1.2 of Chapter 1. The optimal wingspan and chord are $b = 25.48$ m and $c = 0.50$ m, respectively, given the parameters. The contour and the optimal wing shape are shown in Fig. D.6.

Because there are no structural considerations in this problem, the resulting wing has a higher wing aspect ratio than is realistic. This emphasizes the importance of carefully selecting the objective and including all relevant constraints.

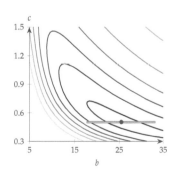

Fig. D.6 Wing design problem with power requirement contour.

D.1 Unconstrained Problems

The parameters for this problem are given as follows:

Parameter	Value	Unit	Description
ρ	1.2	kg/m^3	Density of air
μ	1.8×10^{-5}	kg/(m sec)	Viscosity of air
k	1.2		Form factor
C_L	0.4		Lift coefficient
e	0.80		Oswald efficiency factor
W_0	1,000	N	Fixed aircraft weight
W_s	8.0	N/m^2	Wing area dependent weight
η_{max}	0.8		Peak propulsive efficiency
$\bar{v}$	20.0	m/s	Flight speed at peak propulsive efficiency
σ	5.0	m/s	Standard deviation of efficiency function

D.1.7 Brachistochrone Problem

The brachistochrone problem is a classic problem proposed by Johann Bernoulli (see Section 2.2 for the historical background). Although this was originally solved analytically, we discretize the model and solve the problem using numerical optimization. This is a useful problem for benchmarking because you can change the number of dimensions.

A bead is set on a wire that defines a path that we can shape. The bead starts at some y-position h with zero velocity. For convenience, we define the starting point at $x = 0$.

From the law of conservation of energy, we can then find the velocity of the bead at any other location. The initial potential energy is converted to kinetic energy, potential energy, and dissipative work from friction acting along the path length, yielding the following:

$$mgh = \frac{1}{2}mv^2 + mgy + \int_0^x \mu_k m g \cos\theta \, ds$$

$$0 = \frac{1}{2}v^2 + g(y - h) + \mu_k g x \qquad \text{(D.21)}$$

$$v = \sqrt{2g(h - y - \mu_k x)}.$$

Now that we know the speed of the bead as a function of x, we can compute the time it takes to traverse an differential element of length ds:

$$\Delta t = \int_{x_i}^{x_i + dx} \frac{ds}{v(x)}$$

$$\Delta t = \int_{x_i}^{x_i+dx} \frac{\sqrt{dx^2 + dy^2}}{\sqrt{2g(h - y(x) - \mu_k x)}}$$
$$= \int_{x_i}^{x_i+dx} \frac{\sqrt{1 + \left(\frac{dy}{dx}\right)^2}\,dx}{\sqrt{2g(h - y(x) - \mu_k x)}}. \tag{D.22}$$

To discretize this problem, we can divide the path into linear segments. As an example, Fig. D.7 shows the wire divided into four linear segments (five nodes) as an approximation of a continuous wire. The slope $s_i = (\Delta y / \Delta x)_i$ is then a constant along a given segment, and $y(x) = y_i + s_i(x - x_i)$. Making these substitutions results in

$$\Delta t_i = \frac{\sqrt{1 + s_i^2}}{\sqrt{2g}} \int_{x_i}^{x_{i+1}} \frac{dx}{\sqrt{h - y_i - s_i(x - x_i) - \mu_k x}}. \tag{D.23}$$

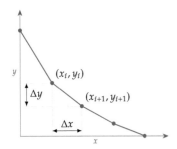

Fig. D.7 A discretized representation of the brachistochrone problem.

Performing the integration and simplifying (many steps omitted here) results in

$$\Delta t_i = \sqrt{\frac{2}{g}} \frac{\sqrt{\Delta x_i^2 + \Delta y_i^2}}{\sqrt{h - y_{i+1} - \mu_k x_{i+1}} + \sqrt{h - y_i - \mu_k x_i}}, \tag{D.24}$$

where $\Delta x_i = (x_{i+1} - x_i)$, and $\Delta y_i = (y_{i+1} - y_i)$. The objective of the optimization is to minimize the total travel time, so we need to sum up the travel time across all of our linear segments:

$$T = \sum_{i=1}^{n-1} \Delta t_i. \tag{D.25}$$

Minimization is unaffected by multiplying by a constant, so we can remove the multiplicative constant for simplicity (we see that the magnitude of the acceleration of gravity has no effect on the optimal path):

$$\text{minimize} \quad f = \sum_{i=1}^{n-1} \frac{\sqrt{\Delta x_i^2 + \Delta y_i^2}}{\sqrt{h - y_{i+1} - \mu_k x_{i+1}} + \sqrt{h - y_i - \mu_k x_i}} \tag{D.26}$$

by varying $y_i, \quad i = 1, \ldots, n$.

The design variables are the $n-2$ positions of the path parameterized by y_i. The endpoints must be fixed; otherwise, the problem is ill-defined, which is why there are $n - 2$ design variables instead of n. Note that x is a parameter, meaning that it is fixed. We could space the x_i any reasonable way and still find the same underlying optimal curve, but it is easiest to just use uniform spacing. As the dimensionality of the problem increases, the solution becomes more challenging. We will use the following specifications:

- Starting point: $(x, y) = (0, 1)$ m.
- Ending point: $(x, y) = (1, 0)$ m.
- Kinetic coefficient of friction $\mu_k = 0.3$.

The analytic solution for the case with friction is more difficult to derive, but the analytic solution for the frictionless case ($\mu_k = 0$) with our starting and ending points is as follows:

$$\begin{aligned} x &= a(\theta - \sin(\theta)) \\ y &= -a(1 - \cos(\theta)) + 1 \end{aligned} \tag{D.27}$$

where $a = 0.572917$ and $\theta \in [0, 2.412]$.

D.1.8 Spring System

Consider a connected spring system of two springs with lengths of l_1 and l_2 and stiffnesses of k_1 and k_2, fixed at the walls as shown in Fig. D.8. An object with mass m is suspended between the two springs. It will naturally deform such that the sum of the gravitational and spring potential energy, E_p, is at the minimum.

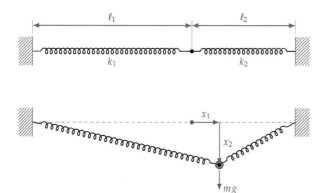

Fig. D.8 Two-spring system with no applied force (top) and with applied force (bottom).

The total potential energy of the spring system is

$$E_p(x_1, x_2) = \frac{1}{2} k_1 (\Delta l_1)^2 + \frac{1}{2} k_2 (\Delta l_2)^2 - m g x_2, \tag{D.28}$$

where Δl_1 and Δl_2 are the changes in length for the two springs. With respect to the original lengths, and displacements x_1 and x_2 as shown, they are defined as

$$\Delta l_1 = \sqrt{(l_1 + x_1)^2 + x_2^2} - l_1$$
$$\Delta l_2 = \sqrt{(l_2 - x_1)^2 + x_2^2} - l_2 .$$
(D.29)

This can be minimized to determine the final location of the object.

With initial lengths of $l_1 = 12$ cm, $l_2 = 8$ cm; spring stiffnesses of $k_1 = 1.0$ N·cm, $k_2 = 10.0$ N·cm; and a force due to gravity of $mg = 7$N, the minimum potential energy is at $(x_1, x_2) = (2.7852, 6.8996)$. The contour of E_k with respect to x_1 and x_2 is shown in Fig. D.9.

The analytic derivatives can also be computed for use in a gradient-based optimization. The derivative of E_p with respect to x_1 is

$$\frac{\partial E_p}{\partial x_1} = \frac{1}{2} k_1 \left(2\Delta l_1 \frac{\partial (\Delta l_1)}{\partial x_1} \right) + \frac{1}{2} k_2 \left(2\Delta l_2 \frac{\partial (\Delta l_2)}{\partial x_1} \right) - mg ,$$
(D.30)

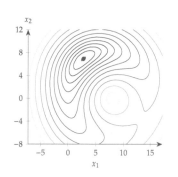

Fig. D.9 Total potential energy contours for two-spring system.

where the partial derivatives of Δl_1 and Δl_2 are

$$\frac{\partial (\Delta l_1)}{\partial x_1} = \frac{l_1 + x_1}{\sqrt{(l_1 + x_1)^2 + x_2^2}}$$
$$\frac{\partial (\Delta l_2)}{\partial x_2} = \frac{l_2 - x_1}{\sqrt{(l_2 - x_1)^2 + x_2^2}} .$$
(D.31)

By letting $\mathcal{L}_1 = \sqrt{(l_1 + x_1)^2 + x_2^2}$ and $\mathcal{L}_2 = \sqrt{(l_2 - x_1)^2 + x_2^2}$, the partial derivative of E_p with respect to x_1 can be written as

$$\frac{\partial E_p}{\partial x_1} = \frac{k_1 (\mathcal{L}_1 - l_1)(l_1 + x_1)}{\mathcal{L}_1} + \frac{k_2 (\mathcal{L}_2 - l_2)(l_2 - x_1)}{\mathcal{L}_2} - mg.$$
(D.32)

Similarly, the partial derivative of E_p with respect to x_2 can be written as

$$\frac{\partial E_p}{\partial x_2} = \frac{k_1 x_2 (\mathcal{L}_1 - l_1)}{\mathcal{L}_1} + \frac{k_2 x_2 (\mathcal{L}_2 - l_2)}{\mathcal{L}_2} .$$
(D.33)

D.2 Constrained Problems

D.2.1 Barnes Problem

224. Barnes, *A comparative study of nonlinear optimization codes*, 1967.

The Barnes problem was devised in a master's thesis[224] and has been used in various optimization demonstration studies. It is a good starter problem because it only has two dimensions for easy visualization while also including constraints.

The objective function contains the following coefficients:

$a_1 = 75.196$ $\quad\quad a_2 = -3.8112$
$a_3 = 0.12694$ $\quad\quad a_4 = -2.0567 \times 10^{-3}$
$a_5 = 1.0345 \times 10^{-5}$ $\quad a_6 = -6.8306$
$a_7 = 0.030234$ $\quad\quad a_8 = -1.28134 \times 10^{-3}$
$a_9 = 3.5256 \times 10^{-5}$ $\quad a_{10} = -2.266 \times 10^{-7}$
$a_{11} = 0.25645$ $\quad\quad a_{12} = -3.4604 \times 10^{-3}$
$a_{13} = 1.3514 \times 10^{-5}$ $\quad a_{14} = -28.106$
$a_{15} = -5.2375 \times 10^{-6}$ $\quad a_{16} = -6.3 \times 10^{-8}$
$a_{17} = 7.0 \times 10^{-10}$ $\quad\quad a_{18} = 3.4054 \times 10^{-4}$
$a_{19} = -1.6638 \times 10^{-6}$ $\quad a_{20} = -2.8673$
$a_{21} = 0.0005$

For convenience, we define the following quantities:

$$y_1 = x_1 x_2, \quad y_2 = y_1 x_1, \quad y_3 = x_2^2, \quad y_4 = x_1^2 \quad\quad (D.34)$$

The objective function is then:

$$\begin{aligned}f(x_1, x_2) = {}& a_1 + a_2 x_1 + a_3 y_4 + a_4 y_4 x_1 + a_5 y_4^2 + a_6 x_2 + a_7 y_1 + \\ & a_8 x_1 y_1 + a_9 y_1 y_4 + a_{10} y_2 y_4 + a_{11} y_3 + a_{12} x_2 y_3 + a_{13} y_3^2 + \\ & \frac{a_{14}}{x_2 + 1} + a_{15} y_3 y_4 + a_{16} y_1 y_4 x_2 + a_{17} y_1 y_3 y_4 + a_{18} x_1 y_3 + \\ & a_{19} y_1 y_3 + a_{20} \exp(a_{21} y_1).\end{aligned} \quad (D.35)$$

There are three constraints of the form $g(x) \leq 0$:

$$\begin{aligned} g_1 &= 1 - \frac{y_1}{700} \\ g_2 &= \frac{y_4}{25^2} - \frac{x_2}{5} \\ g_3 &= \frac{x_1}{500} - 0.11 - \left(\frac{x_2}{50} - 1\right)^2. \end{aligned} \quad (D.36)$$

The problem also has bound constraints. The original formulation is bounded from $[0, 80]$ in both dimensions, in which case the global optimum occurs in the corner at $x^* = [80, 80]$, with a local minimum in the middle. However, for our usage, we preferred the global optimum not to be in the corner and so set the bounds to $[0, 65]$ in both dimensions. The contour of this function is plotted in Fig. D.10.

Global minimum: $f(x^*) = -31.6368$ at $x^* = (49.5263, 19.6228)$.
Local minimum: $f(x) = -17.7754$ at $x = (65, 65)$.

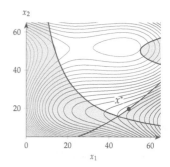

Fig. D.10 Barnes function.

D.2.2 Ten-Bar Truss

225. Venkayya, *Design of optimum structures*, 1971.

The 10-bar truss is a classic optimization problem.[225] In this problem, we want to find the optimal cross-sectional areas for the 10-bar truss shown in Fig. D.11. A simple truss finite-element code set up for this particular configuration is available in the book code repository. The function takes in an array of cross-sectional areas and returns the total mass and an array of stresses for each truss member.

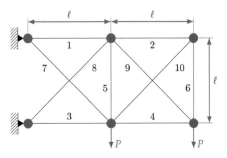

Fig. D.11 Ten-bar truss and element numbers.

The objective of the optimization is to minimize the mass of the structure, subject to the constraints that every segment does not yield in compression or tension. The yield stress of all elements is 25×10^3 psi, except for member 9, which uses a stronger alloy with a yield stress of 75×10^3 psi. Mathematically, the constraint is

$$|\sigma_i| \leq \sigma_{y_i} \quad \text{for } i = 1, \ldots, 10, \tag{D.37}$$

where the absolute value is needed to handle tension and compression (with the same yield strength for tension and compression). Absolute values are not differentiable at zero and should be avoided in gradient-based optimization if possible. Thus, we should put this in a mathematically equivalent form that avoids absolute value. Each element should have a cross-sectional area of at least 0.1 in² for manufacturing reasons (bound constraint). When solving this optimization problem, you may need to scale the objective and constraints.

Although not needed to solve the problem, an overview of the equations is provided. A truss element is the simplest type of structural finite element and only has an axial degree of freedom. The theory and derivation for truss elements are simple, but for our purposes, we skip to the result. Given a two-dimensional element oriented arbitrarily in space (Fig. D.12), we can relate the displacements at the nodes to the forces at the nodes through a stiffness relationship.

In matrix form, the equation for a given element is $K_e d = q$. In

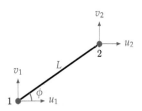

Fig. D.12 A truss element oriented at some angle ϕ, where ϕ is measured from a horizontal line emanating from the first node, oriented in the positive x direction.

D.2 Constrained Problems

detail, the equation is

$$\frac{EA}{L}\begin{bmatrix} c^2 & cs & -c^2 & -cs \\ cs & s^2 & -cs & -s^2 \\ -c^2 & -cs & c^2 & cs \\ -cs & -s^2 & cs & s^2 \end{bmatrix}\begin{bmatrix} u_1 \\ v_1 \\ u_2 \\ v_2 \end{bmatrix} = \begin{bmatrix} X_1 \\ Y_1 \\ X_2 \\ Y_2 \end{bmatrix} \quad (D.38)$$

where the displacement vector is $d = [u_1, v_1, u_2, v_2]$. The meanings of the variables in the equation are described in Table D.1.

Table D.1 The variables used in the stiffness equation.

Variable	Description
X_i	Force in the x-direction at node i
Y_i	Force in the y-direction at node i
E	Modulus of elasticity of truss element material
A	Area of truss element cross section
L	Length of truss element
c	$\cos\phi$
s	$\sin\phi$
u_i	Displacement in the x-direction at node i
v_i	Displacement in the y-direction at node i

The stress in the truss element can be computed from the equation $\sigma = S_e d$, where σ is a scalar, d is the same vector as before, and the element S_e matrix (really a row vector because stress is one-dimensional for truss elements) is

$$S_e = \frac{E}{L}\begin{bmatrix} -c & -s & c & s \end{bmatrix}. \quad (D.39)$$

The global structure (an assembly of multiple finite elements) has the same equations, $Kd = q$ and $\sigma = Sd$, but now d contains displacements for all of the nodes in the structure, $d = [x_1, x_2, \ldots, x_n]$. If we have n nodes and m elements, then q and d are $2n$-vectors, K is a $(2n \times 2n)$ matrix, S is an $(m \times 2n)$ matrix, and σ is an m-vector. The elemental stiffness and stress matrices are first computed and then assembled into the global matrices. This is straightforward because the displacements and forces of the individual elements add linearly.

After we assemble the global matrices, we must remove any degrees of freedom that are structurally rigid (already known to have zero displacement). Otherwise, the problem is ill-defined, and the stiffness matrix will be ill-conditioned.

Given the geometry, materials, and external loading, we can populate the stiffness matrix and force vector. We can then solve for the unknown displacements from

$$Kd = q. \qquad (D.40)$$

With the solved displacements, we can compute the stress in each element using

$$\sigma = Sd. \qquad (D.41)$$

Bibliography

1. Wu, N., Kenway, G., Mader, C. A., Jasa, J., and Martins, J. R. R. A., "PyOptSparse: A Python framework for large-scale constrained nonlinear optimization of sparse systems," *Journal of Open Source Software*, Vol. 5, No. 54, October 2020, p. 2564.
   DOI: 10.21105/joss.02564
   cited on pp. 15, 200

2. Lyu, Z., Kenway, G. K. W., and Martins, J. R. R. A., "Aerodynamic Shape Optimization Investigations of the Common Research Model Wing Benchmark," *AIAA Journal*, Vol. 53, No. 4, April 2015, pp. 968–985.
   DOI: 10.2514/1.J053318
   cited on p. 20

3. He, X., Li, J., Mader, C. A., Yildirim, A., and Martins, J. R. R. A., "Robust aerodynamic shape optimization—From a circle to an airfoil," *Aerospace Science and Technology*, Vol. 87, April 2019, pp. 48–61.
   DOI: 10.1016/j.ast.2019.01.051
   cited on p. 20

4. Betts, J. T., "Survey of numerical methods for trajectory optimization," *Journal of Guidance, Control, and Dynamics*, Vol. 21, No. 2, 1998, pp. 193–207.
   DOI: 10.2514/2.4231
   cited on p. 26

5. Bryson, A. E. and Ho, Y. C., *Applied Optimal Control; Optimization, Estimation, and Control*. Waltham, MA: Blaisdell Publishing, 1969.
   cited on p. 26

6. Bertsekas, D. P., *Dynamic Programming and Optimal Control*. Belmont, MA: Athena Scientific, 1995.
   cited on p. 26

7. Kepler, J., *Nova stereometria doliorum vinariorum (New Solid Geometry of Wine Barrels)*. Linz, Austria: Johannes Planck, 1615.
   cited on p. 34

8. Ferguson, T. S., "Who solved the secretary problem?" *Statistical Science*, Vol. 4, No. 3, August 1989, pp. 282–289.
   DOI: 10.1214/ss/1177012493
   cited on p. 34

9. Fermat, P. de, *Methodus ad disquirendam maximam et minimam (Method for the Study of Maxima and Minima)*. 1636, translated by Jason Ross.
   cited on p. 35

[10] Kollerstrom, N., "Thomas Simpson and 'Newton's method of approximation': An enduring myth," *The British Journal for the History of Science*, Vol. 25, No. 3, 1992, pp. 347–354. (cited on p. 35)

[11] Lagrange, J.-L., *Mécanique analytique*. Paris, France: Jacques Gabay, 1788, Vol. 1. (cited on p. 36)

[12] Cauchy, A.-L., "Méthode générale pour la résolution des systèmes d'équations simultanées," *Comptes rendus hebdomadaires des séances de l'Académie des sciences*, Vol. 25, October 1847, pp. 536–538. (cited on p. 36)

[13] Hancock, H., *Theory of Minima and Maxima*. Boston, MA: Ginn and Company, 1917. (cited on p. 36)

[14] Menger, K., "Das botenproblem," *Ergebnisse eines Mathematischen Kolloquiums*. Leipzig, Germany: Teubner, 1932, pp. 11–12. (cited on p. 36)

[15] Karush, W., "Minima of functions of several variables with inequalities as side constraints," Master's thesis, University of Chicago, Chicago, IL, 1939. (cited on p. 37)

[16] Dantzig, G., *Linear programming and extensions*. Princeton, NJ: Princeton University Press, 1998. ISBN: 0691059136 (cited on p. 37)

[17] Krige, D. G., "A statistical approach to some mine valuation and allied problems on the Witwatersrand," Master's thesis, University of the Witwatersrand, Johannesburg, South Africa, 1951. (cited on p. 37)

[18] Markowitz, H., "Portfolio selection," *Journal of Finance*, Vol. 7, March 1952, pp. 77–91. DOI: 10.2307/2975974 (cited on p. 38)

[19] Bellman, R., *Dynamic Programming*. Princeton, NJ: Princeton University Press, 1957. ISBN: 9780691146683 (cited on p. 38)

[20] Davidon, W. C., "Variable metric method for minimization," *SIAM Journal on Optimization*, Vol. 1, No. 1, February 1991, pp. 1–17, ISSN: 1095-7189. DOI: 10.1137/0801001 (cited on pp. 38, 126)

[21] Fletcher, R. and Powell, M. J. D., "A rapidly convergent descent method for minimization," *The Computer Journal*, Vol. 6, No. 2, August 1963, pp. 163–168, ISSN: 1460-2067. DOI: 10.1093/comjnl/6.2.163 (cited on pp. 38, 126)

[22] Wolfe, P., "Convergence conditions for ascent methods," *SIAM Review*, Vol. 11, No. 2, 1969, pp. 226–235. DOI: 10.1137/1011036 (cited on p. 38)

23 Wilson, R. B., "A simplicial algorithm for concave programming," PhD dissertation, Harvard University, Cambridge, MA, June 1963. — cited on p. 38

24 Han, S.-P., "Superlinearly convergent variable metric algorithms for general nonlinear programming problems," *Mathematical Programming*, Vol. 11, No. 1, 1976, pp. 263–282.
DOI: 10.1007/BF01580395 — cited on p. 38

25 Powell, M. J. D., "Algorithms for nonlinear constraints that use Lagrangian functions," *Mathematical Programming*, Vol. 14, No. 1, December 1978, pp. 224–248.
DOI: 10.1007/bf01588967 — cited on pp. 38, 200

26 Holland, J. H., *Adaptation in Natural and Artificial Systems*. Ann Arbor, MI: University of Michigan Press, 1975. — cited on p. 39

27 Hooke, R. and Jeeves, T. A., "'Direct search' solution of numerical and statistical problems," *Journal of the ACM*, Vol. 8, No. 2, 1961, pp. 212–229.
DOI: 10.1145/321062.321069 — cited on p. 39

28 Nelder, J. A. and Mead, R., "A simplex method for function minimization," *Computer Journal*, Vol. 7, 1965, pp. 308–313.
DOI: 10.1093/comjnl/7.4.308 — cited on pp. 39, 287

29 Karmarkar, N., "A new polynomial-time algorithm for linear programming," *Proceedings of the Sixteenth Annual ACM Symposium on Theory of Computing*. New York, NY: Association for Computing Machinery, 1984, pp. 302–311.
DOI: 10.1145/800057.808695 — cited on p. 39

30 Pontryagin, L. S., Boltyanskii, V. G., Gamkrelidze, R. V., and Mishchenko, E. F., *The Mathematical Theory of Optimal Processes*. New York, NY: Interscience Publishers, 1961, translated by K. N. Triruguf, edited by T. W. Neustadt. — cited on p. 39

31 Bryson Jr, A. E., "Optimal control—1950 to 1985," *IEEE Control Systems Magazine*, Vol. 16, No. 3, June 1996, pp. 26–33.
DOI: 10.1109/37.506395 — cited on p. 39

32 Schmit, L. A., "Structural design by systematic synthesis," *Proceedings of the 2nd National Conference on Electronic Computation*. New York, NY: American Society of Civil Engineers, 1960, pp. 105–122. — cited on pp. 40, 219

33 Schmit, L. A. and Thornton, W. A., "Synthesis of an airfoil at supersonic Mach number," CR 144, National Aeronautics and Space Administration, January 1965. — cited on p. 40

cited on p. 40   34   Fox, R. L., "Constraint surface normals for structural synthesis techniques," *AIAA Journal*, Vol. 3, No. 8, August 1965, pp. 1517–1518.
DOI: 10.2514/3.3182

cited on p. 40   35   Arora, J. and Haug, E. J., "Methods of design sensitivity analysis in structural optimization," *AIAA Journal*, Vol. 17, No. 9, 1979, pp. 970–974.
DOI: 10.2514/3.61260

cited on p. 40   36   Haftka, R. T. and Grandhi, R. V., "Structural shape optimization—A survey," *Computer Methods in Applied Mechanics and Engineering*, Vol. 57, No. 1, 1986, pp. 91–106, ISSN: 0045-7825.
DOI: 10.1016/0045-7825(86)90072-1

cited on p. 40   37   Eschenauer, H. A. and Olhoff, N., "Topology optimization of continuum structures: A review," *Applied Mechanics Reviews*, Vol. 54, No. 4, July 2001, pp. 331–390.
DOI: 10.1115/1.1388075

cited on p. 40   38   Pironneau, O., "On optimum design in fluid mechanics," *Journal of Fluid Mechanics*, Vol. 64, No. 01, 1974, p. 97, ISSN: 0022-1120.
DOI: 10.1017/S0022112074002023

cited on p. 40   39   Jameson, A., "Aerodynamic design via control theory," *Journal of Scientific Computing*, Vol. 3, No. 3, September 1988, pp. 233–260.
DOI: 10.1007/BF01061285

cited on p. 40   40   Sobieszczanski-Sobieski, J. and Haftka, R. T., "Multidisciplinary aerospace design optimization: Survey of recent developments," *Structural Optimization*, Vol. 14, No. 1, 1997, pp. 1–23.
DOI: 10.1007/BF011

cited on pp. 40, 520, 532, 534   41   Martins, J. R. R. A. and Lambe, A. B., "Multidisciplinary design optimization: A survey of architectures," *AIAA Journal*, Vol. 51, No. 9, September 2013, pp. 2049–2075.
DOI: 10.2514/1.J051895

cited on p. 40   42   Sobieszczanski-Sobieski, J., "Sensitivity of complex, internally coupled systems," *AIAA Journal*, Vol. 28, No. 1, 1990, pp. 153–160.
DOI: 10.2514/3.10366

cited on p. 41   43   Martins, J. R. R. A., Alonso, J. J., and Reuther, J. J., "A coupled-adjoint sensitivity analysis method for high-fidelity aero-structural design," *Optimization and Engineering*, Vol. 6, No. 1, March 2005, pp. 33–62.
DOI: 10.1023/B:OPTE.0000048536.47956.62

44 Hwang, J. T. and Martins, J. R. R. A., "A computational architecture for coupling heterogeneous numerical models and computing coupled derivatives," *ACM Transactions on Mathematical Software*, Vol. 44, No. 4, June 2018, Article 37.
DOI: 10.1145/3182393

cited on pp. 41, 498

45 Wright, M. H., "The interior-point revolution in optimization: History, recent developments, and lasting consequences," *Bulletin of the American Mathematical Society*, Vol. 42, 2005, pp. 39–56.
DOI: 10.1007/978-1-4613-3279-4_23

cited on p. 41

46 Grant, M., Boyd, S., and Ye, Y., "Disciplined convex programming," *Global Optimization—From Theory to Implementation*, Liberti, L. and Maculan, N., Eds. Boston, MA: Springer, 2006, pp. 155–210.
DOI: 10.1007/0-387-30528-9_7

cited on pp. 41, 430

47 Wengert, R. E., "A simple automatic derivative evaluation program," *Communications of the ACM*, Vol. 7, No. 8, August 1964, pp. 463–464, ISSN: 0001-0782.
DOI: 10.1145/355586.364791

cited on p. 41

48 Speelpenning, B., "Compiling fast partial derivatives of functions given by algorithms," PhD dissertation, University of Illinois at Urbana–Champaign, Champaign, IL, January 1980.
DOI: 10.2172/5254402

cited on p. 41

49 Squire, W. and Trapp, G., "Using complex variables to estimate derivatives of real functions," *SIAM Review*, Vol. 40, No. 1, 1998, pp. 110–112, ISSN: 0036-1445 (print), 1095-7200 (electronic).
DOI: 10.1137/S003614459631241X

cited on pp. 42, 232

50 Martins, J. R. R. A., Sturdza, P., and Alonso, J. J., "The complex-step derivative approximation," *ACM Transactions on Mathematical Software*, Vol. 29, No. 3, September 2003, pp. 245–262.
DOI: 10.1145/838250.838251

cited on pp. 42, 233, 235, 237

51 Torczon, V., "On the convergence of pattern search algorithms," *SIAM Journal on Optimization*, Vol. 7, No. 1, February 1997, pp. 1–25.
DOI: 10.1137/S1052623493250780

cited on p. 42

52 Jones, D., Perttunen, C., and Stuckman, B., "Lipschitzian optimization without the Lipschitz constant," *Journal of Optimization Theory and Application*, Vol. 79, No. 1, October 1993, pp. 157–181.
DOI: 10.1007/BF00941892

cited on pp. 42, 298

53 Jones, D. R. and Martins, J. R. R. A., "The DIRECT algorithm—25 years later," *Journal of Global Optimization*, Vol. 79, March 2021, pp. 521–566.
DOI: 10.1007/s10898-020-00952-6

cited on pp. 42, 298

cited on p. 42   54   Kirkpatrick, S., Gelatt, C. D., and Vecchi, M. P., "Optimization by simulated annealing," *Science*, Vol. 220, No. 4598, 1983, pp. 671–680.
DOI: 10.1126/science.220.4598.671

cited on p. 42   55   Kennedy, J. and Eberhart, R. C., "Particle swarm optimization," *Proceedings of the IEEE International Conference on Neural Networks*. Institute of Electrical and Electronics Engineers, 1995, Vol. IV, pp. 1942–1948.
DOI: 10.1007/978-0-387-30164-8_630

cited on p. 42   56   Forrester, A. I. and Keane, A. J., "Recent advances in surrogate-based optimization," *Progress in Aerospace Sciences*, Vol. 45, No. 1, 2009, pp. 50–79, ISSN: 0376-0421.
DOI: 10.1016/j.paerosci.2008.11.001

cited on p. 43   57   Bottou, L., Curtis, F. E., and Nocedal, J., "Optimization methods for large-scale machine learning," *SIAM Review*, Vol. 60, No. 2, 2018, pp. 223–311.
DOI: 10.1137/16M1080173

cited on pp. 43, 412   58   Baydin, A. G., Pearlmutter, B. A., Radul, A. A., and Siskind, J. M., "Automatic differentiation in machine learning: A survey," *Journal of Machine Learning Research*, Vol. 18, No. 1, January 2018, pp. 5595–5637.
DOI: 10.5555/3122009.3242010

cited on p. 43   59   Gerdes, P., "On mathematics in the history of sub-Saharan Africa," *Historia Mathematica*, Vol. 21, No. 3, 1994, pp. 345–376, ISSN: 0315-0860.
DOI: 10.1006/hmat.1994.1029

cited on p. 43   60   Closs, M. P., *Native American Mathematics*. Austin, TX: University of Texas Press, 1986.

cited on p. 43   61   Shen, K., Crossley, J. N., Lun, A. W.-C., and Liu, H., *The Nine Chapters on the Mathematical Art: Companion and Commentary*. Oxford University Press on Demand, 1999.

cited on p. 43   62   Hodgkin, L., *A History of Mathematics: From Mesopotamia to Modernity*. Oxford University Press on Demand, 2005.

cited on p. 43   63   Joseph, G. G., *The Crest of the Peacock: Non-European Roots of Mathematics*. Princeton, NJ: Princeton University Press, 2010.

cited on p. 44   64   Hollings, C., Martin, U., and Rice, A., *Ada Lovelace: The Making of a Computer Scientist*. Oxford, UK: Bodleian Library, 2014.

cited on p. 44   65   Osen, L. M., *Women in Mathematics*. Cambridge, MA: MIT Press, 1974.

66   Hodges, A., *Alan Turing: The Enigma*. Princeton, NJ: Princeton University Press, 2014.
ISBN: 9780691164724   cited on p. 44

67   Lipsitz, G., *How Racism Takes Place*. Philadelphia, PA: Temple University Press, 2011.   cited on p. 44

68   Rothstein, R., *The Color of Law: A Forgotten History of How Our Government Segregated America*. New York, NY: Liveright Publishing, 2017.   cited on p. 44

69   King, L. J., "More than slaves: Black founders, Benjamin Banneker, and critical intellectual agency," *Social Studies Research & Practice (Board of Trustees of the University of Alabama)*, Vol. 9, No. 3, 2014.   cited on p. 44

70   Shetterly, M. L., *Hidden Figures: The American Dream and the Untold Story of the Black Women Who Helped Win the Space Race*. New York, NY: William Morrow and Company, 2016.   cited on p. 45

71   Box, G. E. P., "Science and statistics," *Journal of the American Statistical Association*, Vol. 71, No. 356, 1976, pp. 791–799, ISSN: 01621459.
DOI: 10.2307/2286841   cited on p. 49

72   Wilson, G., Aruliah, D. A., Brown, C. T., Hong, N. P. C., Davis, M., Guy, R. T., Haddock, S. H. D., Huff, K. D., Mitchell, I. M., Plumbley, M. D., Waugh, B., White, E. P., and Wilson, P., "Best practices for scientific computing," *PLoS Biology*, Vol. 12, No. 1, 2014, e1001745.
DOI: 10.1371/journal.pbio.1001745   cited on p. 59

73   Grotker, T., Holtmann, U., Keding, H., and Wloka, M., *The Developer's Guide to Debugging*, 2nd ed. New York, NY: Springer, 2012.   cited on p. 60

74   Ascher, U. M. and Greif, C., *A First Course in Numerical Methods*. Philadelphia, PA: SIAM, 2011.   cited on p. 61

75   Saad, Y., *Iterative Methods for Sparse Linear Systems*, 2nd ed. Philadelphia, PA: SIAM, 2003.   cited on pp. 62, 569

76   Higgins, T. J., "A note on the history of mixed partial derivatives," *Scripta Mathematica*, Vol. 7, 1940, pp. 59–62.   cited on p. 85

77   Hager, W. W. and Zhang, H., "A new conjugate gradient method with guaranteed descent and an efficient line search," *SIAM Journal on Optimization*, Vol. 16, No. 1, January 2005, pp. 170–192, ISSN: 1095-7189.
DOI: 10.1137/030601880   cited on p. 99

cited on p. 103   78 Moré, J. J. and Thuente, D. J., "Line search algorithms with guaranteed sufficient decrease," *ACM Transactions on Mathematical Software (TOMS)*, Vol. 20, No. 3, 1994, pp. 286–307.
DOI: 10.1145/192115.192132

cited on pp. 103, 118, 141, 142, 190, 209, 567, 568   79 Nocedal, J. and Wright, S. J., *Numerical Optimization*, 2nd ed. Berlin: Springer, 2006.
DOI: 10.1007/978-0-387-40065-5

cited on p. 126   80 Broyden, C. G., "The convergence of a class of double-rank minimization algorithms 1. General considerations," *IMA Journal of Applied Mathematics*, Vol. 6, No. 1, 1970, pp. 76–90, ISSN: 1464-3634.
DOI: 10.1093/imamat/6.1.76

cited on p. 126   81 Fletcher, R., "A new approach to variable metric algorithms," *The Computer Journal*, Vol. 13, No. 3, March 1970, pp. 317–322, ISSN: 1460-2067.
DOI: 10.1093/comjnl/13.3.317

cited on p. 126   82 Goldfarb, D., "A family of variable-metric methods derived by variational means," *Mathematics of Computation*, Vol. 24, No. 109, January 1970, pp. 23–23, ISSN: 0025-5718.
DOI: 10.1090/s0025-5718-1970-0258249-6

cited on p. 126   83 Shanno, D. F., "Conditioning of quasi-Newton methods for function minimization," *Mathematics of Computation*, Vol. 24, No. 111, September 1970, pp. 647–647, ISSN: 0025-5718.
DOI: 10.1090/s0025-5718-1970-0274029-x

cited on pp. 140, 141, 142, 143   84 Conn, A. R., Gould, N. I. M., and Toint, P. L., *Trust Region Methods*. Philadelphia, PA: SIAM, 2000.
ISBN: 0898714605

cited on p. 141   85 Steihaug, T., "The conjugate gradient method and trust regions in large scale optimization," *SIAM Journal on Numerical Analysis*, Vol. 20, No. 3, June 1983, pp. 626–637, ISSN: 1095-7170.
DOI: 10.1137/0720042

cited on pp. 156, 425   86 Boyd, S. P. and Vandenberghe, L., *Convex Optimization*. Cambridge, UK: Cambridge University Press, March 2004.
ISBN: 0521833787

cited on pp. 156, 547   87 Strang, G., *Linear Algebra and its Applications*, 4th ed. Boston, MA: Cengage Learning, 2006.
ISBN: 0030105676

cited on p. 166   88 Dax, A., "Classroom note: An elementary proof of Farkas' lemma," *SIAM Review*, Vol. 39, No. 3, 1997, pp. 503–507.
DOI: 10.1137/S0036144594295502

89 Gill, P. E., Murray, W., Saunders, M. A., and Wright, M. H., "Some theoretical properties of an augmented Lagrangian merit function," SOL 86-6R, Systems Optimization Laboratory, September 1986. *cited on p. 183*

90 Di Pillo, G. and Grippo, L., "A new augmented Lagrangian function for inequality constraints in nonlinear programming problems," *Journal of Optimization Theory and Applications*, Vol. 36, No. 4, 1982, pp. 495–519
DOI: 10.1007/BF00940544 *cited on p. 183*

91 Birgin, E. G., Castillo, R. A., and Martínez, J. M., "Numerical comparison of augmented Lagrangian algorithms for nonconvex problems," *Computational Optimization and Applications*, Vol. 31, No. 1, 2005, pp. 31–55
DOI: 10.1007/s10589-005-1066-7 *cited on p. 183*

92 Rockafellar, R. T., "The multiplier method of Hestenes and Powell applied to convex programming," *Journal of Optimization Theory and Applications*, Vol. 12, No. 6, 1973, pp. 555–562
DOI: 10.1007/BF00934777 *cited on p. 183*

93 Murray, W., "Analytical expressions for the eigenvalues and eigenvectors of the Hessian matrices of barrier and penalty functions," *Journal of Optimization Theory and Applications*, Vol. 7, No. 3, March 1971, pp. 189–196.
DOI: 10.1007/bf00932477 *cited on p. 187*

94 Forsgren, A., Gill, P. E., and Wright, M. H., "Interior methods for nonlinear optimization," *SIAM Review*, Vol. 44, No. 4, January 2002, pp. 525–597.
DOI: 10.1137/s0036144502414942 *cited on p. 187*

95 Gill, P. E. and Wong, E., "Sequential quadratic programming methods," *Mixed Integer Nonlinear Programming*, Lee, J. and Leyffer, S., Eds., ser. The IMA Volumes in Mathematics and Its Applications. New York, NY: Springer, 2012, Vol. 154.
DOI: 10.1007/978-1-4614-1927-3_6 *cited on p. 190*

96 Gill, P. E., Murray, W., and Saunders, M. A., "SNOPT: An SQP algorithm for large-scale constrained optimization," *SIAM Review*, Vol. 47, No. 1, 2005, pp. 99–131.
DOI: 10.1137/S0036144504446096 *cited on pp. 190, 197, 200*

97 Fletcher, R. and Leyffer, S., "Nonlinear programming without a penalty function," *Mathematical Programming*, Vol. 91, No. 2, January 2002, pp. 239–269.
DOI: 10.1007/s101070100244 *cited on p. 198*

cited on p. 198   98   Benson, H. Y., Vanderbei, R. J., and Shanno, D. F., "Interior-point methods for nonconvex nonlinear programming: Filter methods and merit functions," *Computational Optimization and Applications*, Vol. 23, No. 2, 2002, pp. 257–272.
DOI: 10.1023/a:1020533003783

cited on p. 198   99   Fletcher, R., Leyffer, S., and Toint, P., "A brief history of filter methods," ANL/MCS-P1372-0906, Argonne National Laboratory, September 2006.

cited on p. 200   100   Fletcher, R., *Practical Methods of Optimization*, 2nd ed. Hoboken, NJ: Wiley, 1987.

cited on p. 200   101   Liu, D. C. and Nocedal, J., "On the limited memory BFGS method for large scale optimization," *Mathematical Programming*, Vol. 45, No. 1–3, August 1989, pp. 503–528.
DOI: 10.1007/bf01589116

cited on pp. 200, 208   102   Byrd, R. H., Nocedal, J., and Waltz, R. A., "Knitro: An integrated package for nonlinear optimization," *Large-Scale Nonlinear Optimization*, Di Pillo, G. and Roma, M., Eds. Boston, MA: Springer US, 2006, pp. 35–59.
DOI: 10.1007/0-387-30065-1_4

cited on p. 200   103   Kraft, D., "A software package for sequential quadratic programming," DFVLR-FB 88-28, DLR German Aerospace Center–Institute for Flight Mechanics, Koln, Germany, 1988.

cited on p. 208   104   Wächter, A. and Biegler, L. T., "On the implementation of an interior-point filter line-search algorithm for large-scale nonlinear programming," *Mathematical Programming*, Vol. 106, No. 1, April 2005, pp. 25–57.
DOI: 10.1007/s10107-004-0559-y

cited on p. 208   105   Byrd, R. H., Hribar, M. E., and Nocedal, J., "An interior point algorithm for large-scale nonlinear programming," *SIAM Journal on Optimization*, Vol. 9, No. 4, January 1999, pp. 877–900.
DOI: 10.1137/s1052623497325107

cited on p. 208   106   Wächter, A. and Biegler, L. T., "On the implementation of a primal-dual interior point filter line search algorithm for large-scale nonlinear programming," *Mathematical Programming*, Vol. 106, No. 1, 2006, pp. 25–57.

cited on p. 209   107   Gill, P. E., Saunders, M. A., and Wong, E., "On the performance of SQP methods for nonlinear optimization," *Modeling and Optimization: Theory and Applications*, Defourny, B. and Terlaky, T., Eds. New York, NY: Springer, 2015, Vol. 147, pp. 95–123.
DOI: 10.1007/978-3-319-23699-5_5

108   Kreisselmeier, G. and Steinhauser, R., "Systematic control design by optimizing a vector performance index," *IFAC Proceedings Volumes*, Vol. 12, No. 7, September 1979, pp. 113–117, ISSN: 1474-6670.
DOI: 10.1016/s1474-6670(17)65584-8 — cited on p. 212

109   Duysinx, P. and Bendsøe, M. P., "Topology optimization of continuum structures with local stress constraints," *International Journal for Numerical Methods in Engineering*, Vol. 43, 1998, pp. 1453–1478.
DOI: 10.1002/(SICI)1097-0207(19981230)43:8%3C1453::AID-NME480%3E3.0.CO;2-2 — cited on p. 213

110   Kennedy, G. J. and Hicken, J. E., "Improved constraint-aggregation methods," *Computer Methods in Applied Mechanics and Engineering*, Vol. 289, 2015, pp. 332–354, ISSN: 0045-7825.
DOI: 10.1016/j.cma.2015.02.017 — cited on p. 213

111   Hoerner, S. F., *Fluid-Dynamic Drag*. Bakersfield, CA: Hoerner Fluid Dynamics, 1965. — cited on p. 219

112   Lyness, J. N. and Moler, C. B., "Numerical differentiation of analytic functions," *SIAM Journal on Numerical Analysis*, Vol. 4, No. 2, 1967, pp. 202–210, ISSN: 0036-1429 (print), 1095-7170 (electronic).
DOI: 10.1137/0704019 — cited on p. 232

113   Lantoine, G., Russell, R. P., and Dargent, T., "Using multicomplex variables for automatic computation of high-order derivatives," *ACM Transactions on Mathematical Software*, Vol. 38, No. 3, April 2012, pp. 1–21, ISSN: 0098-3500.
DOI: 10.1145/2168773.2168774 — cited on p. 233

114   Fike, J. A. and Alonso, J. J., "Automatic differentiation through the use of hyper-dual numbers for second derivatives," *Recent Advances in Algorithmic Differentiation*, Forth, S., Hovland, P., Phipps, E., Utke, J., and Walther, A., Eds. Berlin: Springer, 2012, pp. 163–173, ISBN: 978-3-642-30023-3.
DOI: 10.1007/978-3-642-30023-3_15 — cited on p. 233

115   Griewank, A., *Evaluating Derivatives*. Philadelphia, PA: SIAM, 2000.
DOI: 10.1137/1.9780898717761 — cited on pp. 237, 247, 249

116   Naumann, U., *The Art of Differentiating Computer Programs—An Introduction to Algorithmic Differentiation*. Philadelphia, PA: SIAM, 2011. — cited on p. 237

117   Utke, J., Naumann, U., Fagan, M., Tallent, N., Strout, M., Heimbach, P., Hill, C., and Wunsch, C., "OpenAD/F: A modular open-source tool for automatic differentiation of Fortran codes," *ACM Transactions on Mathematical Software*, Vol. 34, No. 4, July 2008, ISSN: — cited on p. 250

0098-3500.
DOI: 10.1145/1377596.1377598

cited on p. 250   118   Hascoet, L. and Pascual, V., "The Tapenade automatic differentiation tool: Principles, model, and specification," *ACM Transactions on Mathematical Software*, Vol. 39, No. 3, May 2013, 20:1–20:43, ISSN: 0098-3500.
DOI: 10.1145/2450153.2450158

cited on p. 250   119   Griewank, A., Juedes, D., and Utke, J., "Algorithm 755: ADOL-C: A package for the automatic differentiation of algorithms written in C/C++," *ACM Transactions on Mathematical Software*, Vol. 22, No. 2, June 1996, pp. 131–167, ISSN: 0098-3500.
DOI: 10.1145/229473.229474

cited on p. 250   120   Wiltschko, A. B., Merriënboer, B. van, and Moldovan, D., "Tangent: Automatic differentiation using source code transformation in Python," arXiv:1711.02712, 2017.
URL: https://arxiv.org/abs/1711.02712.

cited on p. 250   121   Bradbury, J., Frostig, R., Hawkins, P., Johnson, M. J., Leary, C., Maclaurin, D., Necula, G., Paszke, A., VanderPlas, J., Wanderman-Milne, S., and Zhang, Q., "JAX: Composable Transformations of Python+NumPy Programs," 2018.
URL: http://github.com/google/jax.

cited on p. 250   122   Revels, J., Lubin, M., and Papamarkou, T., "Forward-mode automatic differentiation in Julia," arXiv:1607.07892, July 2016.
URL: https://arxiv.org/abs/1607.07892.

cited on p. 250   123   Neidinger, R. D., "Introduction to automatic differentiation and MATLAB object-oriented programming," *SIAM Review*, Vol. 52, No. 3, January 2010, pp. 545–563.
DOI: 10.1137/080743627

cited on p. 250   124   Betancourt, M., "A geometric theory of higher-order automatic differentiation," arXiv:1812.11592 [stat.CO], December 2018.
URL: https://arxiv.org/abs/1812.11592.

cited on pp. 251, 252   125   Giles, M., "An extended collection of matrix derivative results for forward and reverse mode algorithmic differentiation," Oxford, UK, January 2008.
URL: https://people.maths.ox.ac.uk/gilesm/files/NA-08-01.pdf.

cited on p. 252   126   Peter, J. E. V. and Dwight, R. P., "Numerical sensitivity analysis for aerodynamic optimization: A survey of approaches," *Computers and Fluids*, Vol. 39, No. 3, March 2010, pp. 373–391.
DOI: 10.1016/j.compfluid.2009.09.013

127 Martins, J. R. R. A., "Perspectives on aerodynamic design optimization," *Proceedings of the AIAA SciTech Forum*. American Institute of Aeronautics and Astronautics, January 2020.
DOI: 10.2514/6.2020-0043
cited on pp. 257, 444

128 Lambe, A. B., Martins, J. R. R. A., and Kennedy, G. J., "An evaluation of constraint aggregation strategies for wing box mass minimization," *Structural and Multidisciplinary Optimization*, Vol. 55, No. 1, January 2017, pp. 257–277.
DOI: 10.1007/s00158-016-1495-1
cited on p. 260

129 Kenway, G. K. W., Mader, C. A., He, P., and Martins, J. R. R. A., "Effective Adjoint Approaches for Computational Fluid Dynamics," *Progress in Aerospace Sciences*, Vol. 110, October 2019, p. 100 542.
DOI: 10.1016/j.paerosci.2019.05.002
cited on p. 260

130 Curtis, A. R., Powell, M. J. D., and Reid, J. K., "On the estimation of sparse Jacobian matrices," *IMA Journal of Applied Mathematics*, Vol. 13, No. 1, February 1974, pp. 117–119, ISSN: 1464-3634.
DOI: 10.1093/imamat/13.1.117
cited on p. 263

131 Gebremedhin, A. H., Manne, F., and Pothen, A., "What color is your Jacobian? Graph coloring for computing derivatives," *SIAM Review*, Vol. 47, No. 4, January 2005, pp. 629–705, ISSN: 1095-7200.
DOI: 10.1137/s0036144504444711
cited on p. 264

132 Gray, J. S., Hwang, J. T., Martins, J. R. R. A., Moore, K. T., and Naylor, B. A., "OpenMDAO: An open-source framework for multidisciplinary design, analysis, and optimization," *Structural and Multidisciplinary Optimization*, Vol. 59, No. 4, April 2019, pp. 1075–1104.
DOI: 10.1007/s00158-019-02211-z
cited on pp. 264, 494, 501, 508, 533

133 Ning, A., "Using blade element momentum methods with gradient-based design optimization," *Structural and Multidisciplinary Optimization*, May 2021
DOI: 10.1007/s00158-021-02883-6
cited on p. 265

134 Martins, J. R. R. A. and Hwang, J. T., "Review and unification of methods for computing derivatives of multidisciplinary computational models," *AIAA Journal*, Vol. 51, No. 11, November 2013, pp. 2582–2599.
DOI: 10.2514/1.J052184
cited on p. 266

135 Yu, Y., Lyu, Z., Xu, Z., and Martins, J. R. R. A., "On the influence of optimization algorithm and starting design on wing aerodynamic shape optimization," *Aerospace Science and Technology*, Vol. 75, April
cited on p. 283

2018, pp. 183–199.
DOI: 10.1016/j.ast.2018.01.016

cited on pp. 283, 284   136   Rios, L. M. and Sahinidis, N. V., "Derivative-free optimization: A review of algorithms and comparison of software implementations," *Journal of Global Optimization*, Vol. 56, 2013, pp. 1247–1293.
DOI: 10.1007/s10898-012-9951-y

cited on p. 285   137   Conn, A. R., Scheinberg, K., and Vicente, L. N., *Introduction to Derivative-Free Optimization*. Philadelphia, PA: SIAM, 2009.
DOI: 10.1137/1.9780898718768

cited on p. 285   138   Audet, C. and Hare, W., *Derivative-Free and Blackbox Optimization*. New York, NY: Springer, 2017.
DOI: 10.1007/978-3-319-68913-5

cited on p. 285   139   Kokkolaras, M., "When, why, and how can derivative-free optimization be useful to computational engineering design?" *Journal of Mechanical Design*, Vol. 142, No. 1, January 2020, p. 010301.
DOI: 10.1115/1.4045043

cited on pp. 286, 312   140   Simon, D., *Evolutionary Optimization Algorithms*. Hoboken, NJ: John Wiley & Sons, June 2013.
ISBN: 1118659503

cited on p. 297   141   Audet, C. and J. E. Dennis, J., "Mesh adaptive direct search algorithms for constrained optimization," *SIAM Journal on Optimization*, Vol. 17, No. 1, July 2006, pp. 188–217.
DOI: 10.1137/040603371

cited on p. 298   142   Le Digabel, S., "Algorithm 909: NOMAD: Nonlinear optimization with the MADS algorithm," *ACM Transactions on Mathematical Software*, Vol. 37, No. 4, 2011, pp. 1–15.
DOI: 10.1145/1916461.1916468

cited on pp. 298, 304   143   Jones, D. R., "Direct global optimization algorithm," *Encyclopedia of Optimization*, Floudas, C. A. and Pardalos, P. M., Eds. Boston, MA: Springer, 2009, pp. 725–735, ISBN: 978-0-387-74759-0.
DOI: 10.1007/978-0-387-74759-0_128

cited on p. 303   144   Jarvis, R. A., "On the identification of the convex hull of a finite set of points in the plane," *Information Processing Letters*, Vol. 2, No. 1, 1973, pp. 18–21.
DOI: 10.1016/0020-0190(73)90020-3

cited on pp. 304, 415   145   Jones, D. R., Schonlau, M., and Welch, W. J., "Efficient global optimization of expensive black-box functions," *Journal of Global Optimization*, Vol. 13, 1998, pp. 455–492.
DOI: 10.1023/A:1008306431147

146 Barricelli, N., "Esempi numerici di processi di evoluzione," *Methodos*, 1954, pp. 45–68. cited on p. 306

147 Jong, K. A. D., "An analysis of the behavior of a class of genetic adaptive systems," PhD dissertation, University of Michigan, Ann Arbor, MI, 1975. cited on p. 306

148 Deb, K., Pratap, A., Agarwal, S., and Meyarivan, T., "A fast and elitist multiobjective genetic algorithm: NSGA-II," *IEEE Transactions on Evolutionary Computation*, Vol. 6, No. 2, April 2002, pp. 182–197.
DOI: 10.1109/4235.996017 cited on pp. 308, 364

149 Deb, K., *Multi-Objective Optimization Using Evolutionary Algorithms*. Hoboken, NJ: John Wiley & Sons, 2001.
ISBN: 047187339X cited on p. 313

150 Eberhart, R. and Kennedy, J. A., "New optimizer using particle swarm theory," *Proceedings of the Sixth International Symposium on Micro Machine and Human Science*. Institute of Electrical and Electronics Engineers, 1995, pp. 39–43.
DOI: 10.1109/MHS.1995.494215 cited on p. 316

151 Zhan, Z.-H., Zhang, J., Li, Y., and Chung, H. S.-H., "Adaptive particle swarm optimization," *IEEE Transactions on Systems, Man, and Cybernetics, Part B (Cybernetics)*, Vol. 39, No. 6, April 2009, pp. 1362–1381.
DOI: 10.1109/TSMCB.2009.2015956 cited on p. 317

152 Gutin, G., Yeo, A., and Zverovich, A., "Traveling salesman should not be greedy: Domination analysis of greedy-type heuristics for the TSP," *Discrete Applied Mathematics*, Vol. 117, No. 1–3, March 2002, pp. 81–86, ISSN: 0166-218X.
DOI: 10.1016/s0166-218x(01)00195-0 cited on p. 338

153 Kirkpatrick, S., Gelatt, C. D., and Vecchi, M. P., "Optimization by simulated annealing," *Science*, Vol. 220, No. 4598, May 1983, pp. 671–680, ISSN: 1095-9203.
DOI: 10.1126/science.220.4598.671 cited on p. 347

154 Černý, V., "Thermodynamical approach to the traveling salesman problem: An efficient simulation algorithm," *Journal of Optimization Theory and Applications*, Vol. 45, No. 1, January 1985, pp. 41–51, ISSN: 1573-2878.
DOI: 10.1007/bf00940812 cited on p. 347

155 Metropolis, N., Rosenbluth, A. W., Rosenbluth, M. N., Teller, A. H., and Teller, E., "Equation of state calculations by fast computing machines," *Journal of Chemical Physics*, March 1953.
DOI: 10.2172/4390578 cited on p. 347

cited on p. 348 **156** Andresen, B. and Gordon, J. M., "Constant thermodynamic speed for minimizing entropy production in thermodynamic processes and simulated annealing," *Physical Review E*, Vol. 50, No. 6, December 1994, pp. 4346–4351, ISSN: 1095-3787.
DOI: 10.1103/physreve.50.4346

cited on p. 349 **157** Lin, S., "Computer solutions of the traveling salesman problem," *Bell System Technical Journal*, Vol. 44, No. 10, December 1965, pp. 2245–2269, ISSN: 0005-8580.
DOI: 10.1002/j.1538-7305.1965.tb04146.x

cited on p. 351 **158** Press, W. H., Wevers, J., Flannery, B. P., Teukolsky, S. A., Vetterling, W. T., Flannery, B. P., and Vetterling, W. T., *Numerical Recipes in C: The Art of Scientific Computing*. Cambridge, UK: Cambridge University Press, 1992.
ISBN: 0521431085

cited on p. 360 **159** Haimes, Y. Y., Lasdon, L. S., and Wismer, D. A., "On a bicriterion formulation of the problems of integrated system identification and system optimization," *IEEE Transactions on Systems, Man, and Cybernetics*, Vol. SMC-1, No. 3, July 1971, pp. 296–297.
DOI: 10.1109/tsmc.1971.4308298

cited on p. 360 **160** Das, I. and Dennis, J. E., "Normal-boundary intersection: A new method for generating the Pareto surface in nonlinear multicriteria optimization problems," *SIAM Journal on Optimization*, Vol. 8, No. 3, August 1998, pp. 631–657.
DOI: 10.1137/s1052623496307510

cited on p. 362 **161** Ismail-Yahaya, A. and Messac, A., "Effective generation of the Pareto frontier using the normal constraint method," *Proceedings of the 40th AIAA Aerospace Sciences Meeting & Exhibit*. American Institute of Aeronautics and Astronautics, January 2002.
DOI: 10.2514/6.2002-178

cited on p. 362 **162** Messac, A. and Mattson, C. A., "Normal constraint method with guarantee of even representation of complete Pareto frontier," *AIAA Journal*, Vol. 42, No. 10, October 2004, pp. 2101–2111.
DOI: 10.2514/1.8977

cited on p. 362 **163** Hancock, B. J. and Mattson, C. A., "The smart normal constraint method for directly generating a smart Pareto set," *Structural and Multidisciplinary Optimization*, Vol. 48, No. 4, June 2013, pp. 763–775.
DOI: 10.1007/s00158-013-0925-6

cited on p. 363 **164** Schaffer, J. D., "Some experiments in machine learning using vector evaluated genetic algorithms." PhD dissertation, Vanderbilt University, Nashville, TN, 1984.

165 Deb, K., "Introduction to evolutionary multiobjective optimization," *Multiobjective Optimization*. Berlin: Springer, 2008, pp. 59–96.
DOI: 10.1007/978-3-540-88908-3_3
   cited on p. 364

166 Kung, H. T., Luccio, F., and Preparata, F. P., "On finding the maxima of a set of vectors," *Journal of the ACM*, Vol. 22, No. 4, October 1975, pp. 469–476.
DOI: 10.1145/321906.321910
   cited on p. 364

167 Faure, H., "Discrépance des suites associées à un systéme de numération (en dimension s)." *Acta Arithmetica*, Vol. 41, 1982, pp. 337–351.
DOI: 10.4064/aa-41-4-337-351
   cited on p. 384

168 Faure, H. and Lemieux, C., "Generalized Halton sequences in 2008: A comparative study," *ACM Transactions on Modeling and Computer Simulation*, Vol. 19, No. 4, October 2009, pp. 1–31.
DOI: 10.1145/1596519.1596520
   cited on p. 384

169 Sobol, I. M., "On the distribution of points in a cube and the approximate evaluation of integrals," *USSR Computational Mathematics and Mathematical Physics*, Vol. 7, No. 4, 1967, pp. 86–112.
DOI: 10.1016/0041-5553(67)90144-9
   cited on p. 384

170 Niederreiter, H., "Low-discrepancy and low-dispersion sequences," *Journal of Number Theory*, Vol. 30, No. 1, 1988, pp. 51–70.
DOI: 10.1016/0022-314X(88)90025-X
   cited on p. 384

171 Bouhlel, M. A., Hwang, J. T., Bartoli, N., Lafage, R., Morlier, J., and Martins, J. R. R. A., "A Python surrogate modeling framework with derivatives," *Advances in Engineering Software*, 2019, p. 102 662, ISSN: 0965-9978.
DOI: https://doi.org/10.1016/j.advengsoft.2019.03.005
   cited on p. 399

172 Bouhlel, M. A. and Martins, J. R. R. A., "Gradient-enhanced kriging for high-dimensional problems," *Engineering with Computers*, Vol. 1, No. 35, January 2019, pp. 157–173.
DOI: 10.1007/s00366-018-0590-x
   cited on pp. 399, 408

173 Jones, D. R., "A taxonomy of global optimization methods based on response surfaces," *Journal of Global Optimization*, Vol. 21, 2001, pp. 345–383.
DOI: 10.1023/A:1012771025575
   cited on p. 403

174 Sacks, J., Welch, W. J., Mitchell, T. J., and Wynn, H. P., "Design and analysis of computer experiments," *Statistical Science*, Vol. 4, No. 4, 1989, pp. 409–423, ISSN: 08834237.
DOI: 10.2307/2245858
   cited on p. 404

cited on p. 408 175 Han, Z.-H., Zhang, Y., Song, C.-X., and Zhang, K.-S., "Weighted gradient-enhanced kriging for high-dimensional surrogate modeling and design optimization," *AIAA Journal*, Vol. 55, No. 12, August 2017, pp. 4330–4346.
DOI: 10.2514/1.J055842

cited on p. 408 176 Forrester, A., Sobester, A., and Keane, A., *Engineering Design via Surrogate Modelling: A Practical Guide*. Hoboken, NJ: John Wiley & Sons, 2008.
ISBN: 0470770791

cited on p. 414 177 Ruder, S., "An overview of gradient descent optimization algorithms," arXiv:1609.04747, 2016.
URL: http://arxiv.org/abs/1609.04747.

cited on p. 414 178 Goh, G., "Why momentum really works," *Distill*, 2017.
DOI: 10.23915/distill.00006

cited on p. 423 179 Diamond, S. and Boyd, S., "Convex optimization with abstract linear operators," *Proceedings of the 2015 IEEE International Conference on Computer Vision (ICCV)*. Institute of Electrical and Electronics Engineers, December 2015.
DOI: 10.1109/iccv.2015.84

cited on p. 425 180 Lobo, M. S., Vandenberghe, L., Boyd, S., and Lebret, H., "Applications of second-order cone programming," *Linear Algebra and Its Applications*, Vol. 284, No. 1–3, November 1998, pp. 193–228.
DOI: 10.1016/s0024-3795(98)10032-0

cited on p. 425 181 Parikh, N. and Boyd, S., "Block splitting for distributed optimization," *Mathematical Programming Computation*, Vol. 6, No. 1, October 2013, pp. 77–102.
DOI: 10.1007/s12532-013-0061-8

cited on p. 425 182 Vandenberghe, L. and Boyd, S., "Semidefinite programming," *SIAM Review*, Vol. 38, No. 1, March 1996, pp. 49–95.
DOI: 10.1137/1038003

cited on p. 425 183 Vandenberghe, L. and Boyd, S., "Applications of semidefinite programming," *Applied Numerical Mathematics*, Vol. 29, No. 3, March 1999, pp. 283–299.
DOI: 10.1016/s0168-9274(98)00098-1

cited on p. 436 184 Boyd, S., Kim, S.-J., Vandenberghe, L., and Hassibi, A., "A tutorial on geometric programming," *Optimization and Engineering*, Vol. 8, No. 1, April 2007, pp. 67–127.
DOI: 10.1007/s11081-007-9001-7

185 Hoburg, W., Kirschen, P., and Abbeel, P., "Data fitting with geometric-programming-compatible softmax functions," *Optimization and Engineering*, Vol. 17, No. 4, August 2016, pp. 897–918.
DOI: 10.1007/s11081-016-9332-3

cited on p. 436

186 Kirschen, P. G., York, M. A., Ozturk, B., and Hoburg, W. W., "Application of signomial programming to aircraft design," *Journal of Aircraft*, Vol. 55, No. 3, May 2018, pp. 965–987.
DOI: 10.2514/1.c034378

cited on p. 437

187 York, M. A., Hoburg, W. W., and Drela, M., "Turbofan engine sizing and tradeoff analysis via signomial programming," *Journal of Aircraft*, Vol. 55, No. 3, May 2018, pp. 988–1003.
DOI: 10.2514/1.c034463

cited on p. 437

188 Stanley, A. P. and Ning, A., "Coupled wind turbine design and layout optimization with non-homogeneous wind turbines," *Wind Energy Science*, Vol. 4, No. 1, January 2019, pp. 99–114.
DOI: 10.5194/wes-4-99-2019

cited on p. 444

189 Gagakuma, B., Stanley, A. P. J., and Ning, A., "Reducing wind farm power variance from wind direction using wind farm layout optimization," *Wind Engineering*, January 2021.
DOI: 10.1177/0309524X20988288

cited on p. 444

190 Padrón, A. S., Thomas, J., Stanley, A. P. J., Alonso, J. J., and Ning, A., "Polynomial chaos to efficiently compute the annual energy production in wind farm layout optimization," *Wind Energy Science*, Vol. 4, May 2019, pp. 211–231.
DOI: 10.5194/wes-4-211-2019

cited on p. 444

191 Cacuci, D., *Sensitivity & Uncertainty Analysis*. Boca Raton, FL: Chapman and Hall/CRC, May 2003, Vol. 1.
DOI: 10.1201/9780203498798

cited on p. 450

192 Parkinson, A., Sorensen, C., and Pourhassan, N., "A general approach for robust optimal design," *Journal of Mechanical Design*, Vol. 115, No. 1, 1993, p. 74.
DOI: 10.1115/1.2919328

cited on p. 450

193 Golub, G. H. and Welsch, J. H., "Calculation of Gauss quadrature rules," *Mathematics of Computation*, Vol. 23, No. 106, 1969, pp. 221–230, ISSN: 00255718, 10886842.
DOI: 10.1090/S0025-5718-69-99647-1

cited on p. 455

194 Wilhelmsen, D. R., "Optimal quadrature for periodic analytic functions," *SIAM Journal on Numerical Analysis*, Vol. 15, No. 2, 1978, pp. 291–296, ISSN: 00361429.
DOI: 10.1137/0715020

cited on p. 457

cited on p. 457   195   Trefethen, L. N. and Weideman, J. A. C., "The exponentially convergent trapezoidal rule," *SIAM Review*, Vol. 56, No. 3, 2014, pp. 385–458, ISSN: 00361445, 10957200.
DOI: 10.1137/130932132

cited on p. 457   196   Johnson, S. G., "Notes on the convergence of trapezoidal-rule quadrature," March 2010.
URL: http://math.mit.edu/~stevenj/trapezoidal.pdf.

cited on p. 458   197   Smolyak, S. A., "Quadrature and interpolation formulas for tensor products of certain classes of functions," *Proceedings of the USSR Academy of Sciences*, 5. 1963, Vol. 148, pp. 1042–1045.
DOI: 10.3103/S1066369X10030084

cited on p. 462   198   Wiener, N., "The homogeneous chaos," *American Journal of Mathematics*, Vol. 60, No. 4, October 1938, p. 897.
DOI: 10.2307/2371268

cited on p. 465   199   Eldred, M., Webster, C., and Constantine, P., "Evaluation of non-intrusive approaches for wiener–askey generalized polynomial chaos," *Proceedings of the 49th AIAA Structures, Structural Dynamics, and Materials Conference*. American Institute of Aeronautics and Astronautics, April 2008.
DOI: 10.2514/6.2008-1892

cited on p. 466   200   Adams, B. M., Bohnhoff, W. J., Dalbey, K. R., Ebeida, M. S., Eddy, J. P., Eldred, M. S., Hooper, R. W., Hough, P. D., Hu, K. T., Jakeman, J. D., Khalil, M., Maupin, K. A., Monschke, J. A., Ridgway, E. M., Rushdi, A. A., Seidl, D. T., Stephens, J. A., Swiler, L. P., and Winokur, J. G., "Dakota, a multilevel parallel object-oriented framework for design optimization, parameter estimation, uncertainty quantification, and sensitivity analysis: Version 6.14 user's manual," May 2021.
URL: https://dakota.sandia.gov/content/manuals.

cited on p. 466   201   Feinberg, J. and Langtangen, H. P., "Chaospy: An open source tool for designing methods of uncertainty quantification," *Journal of Computational Science*, Vol. 11, November 2015, pp. 46–57.
DOI: 10.1016/j.jocs.2015.08.008

cited on pp. 481, 498   202   Jasa, J. P., Hwang, J. T., and Martins, J. R. R. A., "Open-source coupled aerostructural optimization using Python," *Structural and Multidisciplinary Optimization*, Vol. 57, No. 4, April 2018, pp. 1815–1827.
DOI: 10.1007/s00158-018-1912-8

203 Cuthill, E. and McKee, J., "Reducing the bandwidth of sparse symmetric matrices," *Proceedings of the 1969 24th National Conference.* New York, NY: Association for Computing Machinery, 1969, pp. 157–172.
DOI: 10.1145/800195.805928
cited on p. 486

204 Amestoy, P. R., Davis, T. A., and Duff, I. S., "An approximate minimum degree ordering algorithm," *SIAM Journal on Matrix Analysis and Applications*, Vol. 17, No. 4, 1996, pp. 886–905.
DOI: 10.1137/S0895479894278952
cited on p. 486

205 Lambe, A. B. and Martins, J. R. R. A., "Extensions to the design structure matrix for the description of multidisciplinary design, analysis, and optimization processes," *Structural and Multidisciplinary Optimization*, Vol. 46, August 2012, pp. 273–284.
DOI: 10.1007/s00158-012-0763-y
cited on p. 486

206 Irons, B. M. and Tuck, R. C., "A version of the Aitken accelerator for computer iteration," *International Journal for Numerical Methods in Engineering*, Vol. 1, No. 3, 1969, pp. 275–277.
DOI: 10.1002/nme.1620010306
cited on p. 490

207 Kenway, G. K. W., Kennedy, G. J., and Martins, J. R. R. A., "Scalable parallel approach for high-fidelity steady-state aeroelastic analysis and derivative computations," *AIAA Journal*, Vol. 52, No. 5, May 2014, pp. 935–951.
DOI: 10.2514/1.J052255
cited on pp. 490, 504

208 Chauhan, S. S., Hwang, J. T., and Martins, J. R. R. A., "An automated selection algorithm for nonlinear solvers in MDO," *Structural and Multidisciplinary Optimization*, Vol. 58, No. 2, June 2018, pp. 349–377.
DOI: 10.1007/s00158-018-2004-5
cited on p. 490

209 Kenway, G. K. W. and Martins, J. R. R. A., "Multipoint high-fidelity aerostructural optimization of a transport aircraft configuration," *Journal of Aircraft*, Vol. 51, No. 1, January 2014, pp. 144–160.
DOI: 10.2514/1.C032150
cited on p. 504

210 Hwang, J. T., Lee, D. Y., Cutler, J. W., and Martins, J. R. R. A., "Large-scale multidisciplinary optimization of a small satellite's design and operation," *Journal of Spacecraft and Rockets*, Vol. 51, No. 5, September 2014, pp. 1648–1663.
DOI: 10.2514/1.A32751
cited on p. 512

211 Biegler, L. T., Ghattas, O., Heinkenschloss, M., and Bloemen Waanders, B. van, Eds., *Large-Scale PDE-Constrained Optimization.* Berlin: Springer, 2003.
cited on p. 517

cited on pp. 521, 522   212   Braun, R. D. and Kroo, I. M., "Development and application of the collaborative optimization architecture in a multidisciplinary design environment," *Multidisciplinary Design Optimization: State of the Art*, Alexandrov, N. and Hussaini, M. Y., Eds. Philadelphia, PA: SIAM, 1997, pp. 98–116.
DOI: 10.5555/888020

cited on p. 524   213   Kim, H. M., Rideout, D. G., Papalambros, P. Y., and Stein, J. L., "Analytical target cascading in automotive vehicle design," *Journal of Mechanical Design*, Vol. 125, No. 3, September 2003, pp. 481–490.
DOI: 10.1115/1.1586308

cited on p. 524   214   Tosserams, S., Etman, L. F. P., Papalambros, P. Y., and Rooda, J. E., "An augmented Lagrangian relaxation for analytical target cascading using the alternating direction method of multipliers," *Structural and Multidisciplinary Optimization*, Vol. 31, No. 3, March 2006, pp. 176–189.
DOI: 10.1007/s00158-005-0579-0

cited on p. 524   215   Talgorn, B. and Kokkolaras, M., "Compact implementation of non-hierarchical analytical target cascading for coordinating distributed multidisciplinary design optimization problems," *Structural and Multidisciplinary Optimization*, Vol. 56, No. 6, 2017, pp. 1597–1602
DOI: 10.1007/s00158-017-1726-0

cited on p. 527   216   Sobieszczanski–Sobieski, J., Altus, T. D., Phillips, M., and Sandusky, R., "Bilevel integrated system synthesis for concurrent and distributed processing," *AIAA Journal*, Vol. 41, No. 10, 2003, pp. 1996–2003.
DOI: 10.2514/2.1889

cited on p. 533   217   Tedford, N. P. and Martins, J. R. R. A., "Benchmarking multidisciplinary design optimization algorithms," *Optimization and Engineering*, Vol. 11, No. 1, February 2010, pp. 159–183.
DOI: 10.1007/s11081-009-9082-6

cited on p. 534   218   Golovidov, O., Kodiyalam, S., Marineau, P., Wang, L., and Rohl, P., "Flexible implementation of approximation concepts in an MDO framework," *Proceedings of the 7th AIAA/USAF/NASA/ISSMO Symposium on Multidisciplinary Analysis and Optimization*. American Institute of Aeronautics and Astronautics, 1998.
DOI: 10.2514/6.1998-4959

cited on p. 534   219   Balabanov, V., Charpentier, C., Ghosh, D. K., Quinn, G., Vanderplaats, G., and Venter, G., "Visualdoc: A software system for general purpose integration and design optimization," *Proceedings of the 9th AIAA/ISSMO Symposium on Multidisciplinary Analysis and Optimiza-*

tion. American Institute of Aeronautics and Astronautics, 2002.
DOI: 10.2514/6.2002-5513

220 Trefethen, L. N. and Bau III, D., *Numerical Linear Algebra*. Philadelphia, PA: SIAM, 1997.
ISBN: 0898713617
cited on pp. 559, 569

221 Saad, Y. and Schultz, M. H., "GMRES: A generalized minimal residual algorithm for solving nonsymmetric linear systems," *SIAM Journal on Scientific and Statistical Computing*, Vol. 7, No. 3, 1986, pp. 856–869.
DOI: 10.1137/0907058
cited on p. 570

222 Broyden, C. G., "A class of methods for solving nonlinear simultaneous equations," *Mathematics of Computation*, Vol. 19, No. 92, October 1965, pp. 577–593.
DOI: 10.1090/S0025-5718-1965-0198670-6
cited on p. 571

223 Rosenbrock, H. H., "An automatic method for finding the greatest or least value of a function," *The Computer Journal*, Vol. 3, No. 3, January 1960, pp. 175–184, ISSN: 0010-4620.
DOI: 10.1093/comjnl/3.3.175
cited on p. 579

224 Barnes, G. K., "A comparative study of nonlinear optimization codes," Master's thesis, University of Texas at Austin, 1967.
cited on p. 586

225 Venkayya, V., "Design of optimum structures," *Computers & Structures*, Vol. 1, No. 1–2, August 1971, pp. 265–309, ISSN: 0045-7949.
DOI: 10.1016/0045-7949(71)90013-7
cited on p. 588

Index

absolute value function
- complex-step method, 236
- smoothing, 145

accuracy, 49

activation functions, 410
- rectified linear unit (ReLU), 410
- sigmoid, 410

active constraints, 165, 190

active-set method, 190

acyclic graph, 486

adjoint method, 39, 40, 255
- AD partial derivatives, 260
- constraint aggregation, 211, 260
- coupled, 504
- equations, 256
- structural problem, 259
- variables, 243, 256
- vector, 256
- verification, 262

aerodynamic shape optimization, 20, 40, 283, 444

aerostructural
- analysis, 477
- model, 481

affine function, 424, 435

aggregation functions
- p-norm, 213
- induced exponential, 213
- induced power, 214
- Kreisselmeier–Steinhauser (KS), 212

aircraft fuel tank problem, 218

airfoil optimization, 444

Aitken acceleration, 490, 491, 498

algorithmic differentiation (AD), 41, 225, 237, 497
- adjoint variables, 243
- checkpointing, 247
- computational cost, 246
- computational graph, 242
- connection to complex-step method, 249
- coupled systems, 502
- directional derivative, 242
- forward mode, 238, 239, 502
- forward vs. reverse, 246
- matrix operations, 251
- operator overloading, 247, 248, 250
- partial derivatives, 260
- reverse mode, 43, 239, 243, 412, 502
- scaling, 246
- seed, 243, 246
- shortcuts, 251
- software, 250
- source code transformation, 247, 250
- taping, 249
- verification, 262

analysis, 3, 6, 70

analytic
- function, 232
- methods, *see* implicit analytic methods

analytical target cascading (ATC), 524
anchor points, 360
approximate Hessian, 124, 572
approximate minimum degree (AMD) ordering, 486
Armijo condition, *see* sufficient decrease condition
artificial intelligence (AI), 39, 43
artificial minimum, 282
asymmetric subspace optimization (ASO), 529
asymptotic error constant, 63
augmented Lagrangian method, 176, 181, 198, 315
automatic differentiation, *see* algorithmic differentiation (AD)

back substitution, 245, 561
backpropagation, 43, *see also* algorithmic differentiation (AD)
backtracking, 99, 100, 185
backward difference, 228
banana function, *see* Rosenbrock function
Barnes problem, 447, 586
barrier methods, *see* interior penalty methods
basis functions, 384, 385, 398, 462
    Gaussian, 399
    radial, *see* radial basis function (RBF)
bean function, 98, 113, 119, 122, 130, 133, 291, 314, 319, 580
Bellman
    equation, 38, 342
    principle of optimality, 342
benchmarking, 320

gradient-based algorithms, 133, 135
gradient-free algorithms, 284
MDO architectures, 533
stochastic algorithms, 286
BFGS method, 38, 126, 129, 550, 572, 574, 576
    damped, 200
    derivation, 126
    Hessian reset, 129
    limited memory, *see* L-BFGS method
    SQP, 199
    update, 128
bilevel integrated system synthesis (BLISS), 527
binary
    decoding, 308
    encoding, 308
    representation, 308
    variables, 327, 331
binding-direction method, 191
biological reproduction, 307
bisection, 108, 110
black-box model, 18, 227, 232, 281, 484, 501, 519
    derivatives, 225
    solver, 507
blocking constraint, 194
Boltzmann distribution, 347
bound constraints, 7, 154, 155, 319
    artificial, 156
brachistochrone problem, 35, 152, 583
bracketing, 103
branch-and-bound method, 285, 330, 333
    integer variables, 336
    relaxation, 331
breadth-first tree search, 333
Broyden's method, 69, 496, 571

bugs, 58

calculus of variations, 34–36
callback functions, 15
Cauchy–Riemann equations, 233
ceiling, 320
central difference, 228, 230
central limit theorem, 389
chain rule, 237, 238, 541
    forward, 239
    multivariable, 542
    reverse, 243
characteristic equation, 554
checkerboard pattern, 320
checkpointing, 247
Cholesky factorization, 61, 561
chromosome, 307, 311
    encoding, 308
classification
    convex problems, 425
    gradient-free algorithms, 284
    MDO architectures, 534
    optimization algorithms, 21
    optimization problems, 17
    problem, 432
    stationary points, 92
Clenshaw–Curtis quadrature, 457
collaborative optimization (CO), 521
collocation points, 466
column space, 547
combinatorial optimization, 36, 37, *see also* discrete optimization
complementary slackness condition, 168, 190
complex-step method, 42, 225, 232, 233, 502
    absolute value function, 236
    accuracy, 234
    connection to AD, 249
    implementation, 235
    step size, 234
    testing, 237
    trigonometric functions, 237
component, 269, 475, 478, 480, 487, 488
    explicit, 480, 503, 509
    group, 483, 487
    implicit, 480, 503
    multiple, 225
composite function, 541
computational cost, 50, 63, 374
    AD, 237, 247
    adjoint method, 257
    analysis, 60
    budget, 93, 315
    complex step, 232, 233
    derivatives, 223
    direct method, 256
    direct vs. adjoint, 257
    finite difference, 228
    forward AD, 240
    linear solvers, 62
    optimization, 22, 48, 283
    reverse AD, 244
    solvers, 12, 61, 253
computational differentiation, *see* algorithmic differentiation (AD)
computational fluid dynamics (CFD), 40, 283, 444, 531
computational graph, 242, 246
computer code, *see* source code
conceptual design, 3
concurrent subspace optimization (CSSO), 532
condition number, 560, 570
cone programming, 425
confidence interval, 404
conjugacy, 116, 117, 565
conjugate gradient method, 115, 117, 565
    Fletcher–Reeves formula, 117

linear, 115, 565, 568
nonlinear, 117, 119
Polak–Ribière formula, 118
reset, 118, 119
consistency constraints, 515, 521, 525
constrained optimization, 36, 153
    graphical solution, 154
    problem statement, 154
constraint qualification, 169, 523
constraints, 12, 314
    active, 13, 165, 190
    aggregation, 204, 211, 260
    blocking, 194
    bound, 7, 154
    consistency, 515, 521, 525
    equality, 12, 154
    equality versus inequality, 154, 194
    functions, 12
    handling, 153
    inactive, 13, 165, 190
    inequality, 12, 154
    infeasible, 13
    Jacobian, 188
    reformulation, 197
    scaling, 181
    working set, 190
continuity, 18
continuous parameterization, 330
contour
    perpendicular, 81
    tangent, 157
control law, 429
convergence, 315
    criterion, 237
    failure, 48, 96, 138, 274
    order of, 64, 66
    plot, 66
    quadratic, 121
    rate, 63, *see* rate of convergence, 287
    residuals, 66
    tolerance, 96, 138, 225
convex
    function, 20, 424
    hull, 302, 303
    optimization, 20, 41, 423
    problem, 389
convexity, 20, 27
coordinate
    descent, 476
    search, 39, 115, 116
    search algorithm, 292
correlation, 400, 558
    matrix, 406
coupled
    adjoint, 506
    Broyden method, 499
    derivatives, 509
    model, 478
    Newton's method, 499
    solver, 500
    system, 484, 488
coupling variables, 478, 482, 484, 486, 487, 489, 492, 506, 515
covariance, 401, 449, 450, 558
    matrix, 450
cross validation, 395, 397, 412
    k-fold, 397
    leave-one-out, 397
    simple, 396
crossover, 307, 311, 313
    linear, 313
    point, 311
    single-point, 311
crowded tournament selection, 368
crowding distance, 366
cubature, 457
cubic interpolation, 109, 145
cuboid, 366

cumulative distribution function (CDF), 379, 441, 556
curse of dimensionality, 374, 376, 458
curvature, 22, 85, 112, 119
    approximation, 123
    condition, 126, 573
    directional, 86, 124
    maximum, 87
    principal directions, 86, 116
curve fit, 373
Cuthill–McKee ordering, 486
CVX, 432

damped BFGS update, 200
data
    dependencies, 486
    fitting, 385
    model, 384
    transfer, 343, 486, 487
debugging, 59
decaying sinusoid, 404
decision variables, *see* design variables, 327
decomposition, *see* factorization
deep neural networks, 43, 409
dense matrix, 559
dependence
    implicit, 253
dependency structure matrix, *see* design structure matrix (DSM)
depth-first tree search, 332
derivative-free optimization (DFO), 24, 285
derivatives, 223
    accuracy, 138, 275
    backward propagation, 245
    black box, 225
    computational cost, 275
    coupled, 41, 501, 503, 509
    definition, 228

    directional, 82, 98, 229, 242
    ease of implementation, 275
    eigenvalues, 252
    eigenvectors, 252
    explicit, 266
    first-order, 223
    implicit, 266
    implicit analytic methods, 252
    matrix, 552
    matrix operations, 251, 252
    methods, 225, 275
    mixed partial, 85
    partial, 80, 239, 254, 260, 266, 504, 542
    physical interpretation, 82
    post-optimality, 530, 532
    propagation, 238, 239, 242, 243, 253
    relative, 82
    scalability, 275
    scaling, 274
    second-order, 85
    singular value decomposition, 252
    sparse, 262, 264
    total, 239, 254, 542
    verification, 242, 274
    weighted function, 246
descent direction, 98, 165
design
    constraints, 12
    cycle, 3
    optimal vs. conventional, 4, 5
    optimization, 3
    phases, 2
    process, 2
    sensitivities, *see* derivatives
    space visualization, 10, 12, 14
    specifications, 3

design structure matrix (DSM), 485, 504
design variables, 6, 253, 479
    binary, 327, 331
    bounds, 154, 155, 319
    continuous, 7, 17
    converting integer to binary, 331
    discrete, 8, 17, 27, 283, 312, 327
    integer, 312, 327, 336
    mixed, 17
    parameterization, 9, 330
    scalability, 22, 283
    scaling, 114, 137
    shared, 476
    units, 114
detailed design, 3
determinant, 551
deterministic function, 20
DFP method, 38, 126, 573, 576
    update, 573
diagonal matrix, 550
Dido's problem, 33
differential, 174, 254, 266, 543
differentiation, *see also* derivatives
    algorithmic, *see* algorithmic differentiation (AD)
    chain rule, *see* chain rule
    numerical, 227
    symbolic, *see* symbolic differentiation
DIRECT algorithm, 42, 285, 298
    n-dimensional, 304, 305
    one-dimensional, 301
direct linear solver, 560
direct method, 255
    coupled, 503
    structural problem, 259
    verification, 262
direct quadrature, 448, 452

directed graph, 486
    weighted, 338
directional
    curvature, 86, 124
    derivative, 82, 98, 229, 242
disciplinary subproblem, 519, 521
discipline, 475, 480
disciplined convex optimization, 424, 430
    software, 432
discontinuity, 7, 19, 231, 282, 320, 321
    smoothing, 145
discrete optimization, 26, 36, 327
    dynamic programming, 339
    dynamic rounding, 329
    genetic algorithm, 351
    greedy algorithms, 337
    rounding, 329
    simulated annealing, 347
discrete variables, 27, 283, 312, 327
    avoiding, 328
discretization, 48, 50, 63
    error, 49, 57
    methods, 52
divergence, 64
diversity, 43, 310, 312
dominance, 198, 357
    depth, 365
dominated point, 358
dot product, 545
    2-norm, 549
    test, 262
double-precision
    floating-point format, 54
    number, 55
dual number, 248
dynamic
    polling, 294

programming, 26, 38, 339, 344
rounding, 329
system, 429

efficient global optimization (EGO), 286, 415, 416
eigenvalues, 86, 90, 252, 552, 553
eigenvectors, 86, 116, 252, 553, 566
elitism, 316, 368
engineering design, 2, 45, 475, 476
enhanced collaborative optimization (ECO), 532
equality constraints, 12
equality of mixed partials, 85
error, 58, 225
    absolute, 54
    constant, *see* asymptotic error constant
    discretization, 49, 57
    iterative solver tolerance, 57
    modeling, 48
    numerical, 48, 49, 53
    programming, 58
    propagation, 55
    relative, 54
    roundoff, 49, 54, 57, 58, 561
    truncation, 57, 228
Euclidean
    norm, 549
    space, 545
Euler–Lagrange equation, 36
evolution, 307
evolutionary algorithms, 39, 42, 286, 363
    GA, 306
    PSO, 316
exact penalty decomposition (EPD), 532
exhaustive search, 298, 328, 329

exit conditions, 93
expected
    improvement, 415
    value, 412, 416, 441, 442
expected value, *see* mean
experimental data, 374
experiments, 5
explicit
    component, 480, 503, 509
    equation, 266
    function, 50, 482
    model, 51
exploitation, 319, 414
exploration, 286, 319, 415
exponential
    convergence, 457
    distribution, 557
    function, 212
expression swell, 227
exterior penalty, 176
extrapolation, 313

factorization
    Cholesky, 561
    LU, 560
Farkas' lemma, 36, 166
Faure sequence, 384
feasibility tolerance, 201, 208
feasible
    descent direction, 166
    direction, 159
    region, 12
    space, 185
feedback, 491
Fibonacci sequence, 340
file input and output, 225, 479, 487
filter methods, 198, 315
finite-difference derivatives, 225, 227, 253, 274, 281, 502, 512
    accuracy, 230

backward difference, 228
central difference, 228, 230
coupled, 502
forward difference, 228, 230
higher-order, 229
implementation, 231
optimal step size, 230
step, 228
step-size dilemma, 229, 230
step-size study, 230
finite-difference discretization, 52
finite-element
discretization, 50, 52, 588
structural model, 531
finite-precision arithmetic, 54, 56, 93, 229, 309
finite-volume discretization, 52
first-order
derivatives, 223
perturbation methods, 448, 449
fitness, 286, 309, 310
fixed-point iteration, 62, 226, 487, 534, 562
Fletcher–Reeves formula, 117
floating-point format, 54
food shopping problem, 426
forward difference, 228, 230
forward propagation, 448
direct quadrature, 452
first-order perturbation, 449
Monte Carlo, 459
polynomial chaos, 462
forward substitution, 242, 486
four fundamental subspaces, 547
Frobenius norm, 550
full factorial sampling, 375
full-space hierarchical Newton's method, 493
full-space optimization, *see* simultaneous analysis and design (SAND)
function
blending, 145
constraint, 12
explicit, 482
implicit, 483
of interest, 224, 253
objective, 9
smoothness, 282
functional form, 484, 487, 505
Jacobian, 507

Gauss–Hermite quadrature, 455–457, 468
Gauss–Konrod quadrature, 457
Gauss–Legendre quadrature, 455
Gauss–Newton algorithm, 391
Gauss–Seidel method
linear, 563
nonlinear block, 489, 491, 498, 499, 511
Gaussian
basis, 399
distribution, *see* normal probability distribution
elimination, 560
kernel, 401
multivariate distribution, 401
process, *see* kriging, 416
quadrature, 453
gene, 307
generalization error, 397
generalized minimum residual (GMRES) method, 569
generalized pattern search (GPS), 285, 292, 295, 296
genetic algorithm (GA), 39, 306, 308, 363, 376
binary-encoded, 307, 308, 351
constraints, 314
crossover, 311

discrete optimization, 351
multiobjective, 363, 367
mutation, 311
real-encoded, 307, 312
selection, 309
geometric programming (GP), 41, 425, 434
software, 437
Gibbs distribution, *see* Boltzmann distribution
global
optimization, 42
optimum, 19, 20, 24, 146, 287, 299, 322, 423
search, 23, 146, 282, 284, 286, 322
globalization strategy, 69, 95, 492
governing equations, 50, 61, 70, 252, 254, 479
GPkit, 437
gradient, 79, 80, 121
normalization, 111
scaling, 137
gradient-based algorithms, 22, 28, 79, 376, 412
comparison, 133, 135
constrained, 153
efficiency, 223
unconstrained, 79, 97
gradient-descent method, *see* steepest-descent method
gradient-enhanced kriging, 405
predictor, 405
gradient-free algorithms, 22, 28, 281
graph, 486
acyclic, 486
coloring, 262–265, 504, 508
cyclic, 486
directed, 486
weighted directed, 338

graph form programming, 425
graphical solution, 14
greedy algorithms, 337
grocery store shopping, 339

H-section beam problem, 217
Hadamard product, 167
half-space, 157, 159, 165
intersection, 165
Halton sequence, 382
scrambled, 383
Hammersley sequence, 383, 407
Hamming cliff, 312
Hartmann function, 580
Hermite polynomials, 456, 467
Hessian, 85, 110, 115, 121, 144
approximation, 110, 123, 124, 127, 572
directional curvature, 86
eigenvalues, 86
eigenvectors, 86, 116
Gauss–Newton algorithm, 392
initial approximation, 124
interpretation, 86
inverse approximation, 128, 573, 574
inverse update, 573
Lagrangian, 161, 162, 191
positive-definite, 121, 126
positive-semidefinite, 90
symmetry, 85
update, 124, 572
vector product, 86
heuristics, 24, 115, 287
hierarchical solvers, 41, 499, 503
hierarchy, 481, 483, 501
higher-order moments, 450
histogram, 377, 452
history of optimization, 33
hit-and-run algorithms, 286
human expertise, 4

hybrid adjoint, 260
hypercube, 304
hyperplane, 157
    intersection, 159
    tangent, 157, 159, 165
hyperrectangle, 304
    potentially optimal, 304
    trisection, 304
hypersurface, 11, 254

identity matrix, 206, 550
    scaled, 132
ill-conditioning, 56, 351, 560, 589
    aggregation function, 212
    collaborative optimization (CO), 523
    interpolation, 110
    least squares, 387
    line search, 107
    Newton's method, 69
    penalty function, 176, 182, 187
imaginary step, 232
implicit
    component, 480, 503
    dependence, 253
    equation, 50, 266, 544
    filtering, 285, 322
    function, 50, 71, 253, 483
    model, 51
implicit analytic methods, 225, 252
    adjoint, 211, 255, 260
    coupled, 503
    direct, 255
    direct vs. adjoint, 257
    forward mode, 255
    reverse mode, 256
    structural problem, 259
    verification, 262
inactive constraints, 165, 190

indefinite matrix, 552
individual discipline feasible (IDF), 515
induced functions, 213
inequality
    constraints, 12, 154
    quadratic penalty, 180
inertia, 316
inexact penalty decomposition (IPD), 532
infeasibility, 198
infeasible
    directions, 165
    region, 13
infill, 375, 414
initial design, 3, 8, 79
inner product, 454, 545
    weighted, 455
input and output conversion, 481
inputs, 6, 225, 374, 448, 482, 483
integer
    overflow, 54
    programming, *see* discrete optimization
    variables, 327, 336
integer variables, 312
interior penalty methods, 184
interior-point methods, 39, 41, 153, 187, 204
    line search, 206
    with quasi-Newton approximation, 208
interpolation, 108, 384
    cubic, 109, 145
    ill-conditioning, 110
    non-smooth, 145
    quadratic, 108, 109
intuition, 3, 10, 12, 29
invasive weed optimization, 286
inverse
    barrier, 184

cumulative distribution, 379
inversion sampling, 379
investment portfolio selection, 343
isosurface, 11, 80, 83
    tangent, 157
iterations
    major, 96
    minor, 96
iterative
    linear solvers, 560
    solvers, 57, 62, 63, 226, 237

Jacobi method
    linear, 563
    nonlinear block, 487–489, 499
Jacobian, 69, 121, 155, 224, 228, 232, 244, 246, 254, 495, 571
    compressed, 263
    constraints, 188
    coupled, 504
    diagonal, 263
    inverse, 572
    nullspace, 165
    size, 155, 257
    sparse, 261–264, 504, 508
    square, 261
    structure, 504, 507
    transpose, 504
Jones function, 146, 297, 306, 320, 580
    discontinuous, 320

Kepler's equation, 35, 76, 226
kernel, 400
KKT conditions, 37, 168, 187, 282, 523, 574
knapsack problem, 339, 343
    dynamic programming, 346
    tabulation, 345

Kreisselmeier–Steinhauser (KS) function, 212
kriging, 37, 42, 286, 399, 400, 416
    gradient-enhanced, 405
    kernel, 400, 405
    ordinary, 400
    predictor, 403
    regression-based, 408
Krylov subspace methods, 62, 69, 499, 562, 569
kurtosis, 450, 556

L-BFGS method, 131, 132
Lagrange multipliers, 36, 182, 188, 198, 528
    adjoint interpretation, 257
    equality constraints, 160
    inequality constraints, 167
    interior-point method, 207
    meaning of, 173
Lagrangian
    function, 161, 187, 205
    mechanics, 36
Latin hypercube sampling (LHS), 146, 377, 379, 380, 460
law of large numbers, 459
law of reflection, 34
law of refraction, 34
leading principal
    minor, 551
    submatrix, 551
learning rate, 413
least squares, 36, 386
    constrained, 428
    linear, 386
    nonlinear, 391
    regularized, 388
left nullspace, 548
legacy codes, 226
Legendre polynomial, 454
Levenberg–Marquardt algorithm, 391

likelihood function, 390
    concentrated, 402
line search, 38, 94, 96, 116
    algorithm, 97
    backtracking, 99, 100
    bracketing, 103, 104
    comparison with trust region, 95, 144
    exact, 98, 102, 117
    ill-conditioning, 107
    interior-point methods, 206
    interpolation, 108
    Newton's method, 121
    overview, 95
    pinpointing, 103, 106
    plot, 108, 123, 136
    quasi-Newton method, 124
    SQP, 188, 197
    step length, 99
    sufficient decrease, 99
    unit step, 136
linear
    conjugate gradient, 568
    convergence, 64
    direct solvers, 560
    function, 424
    independence constraint qualification, *see* constraint qualification
    iterative solvers, 560
    least squares, 424
    mixed-integer programming, 330, 331
    programming (LP), 19, 37, 331, 425
    regression, 385
    solvers, 559
    system, 559
linear-quadratic regulator (LQR), 429
Lipschitz
    constant, 299
    continuity, 299
local
    constraints, 520
    design variables, 520
    optimum, 19, 287
    search, 23, 79, 284
log likelihood function, 390
logarithmic
    barrier, 185
    scale, 66
logical operators, 236
lognormal distribution, 557
loops, 138, 226, 238
    unrolling, 238
low-discrepancy sequence, 380, 460
lower
    convex hull, 302, 303
    triangular matrix, 242, 273, 486, 560
LU factorization, 61, 497, 560

machine learning, 2, 5, 43
    deep neural networks, 409
    hidden layers, 408
    input layer, 408
    maximum likelihood, 389
    minibatch, 412
    neural networks, 408
    output layer, 408
    support vector machine, 433
machine precision, 55, 312
machine zero, *see* machine precision
major iterations, 96
manifold, 254
manufacturing, 447
Markov
    chain, 339, 344
        variable-order, 339
    process, 339
mating pool, 309

matrix
- bandwidth, 486
- block diagonal, 497
- column space, 547
- condition number, 560, 570
- dense, 559
- derivatives, 552
- determinant, 551
- diagonal, 550
- factorization, 62, 560
- Hadamard product, 167
- identity, 206, 550
- ill-conditioned, 560
- indefinite, 552
- inverse, 62, 551, 559
- inverse product, 251
- Jacobian, 224
- leading principal minor, 551
- lower triangular, 242, 273, 486, 560
- multiplication, 544
- negative-definite, 552
- norm, 550
- nullspace, 156, 160, 548
- positive-definite, 90, 551
- positive-semidefinite, 90, 552
- rank, 156, 547
- reordering, 486, 491
- row space, 547
- scaled identity, 132
- size notation, 544
- sparse, 486, 559
- splitting, 562
- stiffness, 589
- symmetric, 551
- symmetric positive-definite, 561, 565
- transpose, 485, 550
- upper triangular, 246, 274, 560
- vector product, 546
- well-conditioned, 560

MAUD, *see* modular analysis and unified derivatives (MAUD)

maximization as minimization, 10

maximum
- curvature, 87
- likelihood, 389, 402
- log likelihood, 402
- point, 88, 93
- step, 100

MDO architectures, 40, 475, 533
- ASO, 529
- ATC, 524
- BLISS, 527
- BLISS-2000, 532
- CO, 521
- CSSO, 532
- distributed, 519, 533
- ECO, 532
- EPD, 532
- IDF, 515
- IPD, 532
- MAUD, 533
- MDF, 511, 531
- MDOIS, 532
- monolithic, 510, 533
- QSD, 532
- SAND, 517, 519

MDO frameworks, 534

MDO of independent subspaces (MDOIS), 532

mean value, 554, 555

memoization, 340

merit function, 198

mesh refinement, 57

mesh-adaptive direct search (MADS), 285

metamodel, *see* surrogate models

method of lines, 53

minibatch, 412

minimum, 93
    global vs. local, 19
    strong, 90
    weak, 19
minor iterations, 96
mixed-integer programming, 327
    linear, 331
model, 6, 475, 487
    data-driven, 384
    explicit, 51
    implicit, 51
    inputs and outputs, 483
    multidisciplinary, 479
    optimization considerations, 70
    physics-based, 384
    statistical, 400, 401
modeling error, 48
modular analysis and unified derivatives (MAUD), 498, 499, 501, 503, 508, 512, 533
modularity, 479, 481
monolithic solver, 499
monomial, 434
Monte Carlo simulation, 376, 448, 452, 459, 473
multidisciplinary model, 479
multidisciplinary analysis (MDA), 486, 488, 502, 511, 534
multidisciplinary design feasible (MDF), 511, 513, 531
multidisciplinary design optimization (MDO), 2, 28, 40, 475
multidisciplinary model, 39
multifidelity models, 374
multilevel coordinate search (MCS), 285

multimodality, 19, 20, 23, 79, 138, 146, 282, 350
multiobjective optimization, 10, 18, 28, 198, 283, 355, 443
    NBI method, 360
    epsilon constraint method, 360
    weighted-sum method, 358
    evolutionary algorithms, 363
    GA, 363
    objectives versus constraints, 155
    problem statement, 357
multiphysics, 475
multiple local minima, *see* multimodality
multipoint optimization, 444
multistart, 146, 282, 321, 375, 376
multivariate Gaussian distribution, 401
mutation, 307, 311

N^2 matrix, *see* design structure matrix (DSM)
natural selection, 307, 310
negative-definite matrix, 552
neighboring design, 348
Nelder–Mead algorithm, 39, 285, 287, 290, 314, 351
    convergence, 289
    operations, 288
    simplex, 287, 288
neural networks, 43, 408
    deep, *see* deep neural networks
    depth, 409
    feedforward, 408
    node, 408
    recurrent, 408
    weights, 411

Newton's method, 23, 35, 139, 187, 391, 559
    computational cost, 136
    convergence rate, 68
    coupled, 492
    full-space hierarchical, 493
    globalization, 69, 492
    ill-conditioning, 69
    issues, 121
    linear system, 69, 121
    minimization, 119
    monolithic, 492
    preconditioning, 69
    reduced-space hierarchical, 494, 496
    root finding, 62
    scale invariance, 121, 145
    solver, 66
    step, 69, 121
Newton–Cotes formulas, 453
Newton–Krylov method, 69
Niederreiter sequence, 384
noisy
    data, 388
    function, 24, 93, 231, 423
    model, 28, 374
NOMAD, 298
nondominated
    point, 358
    set algorithm, 364
    sorting, 365
nonlinear
    block methods, 488
    least squares, 391
    simplex algorithm, *see* Nelder–Mead algorithm
nonsingular matrix, 559
normal probability distribution, 314, 378, 389, 400, 450, 473, 556
    uncorrelated, 468
norms, 548

∞-norm, 93, 140, 549
p-norm, 213, 549
1-norm, 176, 549
2-norm, 140, 545, 549
    Frobenius, 550
    matrix, 550
    visualization, 549
    weighted, 549
    weighted Frobenius, 574
NP-complete, *see* polynomial-time complete
NSGA-II, 308, 364
nullspace, 156, 160, 548
    Jacobian, 165
    left, 548
numerical
    conditioning, *see* ill-conditioning
    errors, 48, 49, 53, 479
    integration, *see* quadrature
    models, *see* model
    noise, 28, 48, 58, 94, 138, 225, 231, 282, 321, 388
    optimization, 45
    stability, 56

objective function, 9, 79
    multiple, 283, *see* multiobjective optimization
    scaling, 114, 137
    selecting, 9, 11
    separable, 357
    units, 114
offspring, 307
one-shot optimization, 72
OpenAeroStruct, 498, 508
OpenMDAO, 498, 501, 508, 512
operations research, 2, 19, 41, 45
operator overloading, 237, 247, 248
opportunistic polling, 293
optimal control, 2, 5, 26, 40, 41, 429

optimal-stopping problem, 34
optimality, 4, 287
    criteria, 22, 24
    dynamic programming, 342
    Farkas' lemma, 166
    first-order equality constrained, 161
    first-order inequality constrained, 166
    KKT conditions, 168
    second-order constrained, 162, 169
    tolerance, 201, 208
    unconstrained, 90, 91
optimization
    algorithm selection, 26
    difficulties, 136, 138, 274
    problem classification, 17
    problem formulation, 4, 6, 17
    problem reformulation, 5
    problem statement, 14
    software, 15, 41, 94, 200
    under uncertainty (OUU), 28, 441
optimum, *see* minimum
order of convergence, 64, 66
ordinary
    differential equation (ODE), 52
    kriging, 400
orthogonal, 454
    columns, 264
    polynomials, 454, 463
    search directions, 113
    vectors, 547
outer product, 545
    self, 127, 571, 575
outputs, 6, 18, 225, 374, 482
overdetermined system, 387
overfitting, 395, 396
overflow, 55
    integer, 54

parallel computation, 309, 488, 489, 495, 498, 499, 522
parameterization, 9, 330
parents, 307
Pareto
    anchor points, 360
    front, 358, 368, 443
    optimal, 358
    optimality, 357
    set, 358
    utopia point, 361
partial
    derivatives, 80, 239, 254, 260, 266, 504, 542
    differential equation (PDE), 52, 63
    pivoting, 561
particle swarm optimization (PSO), 42, 316, 318, 376
    convergence, 319
    initial population, 318
    particle position update, 317
partitioning, 479, 480
pattern-search algorithms, *see also* generalized pattern search (GPS)
PDE-constrained optimization, 517
penalty function, 176, 198
    ATC, 524
    exterior, 176
    interior, 184
    methods, 38, 153, 175, 291, 314, 322
    parameter, 176, 177, 185, 198
    quadratic, 176, 179, 525
    relaxation, 525
percent-point function, 379
physics-based model, 37, 384

pinpointing, 103
plane, 157
Polak–Ribière formula, 118
polar cone, 166
politics, 477
polling, 293
polyhedral cone, 158, 165
polynomial chaos, 448, 462
    intrusive, 470
    nonintrusive, 470
    software, 466
polynomial-time complete, 328
polynomials, 398
    Hermite, 456, 467
    Legendre, 454
    orthogonal, 454, 463
    quadratic, 399
population, 306, 307, 310
    initial, 307, 309
portfolio optimization, 38
positive-definite matrix, 90, 127, 551
positive-semidefinite matrix, 90, 552
positive spanning
    directions, 292
    set, 292
post-optimality
    derivatives, 530, 532
    sensitivity, 5, 175
    studies, 5
posynomial, 434
potentially optimal rectangles, 304
precision, 20, 48, 49, 54, 55, 57, 61, 154, 225, 237, 252, 479, 487, 501
preconditioner, 493, 569
preconditioning, 69, 569
principal curvature directions, 86, 116
principle of least time, 34

principle of minimum energy, 1
probability density function (PDF), 379, 401, 441, 555, 556
probability distribution, 441, 442
    exponential, 557
    Gaussian, *see* normal probability distribution
    lognormal, 557
    uniform, 313, 377, 557
    Weibull, 557
programming, 59
    bugs, 58
    errors, 58
    language, 15, 42, 44, 49, 54, 236, 237, 250, 387, 487, 502
    modular, 59
    profiling, 60
    testing practices, 60
propagated error, 55
pruning, 331
pseudo-load, 260

QR factorization, 387
quadratic
    approximation, 120, 123, 124
    convergence, 64, 121
    form, 546, 552
    function, 115, 139
    interpolation, 108, 109
    penalty, 176, 179, 180
    programming (QP), 188, 386, 425, 427
quadratically constrained quadratic programming (QCQP), 140, 429, 430
quadrature, 452, 457
    Clenshaw–Curtis, 457
    direct, 452
    Gauss–Hermite, 455–457, 468
    Gauss–Konrod, 457
    Gauss–Legendre, 455

sparse grid, 458
quantile function, 379
quasi-Monte Carlo method, 460
quasi-Newton methods, 38, 121, 123, 124, 572
    BFGS, 38, 126, 572, 574, 576
    Broyden, 571
    condition, 126
    curvature condition, 126
    DFP, 38, 126, 573, 576
    Hessian reset, 129
    L-BFGS, 131, 132
    SR1, 130, 575, 576
    unification, 576
quasi-random sequences, 381
quasi-separable decomposition (QSD), 532

radial basis function (RBF), 286, 399
radical inverse function, 381
random
    sampling, 294, 309, 312, 377, 459
    variable, 400, 401, 442, 554
rank, 156, 387, 547
rate of convergence, 63
    linear, 64
    plot, 65
    quadratic, 64
    superlinear, 65
real-encoded GA
    crossover, 313
    initial population, 312
    mutation, 313
    selection, 312
rectified linear unit (ReLU), 410
recursion, 339
recursive solver, 499
reduced-space
    Newton's method, 499
    optimization, *see* multidisciplinary design feasible (MDF)
regression, 5, 384
    linear, 385, 388
    nonlinear, 391
    testing, 60
regular point, 161, 169
regularization, 393, 399, 433
regularized least squares, 388
relative
    derivatives, 82
    error, 54
    step size, 231
relaxation, 331, 489
    factor, 490, 564
reliability metric, 444
reliable design, 441, 447
reordering, 486, 491
residual form, 50, 484
residuals, 50, 51, 121, 225, 253, 266, 479–481, 560
    derivatives, 223
    norm, 66
response surface model, *see* surrogate models
restricted-step methods, *see* trust region
reverse chain rule, 243
reverse Cuthill–McKee (RCM) ordering, 486
reward, 443
risk, 443
robust design, 441, 442
Rosenbrock function, 579
    n-dimensional, 283, 579
    two-dimensional, 135, 137, 144, 395, 579
roulette wheel selection, 310
rounding, 329
roundoff error, 49, 54, 57, 58, 230, 561

row space, 547

saddle point, 92, 93, 580
safety factor, 513
sampling, 374
    full factorial, 375
    inversion, 379
    plan, 375
    random, 294, 309, 312, 377, 459
scalar product, 545
scale invariance, 121, 145
scaled identity matrix, 132
scaling, 69, 93, 94
    constraints, 181
    design variables, 114, 137
    gradient, 137
    logarithm, 115
    objective function, 114, 137
    trust-region method, 145
search direction, 110
    conjugate, 116
    method comparison, 133, 135
    normalization, 118
    steepest descent, 110
    zigzagging, 112, 113
secant
    equation, 125, 574, 575
    inverse equation, 573
    method, 67, 123, 571
second-order cone programming (SOCP), 41, 425, 429
seed, *see* algorithmic differentiation (AD)
self influence, 318
semidefinite programming (SDP), 41, 425
sensitivities, *see* derivatives
separable objectives, 357
sequence
    Faure, 384
    Fibonacci, 340
    Halton, 382
    Hammersley, 383, 407
    low-discrepancy, 380
    Niederreiter, 384
    scrambled Halton, 383
    Sobol, 384
    van der Corput, 381
sequential optimization, 5, 476, 514, 532, 533
sequential quadratic programming (SQP), 38, 41, 153, 187
    equality constrained, 187
    active set, 190
    inequality constrained, 190
    line search, 188, 197
    meaning, 189
    quasi-Newton, 201
    system, 188
shape optimization, 35, 40
shared design variables, 476, 520
Sherman–Morrison–Woodbury formula, 128, 572, 574, 576
shipping, 343
Shubert's algorithm, 299
side constraints, *see* bound constraints
sigmoid function, 145, 410
signomial programming, 436
simplex, 287
simplex algorithm, 37
simulated annealing, 42, 347, 349
simulation, 3, 70
simultaneous analysis and design (SAND), 72, 73, 517, 519
singular matrix, 559
skewness, 450, 556
slack variables, 167, 205

slanted quadratic function, 579
slope, 80, 84, 98
smooth functions, 153
smoothing discontinuities, 145
smoothness, 18, 24
Sobol sequence, 384
social influence, 317, 318
software, 40, 41
    AD, 250
    engineering, 44
    geometric programming, 437
    MDO frameworks, 534
    optimization, 15, 200
    stochastic gradient descent, 414
    surrogate modeling, 399
solvers, 50
    hierarchical, 41, 499, 501
    iterative, 225, 226, 237
    linear, 61, 62, 559
    monolithic, 499
    Newton, 62, 66
    nonlinear, 62
    overview, 61
    recursive, 499
source code, 225, 232, 236, 249
    transformation, 247, 248, 250
span, 156, 547
sparse
    Jacobian, 261–264
    linear systems, 62, 246, 559
    matrix, 486
spectral
    expansions, *see* polynomial chaos
    projection, 465
splines, 9
splitting matrix, 563
spring system problem, 133, 143, 210, 585
SQP, *see* sequential quadratic programming (SQP)

SR1 method, 576
    update, 575
stability, 56
standard
    deviation, *see* variance
    error, 404
standard deviation, 314
state variables, 50, 70, 225, 253, 478, 487, 492
stationary point, 91, 93, 94, 167
statistical model, 400, 401
steepest-descent method, 36, 110, 111
step length, 99, 136
step-size dilemma, 229, 230
stiffness matrix, 50, 253, 259, 481, 589
stochastic
    algorithms, 25, 286
    collocation, 465
    function, 20
    gradient descent, 43, 413
strong Wolfe conditions, 102, 106, 124
structural
    design problem, 71, 73, 253, 259, 588
    model, 49, 253, 482
    optimization, 40, 71, 447
structurally orthogonal columns, 263
subspace, 157, 547
subsystem, 499
subtractive cancellation, 56, 230
successive over-relaxation (SOR), 62, 564
sufficient curvature condition, 102
sufficient decrease condition, 99, 100
sum of squared errors, 411
superlinear convergence, 65

supervised learning, 433
support vector machine, 433
surrogate modeling toolbox (SMT), 399
surrogate models, 25, 28, 37, 285, 294, 373, 375, 532
    interpolatory, 384
    kriging, 37, 400
    linear regression, 385
    polynomial, 385
    regression, 384
surrogate-assisted optimization, 373
surrogate-based optimization (SBO), 28, 37, 42, 373, 414, 532
swarm, 316
symbolic differentiation, 81, 226, 238, 239, 247
    toolbox, 226
symmetric rank 1 (SR1) method, 130, 575
symmetry of second derivatives, 85
system-level
    optimization, 532
    representation, 484, 505
    solver, 487, 500
    subproblem, 519

tabulation, 341, 342, 345
tangent
    hyperplane, 157, 159, 165
    Jacobian, 224, 240
taping, 249
target variables, 515
Taylor series
    approximation, 88, 95, 120, 449
    complex step, 232
    constraint, 159
    finite differences, 227
    multivariable, 540

    Newton's method, 68
    single variable, 539
    Taylor series expansion, 539
ten-bar truss problem, 221, 588
three-bar truss problem, 219
time
    dependence, 26
    horizon, 341
    integration, 63
topology optimization, 40
total
    derivatives, 239, 254, 503, 504, 542
    differential, *see also* differential
    potential energy, 133, 585
tournament selection, 310, 315
    multiobjective, 367
trade-offs
    cost vs. performance, 10
    direct vs. adjoint, 257
    forward vs. reverse mode AD, 246
    multidisciplinary, 477
    performance vs. robustness, 444
    risk vs. reward, 355, 443
    weight vs. drag, 356
training data, 374, 384, 396, 411
trajectory optimization, 1, 26, 39, 375
transportation problem, 36
traveling salesperson problem, 37, 328, 349
tree
    breadth-first search, 333
    data structure, 498, 500
    depth-first search, 332
    pruning, 331
trisection, 304
truncation error, 57, 228, 229, 234

trust region, 96, 139, 528
    comparison with line search, 95, 144
    methods, 122, 139, 142–144, 208, 285
    overview, 95
type casting, 479

uncertainty, 5, 20, 357, 389, 404, 441, 442, 447, 476
    quantification, 443, 444, 448
unconstrained optimization, 79
underdetermined system, 466
underfitting, 396
underflow, 55, 233, 234
unified derivatives equation (UDE), 265, 498, 507, 544
    AD, 272
    adjoint method, 270
    derivation, 266
    direct method, 270
    forward, 267
    reverse, 268, 509
uniform distribution, 313, 377, 557
unimodal function, 19, 282
unimodality, 20, 79
unit testing, 60
units, 7
unsupervised learning, 433
upper triangular matrix, 246, 274, 560
utopia point, 361

validation, 5, 49
van der Corput sequence, 381
variable-order Markov chain, 339
variables
    bounds, 528
    coupling, 478, 482, 484, 486, 487, 489, 492, 506, 515
    design, 6, 253, 479
    input, 374, 448, 482, 483
    output, 374, 482
    random, 400
    state, 253, 478, 487, 492
    target, 515
variance, 400, 441, 443, 451, 554, 555
vector, 545
    operations, 545
    space, 547
verification, 49, 242, 262, 274
visualization, 10

warm start, 152, 562
weak minimum, 19
Weibull distribution, 473, 557
weighted
    directed graph, 338
    Frobenius norm, 574
    function, 246
    inner product, 455
    norm, 549
    sum, 358
wind
    farm problem, 27, 40, 330, 358, 444, 445, 472
    rose, 445
wine barrel problem, 34, 150
wing design problem, 8, 40, 82, 477, 497, 508, 513, 517, 519, 523, 531, 581
Wolfe conditions, *see* strong Wolfe conditions
Woodbury matrix identity, *see* Sherman–Morrison–Woodbury formula
working set, 190

XDSM diagram, 475, 486
    data dependency lines, 486
    iterator, 488
    process lines, 488

zero-one variables, *see* binary variables
zigzagging, 112, 113